PICTURING QUANTUM PROCESSES

The unique features of the quantum world are explained in this book through the language of diagrams, setting out an innovative visual method for presenting complex theories. Requiring only basic mathematical literacy this book employs a unique formalism that builds an intuitive understanding of quantum features while eliminating the need for complex calculations. This entirely diagrammatic presentation of quantum theory represents the culmination of 10 years of research, uniting classical techniques in linear algebra and Hilbert spaces with cutting-edge developments in quantum computation and foundations.

Written in an entertaining and user-friendly style and including more than 100 exercises, this book is an ideal first course in quantum theory, foundations, and computation for students from undergraduate to PhD level, as well as an opportunity for researchers from a broad range of fields, from physics to biology, linguistics, and cognitive science, to discover a new set of tools for studying processes and interaction.

BOB COECKE is Professor of Quantum Foundations, Logic and Structures at Oxford University, where he also heads the multidisciplinary Quantum Group. His pioneering research stretches from categorical quantum mechanics to the compositional structure of natural language meaning, and recent interests include causality and cognitive architecture.

ALEKS KISSINGER is an Assistant Professor of Quantum Structures and Logic at Radboud University. His research focuses on diagrammatic language, rewrite theory, category theory, and applications to quantum computation and the foundations of physics.

PICTURING QUANTUM PROCESSES

A First Course in Quantum Theory and Diagrammatic Reasoning

BOB COECKE

University of Oxford

ALEKS KISSINGER

Radboud University

CAMBRIDGE
UNIVERSITY PRESS

CAMBRIDGE
UNIVERSITY PRESS

University Printing House, Cambridge CB2 8BS, United Kingdom

One Liberty Plaza, 20th Floor, New York, NY 10006, USA

477 Williamstown Road, Port Melbourne, VIC 3207, Australia

314-321, 3rd Floor, Plot 3, Splendor Forum, Jasola District Centre, New Delhi - 110025, India

79 Anson Road, #06-04/06, Singapore 079906

Cambridge University Press is part of the University of Cambridge.

It furthers the University's mission by disseminating knowledge in the pursuit of education, learning and research at the highest international levels of excellence.

www.cambridge.org
Information on this title: www.cambridge.org/9781107104228
10.1017/9781316219317

First published 2017

A catalogue record for this publication is available from the British Library

Library of Congress Cataloging in Publication data
Names: Coecke, Bob, author. | Kissinger, Aleks, author.
Title: Picturing quantum processes : a first course in quantum theory and diagrammatic reasoning / Bob Coecke (University of Oxford), Aleks Kissinger (Radboud University).
Description: Cambridge, United Kingdom ; New York, NY : Cambridge University Press, 2017. | Includes bibliographical references and index.
Identifiers: LCCN 2016035537 | ISBN 9781107104228 (hardback ; alk. paper) | ISBN 110710422X (hardback ; alk. paper)
Subjects: LCSH: Quantum theory. | Quantum computing. | Logic, Symbolic and mathematical.
Classification: LCC QC174.12 .C57 2017 | DDC 530.12–dc23 LC record available at https://lccn.loc.gov/2016035537

ISBN 978-1-107-10422-8 Hardback

Contents

Preface

Glad you made it here! This book is about telling the story of quantum theory entirely in terms of pictures. Before we get into telling the story itself, it's worth saying a few words about how it came about. On the one hand, this is a very new story, in that it is closely tied to the past 10 years of research by us and our colleagues. On the other hand, one could say that it traces back some 80 years when the amazing John von Neumann denounced his own quantum formalism and embarked on a quest for something better. One could also say it began when Erwin Schrödinger addressed Albert Einstein's concerns about 'spooky action at a distance' by identifying the structure of composed systems (and in particular, their non-separability) as the beating heart of quantum theory.

From a complementary perspective, it traces back some 40 years when an undergraduate student named Roger Penrose noticed that pictures out-classed symbolic reasoning when working with the tensor calculus.

But 80 years ago the authors weren't around yet, at least not in human form, and 40 years ago there wasn't really that much of us either, so this preface will provide an egocentric take on the birth of this book. This also allows us to wholeheartedly acknowledge all of those without whom this book would never have existed (as well as some who nearly succeeded in killing it).

Things started out pretty badly for Bob, with a PhD in the 1990s on a then completely irrelevant topic of contextual 'hidden variable representations' of quantum theory – which recently have been diplomatically renamed to *ontological models* (Harrigan and Spekkens, 2010; Pusey et al., 2012). After a period of unemployment and a failed attempt to become a rock star, Bob ventured into the then even more irrelevant topic of von Neumann's quantum logic (Birkhoff and von Neumann, 1936) in the vicinity of the eccentric iconoclast Constantin Piron (1976).

It was there that category theory entered the picture, as well as serious considerations on the fundamental status of composition in quantum systems – something that went hand-in-hand with bringing quantum processes (rather than quantum states) to the forefront ...

⚠️ In the case you are suffering from some kind of category theory phobia, do not stop reading here! Though it has influenced many of the ideas in this book, this is by no means a book about category theory!

… these considerations would ultimately provide the formal and the conceptual backbone for a pictorial approach to quantum theory. The categorical push in quantum foundations came initially from David Moore (1995), a very gifted researcher who suffered his academic end in the late 1990s, an era in which conceptually oriented physics suffered from widespread prohibition. In collaboration with Moore and Isar Stubbe, Bob made some early attempts towards a categorical reformulation of quantum theory (Coecke et al., 2001), which unfortunately inherited too many deficiencies from old-fashioned quantum logic. The main problem with quantum logic was its implicit assumption that the physical systems under consideration are always: 'some part of the ostensibly external phenomenal world, supposed separated from its surroundings in the sense that its interactions with the environment can either be ignored or effectively modelled in a simple way' (Moore, 1999). However, interactions with the environment happen to be something one should really care about!

After being kicked out of his university (cf. bureaucrats, village politics, and lots of hypocrisy), a second failed attempt in the arts, and looming unemployment, a bit of a miracle happened to Bob when two complete strangers, Prakash Panangaden and Samson Abramsky, arranged a 'trial' postdoc in the Computing Laboratory at Oxford, which was back then affectionately known as the Comlab. Despite knowing nothing about computer science and thinking of computer scientists as a bunch of nerds staring at screens all day, Bob found a home in this department and quickly discovered that, unlike quantum logicians, computer scientists had for a long time already studied the structure of interacting systems and were able to describe such systems elegantly in the language of category theory. In fact, in this particular computer science department, category theory was even taught at undergraduate level.

It was here that the second author entered the picture. While on a two-month exchange to Oxford from his homeland of Tulsa, Oklahoma (Fig. 0.1), Aleks happened to take the aforementioned undergraduate course in category theory, which was at the time taught by Samson. The mind-expanding nature of that course (including a guest lecture about weird-looking pictures of monoidal categories by an equally weird-looking guy) got Aleks interested enough to get involved in the subject. At Samson's prompting, he started coming along to the group's Quantum Lunch seminar. The seminar format consisted of a large pub lunch followed by a talk where a drunk and drowsy speaker addressed an equally drunk and drowsy audience on topics in the newborn subject of categorical quantum mechanics. It was great.

Two months turned into nine years, the Comlab became the 'Department of Computer Science', and though it seemed like nobody could remember when Aleks first starting hanging around, he ended up doing a master's, PhD, and a postdoc.

Figure 0.1 Some typical sights of Tulsa, Oklahoma.

Without the surprising wealth of both mathematical machinery and conceptual thinking in this unique computer science environment, this book simply would not have existed. In sharp contrast to the prohibition in the 1990s of the words 'foundational' and 'conceptual' in physics, in this new environment 'foundational' and 'conceptual' were (and still are) big virtues! This led to the birth of a new research community, in which computer scientists, pure mathematicians, philosophers, and researchers in the now-resurgent area of quantum foundations closely interact. It is probably even fair to say that this unique atmosphere contributed to the resurrection of the quantum foundations community as a whole, and along the way several of its highly respected practitioners have adopted the diagrammatic paradigm, notably Chiribella et al. (2010) and Hardy (2013a).

The conference series Quantum Physics and Logic (QPL), founded by Peter Selinger in 2003 under a different name (but with the same abbreviation!), was a particularly important forum for the development of the key results leading up to this book. In fact, the first paper about diagrammatic reasoning for novel quantum features (Coecke, 2003) was presented at the first QPL. The categorical formalisation of this result (Abramsky and

Figure 0.2 Yanking for cash.

Coecke, 2004), now referred to as categorical quantum mechanics, became a hit within the computer science semantics community, and ultimately allowed for several young people to establish research careers in this area. Top computer science conferences (e.g. LiCS and ICALP) indeed regularly accept papers on categorical quantum mechanics, and more recently leading physics journals (e.g. PRL and NJP) have started to do so too.

We are very grateful to have received a healthy flow of research funding (cf. Fig. 0.2), by the UK Engineering and Physical Sciences Research Council (EPSRC), the European Commission (FP6 FET Open), the US Office of Naval Research (ONR), the US Air Force Office of Scientific Research (AFOSR), the Foundational Questions Institute (FQXi), and the John Templeton Foundation (JTF). In particular, during the writing of this book both authors were generously supported by the latter. As a result, the Quantum Group at Oxford has grown over the past 10 years from 5 members in 2004 to 50 members now, and several of its former members have even started to build groups elsewhere, spreading the Gospel of pictures and processes. The constant interaction with the Quantum Group (and its numerous diaspora) has of course been absolutely essential in the development of this book.

So where did the idea of actually writing this book come from? Here at Oxford one typically doesn't like to mention 'that other place'. But, in the summer of 2012 Cambridge University Press contacted us to ask whether the diagrammatic language may be ready for a book. This generated the idea of teaching the Quantum Computer Science course that fall entirely with diagrams. The lecture notes would then provide the basis for a book, which would most certainly be ready by spring 2013. A quick glance at the first couple of pages in this book should tell you that this plan failed. The lecture notes were entirely dumped, and in fall 2013 we started again from scratch, leading up to what you are reading now.

Quite a number of things did happen from start to finish, like one of the authors relaunched a music career, met a girl, got married, made a baby, got a baby, and took a baby

Figure 0.3 Aleks' beard growth, as correlated with textbook completion.

to Beijing to watch him play a metal show. Meanwhile, the other author also got married, had a brief foray into stand-up comedy, got a position at Radboud University (amongst Bob's mortal enemies: the Dutch), and grew a humongous beard (Fig. 0.3). Both authors became well known at a local pub for being beaten up outside (rumour has it over a dispute on the interpretations of quantum theory, but neither author remembers too well). They also formed a southern country folk industrial noise band called the Quantum Dagger Orchestra.

The students who have taken this course over the past several years have been an invaluable source of inspiration and practical guidance in terms of what works and what doesn't when teaching this drastically new approach to quantum theory. In particular, we would like to thank students Jiannan Zhang, William Dutton, Jacob Cole, Pak Choy, and Craig Hull in the class of 2013, Tomas Halgas in the class of 2014, and Ernesto Ocampo, Matthew Pickering, Callum Oakley, Ashok Menon, Ignacio Funke Prieto, and Benjamin Dawes in the class of 2015, who have all contributed to this finished product by pointing out typos in those original (and revised) lecture notes.

We would especially like to thank Yaared Al-Mehairi, Daniel Marsden, Katie Casey, John-Mark Allen, Fabrizio Genovese, Maaike Zwart, Hector Miller-Bakewell, Joe Bolt, John van de Wetering, and Adrià Garriga Alonso, who all provided substantial qualities and/or quantities of corrections, and the class tutors Miriam Backens, Vladimir Zamdhiev, Will Zeng, John-Mark Allen, and Ciaran Lee for their valuable overall input, model solutions, and for bearing the brunt of the students' frustration when some of the more 'experimental' exercises went wrong.

Detailed corrections on the final manuscript were provided by booze brothers John Harding and Frank Valckenborgh, and the pictures of angry Bob were all taken by yet another booze brother, Ross Duncan. Consistent reminders for Aleks to eat and avoid head trauma were graciously provided by his wife Claire. Bob failed to take account of these reminders by his wife Selma.

All of the diagrams in this book were created using PGF/TikZ package for LaTeX and the TikZiT software. Grab the latest version from tikzit.github.io.

Finally: Why Cambridge University Press and not Oxford University Press? Because a CUP causes a whole lot of magic in this book while an OUP is ... no clue.

1

Introduction

> Under normal conditions the research scientist is not an innovator but a
> solver of puzzles, and the puzzles upon which he concentrates are just
> those which he believes can be both stated and solved within the existing
> scientific tradition.
>
> — *Thomas Kuhn, The Essential Tension, 1977.*

Quantum theory has been puzzling physicists and philosophers since its birth in the early
20th century. However, starting in the 1980s, rather than asking why quantum theory is so
weird, many people started to ask the question:

What can we do with quantum weirdness?

In this book we not only embrace this perspective shift, but challenge the quantum icons
even more. We contend that one should not only change the kinds of questions we ask about
quantum theory, but also:

change the very language we use to discuss it!

Before meeting this challenge head-on, we will tell a short tale to demonstrate how the
quantum world defies conventional intuitions ...

1.1 The Penguins and the Polar Bear

Quantum theory is about very special kinds of physical systems – often very small
systems – and the ways in which their behaviour differs from what we observe in everyday
life. Typical examples of physical systems obeying quantum theory are microscopic
particles such as photons and electrons. We will ignore these for the moment, and begin by
considering a more 'feathered' quantum system. This is Dave:

He's a dodo. Not your typical run-of-the-mill dodo, but a *quantum dodo*. We will assume that Dave behaves in the same manner as the smallest non-trivial quantum system, a two-level system, which these days gets referred to as a quantum bit, or *qubit*. Let's compare Dave's state to the state of his classical counterpart, the *bit*. Bits form the building blocks of classical computers, whereas (we will see that) qubits form the building blocks of quantum computers. A bit:

1. admits two states, which we tend to label 0 and 1,
2. can be subjected to any function, and
3. can be freely read.

Here, 'can be subjected to any function' means that we can apply any function on a bit to change its state. For example, we can apply the 'NOT' function to a bit, which interchanges the states 0 and 1, or the 'constant 0' function which sends any state to 0. What we mean by 'can be freely read' is that we can read the state of any bit in a computer's memory without any kind of obstruction and without changing that state.

The fact that we even mention all of this may sound a bit odd...until we compare this to the quantum analogue. A qubit:

1. admits an entire sphere of states,
2. can only be subjected to rotations of the sphere, and
3. can only be accessed by special processes called *quantum measurements*, which only provide limited access, and are moreover extremely invasive.

The set of states a system can occupy is called the *state space* of that system. For classical bits, this state space contains just two states, whereas a qubit can be in infinitely-many states, which we can visualise as a sphere. In the context of quantum theory, this state space is called the *Bloch sphere*. For the sake of explanation, any sphere will do, so we'll just take the Earth. There's plenty of space on Earth for two states of a bit, so put 0 on the North Pole and 1 on the South Pole:

If we ask a polar bear whether she lives at the North Pole or the South Pole, then she'll say 'the North Pole'. If we ask again, she'll say 'the North Pole' again, because that's just where polar bears are from. Similarly, if we ask a penguin, he'll keep saying 'the South Pole', as long as we keep asking.

On the other hand, what will Dave say if we ask him whether he lives at the North Pole or the South Pole? Now, Dave doesn't really understand the question, but since dodos are a bit thick, he'll give an answer anyway. However, assumption 2 was that all animals will answer correctly. Consequently, as soon as Dave says 'the North Pole', his statement is correct: he actually <u>is</u> at the North Pole!

Now, if we ask him again, he'll say 'the North Pole' again, and he'll keep answering thus until he's eaten by a polar bear (Fig. 1.1). Alternatively, if he had initially said 'the South Pole', he would immediately have been at the South Pole.

The particular choice of North Pole/South Pole is not important, but it is important that they are *antipodal* points on the sphere.

Since we can only apply rotations to the sphere of qubit-states, we cannot map both 0 and 1 to 0 (as we could with classical bits), simply because there is no rotation that does that. On the other hand, there are lots of ways to interchange 0 and 1, since there are many (different!) rotations that will turn a sphere upside-down.

So what are quantum measurements? Just like when we read a normal bit, measuring a qubit will produce one of two answers (e.g. 0 or 1, hence the name qu<u>bit</u>). However, this act of 'measuring' is not quite as innocent as simply reading a bit to get its value. To get a feel for this, we return to Dave. Since qubits can live anywhere in the world, Dave – like one particularly famous (classical) dodo – lives in Oxford:

Now, suppose we wish to ascertain where in the world certain animals live, subject to the following assumptions:

1. we are only allowed to ask whether an animal lives at a specific location on Earth or its antipodal location;
2. all animals can talk and will always answer 'correctly'; and
3. predatory animals will refrain from eating the questioner.

Figure 1.1 A polar bear attempting a 'demolition measurement' on Dave.

So, no matter what answer Dave gives, his state has changed. The fact that he was originally in Oxford is permanently lost. This phenomenon, known as the *collapse* of the quantum state, happens for almost all questions (i.e. measurements) we might perform. Crucially, this collapse is almost always *non-deterministic*. We almost never know until we measure Dave whether he'll be at the North Pole or the South Pole. We say 'almost', because there is one exception: if we ask whether Dave is in Oxford or the Antipodes Islands, he'll say 'Oxford' and stay put.

While quantum theory cannot predict with certainty the fate of Dave, what it does provide are the *probabilities* for Dave to either collapse to the North Pole or to the South Pole. In this case, quantum theory will tell us that Dave is more likely to go to the North Pole and get eaten by a polar bear than to go to the South Pole and chill with some penguins. The dodo is extinct for a reason after all ...

1.2 So What's New?

Almost a century has passed since Dave's unfortunate travels to the North Pole. In particular, the past two decades have seen a humongous surge in new kinds of research surrounding quantum theory, ranging from re-considering basic concepts (Fig. 1.2) to envisioning radically new technologies. A paradigmatic example is *quantum teleportation*, whereby the non-local features of quantum theory are exploited to send a quantum state across (sometimes) great distances, using nothing but a little bit (actually two little bits ...) of classical communication. Quantum teleportation exposes a delicate interaction between quantum theory and the structure of spacetime at the most fundamental level. At the same time, it is also a template for an important quantum computational model (measurement-based quantum computing), as well as a component in many quantum communication protocols.

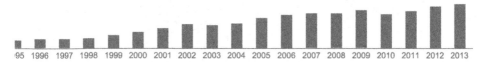

95 1996 1997 1998 1999 2000 2001 2002 2003 2004 2005 2006 2007 2008 2009 2010 2011 2012 2013

Figure 1.2 The paper by Einstein, Podolsky, and Rosen, which was the first to identify quantum non-locality, has enjoyed a huge surge in citations over the past two decades according to Google Scholar, now making it Albert Einstein's most cited paper. And considering the competition, that's saying something.

Quantum theory as we now know it – that is to say, its formulation in terms of *Hilbert spaces* – first saw daylight in 1932 with John von Neumann's book *Mathematische Grundlagen der Quantenmechanik*. On the other hand, quantum teleportation was only discovered in 1992. Hence the question:

Why did it take 60 years for quantum teleportation to be discovered?

A first explanation is that within the tradition of physics research during those 60 years, the question of whether something like quantum teleportation would be possible was simply never asked. It only became apparent when researchers stepped outside the existing scientific tradition and asked a seemingly bizarre question:

What are the information processing features of quantum theory?

However, one could go a step further and ask why it was even necessary to first pose such a question for teleportation to be discovered. Why wasn't it plainly obvious that quantum theory allowed for quantum teleportation, in the same way that it is plainly obvious that hammers are capable of hitting nails? Our answer to this question is that the traditional language of Hilbert spaces just isn't very good at exposing many of the features of quantum theory, and in particular, those features such as teleportation that involve the interaction of multiple systems across time and space. Thus, we pose a new question:

What is the most appropriate language to reason about quantum theory?

The answer to this question is what this book is all about. The reader will learn about many important new quantum features that rose to prominence within the emerging fields of quantum computation, quantum information, and quantum technologies, and how these developments went hand-in-hand with a revival of research into the foundations of quantum theory. All of this will be done by using a novel presentation of quantum theory in a purely diagrammatic manner. This not only consists of developing a two-dimensional notation for describing and reasoning about quantum processes, but also of a unique methodology that treats quantum processes, and most importantly *compositions* of processes, as first-class citizens.

1.2.1 A New Attitude to Quantum Theory: 'Features'

Since its inception, many prominent thinkers were deeply unsettled by quantum theory. A great deal of effort and ingenious mathematics in the early twentieth century went into demonstrating the *bugs* in quantum theory, starting with the now famous EPR paper by Einstein, Podolsky, and Rosen (EPR) in 1935, which claimed that the quantum state provided an 'incomplete description' of physical reality. Roughly speaking, they claimed that something must be missing in order to make sense of quantum theory in a manner compatible with our conventional intuitions. However, John Bell showed in 1964 that any attempt to 'complete' quantum theory to EPR's standards was doomed to failure and thereby binned our conventional intuitions as far as quantum theory is concerned. Bell showed that quantum theory contains at its heart a fundamental, irreducible non-locality (Fig. 1.3).

While relativity theory led Einstein to a beautiful and elegant description of the universe in-the-large, quantum theory seemed to muddy the waters. And this more or less characterises how most scientists perceived quantum theory. There were essentially two ways of dealing with this discomfort with 'quantum weirdness'. One way is to simply ignore any conceptual considerations. This has been the main attitude within the particle physics community, who exemplify the motto 'shut up and calculate'. Alternatively, one can be obsessively concerned with the conceptual problems surrounding

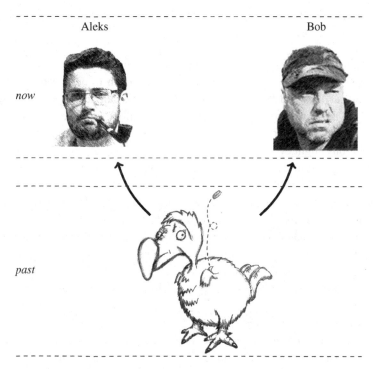

Figure 1.3 Non-locality of quantum theory means that quantum features cannot be explained by means of a classical probabilistic model. In other words, there are situations (unlike the one above) where distantly located observers can experience statistical correlations when they make quantum measurements that cannot be explained by a common cause.

Figure 1.4 Alice queries Dave about Aleks having been partnered with Bob.

quantum theory, sacrificing most of one's life (not to mention sanity) trying to 'fix' them.

Then, starting in the early 1980s, there was an important attitude change, which could be summed up in a simple question:

> *What if the purported <u>bugs</u> of quantum theory are actually <u>features</u>?*

In other words, people began to realise that there was much to be gained by embracing quantum theory as it is and trying to figure out how one can actually exploit 'quantum weirdness'. One may even hope that by doing so, we will become more acquainted with quantumness, get more comfortable with its quirkiness, and maybe the resulting less conventional intuitions might even start to make a lot of sense.

And indeed, quantum non-locality, once perceived by Einstein as some unwanted 'spooky action at a distance', suddenly became a key resource. In fact, decades before software developers started using the motto above to excuse their lazy debugging practices ('It's not a bug, it's a feature!'), Richard Feynman had already pointed out that there was at least one thing that quantum systems were really good at: simulating quantum systems! As it turns out, this problem is pretty difficult using a normal, classical computer. Over the next few decades, scientists discovered lots of weird and wonderful things that quantum systems can do: send secure messages, teleport physical systems, and efficiently factor large numbers.

The new focus on quantum features gave birth to several new fields: quantum computing, which studies how quantum systems can be used to compute; quantum information theory, which studies the implications of incorporating quantum phenomena into gathering and sharing information; and quantum technologies, which concerns the actual business of building devices that exploit quantum effects to make our lives better.

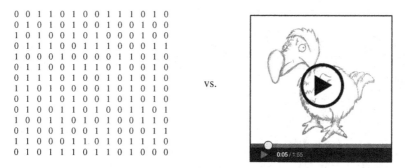

Figure 1.5 Contrasting a low-level and a high-level representation of the digital data that one may find within a computational device.

1.2.2 A New Form of Mathematics: 'Diagrams'

It should be emphasised that discovering these new quantum features wasn't trivial and involved some very smart people. Our bold claim is that when one adopts the appropriate language for quantum theory, these features jump right off the page. Conversely, the traditional, Hilbert space–based language of quantum theory forms a major obstruction to discovering such features. To give some idea of why this is the case, we will make use of some simple metaphors.

Imagine that you were trying to determine what was happening in a video just by looking at its digital encoding (Fig. 1.5). Obviously this is a more or less impossible task. While digital data, i.e. strings of 0s and 1s, is the workhorse of digital technology, and while it is possible to understand 'in principle' how they encode all of the media stored on your hard drive, asking a person to decode a particular string of binary by hand is more suitable for punishing greedy bankers and corrupt politicians than solving interesting problems.

Of course, even skilled computer programmers wouldn't be expected to interact directly with binary data. Somewhere along the way to modern computer programming came the advent of assembly language, which gives a (somewhat) human-readable translation for individual instructions sent to a computer processor. While this made it more practical to write programs to drive computers, it still takes a lot of head-scratching to figure out what any particular piece of assembly code does. Using *low-level languages* such as assembly language creates an artificial barrier between programs and the concepts that they represent and places practical limits on the complexity of problems those programs can solve. For this reason, virtually every programmer today uses *high-level languages* in their day-to-day work (Fig. 1.6).

Similarly, 'detecting new quantum features' in terms of the traditional (i.e. low-level) language for quantum theory, namely 'strings of complex numbers' (rather than 'strings of 0s and 1s'), isn't that easy either. This could explain why it took six highly esteemed researchers to discover quantum teleportation, some 60 years since the actual birth of the quantum theoretical formalism. By contrast, the diagrammatic language we use in this book is a *high-level* language for exploring quantum features (Fig. 1.7). We will soon see

```
.LC0:
        .string "QUANTUM!"
        .text
        .globl  main
        .type   main, @function
main:
.LFB0:
        .cfi_startproc
        pushq   %rbp
        .cfi_def_cfa_offset 16
        .cfi_offset 6, -16
        movq    %rsp, %rbp
        .cfi_def_cfa_register 6
        subq    $16, %rsp
        movl    $0, -4(%rbp)
        jmp .L2
.L3:
        movl    $.LC0, %edi
        movl    $0, %eax
        call    printf
        addl    $1, -4(%rbp)
.L2:
        cmpl    $4, -4(%rbp)
        jle .L3
        leave
        .cfi_def_cfa 7, 8
        ret
        .cfi_endproc
```

vs.

```
5.times do
    print "QUANTUM!"
end
```

Figure 1.6 Contrasting a low-level and a high-level language for computer programs. The programs on the left and right perform the same task, but one is written in the low-level x86 assembly language and one in the high-level language Ruby.

Figure 1.7 Contrasting a low-level and a high-level language for quantum processes, just like we contrasted the low-level and a high-level representation for digital data in Fig. 1.5 and a low-level and a high-level programming language in Fig. 1.6.

that by embracing the diagrammatic language for quantum theory, features like quantum teleportation are pretty much staring you in the face!

Although it goes beyond the scope of this book, it is worth mentioning that the diagrammatic language we use has found applications in other areas as well, such as modelling meaning in natural language (Fig. 1.8), doing proofs in formal logic, control theory, and modelling electrical circuits.

Diagrams are also becoming increasingly important in some fancy research areas of pure mathematics, such as knot theory, representation theory, and algebraic topology. By using diagrams we eliminate a huge amount of redundant syntactic garbage in representing

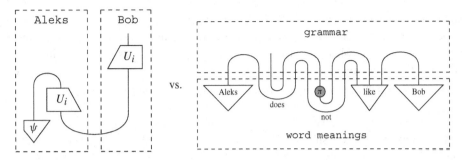

Figure 1.8 Comparing diagrammatic representations of quantum processes to those of 'the flow of meaning' in natural language. While these are two very different contexts, Aleks and Bob feel well at home in both due to their diagrammatic similarity. In the diagram representing natural language, the upper half represents the grammatical structure, while the bottom half represents meaning of individual words, and the overall wiring exposes how the meanings of these words interact in order to produce the meaning of the entire sentence.

$$(g_1 \otimes g_2) \circ (f_1 \otimes f_2) = (g_1 \circ f_1) \otimes (g_2 \circ f_2) \qquad \text{vs.}$$

Figure 1.9 Two distinct syntactic descriptions corresponding to the same diagram. In terms of the symbolic language that is used on the left, two syntactically non-equal expressions might mean the same thing. On the other hand, in terms of the graphical language that is used on the right, there is only one representation. This example is explained in great detail in Section 3.2.4.

mathematical objects (Fig. 1.9), freeing us to concentrate on the important features of the mathematical objects themselves.

There are clear indications that diagrammatic reasoning will become increasingly important in the sciences in general, and this book represents the first attempt to comprehensively introduce a big subject like quantum theory entirely in this new language. By reading this book – or even more, taking a course based on this book – you, like the monkeys launched into space in the 1960s, are the 'early adopters' (a.k.a. 'test subjects') in a totally new enterprise.

1.2.3 A New Foundation for Physics: 'Process Theories'

By taking diagrammatic language as a formal backbone for describing quantum theory (or any other physical theory, for that matter) one also subscribes to a new perspective on physical theories.

First, traditional physical theories take the notion of a 'state of a system' as the primary focus, whereas in diagrammatic theories, it is natural to treat arbitrary processes on equal footing with states. States are then treated just as a special kind of process, a 'preparation' process. In other words, there is a shift from focussing on 'what is' to 'what happens', which is clearly a lot more fun. This is very much in line with the concerns of computer science, where the majority of time and energy goes into reasoning about processes (i.e. programs), and states (i.e. data) only exist to be used and communicated by programs. It is also becoming clear that one should focus not just on single programs but on collections of *interacting* programs to understand the complex, distributed computer systems that are becoming increasingly prevalent in the modern world.

Another example where studying interaction is crucial to understanding a system comes from biology. While one can (in principle) deduce the coat of an animal from its genetic code, this does not explain <u>why</u> that animal has such a coat. On the other hand, if we look at where an animal lives or how it attracts a mate, for example, this can immediately become clear. Similarly, rather than concentrating on systems in isolation, our approach to physics looks at the overall structure of many systems and processes and how they compose. We call such a structure consisting of all the 'allowed processes' and how these interact a *process theory*.

Schrödinger realised early on that the most startlingly non-classical features of quantum theory came from looking not at a single system, but rather at how multiple systems behave together. Rather than acting as a collection of individuals, quantum systems establish complex relationships, and it is these relationships that suddenly enable amazing new things:

> When two systems, of which we know the states by their respective representatives, enter into temporary physical interaction due to known forces between them, and when after a time of mutual influence the systems separate again, then they can no longer be described in the same way as before, viz. by endowing each of them with a representative of its own. I would not call that *one* but rather *the* characteristic trait of quantum mechanics, the one that enforces its entire departure from classical lines of thought.

Schrödinger says that the most important trait of quantum mechanics only becomes apparent when we study the interactions of two systems. So, one might expect any approach to quantum theory to start from a point of view that emphasises compositionality from page 1. But oddly, if you pick up a random textbook on quantum theory, this is not the point of view you will see. It was not until the late 1990s – largely prompted by the discoveries we discussed in Section 1.2.1 – that this idea made it back into the mainstream.

The concept of a process theory does of course put composition at the forefront, and it suggests a natural way of reasoning about processes. One should pare down the nitty-gritty aspects of how processes are mathematically defined and seek out high-level principles that govern their interactions. These principles taken together constitute what can be called the *logic of interaction* for a process theory. Von Neumann also thought that quantum theory should be understood in terms of logical principles. Three years after he published *Mathematische Grundlagen*, von Neumann wrote:

> I would like to make a confession which may seem immoral: I do not believe absolutely in Hilbert space no more. [sic]

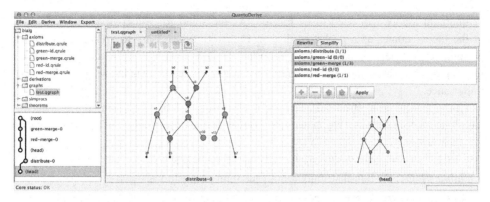

Figure 1.10 Quantomatic: a diagrammatic proof assistant.

He went on to say that it is not the Hilbert-space structure of quantum theory that is physically relevant, but rather the quantum analogue of 'logical propositions', namely those properties verifiable by means of quantum measurements. However, this new kind of logic, called *quantum logic*, ultimately failed to replace Hilbert space as a conceptual basis for quantum theory. Its biggest stumbling block was its complete focus on systems in isolation and its inability to obtain any conceptual account of composed systems. More pragmatically, the passage from Hilbert space to quantum logic seems to make it <u>more</u> difficult to establish new facts or discover new features of quantum theory, typically requiring extreme cleverness on the part of its practitioners to establish even basic facts.

In contrast, this new kind of interaction logic via process theories has very quickly become a practical tool for high-level reasoning about quantum systems and beyond, not in the least due to its intuitive diagrammatic language. It has even formed the basis of a diagrammatic *proof assistant* – i.e. an interactive software tool that constructs proofs (semi-)automatically – called Quantomatic (Fig. 1.10).

So, what should we call this new way of reasoning about quantum theory entirely with diagrams, focussing crucially on its processes and the logic of their interactions? Since the term 'quantum logic' is already trademarked, we'll have to use something a bit more descriptive ...

1.2.4 A New Paradigm: 'Quantum Picturalism'

In Section 1.2.1 we said that the dissatisfaction with quantum theory obstructed people from realising what the actual features of quantum theory were and how those features could be put to good use. Then, by asking the right 'positive' question, many new features were discovered. We went on to argue that in an adequate mathematical language, the features of quantum theory should be plainly obvious. Taking things one step further, one could wonder if this adequate mathematical language isn't just easier to work with, but is also closer to what the world is made up of!

The Holy Grail of theoretical physics is to come up with a theory of quantum gravity. It is likely that to develop a consistent theory of quantum gravity, some of the core assumptions of quantum theory will need to be relaxed. As the standard, Hilbert-space presentation of the theory comes as a packaged deal, it is therefore necessary to seek an alternative presentation that lets us tease out the important features from the incidental ones. Hence, figuring out the presentation that matches what is actually out there in the world is of crucial importance.

Until recently, most if not all attempts to do so suffered from the same obsessions we mentioned before. That is, they take as their starting points some <u>failure</u> of quantum theory:

- *C*-algebras*: the <u>non</u>-commutativity of 'quantum observables'
- *quantum logic*: the <u>non</u>-distributivity of 'quantum propositions'
- *quantum measure theory*: the <u>non</u>-additivity of 'quantum measures'

(Sorry about all the jargon.) It doesn't matter what all of these exactly mean, but the key thing to observe is that they all emphasise something that quantum theory fails to be. What can you do with that? How useful is it to know that a fish is <u>not</u> a dodo? Not much, since a screwdriver is also not a dodo.

Instead of highlighting properties that quantum theory fails to satisfy, one should instead seek out the unique new possibilities highlighted by quantum theory. We contend that the most interesting features of quantum theory are diagrammatic ones, which brings us to the first 'definition' in this book.

Definition 1.1 *Quantum picturalism* refers to the use of diagrams to capture and reason about all of the essential features of interacting quantum processes, in a manner that these diagrammatic equations become the very foundation of quantum theory.

Now, let's see quantum picturalism in action. Consider the following failure of quantum theory: 'A state of two quantum systems fails in general to separate into distinct states of systems *A* and *B*'. Flipping this around, we can 'witness' non-separability by noting that there exist *maximally entangled states* for two systems, which can be succinctly characterised by a single (very useful!) graphical equation (see Fig. 1.11).

A second 'positive' feature we explore in this book is the *existence of complementary measurements*. Here the corresponding 'negative' statement is that most quantum measurements fail to be *compatible* (i.e. they cannot both be performed simultaneously). In fact, the aforementioned conditions of non-commutativity and non-distributivity both aimed to capture the incompatibility of measurements. By contrast, *maximum incompatibility*

Figure 1.11 The two quantum features, 'existence of maximally entangled states' and 'existence of complementary measurements', in terms of graphical primitives. Chapters 4 and 9 are devoted to these two features.

captured by the second equation in Fig. 1.11 represents an actual quantum behaviour that we can observe in experiments and exploit, e.g. in the design of quantum security protocols.

One can now even ask whether such features are enough to completely characterise quantum theory. Can we craft a new formalism for quantum theory for which the defining axioms represent essential physical features through elegant diagrammatic properties? Though this story is by no means complete, the answer seems to be 'yes'. We hope that during the course of this book you will enjoy discovering this fact as much as we did.

1.3 Historical Notes and References

At the end of each chapter, including this one, we will present a brief overview of the history of its main content. This particular section, due to the very nature of an introductory chapter, will cover a lot of ground, and many things touched on here will be discussed with greater care at the end of later chapters.

First, let us say a word about dodos. Dodos became extinct in 1680. The world's only preserved dodo remains are resting at Oxford University's Museum of Natural History. Thus the appearance of Dave the Dodo is an homage both to the 'Oxford Dodo' and to another dodo famously appearing in a fellow Oxonian and local logician's hallucinogenic trip *Alice in Wonderland* (Carroll, 1942). It is fortuitous that our hero is entombed less than 100 metres from both the place where the first experimental demonstration of a quantum algorithm took place (Jones et al., 1998) and the offices where this book was written.

Actual quantum systems, rather than imaginary quantum dodos, were first identified as such by Max Planck (1900), which initiated some 30 years of constructing the formalism of quantum theory as we know it now, ultimately yielding von Neumann's formulation of quantum theory based on Hilbert space and linear maps (von Neumann, 1932). Since then, pretty much any standard textbook of quantum theory still very much resembles the original, with the exception of this one of course!

The paper most associated with Einstein's discomfort with quantum theory is the EPR paper (Einstein et al., 1935). Einstein himself never subscribed to the precise wording of the EPR paper and republished a single-authored paper (Einstein, 1936). As we mentioned earlier, the EPR paper is now considered to be the first paper pointing in the direction of quantum non-locality, by highlighting a conflict between quantum theory and the assumption of 'local realism'. John Bell (1964) strengthened this claim by proving a general theorem about local realistic models and showed that if quantum theory is correct, then it must violate local realism. The modern conception of local realism is that the probabilistic correlations between events observed at distinct locations can be explained by a causal, classical probabilistic model (see e.g. Pearl, 2000). Even today, seeking refinements and generalisations of Bell's theorem is a topic of active research (see e.g. Wood and Spekkens, 2012).

Importantly, Bell's version was directly experimentally verifiable, and the violation of local realism has been experimentally verified many times, starting with Aspect et al. (1981, 1982). Hence it has been experimentally established that the world is indeed 'non-local', exactly as predicted by quantum theory. It should be mentioned that there were a handful of objections to that initial experiment, in the form of 'loopholes' in that particular demonstration of non-locality. All of these have meanwhile been closed by other experiments (Weihs et al., 1998; Rowe et al., 2001; Hensen et al., 2015). On top of that, the huge variety of experiments that have been done confirming some form or another of non-locality for quantum theory is pretty compelling, to say the least (see e.g. Rauch et al., 1975; Zeilinger, 1999; Pan et al., 2000; Gröblacher et al., 2007).

Another development prompted by the EPR paper was the quest for interpretations of quantum theory. Given EPR's claim of incompleteness of quantum theory, one family of interpretations was all about completing quantum theory, although due to Bell's theorem, any such an attempt is bound to be non-local. The most famous one of these is the hidden variable interpretation due to David Bohm (1952a,b). Another one loved by Hollywood is the many-worlds interpretation due to Hugh Everett III (1957). The official default interpretation of quantum theory is the Copenhagen interpretation due to Niels Bohr and Werner Heisenberg, which in the eyes of many is a non-interpretation, given that at first sight it provides nothing more than a recipe to compute probabilities. A detailed survey and extensive discussion of the interpretations of quantum theory can be found in Bub (1999).

The shut-up-and-calculate slogan is often associated with Richard Feynman, who did in many ways embody this way of working, but in fact it was coined by David Mermin (May 2004), who very much did <u>not</u> skirt around foundational questions in quantum theory. He used the term not to refer to this common practice in particle physics, but rather to give his view on the Copenhagen interpretation (Mermin, April 1989).

While on the topic of Feynman, it is worth mentioning that he was the first to realise that there was something quantum systems are really good at: simulating themselves (Feynman, 1982). Thus, his notion of a quantum simulator contained the first seeds of the idea of quantum computation. The discovery of less self-referential applications for quantum features in information processing began a few years later with the advent of quantum key distribution (Bennett and Brassard, 1984). A year later, at the University of Oxford, David Deutsch (1985) gave a formulation for a universal quantum computer, the quantum analogue to Turing's universal machine (Turing, 1937). This led to the discovery of quantum algorithms that substantially outperformed any classical algorithm (Deutsch and Jozsa, 1992; Shor, 1994; Grover, 1996; Simon, 1997). The term 'qubit' was coined by Schumacher (1995).

Quantum teleportation was proposed by Bennett et al. (1993), and its first experimental realisation was by Bouwmeester et al. (1997). The question of why it took 60 years for quantum teleportation to be discovered was asked by one of the authors in a seminar at the Perimeter Institute of Theoretical Physics and was immediately answered by Gilles Brassard, co-inventor of quantum teleportation and a pioneer of the quantum information endeavour as a whole, who happened to be in the audience. He said that no one before had considered the information-processing features of quantum theory and had therefore simply not thought to ask the question. This exchange is reported in Coecke (2005).

The diagrams used in this book are an extension of those used by Roger Penrose (1971), who introduced them as an alternative for ordinary tensor notation (see Section 3.6.1).

However, many similar diagrammatic languages were invented prior to this or reinvented later.

In programming language theory, flow charts were among the first abstract presentations of programs and algorithms. These flow charts, introduced in Gilbreth and Gilbreth (1922) under the name 'process charts', are widely used in many other disciplines too. In quantum information the use of diagrammatic representations started with quantum circuits, a notation borrowed from circuits made up of Boolean logic gates, to which a number of new properly quantum gates were added (see e.g. Nielsen and Chuang, 2010).

A diagrammatic notation specifically tailored towards the processes responsible for quantum weirdness was first introduced in Coecke (2003, 2014a) and, independently, also in Kauffman (2005). These diagrams were provided with an axiomatic underpinning in Abramsky and Coecke (2004), and independently in Baez (2006), paving the way to a diagrammatic approach for quantum theory as a whole. The main pillars supporting the story outlined in this book are the diagrammatic representation of mixed states and completely positive maps by Selinger (2007), the diagrammatic representation of classical data as 'spiders' in (Coecke and Pavlovic, 2007; Coecke et al., 2010a), the diagrammatic representation of phases, complementarity, and the introduction of strong complementarity by Coecke and Duncan (2008, 2011), again in terms of spiders, and the causality postulate introduced by Chiribella et al. (2010). 'Quantum picturalism' was coined in Coecke (2009).

Important topics in the area of quantum computing and related areas that were diagrammatically explored are quantum circuits (Coecke and Duncan, 2008), (topological) measurement-based quantum computing (Coecke and Duncan, 2008; Duncan and Perdrix, 2010; Horsman, 2011), quantum error correction (Duncan and Lucas, 2013), quantum key exchange (Coecke and Perdrix, 2010; Coecke et al., 2011a), non-locality (Coecke et al., 2011b, 2012), and quantum algorithms (Vicary, 2013; Zeng and Vicary, 2014).

Structural theorems about quantum theory emerging from the diagrammatic approach include a number of completeness theorems (Selinger, 2011a; Duncan and Perdrix, 2013; Backens, 2014a; Kissinger, 2014b) and some representation theorems (Kissinger, 2012a; Coecke et al., 2013c).

For applications of the kinds of diagrams that we consider here to other scientific disciplines, there are, for example, Coecke et al. (2010c) and Sadrzadeh et al. (2013) for applications to natural language, Mellies (2012) for logic in computer science, Pavlovic (2013) for computability, Hinze and Marsden (2016) for programming, Baez and Fong (2015) for applications to electrical circuits, Bonchi et al. (2014a) and Baez and Erbele (n.d.) for applications in control theory, Hedges et al. (2016) for applications in economic game theory, Baez and Lauda (2011) for a prehistory, Baez and Stay (2011) for a Rosetta Stone, and Coecke (2013) for an alternative Gospel.

A discussion of von Neumann's discontent with Hilbert space is in Redei (1996), from which we also took the second quote in Section 1.2.3. Attempted modifications/ generalisations/axiomatisations of quantum theory, all to a great extent inspired by quantum logic (Birkhoff and von Neumann, 1936), were pioneered by Mackey (1963), Jauch (1968), Foulis and Randall (1972), Piron (1976), and Ludwig (1985). Coecke et al. (2000) provides a survey of these approaches. A tutorial on how property lattices yield Hilbert spaces is in Stubbe and van Steirteghem (2007). A survey of Foulis and Randall's 'test space' formalism (or 'manuals' formalism) is in Wilce (2000). Ludwig's approach has recently become very prominent again under the name 'generalised probabilistic theories' (Barrett, 2007). Also, several researchers have tried to combine the earlier axiomatic approaches with diagrams and/or compositional structure (Harding, 2009; Heunen and Jacobs, 2010; Jacobs, 2010; Vicary, 2011; Abramsky and Heunen, 2012; Coecke et al., 2013a,b; Tull, 2016).

Giving processes a privileged role in quantum theory was already present in the work of Whitehead (1957) (cf. the quotation at the beginning of Chapter 6) and in Bohr (1961) and became more prominent in Bohm (1986). Process ontologies trace back to the pre-Socratics, most notably to Heraclitus of Ephesus in the sixth century BC (cf. the quotation at the beginning of Chapter 2). Diagrams, and hence a privileged role for processes and composition thereof, are used as a canvas for drafting theories of physics in Chiribella et al. (2010), Coecke (2011), and Hardy (2011, 2013b).

The second quote in Section 1.2.3 on the importance of the role of composition in quantum theory is taken from Schrödinger (1935). The first proper 'interaction logic' was *Geometry of Interaction* due to Jean-Yves Girard (1989), which was recast in a form more resembling the language used in this book in Abramsky and Jagadeesan (1994), and even more so in Duncan (2006). That quantum picturalism can be seen as a logic of interaction is argued in Coecke (2016), on the basis that the roots of logic are language and that the use of logic is artificial reasoning. The diagrammatic proof assistant Quantomatic is described in (Kissinger and Zamdzhiev, 2015) and at the time of this writing is still being actively developed. It is available from the project Website quantomatic.github.io.

2

Guide to Reading This Textbook

Nothing endures but change.

– Heraclitus of Ephesus, 535–475 BC

2.1 Who Are You and What Do You Want?

While there is already a plethora of textbooks on quantum theory and its features, this one is unique because it is based on quantum picturalism.

Prerequisites. There are hardly any prerequisites to this textbook. We do not expect our readers to have a background in physics or computer science or to have any profound background in mathematics. In principle, some basic secondary-school mathematics should be sufficient.

For example, linear algebra (and of course quantum theory) is presented from scratch in a diagrammatic manner. However, this does not mean that the first half of this book will be a boring read for the specialist, since these presentations from scratch are radically different from the usual ones.

Target audience. Given its low entrance fee, as well as its unique form and content, this textbook should appeal to a broad audience, ranging from students to experts from a wide range of disciplines including physicists, computer scientists, mathematicians, logicians, philosophers of science, and researchers from other areas with a multidisciplinary interest, such as biologists, engineers, cognitive scientists, and educational scientists.

One particular target audience for this book consists of students and researchers in quantum computation and quantum information, as we will apply the tools from quantum picturalism directly to these areas. A practising quantum computing researcher may discover a new set of tools to attack open problems where traditional methods have failed, and a student may find some subjects explained in a manner that is much easier to grasp.

Another target audience consists of students and experts with an interest in foundations and/or philosophy of physics, who can read in this book about a process-oriented approach to physics that takes composition of systems as a first-class citizen, rather than a derived notion. In particular, this is the first book that uses diagrammatic language to capture the idea of a process theory and puts such process theories forward as a new foundation for quantum theory, in which all standard quantum theoretical notions can be expressed.

Yet another target audience consists of logicians and computer scientists, who may want to learn about diagrams as a new kind of logical paradigm, which emphasises 'composition' over 'proposition'. For them, learning quantum theory may just be an added bonus to learning about this new paradigm for theory development.

But since this textbook covers the essential ingredients of a standard textbook on quantum computation and quantum foundations, it can just as well be used as a first introduction to those fields. Even though our notation is different from the one used in other textbooks, we cover the core curriculum of a first course in quantum computation and make a continual effort to relate the concepts and notations we introduce to those used more commonly in the literature.

Similarly, this textbook can be used as a first introduction to diagrammatic reasoning, being pretty much the first textbook that does that too.

2.2 The Menu

While we cannot yet offer dodo steaks, with recent advances in science, it may not be too long before pigeons give birth to new Daves and Davettes.

2.2.1 How Diagrams Evolve in This Book

In this book two stories more or less evolve in parallel:

- the development of the diagrammatic language and
- the presentation of quantum theory as a process theory.

Quantum picturalism is indeed all about how these two are closely intertwined. We begin with a very general diagrammatic language and gradually add features to increase its expressiveness. Thus, we start with a language that is general enough to describe many different kinds of processes and gradually home in on quantum processes. Along the way, we present quantum features as and when the language is rich enough to discuss them. As a result, we will encounter features such as quantum teleportation way before more concrete notions such as qubits. All together there are five major jumps in the expressiveness of diagrams on the way to capturing full-blown quantum theory.

1. We first introduce a very basic diagrammatic language in Chapter 3 consisting of nothing but *boxes* and *wires*. This gives us a natural way to express compositions of processes in any process theory:

2. *String diagrams*, defined in Chapter 4, single out special kinds of process theories for which wires can be cup- and cap-shaped, and each box can be both horizontally and vertically reflected:

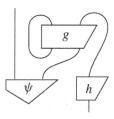

They already expose several quantum-like features, such as non-separability, unitarity, and the impossibility to clone arbitrary states. Following our discussion at the beginning of Section 1.2 on why it took 60 years for teleportation to be discovered, the language of string diagrams makes it plainly obvious that something like quantum teleportation is possible.

3. Next, we allow for two kinds of boxes and wires in diagrams, 'thin' and 'thick' ones, which allows us to distinguish quantum systems from classical ones. Thick boxes and wires, defined in Chapter 6, arise by *doubling* their thin counterparts:

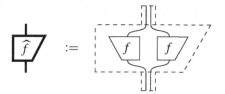

This doubling guarantees that numbers produced by the theory are positive, so these can be interpreted as probabilities. It also allows us to define *discarding*:

$$\overline{\overline{\top}} \; := \; \overline{}$$

a special process that plays a crucial role in our presentation of quantum theory and the causality postulate, which imposes compliance with the theory of relativity.

4. *Spiders*, defined in Chapter 8, are a funky generalisation of wires. Whereas wires have just one input and one output, spiders represent a connection between any number of inputs/outputs. They are governed by a 'spider fusion' rule, which states that when two spiders are connected, they fuse together into a single spider.

Among other things, spiders are used to capture the unique behaviour of classical data, in that it can be copied and deleted. They also allow us to represent the interaction of classical and quantum systems directly in our diagrams, via measurement and encoding operations:

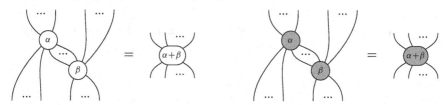

5. For the final jump, we allow for diversification of spiders in Chapter 9. Spiders can now be decorated by *different colours* and *phases*:

This extra data allows us to define all of the processes we need in a purely diagrammatic way (without invoking e.g. matrices) and gives elegant expressions of extremely important quantum features, such as complementarity:

The diagrammatic language now becomes rich enough to unambiguously write down any process from *m* qubits to *n* qubits; that is, it becomes *universal* for qubits. The rules for manipulating these diagrams are called the *ZX-calculus*. In fact, this calculus is not just universal, but also complete for important fragments of qubit quantum theory. This means that all equations that can be derived using matrices can also be derived diagrammatically.

Once we have the full diagrammatic language available, we give a succinct and elegant picture of quantum theory in Chapter 10, 'Quantum Theory: The Full Picture'. Here's a preview of what this looks like in the form of a ...

2.2.2 Hollywood-Style Trailer

The coolest and at the same time least understood quantum feature is most certainly quantum non-locality. Near the end of this book, we provide a detailed account (and proof) of the existence of non-locality. Here, we'll give the reader a glimpse into the mysteries that will be revealed.

While the account of non-locality in standard textbooks involves pages mixing up words and formulas, for us it simply boils down to two diagrammatic computations, one about quantum theory and one about local theories, which yield contradictory results:

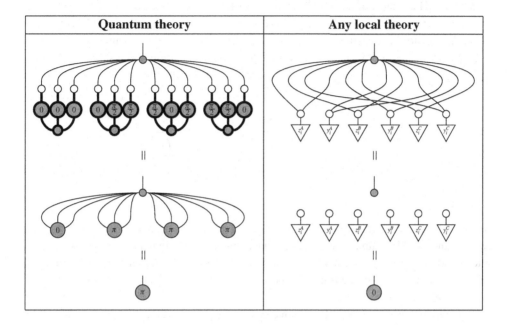

| **Quantum theory** | **Any local theory** |

We will learn that the diagram on the left models four measurements performed on a GHZ-state, followed by computing the *parity* of the measurement outcomes. In other words, our theory is telling us whether to expect an even or an odd number of clicks (or beeps, flashes, etc.) coming from our measurement devices. We will see that the reduction to π on the left says that, according to quantum theory, the parity will be odd. Compare this with the derivation on the right, which assumes that there are some pre-established correlations that determine the measurement outcomes. This is always the case with a local theory, since all correlations of distant events can be traced back to some common cause. The reduction to 0 on the right says that any local theory predicts an even number of clicks, and hence there is a contradiction between quantum theory and locality.

This example is taken from Chapter 11, 'Quantum Foundations', in which we also present a toy theory that looks very much like qubit quantum theory, but fails to be non-local. There are two more themed chapters: Chapter 12, 'Quantum Computation', where we address standard topics such as the circuit model of quantum computation and quantum algorithms, as well as less standard (but increasingly important) topics such as measurement-based quantum computation. In Chapter 6, 'Quantum Resources', we present a general framework to study resources in quantum theory, with quantum entanglement as a particularly important example. We also show how qualitatively distinct types of quantum entanglement can be thought of as very differently behaving spiders.

2.2.3 Some Intermediate Symbolic Pollution

We haven't mentioned Chapter 5 yet. This chapter doesn't contribute to the development of quantum picturalism, but makes the connection with the usual quantum theoretical formalism. The main question addressed is the following. Given a process theory with string diagrams, when do wires represent Hilbert spaces and boxes represent linear maps? As an answer to this question, we adjoin some symbols to string diagrams, resulting in a hybrid diagram–symbol formalism in which one encounters computations like this one:

$$\Big(\boxed{f}\ =\ \sum_i\ \ \ =\ \sum_i \boxed{f}\Big)$$

Such a hybrid formalism is in itself useful, and is widely used in areas of mathematics such as knot theory. More importantly, it introduces the usual formalism to those who don't know it already, and for those who are familiar with it, it makes clear how the diagrams relate to it. Finally, and perhaps most importantly, it allows us to state what exactly one can compute with string diagrams.

Something this allows us to do, which cannot be found in standard textbooks, is to make a smooth transition from linear maps to quantum processes and ultimately to processes that model quantum non-determinism and the classical-quantum interaction in a purely graphical way.

Along the way, we, like Dante, will proceed from linear algebraic Hell:

$$\boxed{U_i}\circ\left(\boxed{U_i}\otimes 1_{\mathbb{C}^2}\right)\circ\left(\boxed{\psi}\otimes\left(\boxed{0}\,\boxed{0}+\boxed{1}\,\boxed{1}\right)\right)$$

through Purgatory:

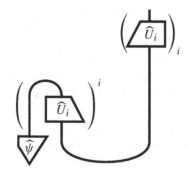

and ultimately to a purely diagrammatic Paradise:

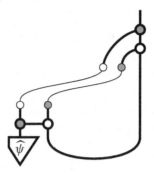

2.2.4 Summaries, Historical Notes, References, Epigraphs

Each chapter (with the exception of Chapter 10, which is already a summary) contains a short section entitled 'Summary: What to Remember' listing the essential material to be taken from it.

At the end of each chapter there is also a short section entitled 'Historical Notes and References' in which we sketch the historical development of the material covered in that chapter, list some key references, and suggest some further reading.

The epigraphs appearing at the beginning of each chapter are all relevant to the contents of that chapter. For some it will be immediately clear from the text. Determining the relevance of the others is left as an exercise for the reader.

2.2.5 Starred Headings and Advanced Material Sections

Any sections, theorems, remarks, examples, or exercises that have a star (*) as superscript, e.g. 'Remark* x.y.z', are to be considered as optional. Typically they require some knowledge that only a fraction of our readers would either know about or be interested in, and hence they are only intended for that particular fraction of readers. For instance, a starred remark may require knowledge of some advanced concepts from linear algebra, quantum theory, or programming. Notably, each chapter has a section entitled 'Advanced Material', which contains material that particularly advanced students or specialists may find interesting. These in particular include clarification of the connection between diagrams and monoidal categories. Some attention is also given to currently ongoing research in quantum picturalism, how it is related to recent developments in pure mathematics, and the surprising connection with natural language meaning.

2.3 FAQ

Over the years, we have noticed that people ask a few questions all of the time. We also anticipate a couple of new questions arising about this book in particular. We'll try to address both of these here.

Q1: *Why does it take X pages to get to some basic stuff such as Y?*

A: There are a few reasons for this:

- As the title suggests, this is a first course not only in quantum theory, but also in diagrammatic reasoning. Thus, we introduce features at the exact points where we have a rich enough language to talk about them. This doesn't always happen in the order you might expect.

- We assume no preliminaries and construct as much as possible from first principles, anticipating a very broad spectrum of potential readers. This means introducing many things, such as linear algebra, that will be old hat for many readers. However, we develop these basic concepts in such a drastically different, diagrammatic way that we think there is something for everyone in each chapter.

- It turns out that diagrams take up a lot of space. Sorry, trees.

Q2: *Where's the beef (i.e. the numbers)?*

A: A traditionally held belief is that the predictive content of a physical theory lies in its ability to produce numbers, such as probabilities. Many find it difficult to reconcile this idea with a diagram, which seems like a discrete, logical object. However, we will see in Section 3.4.1 that numbers arise naturally as special kinds of diagrams. That being said, the most interesting features we will highlight in this book are qualitative, not quantitative. As previewed in the Hollywood-style trailer, we will use diagrams continuously to see that quantum theory exhibits behaviours that are simply not possible in classical physics.

Q3: *What about infinite-dimensional Hilbert spaces?*

A: The lesson of quantum computing and quantum information processing is that we can access many new, revolutionary features of quantum theory by restricting to finite (and often just two!) dimensions. In fact, the long-held belief that 'real physics' only happens in the infinite-dimensional realm, with all of its associated difficulties, could have contributed to the blindness to new quantum features that were not discovered until the past couple of decades. Of course, we are not claiming that infinite dimensions should therefore be ignored, but it does present a unique set of difficulties for quantum picturalism. Most notably, the main diagrammatic workhorses in this book (cups, caps, and spiders) simply don't exist as bounded operators between infinite-dimensional Hilbert spaces. So, for many years, we thought one had to choose between the power (and complexity) of infinite dimensions or the elegance of diagrammatic reasoning with cups, caps, and spiders. However, certain constructions in this book that seem to rely on caps and cups can be done in ways to avoid them altogether (Coecke and Heunen, 2011), and recent results of Gogioso and Genovese (2016) suggest it is possible for us to have our caps and eat them too! Namely, they showed using techniques from non-standard analysis that, while cups, caps, and spiders don't actually <u>exist</u> in infinite dimensions, reasoning with them is still sound, as long as they don't appear in the final answer that comes out.

While this new way of working with infinite-dimensional systems is still in its infancy, it shows promise for the advent of a true infinite-dimensional quantum picturalism.

Q4: *What about Schrödinger's equation?*

A: You will notice that we are careful to use the term 'quantum theory', as opposed to 'quantum mechanics', which we take to mean the core of quantum mechanics that ignores things like positions, momenta, and continuous time-evolution. For our purposes, it suffices to just consider the overall change of systems between some time t_1 and some time t_2, without expounding on the details of what exactly happens in between these times. As in the case of finite dimensions, the huge advances in quantum information/computation have shown us that, even working at this level, we can access many fascinating features of the theory. That being said, there has recently been some exciting new research to accommodate dynamics in quantum picturalism by Gogioso (2015b,c), and it seems likely that many of the features we discuss in this book (e.g. strong complementarity) play a major role.

Q5: *Has quantum picturalism ever produced anything new?*

A: What people posing this question usually mean is: Has quantum picturalism helped solve problems that were already out there and that one couldn't solve with other existing methods? The answer is yes, and some examples of that are Duncan and Perdrix (2010), Coecke et al. (2011b), Horsman (2011), and Boixo and Heunen (2012). However, in science, coming up with new, interesting questions is sometimes more important than answering old ones. Issues such as completeness of calculi is a question that to our knowledge has never been asked before in physics, and quantum picturalism has meanwhile produced a string of results in that area (Backens, 2014a,b; Schröder de Witt and Zamdzhiev, 2014; Hadzihasanovic, 2015). It would be difficult to see how any answer could arise when sticking to the traditional Hilbert space formalism. Another new question is whether we can automate reasoning about physics, for which the Quantomatic software discussed in Chapter 14 has been produced (Kissinger and Zamdzhiev, 2015). Finally, there is the communality of structure and use of quantum picturalism methods elsewhere, like in developing a compositional distributional model of natural language meaning (Clark et al., 2014; Coecke, 2016), which has actually outperformed other existing methods when applied to empirical data (Grefenstette and Sadrzadeh, 2011; Kartsaklis and Sadrzadeh, 2013).

3

Processes as Diagrams

We haven't really paid much attention to thought as a process. We have engaged in thoughts, but we have only paid attention to the content, not to the process.

– David Bohm and David Peat, 1987

In this chapter we provide a practical introduction to basic diagrammatic reasoning, namely how to perform computations and solve problems using diagrams. We also demonstrate why diagrams are far better in many ways than traditional mathematical notation. The development and study of diagrammatic languages is a very active area of research, and intuitively obvious aspects of diagrammatic reasoning have actually taken many years to get right. Luckily, the hard work needed to formalise the diagrams in this book has already been done! So, all that remains to do is reap the benefits of a nice, graphical language.

Along the way, we will encounter *Dirac notaton*. Readers who have previously studied quantum mechanics or quantum information theory may have already seen Dirac notation used in the context of linear maps. Here, we'll explain how it arises as a one-dimensional fragment of the two-dimensional graphical language. Thus, readers not familiar with Dirac notation will learn it as a special case of the graphical notation we use throughout the book.

We also introduce the notion of a *process theory*, which provides a means of interpreting diagrams by fixing a particular collection of (physical, computational, mathematical, edible, etc.) systems and the processes that these systems might undergo (being heated up, sorted, multiplied by two, cooked, etc.).

As we pointed out in Chapter 1, taking process theories as our starting point represents a substantial departure from standard practice in many disciplines. Rather than forcing ourselves to totally understand single systems before even thinking about how those systems compose and interact, we will seek to understand systems primarily in terms of their interactions with others. Rather than trying to understand Dave the dodo by dissecting him (at which point, he'll look pretty much like any other fat bird), we will turn him loose in the world and see what he does.

This turns out to be very close in spirit to the aims of *category theory*, which we will meet briefly in the advanced material at the end of this chapter.

3.1 From Processes to Diagrams

Let's have a look at how diagrammatic language gives us a general way to speak about processes and how they compose, and show that these diagrams provide a rigorous mathematical notation on par with traditional mathematical formulas.

3.1.1 Processes as Boxes and Systems as Wires

We shall use the term *process* to refer to anything that has zero or more inputs and zero or more outputs. For instance, the function

$$f(x, y) = x^2 + y \tag{3.1}$$

is a process that takes two real numbers as input and produces one real number as output. We represent such a process as a *box* with some *wires* coming in the bottom to represent input systems and some wires coming out the top to represent output systems. For example, we could write the function (3.1) like this:

$$\tag{3.2}$$

The labels on wires are called *system-types* or simply *types*.

Similarly, a computer program is a process that takes some data (e.g. from memory) as input and produces some new data as output. For example, a program that sorts lists might look like this:

The following are also perfectly good processes:

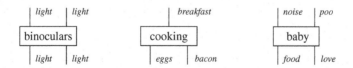

Clearly the world around us is packed with processes!

Note how sometimes a box is actually labelled with a process that occurs over time (e.g. 'the process of cooking breakfast'), while other times we label a box with an apparatus (e.g. 'binoculars' or 'baby'). Typically this means 'the process of using this apparatus to convert input systems into output systems'.

In some cases, wires in the diagram may even correspond to actual physical wires, for example:

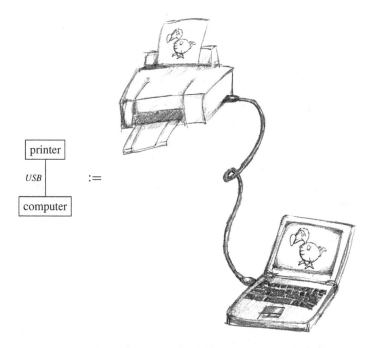

Though, of course, this needs not always be the case. Wires can also represent e.g. 'aligning apertures' on lab equipment, or shipping something by boat. The important thing is that wires represent the flow of data (or more general 'stuff') from one process to another.

As we have just seen, we can *wire together* simple processes to make more complicated processes, which are described by *diagrams*:

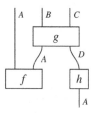

In such a diagram outputs can be wired to inputs only if their types match. For example, the following two processes:

can be connected in some ways, but not in others, depending on the types of their wires:

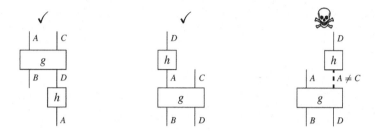

This restriction on which wirings are allowed is an essential part of the language of diagrams, in that it tells us when it makes sense to apply a process to a certain system and prevents occurrences like this:

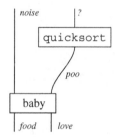

which probably wouldn't be very good for your computer! Much like *data types* in computer science, the types on wires tell us what sort of data (or stuff) the process expects as input and what it produces as output. For example, a calculator program expects numbers as inputs and produces numbers as outputs. Thus, it can't make sense of 'Dave' as input, nor will it ever produce, say, 'carrots' as an output.

Another useful example is that of electrical appliances. Suppose we consider processes to be appliances that have plugs as inputs (i.e. where electricity is 'input') and sockets as outputs:

Then, system-types are just the shape of the plug (UK, European, etc.). Clearly we can't connect a plug to a socket of the wrong shape, so the type information on wires gives us precisely the information we need to determine which wirings are possible. For example, one possible wiring is:

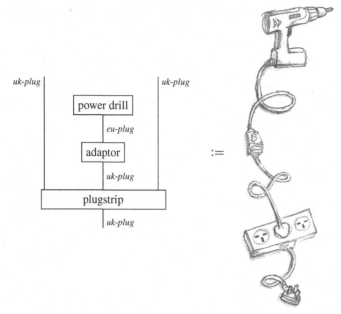

Though this is a toy example, it is often very useful to draw diagrams whose boxes refer to devices that actually exist in the real world (e.g. in a lab). We will refer to such diagrams as *operational*.

3.1.2 Process Theories

Usually one is not interested in all possible processes, but rather in a certain class of related processes. For example, practitioners of a particular scientific discipline will typically only study a particular class of processes: physical processes, chemical processes, biological processes, computational processes, mathematical processes, etc. For that reason, we organise processes into *process theories*. Intuitively, a process theory tells us how to interpret wires and boxes in a diagram (cf. (i) and (ii) below) and what it means to form diagrams, that is, what it means to wire boxes together (cf. (iii) below).

Definition 3.1 A *process theory* consists of:

  (i) a collection T of *system-types* represented by wires,
 (ii) a collection P of *processes* represented by boxes, where for each process in P the input types and output types are taken from T, and
(iii) a means of 'wiring processes together', that is, an operation that interprets a diagram of processes in P as a process in P.

In particular, (iii) guarantees that

> *process theories are 'closed under wiring processes together',*

since it is this operation that tells us what 'wiring processes together' means. In some cases this operation consists of literally plugging things together with physical wires, as in the example of the power drill. In other cases this will require some more work, and sometimes there is more than one obvious choice available. We shall see in Section 3.2 that in traditional mathematical practice one typically breaks down 'wiring processes together' in two sub-operations: parallel composition and sequential composition of processes.

Example 3.2 Some process theories we will encounter are:

- **functions** (types = sets)
- **relations** (types = sets, again)
- **linear maps** (types = vector spaces, or Hilbert spaces)
- **classical processes** (types = classical systems)
- **quantum processes** (types = quantum and classical systems)

We will use the first two mainly in various examples and exercises, whereas the remaining three will play a major role in this book.

Remark 3.3 Note how we refer to a particular process theory by saying what the 'processes' are, leaving the types implicit. This tends to be a good practice, because the processes are the important part. For instance, we'll see that the process theory of functions is quite different from that of relations, so it will not do to just call both 'sets'.

Since a process theory tells us how to interpret diagrams as processes, it crucially tells us when two diagrams represent the <u>same</u> process. For example, suppose we define a simple process theory for **computer programs**, where the types are data-types (e.g. integers, booleans, lists) and the processes are computer programs. Then, consider a short program, which takes a list as input and sorts it. It might be defined this way (don't worry if you can't read the code, neither can half of the authors):

$$
\boxed{\text{quicksort}} \quad := \quad
\begin{cases}
\texttt{qs [] = []} \\
\texttt{qs (x :: xs) =} \\
\quad \texttt{qs [y | y <- xs; y < x] ++ [x] ++} \\
\quad \texttt{qs [y | y <- xs; y >= x]}
\end{cases}
$$

Wiring together programs means sending the output of one program to the input of another program. Taking two programs to be equal if they behave the same (disregarding some details like execution time, etc.), our process theory yields equations like this one:

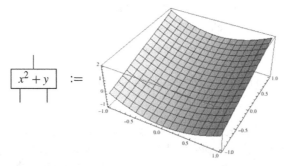

i.e. sorting a list twice has the same effect as sorting it once.

The reason we call a process theory a *theory* is that it comes with lots of such equations, and these equations are precisely what allows us to draw conclusions about the processes we are studying.

To take another example, the function we defined at the beginning of Section 3.1.1 lives in the process theory of **functions**. Types are sets and processes are functions between sets. We consider two functions equal if they behave the same; i.e. they have the same graph:

(Note how two input wires means a function of two variables, hence the three-dimensional plot.) Since they have the same graph, we have:

as well as:

Exercise 3.4 Cook up some of your own process theories. For each one, answer the following questions:

1. What are the system-types?
2. What are the processes?

3. How do processes compose?
4. When should two processes be considered equal?

One thing to note is we haven't yet been too careful to say what a diagram actually i. A complete description of a diagram consists of

1. what boxes it contains and
2. how those boxes are connected.

So the diagram refers to the 'drawing' of boxes and wires without the interpretation within a process theory. However, it makes no reference to where boxes are written on the page. An immediate corollary of this fact is the following.

Corollary 3.5 If two diagrams can be deformed into each other (without changing connections, of course), then they are equal.

Or put more succinctly:

> *Only connectivity matters!*

Example 3.6 Here are some pairs of equal diagrams:

Remark* 3.7 The first pair of equal diagrams in Example 3.6 is called the *Yang–Baxter equation*, which some readers may have already encountered in the mathematical physics literature.

3.1.3 Diagrams Are Mathematics

Diagrams provide a rigorous language for doing mathematics on par with traditional mathematical formulas. Just as traditional formulas can be used to define new concepts and reason about many mathematical structures (e.g. sets, groups, topological spaces, vector spaces), diagrams can be used to define new concepts and reason about process theories.

Perhaps as a result of their upbringing, some people will not accept something as a rigorous mathematical object unless it can be expressed as a formula of some kind. To please those people (and also in order to give another perspective on diagrams) we will briefly introduce a formula-like notation. However, once the correct intuition is in place, we shall drop this notation and (almost) always work with diagrams directly.

We can turn a box into a formula as follows:

$$\begin{array}{c} \boxed{\;f\;} \end{array} \qquad \longleftrightarrow \qquad f_{A_1 B_1}^{B_2 C_1 D_1}$$

In the formula on the right, the subscripts A_1 and B_1 of f represent inputs, while the superscripts B_2, C_1, and D_1 represent outputs. Why did we add numbers to each of these? This numbering doesn't have any meaning in its own right, but it is necessary to eliminate ambiguity whenever we have more than one wire of the same type in a single diagram. For instance, the second input to f is a different wire from the first output, thus we refer to these wires as B_1 and B_2, respectively. We will call a type with some subscript a *wire name*, since it refers to a particular wire in the diagram. The subscripts are of course unnecessary in diagrams because it is already clear that B_1 and B_2 are two distinct wires, since they are in different positions on the page.

The entire expression $f_{A_1 B_1 C_1}^{B_2 D_1}$ is called a *box name*, since it refers to a particular box in the diagram. We can express more than one box by writing down a sequence of box names, where the order we write them down doesn't matter:

$$\begin{array}{cc} \boxed{\;f\;} & \boxed{\;g\;} \end{array} \qquad \longleftrightarrow \qquad f_{A_1 B_1 C_1}^{B_2 D_1} g_{A_2}^{D_2} = g_{A_2}^{D_2} f_{A_1 B_1 C_1}^{B_2 D_1}$$

Note that all of the upper and lower wire names in the formula on the right are distinct. This is because no wires are connected to each other in the diagram. The way we represent connections is by repeating the name of a wire, once as an upper wire name (which corresponds with the output of a box) and once as a lower wire name (which corresponds with the input):

$$\begin{array}{c} \boxed{\;g\;} \\ \boxed{\;f\;} \quad \boxed{\;h\;} \end{array} \qquad \longleftrightarrow \qquad f^{A_1 A_2} g_{A_2 D_1}^{B_1 C_1} h_{A_3}^{D_1}$$

We can give any subscript to a repeated wire name as long as it doesn't clash with any of the other wire names. For example, the following two formulas represent exactly the same diagram:

$$f^{A_1 A_2} g_{A_2}^{B_1} = f^{A_1 A_4} g_{A_4}^{B_1} \tag{3.3}$$

We will also consider a single wire of type A as a 'special' box:

with a corresponding box name $1^{A_2}_{A_1}$. Hence we can write, e.g.:

$$\longleftrightarrow \quad f^{B_3}_{A_1} 1^{A_4}_{A_2}$$

In almost every respect, these special boxes are just boxes like any other and may be interpreted as the process that 'does nothing'. However, when we connect them to the input or output of another box, they vanish:

$$\longleftrightarrow \quad f^{B_1 B_3}_{A_1 A_2} 1^{B_2}_{B_3} = f^{B_1 B_2}_{A_1 A_2} \tag{3.4}$$

The following definition summarises all of above.

Definition 3.8 A *diagram formula* is a sequence of box names such that all wire names are unique, except for matched pairs of upper and lower names. Two diagram formulas are equal if and only if one can be turned into the other by:

(a) changing the order in which box names are written,
(b) adding or removing identity boxes, as in (3.4), or
(c) changing the name of a repeated wire name.

With this concept in hand, we can now provide a definition of a diagram that makes direct reference to a formula-like counterpart.

Definition 3.9 A *diagram* is a pictorial representation of a diagram formula where box names are depicted as boxes (or various other shapes), wire names are depicted as input and output wires of the boxes, and repeated wire names tell us which outputs are connected to which inputs.

Exercise 3.10 Draw the diagrams of the following diagram formulas:

$$f^{C_4}_{B_1 C_2} g^{D_3}_{C_4} \qquad f^{A_1}_{A_1} \qquad g^{A_1}_{B_1} f^{B_1}_{A_1} \qquad 1^{A_6}_{A_1} 1^{A_5}_{A_2} 1^{A_4}_{A_3}$$

Use the convention that inputs and outputs are numbered from left to right.

The above construction shows that all of our work with diagrams could be translated, without ambiguity, into work on diagram formulas. We will mostly stick to diagrams since they are easier to visualise, are easier to work with, and do away with the bureaucratic overhead of extra subscripts. However, we will occasionally use diagram formulas as a handy tool for computing the process of a given diagram, as in Sections 3.3.3 and 5.2.4.

3.1.4 Process Equations

In Example 3.6 above, we wrote down many equations involving diagrams on the left-hand side (LHS) and the right-hand side (RHS). These *diagram equations* always hold, regardless of how we interpret the boxes and wires. That is, they are true in any process theory. On the other hand, within a particular process theory it may be possible to represent the same *process* using two distinct diagrams. We have already seen a couple of examples of this in Section 3.1.2.

Returning to the process theory **functions**, suppose we define two processes: 'minus', which takes two inputs m, n and subtracts them, outputting $m - n$:

and 'times 2', which takes a single input and multiplies it by 2:

where all of the wires have type $\mathbb{R}$. Now, consider the following two diagrams:

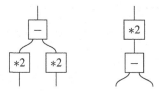

As *diagrams*, these two are not equal. However, if we look at the *processes* they represent, we see that the process on the left multiplies both inputs by 2, then subtracts one from the other. The one on the right first subtracts the inputs, then multiplies the result by 2. Even though they have different diagrams, both processes compute the same function, namely:

$$2m - 2n = 2(m - n)$$

Thus, these distinct diagrams represent the same process when interpreted in the process theory of **functions**; i.e. they are *equal as processes*, which we write simply as:

$$(3.5)$$

This is called a *process equation*. A diagram equation is a (trivial) special case of a process equation, since equal diagrams will always be interpreted as equal processes.

Remark* 3.11 There is a special class of theories where diagram equations and process equations coincide. That is, two processes are equal if and only if they have the same diagram. These are called *free process theories*, in analogy with the use of the term 'free' in algebra. Here, 'free' means 'no extra equations'. In general, imposing extra equations between processes puts a constraint on which interpretations of boxes are allowed (i.e. only those satisfying the extra equations). Free process theories have the property that any interpretation of their boxes as processes (in some other process theory) extends to a consistent interpretation of diagrams.

Exercise 3.12 Give the (diagram) equations that express the algebraic properties of associativity, unitality, and commutativity of a two-input, one-output process. Can you do the same for distributivity of a pair of two-input processes (e.g. 'plus' and 'times')? If not, what's the problem?

The notion of a diagram equation may seem completely obvious (because it is!), but it is important to note that such an equation gives a bit more information than you might first think. To see why, note that equation (3.5) is true, but the following equation is **false**:

$$(3.6)$$

The LHS will map the inputs m, n to $2m - 2n$, whereas the RHS will map those same inputs to $2(n - m) \neq 2m - 2n$. The only difference between this equation and the true equation (3.5) is that the inputs of the RHS are flipped. Even though we do not write down the names of wires in the diagrams (as we do for the diagram formulas of Definition 3.8), these inputs and outputs do have distinct identities. Furthermore, for a diagram equation to be well formed, both sides must have the *same* inputs and outputs, which suggests that there is a correspondence between them:

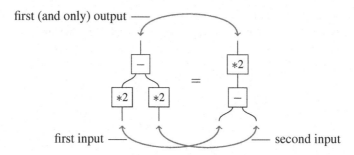

In terms of diagram formulas, this would be reflected by the fact that we call e.g. the first input A_1 and the second input A_2 on both sides of the equation. In a diagram, we show which inputs and outputs are in correspondence by their positions on the page. Bearing this rule in mind, there is clearly a difference between the correct equation (3.5) and the erroneous (3.6).

Equations between processes with only one 'output' should already be familiar from algebra. When we write an equation like:

$$2m - 2n = 2(m - n)$$

we distinguish the inputs m, n of the formulas on the LHS and RHS by giving them names: namely 'm' and 'n'. In algebra, there is always exactly one 'output' of a formula, namely the value computed when all the variables have been substituted by numbers, so we don't bother to give it a name. On the other hand, diagrams can have *many* outputs in general, so we need to distinguish those as well.

We typically think of multivariable functions, such as the algebraic operations above, as taking one or more inputs to a single output:

$$f \qquad :: \qquad (a_1, \ldots, a_m) \mapsto b$$

where the '$\mapsto$' symbol means 'maps to' (cf. Appendix). However, in the process theory **functions** we can also consider functions of this form:

$$f \qquad :: \qquad (a_1, \ldots, a_m) \mapsto (b_1, \ldots, b_n) \tag{3.7}$$

Remark* 3.13 In functional programming, one might encounter a function with many outputs in the sense of (3.7) in expressions like:

$$\text{'}\textbf{let } (b_1, \ldots, b_n) = f(a_1, \ldots, a_m) \textbf{ in } \ldots\text{'}$$

A simple example of a two-output function is *cp*, which takes in a number *n* and sends a copy of *n* down both of its output wires:

$$cp \quad :: \quad n \mapsto (n, n)$$

The function *cp* satisfies the following process equation with minus:

$$= \tag{3.8}$$

The process on the LHS takes in two inputs *m* and *n*, copies each of them, and sends one copy of each to two separate subtraction operations. So, the output is two copies of $m - n$, one on each wire:

$$(m, n) \;\mapsto\; (m, m, n, n) \;\mapsto\; (m, n, m, n) \;\mapsto\; (m - n, m - n)$$

The process on the RHS takes the inputs *m* and *n*, subtracts them, and copies the result. Again, this yields two copies of $m - n$:

$$(m, n) \;\mapsto\; m - n \;\mapsto\; (m - n, m - n)$$

As before, implicit in equation (3.8) is a correspondence between inputs and outputs:

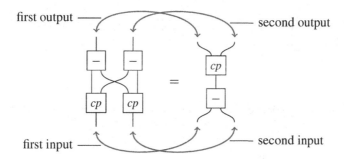

Exercise 3.14 Write (3.8) as an equation between diagram formulas.

Exercise* 3.15 Using *cp*, is it now possible to express the distributivity equation between 'plus' and 'times' mentioned in Exercise 3.12?

3.1.5 Diagram Substitution

We can obtain new process equations from old ones via *diagram substitution*. Along with simple diagram deformation, diagram substitution will be the most common style of calculation in this book.

Diagram substitution refers to the act of replacing some subdiagram of a given diagram with a new subdiagram using a process equation. The resulting diagram will then represent an equivalent process. It is easiest to see how this works by example. Since we are getting a bit bored with our copy map, which produces two identical copies of a number, we define a more exciting one that also multiplies the right copy by two:

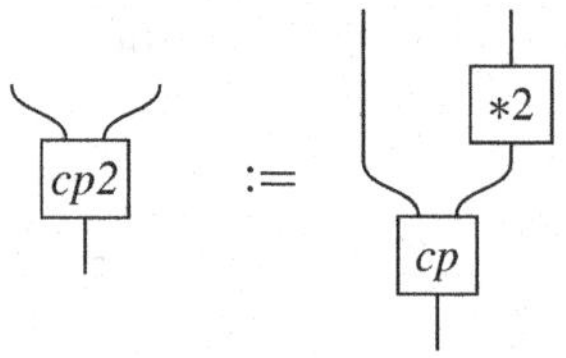

(3.9)

Now, we can ask which equations *cp2* satisfies. We hypothesise that it satisfies a similar equation to (3.8):

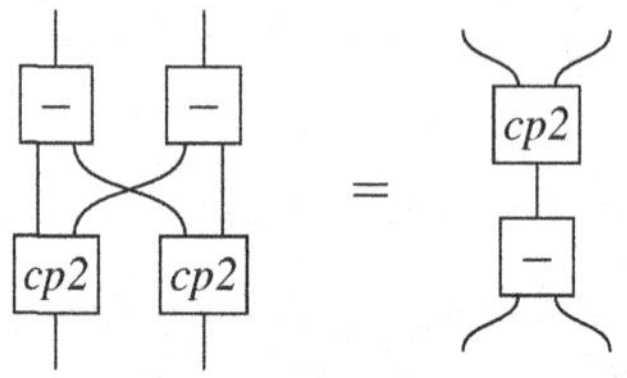

(3.10)

Starting with the LHS of (3.10), we apply equation (3.9) to expand the definition of *cp2*. So, let's cut out the LHS of (3.9) and replace it with the RHS of (3.9):

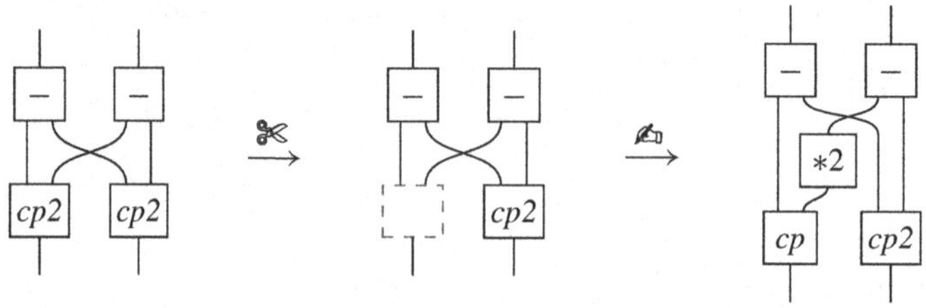

Thus, from (3.9), we can deduce the following equation, by diagram substitution:

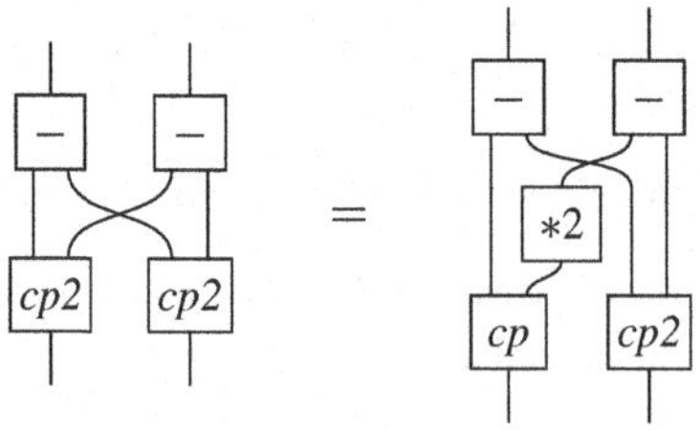

Remark 3.16 An important aspect of diagram substitution is that the RHS should be plugged into the *same place* as the LHS was. For example, the first input of the RHS should be connected to the same wire that the first input of the LHS was previously connected to, and so on. This is where the correspondence between the inputs/outputs in a diagram equation that we discussed in the previous section plays an important role.

Continuing this procedure, we can construct a proof of equation (3.10):

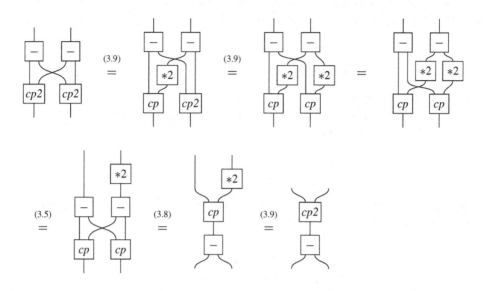

Each of the steps in the proof is marked with the process equation we used, except for step 3, which is just a diagram deformation. Also note that for the last step, (3.9) is being used backwards, which of course is okay.

Remark* 3.17 (algebra vs. coalgebra) While *cp* is very intuitive in what it does, it is nothing like the usual operations that one encounters in algebra, like 'sum' and 'times', exactly for the reason that it has one input and two outputs. The realisation that many interesting operations indeed have multiple outputs resulted in a new research area called *coalgebra*. In terms of diagrams, 'co' can be understood as 'upside-down'. So, for instance, a 'multiplication' operation can be flipped upside-down to produce a 'comultiplication'. Just as one can talk about associativity, unitality and commutativity in algebra, one can talk about coassociativity, counitality and cocommutativity in coalgebra, and these conditions are obtained simply by flipping the equations of algebra upside-down, e.g.:

associativity   coassociativity

3.2 Circuit Diagrams

Recalling that boxes represent processes, we can define two basic *composition operations* on processes with the following interpretations:

$$f \otimes g := \text{'process } f \text{ takes place } \underline{\text{while}} \text{ process } g \text{ takes place'}$$

$$f \circ g := \text{'process } f \text{ takes place } \underline{\text{after}} \text{ process } g \text{ takes place'}$$

These operations allow us to define an important class of diagrams called *circuits*. Despite their importance, we will see in the next chapter that the most interesting processes are in fact those that fail to be circuits.

3.2.1 Parallel Composition

The *parallel composition* operation consists of placing a pair of diagrams side by side. We write this operation using the symbol '$\otimes$':

Any two diagrams can be composed in this manner, since placing diagrams side by side does not involve connecting anything. This reflects the intuition that both processes are happening independently of each other.

This composition operation is associative:

(3.11)

and it has a unit, the empty diagram:

(3.12)

Parallel composition is defined for system-types as well. That is, for types A and B, we can form a new type $A \otimes B$, called the *joint system-type*:

We have already made use of composition of system-types implicitly by drawing boxes that have many inputs and outputs. Formally, a box with three inputs A, B, C and two outputs D and E is the same as a box with a single input $A \otimes B \otimes C$ and a single output $D \otimes E$:

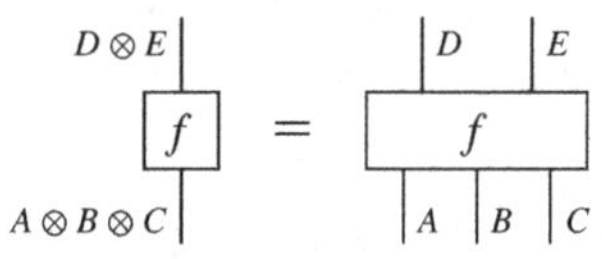

There is also a special 'empty' system-type, symbolically denoted I, which is used to represent 'no inputs', 'no ouputs', or both.

Examples 3.18 Here are some examples of boxes with no input and/or output wires, written both in terms of '$\otimes$' and I and in the usual way:

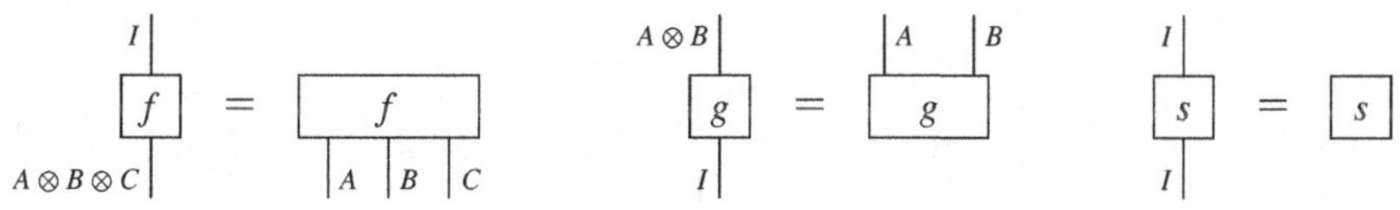

Processes with no inputs and/or outputs play a special role in this book, as we will see starting in Section 3.4.1.

In the diagrammatic notation, we will rarely use the $\otimes$ symbol explicitly, preferring instead to express joint systems as multiple wires.

3.2.2 Sequential Composition

The *sequential composition* operation consists of connecting the outputs of a first diagram to the inputs of a second diagram. We write this operation using the symbol '$\circ$':

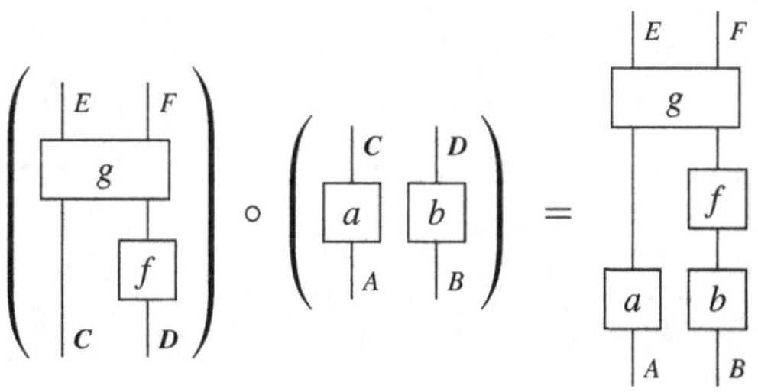

The intuition is that the process on the right happens first, and then the process on the left happens, taking the output of the first process as its input. Clearly not any pair of diagrams can be composed in this manner: the number and type of the inputs of the left process must match the number and type of the outputs of the right process.

We assume that sequential composition will connect outputs to inputs in order. If we wish to vary this order, we can introduce *swaps*:

The LHS of this equation is well-defined of course, since the sequential composition operation is also associative:

$$\left(\boxed{h} \circ \boxed{g} \right) \circ \boxed{f} \;=\; \boxed{g} \;=\; \boxed{h} \circ \left(\boxed{g} \circ \boxed{f} \right) \tag{3.13}$$

and it also has a unit. This time, it's a plain wire of appropriate type:

$$\Big|_B \circ \boxed{f} \;=\; \boxed{f} \circ \Big|_A \;=\; \boxed{f} \tag{3.14}$$

We already encountered these plain wires or *identities* in Section 3.1.3, where we mentioned that as a process, we can think of them as 'doing nothing' to the input. For a system of type A we will denote the identity on A symbolically as 1_A. The empty diagram is also an identity, but on the system-type I, so we denote it as 1_I.

3.2.3 Two Equivalent Definitions of Circuits

Definition 3.19 A diagram is a *circuit* if it can be constructed by composing boxes, including identities and swaps, by means of $\otimes$ and $\circ$.

Every diagram that we have seen in this chapter is in fact a circuit. Here is an example of the assembly of such a circuit:

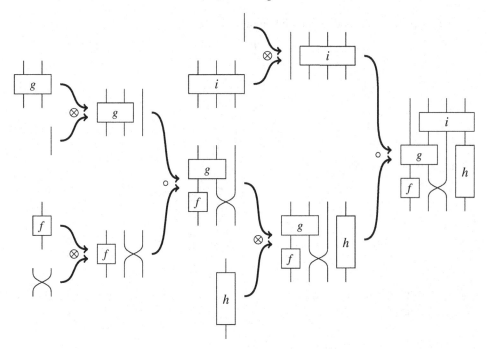

We can still recognise this assembly structure in the resulting diagram:

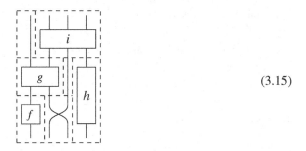

(3.15)

Note however that such a decomposition of a diagram does not uniquely determine the assembly process (e.g. associativity of $\otimes$ and $\circ$ allows for two manners of composing three boxes), and the same diagram may even allow for several distinct decompositions. We will see in the next section how this feature makes a non-diagrammatic treatment of circuits in terms of $\otimes$ and $\circ$ especially unwieldy.

Not all diagrams are circuits. To understand which ones are, we provide an equivalent characterisation that doesn't refer to the manner that one can build these diagrams but instead to a property that has to be satisfied.

Definition 3.20 A *directed path* of wires is a list of wires $(w_1, w_2, \ldots, w_n)$ in a diagram such that for all $i < n$, the wire w_i is an input to some box for which the wire w_{i+1} is an output. A *directed cycle* is a directed path that starts and ends at the same box.

An example of a directed path is shown in bold here:

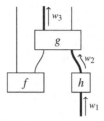

and an an example of a directed cycle is:

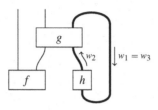

Remark* 3.21 In an area of mathematics called graph theory, graphs without directed cycles are called *directed acyclic*.

Theorem 3.22 The following are equivalent:

- a diagram is a circuit, and
- it contains no directed cycles.

Proof   There exists a (non-unique) way to express any diagram with no directed cycles in terms of $\otimes$, $\circ$, and swap. Since the diagram contains no directed cycles, we divide the diagram into 'layers' $l_1, \ldots, l_n$, where the outputs of each box (including identities and swaps) in layer l_i connect only to boxes in layer l_{i+1}. For example, one way to divide diagram (3.15) into layers is:

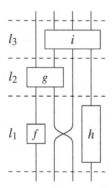

In each layer, $\otimes$ combines boxes into one, and $\circ$ combines the layers together, so we indeed have a circuit. Conversely, any diagram built up using just $\otimes$ and $\circ$ can be decomposed into such layers, which implies the absence of directed cycles. $\square$

Remark* 3.23 In relativity theory, a particular 'division of a process into layers' used in the previous proof is called a *foliation*. We will have a closer look at foliations in Section 6.3.3.

Since circuit diagrams contain no directed cycles, we can give them a 'temporal interpretation' by letting time flow from the bottom of the diagram to the top:

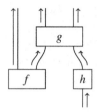

In particular, no wires go 'backwards in time'. This interpretation implicitly assumes that 'inside the boxes' the temporal flow is from the inputs to the outputs:

This is for example the case for identities as well as for swaps:

A canonical example of a diagram that is not a circuit is one involving a *feedback loop*; i.e. a box's output is connected to its own input:

which is indeed a directed cycle consisting of just one wire.

Remark* 3.24 Even more fundamental than foliations, all circuit diagrams admit a *causal structure*, which means that for any two boxes f and g, there are only three possibilities:

(1) f is in the causal past of g,
(2) g is in the causal past of f, or
(3) neither is in the causal past of the other.

We elaborate a bit more on the connection between circuits and the causal structure of spacetime in Section 6.3, when we explore the connections between quantum theory and the theory of relativity.

Circuit diagrams provide the correct language to discuss experimental setups that one can do in a laboratory, which take certain physical systems and/or data as input and which output other physical systems and/or data. 'Parallel' and 'sequential' then directly refer to how the devices are arranged.

Remark 3.25 Some might find the term 'circuit' somewhat oddly chosen for a kind of diagram that <u>excludes</u> feedback loops. This terminology is justified for two reasons. First, it mirrors the terminology used in quantum computing literature (cf. 'quantum circuits'). Second, while it is impossible to 'wire in' feedback loops, we can actually introduce feedback-like behaviour by introducing special boxes called 'caps' and 'cups' to circuits. We will see how this works in Chapter 4.

3.2.4 Diagrams Beat Algebra

In Section 3.1.3 we gave a representation of diagrams in terms of diagram formulas in order to convince the reader that one should think of diagrams as mathematical entities on par with traditional formulas. However, the diagram formulas caused an increase in bureaucracy and are not at all easy to parse. We have now seen that it is possible to represent (at least) circuit diagrams in a more traditional algebraic language using just $\otimes$ and $\circ$. One might now wonder if this is a good alternative to using diagrams. The answer is a resounding NO!

An algebraic syntax involving the $\otimes$- and $\circ$-symbols needs to be subjected to many additional equations, all of which are built in to the diagrammatic language. A first example of this is the fact that associativity and unitality of parallel composition are handled automatically:

- Drawing three boxes side by side implicitly assumes associativity:

$$(f \otimes g) \otimes h = \boxed{f}\ \boxed{g}\ \boxed{h} = f \otimes (g \otimes h)$$

- Composition with empty space does nothing:

$$f \otimes 1_I = \boxed{f}\ \vdots\ \overset{\ulcorner\ \ \ \urcorner}{\underset{\llcorner\ \ \ \lrcorner}{\ }}\ \vdots = \boxed{f} = f$$

Associativity and unitality of sequential composition are also built in: associativity is handled again by throwing away the brackets, and unitality is handled by depicting identities as a piece of wire with no box on it.

A more striking example occurs when we combine the two compositions. Consider the following two expressions:

$$(g_1 \otimes g_2) \circ (f_1 \otimes f_2) \qquad \text{and} \qquad (g_1 \circ f_1) \otimes (g_2 \circ f_2)$$

Evidently, in the absence of any additional equations on the $\otimes$- and $\circ$-symbols these are not equal. Now we compute their corresponding diagrams. For the first expression we obtain:

$$\left(\boxed{g_1} \otimes \boxed{g_2} \right) \circ \left(\boxed{f_1} \otimes \boxed{f_2} \right) = \left(\boxed{g_1}\ \boxed{g_2} \right) \circ \left(\boxed{f_1}\ \boxed{f_2} \right) = \begin{array}{c} \boxed{g_1}\ \boxed{g_2} \\ \boxed{f_1}\ \boxed{f_2} \end{array}$$

and for the second we obtain:

$$\left(\boxed{g_1} \circ \boxed{f_1} \right) \otimes \left(\boxed{g_2} \circ \boxed{f_2} \right) = \left(\begin{array}{c} \boxed{g_1} \\ \boxed{f_1} \end{array} \right) \otimes \left(\begin{array}{c} \boxed{g_2} \\ \boxed{f_2} \end{array} \right) = \begin{array}{c} \boxed{g_1}\ \boxed{g_2} \\ \boxed{f_1}\ \boxed{f_2} \end{array}$$

So we get the same diagram twice!

Since the $\otimes$- and $\circ$-symbols are supposed to be an algebraic syntax for diagrams, we need to impose a new equation:

$$(g_1 \otimes g_2) \circ (f_1 \otimes f_2) = (g_1 \circ f_1) \otimes (g_2 \circ f_2) \tag{3.16}$$

while in the diagrammatic language this is nothing but a tautology! So what is the actual content of (3.16)? It states that the composite process:

process g_1 takes place while process g_2 takes place,

after

process f_1 takes place while process f_2 takes place,

is the same as the composite process:

process g_1 takes place after process f_1 takes place,

while

process g_2 takes place after process f_2 takes place.

Evidently this is true, and it is something we get for free in the diagrammatic language, but in algebraic language we need to explicitly state this.

Given the equations we have stated thus far for the $\otimes$- and $\circ$-symbols, we can derive many new ones. For example:

$$\boxed{g} \circ \boxed{f} = \boxed{g} \otimes \boxed{f}$$

can be derived by combining (3.12), (3.14), and (3.16):

$$\boxed{g} \circ \boxed{f} = \left(\boxed{g} \otimes 1_I \right) \circ \left(1_I \otimes \boxed{f} \right) = \left(\boxed{g} \circ 1_I \right) \otimes \left(1_I \circ \boxed{f} \right) = \boxed{g} \otimes \boxed{f}$$

while in the diagrammatic language it is again a tautology. Another example of an extra equation for the $\otimes$- and $\circ$-symbols involves a crossing:

$$
\begin{array}{c}
\text{(crossing diagram)}
\end{array}
=
\begin{array}{c}
\text{(diagram with } g \text{ and } f\text{)}
\end{array}
\tag{3.17}
$$

One can easily imagine how many more non-trivial equations are derivable when combining (3.11), (3.12), (3.14), (3.16), and (3.17), many of which would look not at all obvious without drawing the diagram.

Exercise* 3.26 Assume the following equation:

$$\sigma_{A \otimes B, C} = (\sigma_{A,C} \otimes 1_B) \circ (1_A \otimes \sigma_{B,C}) \qquad \text{where} \qquad \sigma_{A,B} := \begin{array}{c} \text{(crossing diagram with } B, A \text{ top and } A, B \text{ bottom)} \end{array}$$

Prove the first and the last diagram equation from Example 3.6 as algebraic equations using just the above equation and the other algebraic equations introduced in this section.

So why do things become so complicated? In simple terms, one is trying to squeeze something that wants to live in two dimensions (a diagram) into one dimension (a 'linear' algebraic notation). When we do this, the horizontal and the vertical space on the piece of paper we used to draw our diagram suddenly coincide, and we need a bunch of extra syntax (e.g. brackets), to disambiguate the parallel and sequential compositions in their new, 'compressed' world. We then need to subject this extra syntax to a bunch of extra rules to enable the kinds of deformations we were doing quite naturally with diagrams before. The punchline:

$$\boxed{\textbf{Diagrams rule!}}$$

There is a lot more to say about the precise connection between algebraic equations and diagrammatic reasoning, particularly in the context of an area called *(monoidal) category theory*. Although the latter is not crucial to understanding this book, we refer the interested reader to the advanced material in Section 3.6.2.

3.3 Functions and Relations as Processes

We now introduce two simple process theories: the theory of **functions** and the theory of **relations**. Besides it being useful to have some concrete examples of process theories in hand in order to understand some of the concepts to come, these examples establish two important insights:

- that traditional mathematical structures (sets, in this case) can form the types of a process theory, and

- that, even for some fixed system-types, the choice of processes is in fact far more important in determining the character of a process theory. In particular, while the process theories **functions** and **relations** are both based on sets, we shall see in this and the next chapter that their properties as process theories couldn't be farther apart.

For these reasons, these example process theories and, maybe somewhat surprisingly, **relations** in particular, will prove a useful stepping stone towards the process theories of **linear maps** and **quantum maps**.

3.3.1 Sets

To define a process theory, we need to first say what the system-types are, then what the processes are. For both **functions** and **relations**, the system-types are just sets. We will encounter some familiar sets, like the natural numbers $\mathbb{N}$, the real numbers $\mathbb{R}$, the complex numbers $\mathbb{C}$ (whose properties we review in Section* 5.3.1), and:

$$\mathbb{B} := \{0, 1\}$$

which is called the set of *booleans* or *bit values*. Joint system-types are formed by taking the *Cartesian product* of sets, that is:

$$A \otimes B := A \times B = \{(a, b) \mid a \in A, b \in B\}$$

The definition is similar for three or more sets, in which case we replace pairs of elements by *tuples*:

$$A_1 \times \cdots \times A_n := \{(a_1, \ldots, a_n) \mid a_i \in A_i\}$$

If we ignore brackets on tuples of elements, e.g. letting:

$$((a, b), c) = (a, b, c) = (a, (b, c))$$

parallel composition of systems is associative:

$$A \times (B \times C) = (A \times B) \times C$$

as required.

Example 3.27 The set of *bitstrings* of length n can be expressed as an n-fold Cartesian product:

$$\underbrace{\mathbb{B} \times \cdots \times \mathbb{B}}_{n}$$

We can think of this as the set of natural numbers ranging from 0 to $2^n - 1$ by considering the bit strings as a binary representations of these numbers; e.g. for $n = 4$ we represent 0 as $(0, 0, 0, 0)$, 1 as $(0, 0, 0, 1)$, 2 as $(0, 0, 1, 0)$, ..., up to 15, which we represent as $(1, 1, 1, 1)$. This is a trick often used in computer science, and it is also important in this book.

The trivial type I is defined to be the set $\{*\}$ that contains just a single element, here called '$*$'. If we always drop $*$ from tuples, e.g. letting:

$$(a, *) = a = (*, a)$$

then $\{*\}$ becomes the unit for parallel composition of wires:

$$A \times \{*\} = A = \{*\} \times A$$

Remark* 3.28 Strictly speaking, the sets $(A \times B) \times C$ and $A \times (B \times C)$, as well as the sets $A \times \{*\}$ and A, are not equal, but *isomorphic*; that is, their elements are in one-to-one correspondence. In fact, they are not just isomorphic but isomorphic in a very strong sense, called *naturally isomorphic*, which pretty much means that for all practical purposes they can be treated as equal. The failure of strict equality is mainly due to the sort of 'algebraic bureaucracy' we complained about in Section 3.2.4, so it won't have any effect on working with diagrams. Section* 3.6.2 contain more details on this somewhat delicate issue.

3.3.2 Functions

In the process theory of **functions** every process:

is a function from a set A to a set B. Since joint systems are formed using the Cartesian product, a function with many inputs and outputs, e.g.

is a function from the Cartesian product of sets $A \times B$ to the Cartesian product $B \times C \times D$. So f maps pairs $(a, b) \in A \times B$ to triples $(b', c, d) \in B \times C \times D$. More specifically, a function from $\mathbb{B} \times \mathbb{B}$ to $\mathbb{B} \times \mathbb{B} \times \mathbb{B}$ is a function from bit strings of length 2 to bit strings of length 3, for example:

$$
\begin{cases}
(0,0) \mapsto (0,0,0) \\
(0,1) \mapsto (0,1,1) \\
(1,0) \mapsto (1,0,1) \\
(1,1) \mapsto (0,1,1)
\end{cases}
$$

Now that we know what the system-types and processes are, we need to say what 'wiring processes together' means (cf. Definition 3.1). We will do so by specifying what parallel and sequential composition for functions means. Sequential composition is the usual composition of functions:

$$(g \circ f)(a) := g(f(a)) \tag{3.18}$$

Parallel composition is defined as applying each function to its corresponding element of a pair:

$$(f \otimes g)(a, b) := (f(a), g(b)) \tag{3.19}$$

This tells us everything we need to know to compute the function represented by a circuit diagram, since we can decompose the diagram across $\otimes$ and $\circ$:

$$= (1_A \otimes g) \circ (f \otimes h)$$

then applying equations (3.18) and (3.19). There might be more than one way to decompose a diagram, but the actual function computed will always be the same. Note that this will work for any diagram where the wires do not cross. To handle crossings, we need to define a 'swap' function.

Exercise 3.29 Which function represents the swap:

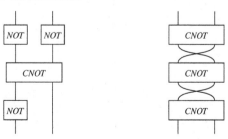

where A and B can be any set?

Exercise 3.30 Compute the functions:

where:

$$\boxed{NOT} \quad :: \quad \begin{cases} 0 \mapsto 1 \\ 1 \mapsto 0 \end{cases} \qquad\qquad \boxed{CNOT} \quad :: \quad \begin{cases} (0,0) \mapsto (0,0) \\ (0,1) \mapsto (0,1) \\ (1,0) \mapsto (1,1) \\ (1,1) \mapsto (1,0) \end{cases}$$

After reading this book, you will be able to straightforwardly write down the answer for each of these two diagrams (not from memory, but because the calculation is so easy with the tools we will provide!).

3.3.3 Relations

In the process theory of **relations** every process:

$$\boxed{R}$$

is a *relation* from a set A to a set B, that is, a subset:

$$R \subseteq A \times B$$

We say a and b are *related* by R if $(a, b) \in R$. It will be useful to use a more function-like notation, writing:

$$R :: a \mapsto b \qquad \text{instead of} \qquad (a,b) \in R \qquad\qquad (3.20)$$

and as a counterpart to the notation $f(a)$ for functions, we can set:

$$R(a) := \{b \mid R : a \mapsto b\}$$

Thus relations that happen to be functions can be expressed in the same notation as before. However, for more general relations, we are allowed to map a single element in the input to any number of elements in the output. For example:

$$\boxed{R} \quad :: \quad \begin{cases} a \mapsto x \\ a \mapsto y \\ b \mapsto z \end{cases}$$

which we can also write as:

$$\boxed{R} \quad :: \quad \begin{cases} a \mapsto \{x, y\} \\ b \mapsto z \\ c \mapsto \emptyset \end{cases}$$

Interpreted as a process, a relation is like a function with added *non-determinism*: a single input can map to several outputs or none at all.

Relations are composed sequentially as follows: an element a is related to c by $(S \circ R)$ if and only if there exists some b such that $R :: a \mapsto b$ and $S :: b \mapsto c$. Symbolically:

$$\boxed{\begin{array}{c} S \\ \hline R \end{array}} :: a \mapsto c \iff \exists b \left(\boxed{R} :: a \mapsto b \text{ and } \boxed{S} :: b \mapsto c \right) \tag{3.21}$$

Note how this 'if and only if' expression is being used to <u>define</u> a new relation; that is, it tells us precisely which elements a and c are related by $S \circ R$. You may have seen a picture of relation composition in school that looks something like this:

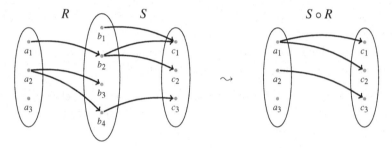

Here, we can see that elements are related by $S \circ R$ if and only if there is a path connecting them in the diagram on the left.

Parallel composition of relations is defined as follows:

$$\boxed{R} \, \boxed{S} :: (a, b) \mapsto (c, d) \iff \left(\boxed{R} :: a \mapsto c \text{ and } \boxed{S} :: b \mapsto d \right) \tag{3.22}$$

As in the case of functions, we can now compute a relation represented by a diagram by decomposing it over $\otimes$ and $\circ$ then applying (3.21) and (3.22), but this can be quite a bit of work. Luckily, there is a general procedure for computing diagrams of relations directly. Here's how it goes: suppose we have some known relations R, S, and T, and we want to compute a new relation P as their composition:

$$\begin{array}{c} A \quad B \quad C \\ \boxed{P} \\ B \quad A \end{array} := \quad \begin{array}{c} A \quad B \quad C \\ \boxed{S} \\ \boxed{R} \quad \boxed{T} \\ B \quad A \end{array} \tag{3.23}$$

Step 1: Write P as a diagram formula:

$$P_{B_1 A_1}^{A_2 B_2 C_1} = R_{B_1}^{A_2 A_3} S_{A_3 D_1}^{B_2 C_1} T_{A_1}^{D_1}$$

Step 2: Change wire names to elements of the appropriate set. For example, write B_2 as an element $b_2 \in B$:

$$P_{b_1 a_1}^{a_2 b_2 c_1} \iff R_{b_1}^{a_2 a_3} S_{a_3 d_1}^{b_2 c_1} T_{a_1}^{d_1}$$

Step 3: Change box names to actual relations that map the element(s) in the subscript to the element(s) in the superscript, and adjoin '$\exists x$' for every repeated element x:

$$P :: (b_1, a_1) \mapsto (a_2, b_2, c_1) \iff \exists a_3 \exists d_1 \left(\begin{array}{l} R :: b_1 \mapsto (a_2, a_3) \text{ and} \\ S :: (a_3, d_1) \mapsto (b_2, c_1) \text{ and} \\ T :: a_1 \mapsto d_1 \end{array} \right)$$

In the absence of inputs and/or outputs, one must use '$*$', i.e. the unique element in the set $I = \{*\}$.

Step 4: The expression in the RHS of '$\iff$' now uniquely characterises all related tuples in the LHS, and hence can now be used to compute P.

For the particular cases of sequential and parallel composition of relations, we recover (3.21) and (3.22), respectively. To see this, just draw the diagrams of $f \otimes g$ and $g \circ f$, apply steps 1–3, and see what comes out.

Exercise 3.31 Suppose A, B, C, and D are the following sets:

$$A = \{a_1, a_2, a_3\}$$
$$B = \mathbb{B}$$
$$C = \{\textbf{red}, \textbf{green}\}$$
$$D = \mathbb{N}$$

Compute P as defined by (3.23), for R, S, T defined as follows:

$$R :: \begin{cases} 1 \mapsto (a_1, a_1) \\ 1 \mapsto (a_1, a_2) \end{cases} \quad S :: \begin{cases} (a_1, 5) \mapsto (0, \textbf{red}) \\ (a_1, 5) \mapsto (1, \textbf{red}) \\ (a_2, 6) \mapsto (1, \textbf{green}) \end{cases} \quad T :: \begin{cases} a_1 \mapsto 200 \\ a_3 \mapsto 5 \end{cases}$$

and for R, S, T defined as follows:

$$R :: \begin{cases} 0 \mapsto A \times \{a_2, a_3\} \\ 1 \mapsto A \times \{a_2, a_3\} \end{cases} \quad S :: \begin{cases} (a_1, 0) \mapsto \mathbb{B} \times \{\textbf{red}, \textbf{green}\} \\ (a_1, 1) \mapsto \mathbb{B} \times \{\textbf{red}, \textbf{green}\} \\ (a_1, 2) \mapsto \mathbb{B} \times \{\textbf{red}, \textbf{green}\} \\ \vdots \end{cases} \quad T :: \begin{cases} a_1 \mapsto \mathbb{N} \\ a_2 \mapsto \mathbb{N} \\ a_3 \mapsto \mathbb{N} \end{cases}$$

Remark* 3.32 Due to the close resemblance between the process theories **relations** and **linear maps**, and in particular, the fact that they both admit a *matrix calculus* (see Chapter 5), this 'trick' for computing diagrams of relations applies also for computing diagrams of linear maps, as we shall see in Section 5.2.4, and hence will also be relevant for **quantum processes**. In fact, if we treat relations as matrices with boolean entries, the computation is essentially identical.

3.3.4 Functions versus Relations

When we restrict the definitions (3.21) and (3.22) to functions, we obtain (3.18) and (3.19), respectively. In both cases, we took advantage of the fact that functions relate each input element to a unique output to obtain a simpler form. For example, in the case of parallel composition of functions $f : A \to C$ and $g : B \to D$, for each $a \in A$ and each $b \in B$ there is a unique $c \in C$ and $d \in D$ such that $f :: a \mapsto c$ and $g :: b \mapsto d$, and hence, building $f \otimes g$ amounts to merely pairing these two elements as we did in (3.19). On the other hand, for relations, $R \otimes S :: (a, b) \mapsto (c, d)$ whenever there exist both a $c \in C$ and a $d \in D$ such that $R :: a \mapsto c$ and $S :: b \mapsto d$, and there could of course be more than one of such pair (c, d).

Now, since:

1. both **functions** and **relations** have sets as system-types, so in particular, the system-types of **relations** include those of **functions**;
2. functions are a special case of relations, and hence the processes of **relations** include the processes of **functions**; and
3. sequential composition ∘ of **functions** is a special case of sequential composition of **relations**, and parallel composition ⊗ of **functions** is a special case of parallel composition of **relations**,

we say that the process theory of **functions** is a *subtheory* of **relations**. We can write this as follows:

$$\text{\textbf{functions} } \subseteq \text{ \textbf{relations}}$$

Of course, there are many more relations than there are functions, and throughout this book we will encounter many important relations that aren't functions. In fact, these extra relations are precisely what make the process theory of **relations** behave much more like **linear maps** than **functions**.

3.4 Special Processes

In this section, we single out several types of process that will play a central role throughout this book.

3.4.1 States, Effects, and Numbers

So far, we have been handling boxes without any inputs and/or without any outputs just like any other boxes. However, these particular boxes have a very special status when we interpret them as processes.

- *States* are processes without any inputs. In operational terms they are 'preparation procedures'. We represent them as follows:

Whereas a system *A* could be in any number of possible different states, a preparation procedure will produce a system of type *A* in a <u>particular</u> state ψ taken from these possibilities. The fact that there is no wire in means we don't know (or care) where this system came from, just that we have it now and it is in a particular state. For example, if we say 'here is a bit initiated in the 0 state', then we don't really care about the previous history of that bit as long as we have the guarantee that it is currently in state 0 and that it is available to perform some further computations. Or, in the case of saying 'here is a fresh raw potato', we don't really care whose farm it came from or which animal's manure was used, as long as we know that we have a fresh raw potato that we can process further into fries.

- *Effects* are processes without any outputs. We have borrowed this terminology from quantum theory, where effects play a key role. From now on we represent them as follows:

The notion of effect is dual to the notion of state, in that one starts with a system and has nothing left afterwards – or, we simply don't care what we have afterwards. The simplest example of an effect consists of *discarding* a system, where the system is destroyed (or simply disregarded). Less trivial effects refer to things that may or may not happen, depending on the actual state of a system. For example, 'the bomb exploded', 'the photon was absorbed', 'all the food was eaten', or 'the dodo became extinct' could all be considered effects, since afterwards there is no bomb, no photon, no food, and no dodo.

Operationally speaking, effects are used to model 'tests'. When we say an effect 'happens', it means we tested whether a system satisfies a certain property and have obtained 'yes' as the answer, after which we discard that system, or – as is often the case – the system is destroyed in the verification process. Altogether, a test consists of:

1. a question about a system,
2. a procedure that enables one to verify this question, and
3. the event of obtaining 'yes' as the answer; i.e. the effect 'happened'.

So, there is more to the notion of a test than just answering a question. For example, when making a drawing, one could ask the question: 'Is this a red pen?' A corresponding test would then consist of taking a piece of paper, then writing with the pen on it, and observing that the colour is indeed red. Of course, in this case one would (usually) not destroy the pen. Genuine destruction takes place when one asks the question: 'Is this a working match?' A corresponding test would then consist of trying to light the match and succeeding, after which we end up with no match. The act of discarding a system is also a test, albeit a trivial one. After verifying some property that holds for any state of a system (e.g. 'does this system exist?'), we discard it. This is of course a very silly test, since no information is gained, but the system is lost.

At first glance, it seems like tests are more than just 'preparation procedures in reverse'. However, if we assume that preparation procedures can fail (as is often the case in the lab), we see that the notions really are symmetric. A preparation procedure consists of:

1. a state we would like our system to be in,
2. a procedure that enables one to put a system in that state, and
3. the event of obtaining 'yes'; i.e. the state was successfully prepared.

Of course, unlike tests, we could just keep retrying until a preparation succeeds (or we run out of funding), so we tend to disregard part 3.

When we compose a state and an effect, a third kind of special box arises:

$$ \overset{\pi}{\triangle} \circ \underset{\psi}{\triangledown} = \begin{array}{c} \overset{\pi}{\triangle} \\ \underset{\psi}{\triangledown} \end{array} $$

- *Numbers* are processes without any inputs or outputs. From now on we represent them as:

$$ \langle\!\langle \lambda \rangle\!\rangle $$

or sometimes simply as:

$$ \lambda $$

At this point, you're probably saying 'Hang on! I already know what numbers are, and this doesn't have anything to do with processes with no inputs and outputs!' Well, the numbers you learned about in school (e.g. whole numbers, real numbers, possibly complex numbers) are all instances of this very general kind of 'number'.

So, what are processes with no inputs or outputs? Well, they are just a set of things that can be 'multiplied', thanks to $\otimes$-composition (or equivalently in this case, $\circ$-composition). That is, if λ and μ are numbers, then:

$$ \lambda \otimes \mu := \langle\!\langle \lambda \rangle\!\rangle \; \langle\!\langle \mu \rangle\!\rangle $$

is also a number. Also, we saw in Section 3.2.4 that diagrams rule, so this 'multiplication' is associative:

$$ (\lambda \otimes \mu) \otimes \nu = \langle\!\langle \lambda \rangle\!\rangle \; \langle\!\langle \mu \rangle\!\rangle \; \langle\!\langle \nu \rangle\!\rangle = \lambda \otimes (\mu \otimes \nu) $$

and has a unit, given by the empty diagram:

$$ \lambda \otimes 1_I = \langle\!\langle \lambda \rangle\!\rangle \; \begin{array}{|c|} \hline \\ \hline \end{array} = \lambda $$

And finally, because of the lack of inputs and outputs numbers can move around liberally within a diagram, and even can do a little dance:

$$
\begin{array}{c} \lambda \\ \\ \mu \end{array} \;=\; \begin{array}{c} \lambda \\ \mu \end{array} \;=\; \lambda \;\; \mu \;=\; \begin{array}{c} \mu \\ \lambda \end{array} \;=\; \begin{array}{c} \mu \\ \\ \lambda \end{array}
\tag{3.24}
$$

This little dance shows that multiplication of numbers is furthermore *commutative*. So, what we call 'numbers' consists of a set of things that can be 'multiplied', and that multiplication operation is associative, commutative, and has a unit (i.e. a chosen number that means '1'). Sound familiar? Mathematicians call this a *commutative monoid*. Other people would call this 'virtually any kind of numbers you can imagine.'

Remark* 3.33 This remarkably simple little proof shown in (3.24) is known as the *Eckmann–Hilton argument*. It shows in general that, whenever we have a pair of associative operations (in this case $\otimes$ and $\circ$, applied specifically to numbers) that share the same unit (1_I in both cases) and satisfy the interchange law (3.16) (which of course comes for free with diagrams), both of these operations are commutative and are in fact the same operation. Here, we used it to prove that $\otimes$ (or equivalently $\circ$) gives us a commutative 'multiplication' operation for the numbers of any process theory.

So, how should we interpret the fact that, when a state meets an effect, a number pops out? One extremely useful interpretation is that this number gives the *probability* that the effect happens, given the system is in a particular state. Or, in terms of tests, it is the probability that when we test the state ψ for an effect π we get 'yes' as the answer:

$$
\left. \begin{array}{l} \text{test} \left\{ \begin{array}{c} \pi \\ \end{array} \right. \\[2mm] \text{state} \left\{ \begin{array}{c} \psi \\ \end{array} \right. \end{array} \right\} \text{probability}
\tag{3.25}
$$

We refer to this as the *generalised Born rule*. Here, we say 'generalised' because it makes sense in any process theory. We will see in Chapter 6 that by restricting to the theory of **quantum maps**, we obtain the rule that is used to compute all probabilities within quantum theory, which is called simply the *Born rule*.

Remark* 3.34 A reader familiar with quantum theory may think that rather than a probability, in (3.25) we obtain a complex number (a.k.a. an 'amplitude'), which we then have to multiply with its conjugate in order to obtain a probability. This is indeed the case if we are dealing with plain old **linear maps**. However, **quantum maps** have this step 'built in', so the numbers are always positive real numbers. In Example 3.37 below we indicate how (3.25) indeed produces the correct probabilities in quantum theory, and in Section 6.1.1 we explain how the passage from complex numbers to probabilities, via multiplication of a number with its conjugate, is nicely built in to the theory of **quantum maps**.

In many physical theories, and quantum theory in particular, states are thought of as elements in some set (or space), and processes as functions between sets and/or spaces.

Numbers are yet again another thing, and effects in many cases don't even seem to have a clear counterpart. However, in the diagrammatic language, we treat states, effects, and numbers all on the same footing as special cases of processes. This is very convenient because we can define many concepts for arbitrary processes that will apply immediately to all of these special cases.

We will now see how all of these special kinds of processes look in the process theories we have encountered so far.

Example 3.35 (functions)

1. *States correspond to the elements of a set.* Since 'no wire' means the single element set {∗} (see Section 3.3.1), a state is a function from {∗} into another set A. This function 'points to' a single element $a \in A$, namely, the image of ∗:

$$\bigtriangledown_a \quad :: * \mapsto a$$

Conversely, for each $a \in A$, there is a unique function that sends ∗ to a, so elements of A and functions from {∗} into A are essentially the same thing, which justifies us simply calling this function a. If we let $A := \mathbb{B}$, there are exactly two states:

$$\bigtriangledown_0 \quad :: * \mapsto 0 \qquad\qquad \bigtriangledown_1 \quad :: * \mapsto 1$$

2. *Effects are boring.* For any set A, there is exactly one function from A to {∗}, namely, the function that sends everything to ∗. Therefore, we shouldn't even bother to give it a name. Again letting $A := \mathbb{B}$ this unique effect is:

$$\bigtriangleup \quad :: \begin{cases} 0 \mapsto * \\ 1 \mapsto * \end{cases}$$

3. *Numbers are boring* – for the same reason that effects are boring. Since a number is a special case of an effect, there is only one number, namely, the function that sends ∗ to ∗. This unique number is already represented by the empty diagram, so we don't even need to draw it!

So in summary, states of type A are the elements of A, but there isn't enough freedom with functions to get interesting effects and numbers.

Let's see what happens if we generalise to relations.

Example 3.36 (relations)

1. *States correspond to subsets of a set (a.k.a. 'non-deterministic' elements).* A state is a relation from the single element set {∗} to another set A, thus it will relate ∗ to zero or more elements of A. Just as we named states in **functions** by set elements, we can name states in **relations** by the subset $B \subseteq A$ that ∗ 'points to':

$$\overset{\displaystyle\downarrow}{\underset{B}{\bigtriangledown}} \;\; :: * \mapsto B$$

When $A := \mathbb{B}$ there are four states in **relations** corresponding to each of the four subsets of $\{0, 1\}$:

$$\overset{\displaystyle\downarrow}{\underset{\emptyset}{\bigtriangledown}} \;\; :: * \mapsto \emptyset \qquad \overset{\displaystyle\downarrow}{\underset{0}{\bigtriangledown}} \;\; :: * \mapsto \{0\} \qquad \overset{\displaystyle\downarrow}{\underset{1}{\bigtriangledown}} \;\; :: * \mapsto \{1\} \qquad \overset{\displaystyle\downarrow}{\underset{\mathbb{B}}{\bigtriangledown}} \;\; :: * \mapsto \mathbb{B}$$

As with **functions**, we have the states 0 and 1, which correspond to the system being definitely in each of these respective states. We can interpret the state $\mathbb{B}$ as the system being in either of the two states 0 or 1. The state $\emptyset$ corresponds to the system being in neither 0 nor 1, i.e. an 'impossible' state of the system.

2. *Effects also correspond to subsets.* Relations from a set A to $\{*\}$ are uniquely defined by the subset $B \subseteq A$ consisting of all $a \in A$ where $a \mapsto *$. For $A := \mathbb{B}$ all the possible effects are:

$$\overset{\bigtriangleup_{\emptyset}}{\underset{\displaystyle\mid}{}} \; :: \begin{cases} 0 \mapsto \emptyset \\ 1 \mapsto \emptyset \end{cases} \quad \overset{\bigtriangleup_{0}}{\underset{\displaystyle\mid}{}} \; :: \begin{cases} 0 \mapsto \{*\} \\ 1 \mapsto \emptyset \end{cases} \quad \overset{\bigtriangleup_{1}}{\underset{\displaystyle\mid}{}} \; :: \begin{cases} 0 \mapsto \emptyset \\ 1 \mapsto \{*\} \end{cases} \quad \overset{\bigtriangleup_{\mathbb{B}}}{\underset{\displaystyle\mid}{}} \; :: \begin{cases} 0 \mapsto \{*\} \\ 1 \mapsto \{*\} \end{cases}$$

The effects 0 and 1 can be interpreted as tests that verify if the system is in the state 0 or 1, respectively. The effect $\mathbb{B}$ can be interpreted as the trivial test that gives 'yes' as the answer for any of the three 'possible' states 0, 1, and $\mathbb{B}$, whereas the effect $\emptyset$ is the 'impossible' effect, which will never happen.

3. *There are two numbers corresponding to 'impossible' and 'possible'.* There are two relations from the set $\{*\}$ to itself, namely the empty relation and the identity relation:

$$\emptyset :: * \mapsto \emptyset \qquad\qquad \boxed{} \;\; :: * \mapsto \{*\}$$

We interpret these two numbers as impossible and possible, which makes perfect sense when we look at what happens when we test whether the system is in a particular state. First consider:

$$\overset{\bigtriangleup_{0}}{\underset{\bigtriangledown_{0}}{}} \; = \; \boxed{}$$

If the system is in the state 0 it should be *possible* to get a positive outcome when testing for effect 0. Of course it's possible, and in fact this should always happen! However, as there is no number 'certain' in **relations**, 'possible' is the most we can get. Next, consider:

$$\overset{\bigtriangleup_{0}}{\underset{\bigtriangledown_{\mathbb{B}}}{}} \; = \; \boxed{}$$

If the state is either 0 or 1, then it is *possible* (but no longer certain) that testing for 0 will yield a positive outcome. Finally consider:

 $= \emptyset$

If the system is in the state 0, then it is indeed *impossible* for the test 1 to give a positive outcome.

The following starred example uses some notation that will be introduced over the next two chapters. However, those already familiar with quantum theory may find it useful. Other readers should skip this example for now.

Example* 3.37 (quantum maps) Quantum maps give a particularly elegant way to express completely positive maps, of which mixed quantum states, quantum effects, and probabilities are all special cases. Most of Chapter 6 is devoted to them, but here we provide a first glimpse. For the benefit of those already familiar with quantum theory, we use traditional terminology and notation here, rather than the more process-theoretic versions introduced in Chapter 6.

1. *States correspond to density operators.* These are depicted as:

The action of a *completely positive map* $\mathcal{E}$ on a state ρ, that is:

$$\mathcal{E} :: \rho \mapsto \sum_i A_i^\dagger \rho A_i$$

is depicted as:

2. *Effects are positive linear functionals.* These are depicted as:

for some positive operator A. Their action on a state ρ, that is:

$$\rho \mapsto \mathrm{tr}(A\rho)$$

(where tr is the trace) is depicted as:

$$\begin{array}{c}\text{[diagram]}\end{array} \tag{3.26}$$

A special case is the trace itself:

$$\rho \mapsto \text{tr}(\rho)$$

which represents 'discarding the system' and is depicted as:

The conditions of being trace-1 (for states) and trace-preserving (for completely positive maps) are written respectively as:

3. *Numbers are positive reals.* The Born rule (3.25) now takes the form (3.26), and it indeed produces positive reals, which are interpreted as probabilities.

What may surprise the reader is that density matrices are depicted as states, i.e. no input, rather than as input-output processes (cf. density 'operator') and that completely positive maps are depicted as input-output processes, rather than as something that takes input-output processes to input-output processes (cf. 'superoperator'). This is a consequence of the manner in which we define *quantum maps*, and this is of course an upshot, since now the diagrams reflect the true nature of quantum states/effects/processes.

3.4.2 Saying the Impossible: Zero Diagrams

An extreme example of (non-function) relations are *zero relations*. These are relations in which 'nothing relates to nothing'; i.e. there are no $a \in A$ and $b \in B$ such that $R :: a \mapsto b$, or equivalently:

$$\boxed{0} :: a \mapsto \emptyset$$

We already encountered a few of these, namely:

$$\bigtriangledown :: * \mapsto \emptyset \qquad \bigtriangleup :: \begin{cases} 0 \mapsto \emptyset \\ 1 \mapsto \emptyset \end{cases} \qquad \emptyset :: * \mapsto \emptyset$$

Each of these represented something 'impossible'.

Since an effect represents a test in which we successfully verify a certain question, it may be the case that for certain states this effect is simply impossible. So what do we get if we compose that state and that effect in a diagrammatic language? We should of course get a number that corresponds to 'impossible'. These impossible numbers, and more generally impossible processes, have an elegant characterisation within a process theory.

For the specific case of **relations**, it is easily seen that both the sequential and parallel composition of a zero relation with any relation is again a zero relation. For general process theories, assuming that an effect π is impossible given a state ψ, the number:

$$0 = \overset{\displaystyle \triangle\!\!\pi}{\underset{\displaystyle \triangledown\!\!\psi}{}}$$

should represent the impossible. Of course, if part of a larger process is impossible, then the entire process is impossible, hence:

$$\boxed{0} := 0\,\boxed{f}$$

should also be impossible for any f. So if we want to accommodate the impossible in our language, for every possible input and output type there should be a *zero process*, which obeys the following composition rules:

$$\frac{\boxed{0}}{\boxed{f}} = \frac{\boxed{f}}{\boxed{0}} = \boxed{0} \qquad \boxed{0}\,\boxed{f} = \boxed{0} \tag{3.27}$$

Exercise 3.38 Prove that if a zero process satisfying (3.27) exists, then for any given types of input and output systems, it is unique.

So just as multiplying by the plain old number zero always yields zero, zero processes 'absorb' any diagram in which they occur. In other words, any diagram containing a zero process is itself a zero process. Due to the uniqueness of the zero process, we will just write it as '0' and ignore its input and output wires:

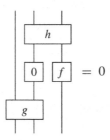 $= 0$

The main point to remember here is that 0 is <u>not</u> the empty diagram, since the empty diagram can be adjoined to any diagram without changing it. Zeros, on the other hand, 'eat' everything around them!

3.4.3 Processes That Are Equal 'Up to a Number'

There will be many instances in this book where numbers are crucially important, most notably when the Born rule (3.25) is used to compute probabilities. However, on other occasions, they are just a bit of a nuisance. For example, we may only be interested in some qualitative aspect of a process, such as whether it separates into disconnected pieces (cf. Section 4.1.1 in the next chapter), in which case numbers play no role. In other cases, we can make life a bit easier by ignoring numbers throughout a calculation and figuring out at the end what they should be. Therefore, we'll introduce some notation to handle these kinds of situations:

Definition 3.39 Two processes are *equal up to a number*, written:

$$\boxed{f} \approx \boxed{g}$$

if there exist non-zero numbers λ and μ such that:

$$\lambda \boxed{f} = \mu \boxed{g} \tag{3.28}$$

We require λ and μ to be non-zero because otherwise everything would be $\approx$-related to everything else:

$$0 \boxed{f} = 0 = 0 \boxed{g}$$

which is not particularly useful. What should instead be true is that the only thing ≈ 0 is zero itself.

Exercise 3.40 Assume a process theory has *no zero-divisors*; i.e. for all processes f and numbers λ we have $\lambda f = 0$ if and only if $\lambda = 0$ or $f = 0$. Then show that:

$$\boxed{f} \approx 0 \quad \Longrightarrow \quad \boxed{f} = 0$$

A fortunate feature of the relation $\approx$ is that it plays well with diagrams. Suppose for example that $h \approx h'$, then:

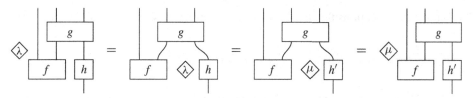

So, for any diagram containing h, we have:

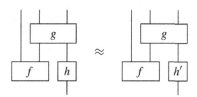

Even if we ignore numbers in a calculation, we might still recover them at the end. Suppose we have established that:

$$\boxed{f} \approx \boxed{g}$$

If there exists a state ψ and an effect π such that:

$$\psi\text{-}f\text{-}\pi \neq 0 \neq \psi\text{-}g\text{-}\pi$$

then by sandwiching both f and g in equation (3.28) we obtain:

$$\lambda\,\psi\text{-}f\text{-}\pi = \mu\,\psi\text{-}g\text{-}\pi$$

where all numbers are non-zero. As a consequence, if the numbers of a process theory are not too crazy, then we can figure out what λ and μ must be in order to obtain an equality.

3.4.4 Dirac Notation

Throughout this chapter we have extolled the virtues of diagrams, most coming from the freedom to write down processes in two dimensions. With modern technology, these diagrams are pretty easy to stick in, say, a textbook. However, things weren't always this good. Let's see now how far we get if we try to turn diagrams into a notation that can easily be churned out on a good old-fashioned `typewriter`.

Our initial conventions for states and effects are as follows:

D1: $\bigtriangledown_\psi$ is written as $|\psi\rangle$ and called a 'Dirac ket'

D2: $\bigtriangleup_\pi$ is written as $\langle\pi|$ and called a 'Dirac bra'

We can now also consider compositions of a state and an effect:

D3: is written as $\langle\pi|\psi\rangle$ and called a 'Dirac bra(c)ket'

As our naming indicates this notation was introduced by a physicist called Paul Dirac, specifically for the purpose of describing quantum theory. It is still widely used in most textbooks on quantum theory today.

Remark 3.41 We have slightly deviated from the usual Dirac notation since we are missing one essential refinement of the diagrammatic language to give a complete account here, namely the ability to turn a ket into a bra. We will do this in Section 4.3.3 where we provide modified versions of **D2** and **D3**.

Note that there's a pretty easy recipe for turning diagrams of states and effects into Dirac notation:

So, up to some cutting and rotating, Dirac notation is actually a (one-dimensional) subset of our diagrammatic notation. With that in mind, let's continue our game of fitting diagrams on a line:

D4: is written as $|\psi\rangle\langle\pi|$

For processes that are not states or effects, we cut the whole box off and write sequential composition just by writing processes from right to left:

D5: 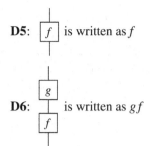 is written as f

D6: is written as gf

Then, exploiting the fact that for numbers λ, μ we have that $\lambda \otimes \mu = \lambda \circ \mu$, we can obtain expressions like:

Ex: 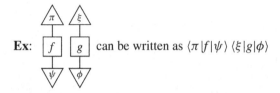 can be written as $\langle \pi |f| \psi \rangle \, \langle \xi |g| \phi \rangle$

But things become of course much trickier when wires and boxes are genuinely side by side, for example, when considering boxes with two wires in and/or out. We can make some progress by writing multiple states or effects inside of a ket or bra to indicate parallel composition:

D8:  is written as $|\psi \phi\rangle$

D9: is written as $\langle \pi \xi |$

Hence we obtain:

Ex: 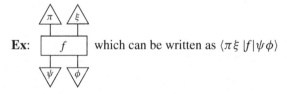 which can be written as $\langle \pi \xi |f| \psi \phi \rangle$

But once the diagrams become more involved, we will have to make parallel composition explicit using '$\otimes$' or employing some other trickery, like letting $f_A := f \otimes 1_B$ and $g_B := 1_A \otimes g$ then setting:

D10: 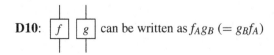 can be written as $f_A g_B \, (= g_B f_A)$

Of course, this is ambiguous unless we already know that we are working with some system $A \otimes B$, and pretty soon we need to start requiring extra equations, as in Section 3.2.4.

By sticking to diagrams all of this is immediately solved, and fortunately, very few of us are still using typewriters. Technology marches on, and now there are some great tools for drawing diagrams (pretty much) as easily as typing out a Dirac-style formula on a keyboard.

3.5 Summary: What to Remember

1. *Diagrams* consist of boxes that are labelled by *processes* and wires that are labelled by *system-types*:

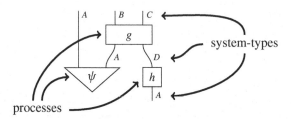

Diagrams are all around us:

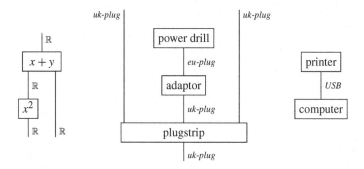

We can obtain new *diagram equations* from old ones in two ways:

1. *Diagram deformation*:

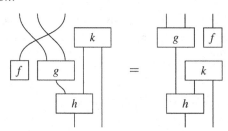

2. *Diagram substitution*:

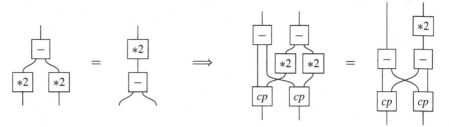

2. A *process theory* is a collection of processes that can be plugged together. It tells us how to interpret the boxes and wires in a diagram, e.g.:

$$
\boxed{\text{quicksort}} \quad := \quad
\begin{cases}
\texttt{qs [] = []} \\
\texttt{qs (x :: xs) =} \\
\quad \texttt{qs [y | y <- xs; y < x] ++ [x] ++} \\
\quad \texttt{qs [y | y <- xs; y >= x]}
\end{cases}
$$

and what it means to 'wire processes together'. By doing so it also tells us which diagrams represent the same process, e.g.:

$$
\begin{array}{c}
\boxed{\text{quicksort}} \\
\boxed{\text{quicksort}}
\end{array}
\quad = \quad \boxed{\text{quicksort}}
$$

Examples are **functions**, **relations**, **linear maps** (defined in Chapter 5), **classical processes** (defined in Chapter 8), **quantum maps** (defined in Chapter 6), and **quantum processes** (also defined in Chapter 8).

3. *Parallel composition* '⊗' and *sequential composition* 'o' are defined as:

$$
\boxed{f} \otimes \boxed{g} = \boxed{f}\ \boxed{g}
\qquad
\boxed{g} \circ \boxed{f} = \begin{array}{c}\boxed{g}\\\boxed{f}\end{array}
$$

4. *Circuit diagrams* are an important class of diagrams. They have the following two equivalent characterisations:

The diagram contains no directed cycles. That is, information flows from bottom to top without 'feeding back' to an earlier process:

The diagram can be constructed by composing boxes, including identities and crossings, by means of the operations $\otimes$ and $\circ$:

5. Processes lacking inputs, outputs, or both have special names:

states: effects: numbers:

These special processes admit a *generalised Born rule*:

6. *Zero processes* 'eat everything':

7. *Dirac notation* is common language for depicting quantum processes. It is a fragment of the diagrammatic language:

$$\psi \leftrightarrow |\psi\rangle \qquad \pi \leftrightarrow \langle\pi| \qquad \frac{\pi}{\psi} \leftrightarrow \langle\pi|\psi\rangle$$

3.6 Advanced Material*

Central to this book are process theories, which tell us how to interpret wires and boxes in a diagram, and what it means to form diagrams, i.e. what it means to wire boxes together. Here we present two alternative ways of defining process theories that make no direct reference to diagrams, namely, *abstract tensor systems* and *symmetric monoidal categories*.

In fact, we already saw a glimpse of each of these, respectively, when we encountered diagram formulas in Section 3.1.3 and when we defined circuits by means of the operations parallel and sequential composition in Section 3.2.

So how does one go about constructing a symbolic counterpart to the definition of process theories? In the first two parts of the definition:

(i) a collection T of *system-types* represented by wires,
(ii) a collection P of *processes* represented by boxes, where for each process in P the input types and output types are taken from T ...

diagrams don't play an essential role and we can just drop the 'represented by ...' bit, provided that we explicitly require that each process comes with a list of input types and a list of output types, all taken from T. The third part of the definition, which concerns what 'forming diagrams' means:

(iii) a means of 'wiring processes together', that is, an operation that interprets a diagram of processes in P as a process in P

is the one that requires some non-trivial effort. In particular, we need to provide a symbolic counterpart to 'wiring processes together' that does not make reference to diagrams. Moreover, the symbolic operation(s) should produce processes of that theory.

3.6.1 Abstract Tensor Systems*

We introduced diagram formulas as a symbolic counterpart to diagrams. Can we define a corresponding notion of process theory? The definition below does exactly that. For notational simplicity we will assume that all system-types are the same, and consequently, distinct wire names are merely distinguishing occurrences of the same system.

Definition 3.42 An abstract tensor system (ATS) consists of:

1. For all wire names $A_1, \ldots, A_m$ and $B_1, \ldots, B_n$, a set of *tensors*:

$$f_{A_1 \ldots A_m}^{B_1 \ldots B_n} \in \mathcal{T}(\{A_1, \ldots, A_m\}, \{B_1, \ldots, B_n\})$$

where $A_1, \ldots, A_m$ are called *inputs* and $B_1, \ldots, B_n$ are called *outputs*.

2. A *unit* tensor:

$$1 \in \mathcal{T}(\{\}, \{\})$$

and for all wire names A and B *identity* tensors:

$$\delta_A^B \in \mathcal{T}(\{A\}, \{B\})$$

3. An operation *tensor product* that combines tensors provided their inputs and outputs are distinct, which is denoted by concatenation:

$$f_{A_1 \ldots A_m}^{B_1 \ldots B_n} g_{C_1 \ldots C_k}^{D_1 \ldots D_l} \in \mathcal{T}(\{A_1, \ldots, A_m, C_1, \ldots, C_k\}, \{B_1, \ldots, B_n, D_1, \ldots, D_l\})$$

4. For any wire names A and B, an operation c_A^B called *tensor contraction*, which is denoted either explicitly or by repeating an upper/lower wire name:

$$f_{A_1\ldots A_m}^{B_1\ldots B_{j-1}A_iB_{j+1}\ldots B_n} := c_{A_i}^{B_j}\left(f_{A_1\ldots A_m}^{B_1\ldots B_n}\right)$$

which, intuitively, 'connects' output B_j to input A_i.

5. *Re-indexing* operations that change wire names:

$$f_{A_1\ldots A_m}^{B_1\ldots B_n}[A_i \mapsto A_i'] = f_{A_1\ldots A_{i-1}A_i'A_{i+1}\ldots A_m}^{B_1\ldots B_n}$$

where these operations satisfy the following conditions:

1. the tensor product is associative, commutative, and has unit 1:

$$(fg)h = f(gh) \qquad\qquad fg = gf \qquad\qquad 1f = f$$

2. the order of tensor contractions/products is irrelevant:

$$c_A^B(c_C^D(f)) = c_C^D(c_A^B(f)) \qquad\qquad c_A^B(f)g = c_A^B(fg)$$

3. contracting with the identity does nothing except change wire names:

$$c_A^B(\delta_C^B f_{\ldots A\ldots}) = f_{\ldots C\ldots} \qquad\qquad c_B^A(\delta_B^C f_{\ldots}^{\ldots A\ldots}) = f_{\ldots}^{\ldots C\ldots}$$

4. re-indexing respects identities, tensor product, and contraction.

The two operations, tensor product and tensor contraction, together play the role of 'wiring processes together'. For example, the symbolic counterpart to the diagram:

is obtained as follows:

$$c_{B'}^B\left(f_A^B g_{B'}^C\right)$$

More generally, we can compute the tensor corresponding to a diagram by first writing it down as a diagram formula and then decomposing it further in terms of the tensor product and tensor contraction. For example:

$$f^{AB}h_C^D g_{BD}^{EF} \quad:=\quad c_{D'}^D\left(c_{B'}^B\left(f^{AB}\left(h_C^D g_{B'D'}^{EF}\right)\right)\right)$$

An expression like $f^{AB}h^D_C g^{EF}_{BD}$ is called *abstract tensor notation*. Just like the decomposition of diagrams into $\otimes$ and $\circ$, the expression above is not unique. However, the requirements in Definition 3.42 guarantee that any two such decompositions will be equal.

Just like the decomposition of diagrams into $\otimes$ and $\circ$ we met in Section 3.2, here 'wiring processes together' is decomposed into two suboperations. This notation has been employed extensively in mathematical physics (and especially general relativity), where one often deals with operations with many inputs or outputs that are connected in ways that would be unwieldy to express using just $\otimes$ and $\circ$.

The reason that abstract tensor notation is better for these applications is that, by simply listing boxes and their connections, it embraces the idea that 'only connectivity matters'. This also explains why the translation between abstract tensor notation (a.k.a. diagram formulas) we saw in Section 3.1.3 is so straightforward.

Exercise* 3.43 Define abstract tensor systems for the more general case that there are distinct system-types.

Exercise* 3.44 How would you define the process theory **relations** as an abstract tensor system? (Hint: see the 'algorithm' for computing relations of a diagram at the end of Section 3.3.3.)

*3.6.2 Symmetric Monoidal Categories**

Another symbolic way to define process theories is what a category-theorist would call a *strict symmetric monoidal category*. In essence this boils down to axiomatising the operations of parallel composition $\otimes$ and sequential composition $\circ$ of Section 3.2. Doing so is a bit much to take in all at once, so let's drop the 'symmetric' part for the moment, and just look at the following definition.

Definition 3.45 A *strict monoidal category* (SMC) $\mathcal{C}$ consists of:

1. a collection ob($\mathcal{C}$) of *objects*,
2. for every pair of objects A, B, a set $\mathcal{C}(A, B)$ of *morphisms*,
3. for every object A, a special identity morphism:

$$1_A \in \mathcal{C}(A, A)$$

4. a sequential composition operation for morphisms:

$$\circ : \mathcal{C}(B, C) \times \mathcal{C}(A, B) \to \mathcal{C}(A, C)$$

5. a parallel composition operation for objects:

$$\otimes : \mathrm{ob}(\mathcal{C}) \times \mathrm{ob}(\mathcal{C}) \to \mathrm{ob}(\mathcal{C})$$

6. a unit object

$$I \in \mathrm{ob}(\mathcal{C})$$

7. and a parallel composition operation for morphisms:

$$\otimes : \mathcal{C}(A, B) \times \mathcal{C}(C, D) \to \mathcal{C}(A \otimes C, B \otimes D)$$

satisfying the following conditions:

1. $\otimes$ is associative and unital on objects:

$$(A \otimes B) \otimes C = A \otimes (B \otimes C) \qquad A \otimes I = A = I \otimes A$$

2. $\otimes$ is associative and unital on morphisms:

$$(f \otimes g) \otimes h = f \otimes (g \otimes h) \qquad f \otimes 1_I = f = 1_I \otimes f$$

3. $\circ$ is associative and unital on morphisms:

$$(h \circ g) \circ f = h \circ (g \circ f) \qquad 1_B \circ f = f = f \circ 1_A$$

4. $\otimes$ and $\circ$ satisfy the *interchange law*:

$$(g_1 \otimes g_2) \circ (f_1 \otimes f_2) = (g_1 \circ f_1) \otimes (g_2 \circ f_2)$$

Wow, that looks like a lot! However, if we stare at this definition for a bit, some of it should start to look fairly familiar. First, note that what we have been calling types or system-types are called *objects* in category theory. Similarly, what we call processes are called *morphisms*. The set $\mathcal{C}(A, B)$ should be thought of as the set of all morphisms with input type A and output type B. Typically we write $f : A \to B$ instead of $f \in \mathcal{C}(A, B)$. A *category* is a collection of such sets with sequential composition. A *monoidal category* is the same, but with sequential and parallel composition. We'll say more about what 'strict' means shortly. First, note that we have almost all of the ingredients we need to build circuit diagrams – all except swaps. This is where the *symmetric* part comes in.

Definition 3.46 A *strict symmetric monoidal category* is a strict monoidal category with a *swap morphism*:

$$\sigma_{A,B} : A \otimes B \to B \otimes A$$

defined for all objects A, B, satisfying:

$$\sigma_{B,A} \circ \sigma_{A,B} = 1_{A \otimes B} \qquad\qquad \sigma_{A,I} = 1_A$$

$$(f \otimes g) \circ \sigma_{A,B} = \sigma_{B',A'} \circ (g \otimes f)$$

$$(1_B \otimes \sigma_{A,C}) \circ (\sigma_{A,B} \otimes 1_C) = \sigma_{A,B \otimes C}$$

Saying that a monoidal category is *strict* means that $\otimes$ is associative and unital 'on the nose', whereas a (non-strict) monoidal category only requires them to be *isomorphic*.

Definition 3.47 An object A is *isomorphic* to an object B, denoted:

$$A \cong B$$

if and only if there exists a pair of morphisms:

$$f : A \to B \qquad\qquad f^{-1} : B \to A$$

such that:

$$f^{-1} \circ f = 1_A \qquad\qquad f \circ f^{-1} = 1_B$$

The morphism f is then called an *isomorphism*.

Most monoidal categories we find in the wild tend to be of the non-strict variety. That is, their parallel composition only satisfies:

$$(A \otimes B) \otimes C \cong A \otimes (B \otimes C) \qquad A \otimes I \cong A \cong I \otimes A$$

We already briefly encountered this situation in Section 3.3.1. Recall that $\otimes$ was defined in terms of the Cartesian product. When we compare the sets $(A \times B) \times C$ and $A \times (B \times C)$ (and the sets A and $A \times \{*\}$), we do not get *equal* sets, but rather sets whose elements are in one-to-one correspondence:

$$((a, b), c) \leftrightarrow (a, (b, c)) \qquad\qquad a \leftrightarrow (a, *)$$

To define a non-strict monoidal category, we must include these correspondences in the definition of the category as so-called *structural isomorphisms*, and we must assume several (more!) equations called *coherence equations* that guarantee that they behave sensibly when they are composed together.

However, we happily glazed over this fact when we defined the process theories of functions and relations in Section 3.3. Why? Because of the following *coherence theorem*.

Theorem 3.48 Every (symmetric) monoidal category $\mathcal{C}$ is equivalent to a strict (symmetric) monoidal category $\mathcal{C}'$.

Equivalent categories are, for all practical purposes, the same. We'll say more about what this means in Section* 5.6.4. The proof of Theorem 3.48 consists of an explicit construction of the category $\mathcal{C}'$ by a procedure called *strictification*. This procedure is used implicitly when, e.g. we decide to treat elements $((a, b), c)$, $(a, (b, c))$ and (a, b, c) as 'the same', which we do without further comment throughout this text.

Furthermore, we are justified in using circuit diagrams to talk about symmetric monoidal categories because of the following theorem.

Theorem 3.49 Circuit diagrams are sound and complete for symmetric monoidal categories. That is, two morphisms f and g are provably equal using the equations of a symmetric monoidal category if and only if they can be expressed as the same circuit diagram.

To summarise, it is possible to translate many concepts between our purely diagrammatic language and category theory, using this correspondence:

$$\text{process theory} \leftrightarrow \text{(strict) symmetric monoidal category}$$

$$\text{process} \leftrightarrow \text{morphism}$$

$$\text{system-type} \leftrightarrow \text{object}$$

3.6.3 General Diagrams versus Circuits*

Some readers may have noted that abstract tensor systems and symmetric monoidal categories aren't exactly one and the same. While symmetric monoidal categories correspond to circuits, abstract tensor systems represent more general kinds of diagrams in which directed cycles (cf. Definition 3.20) are allowed. However, we can make the two concepts coincide.

In one direction, one can force abstract tensor systems to be equivalent to strict symmetric monoidal categories by only requiring the contraction operation c_A^B to be well-defined if it does not introduce a directed cycle.

Conversely, it is possible to introduce 'feedback loops' into symmetric monoidal categories, obtaining so-called *traced symmetric monoidal categories*. These categories assume an additional operation called the *trace*:

$$\begin{array}{c}\begin{array}{c}\text{(diagram of } f \text{ with feedback loop from } A, \text{ output } C, \text{ output } B)\end{array}\end{array} := \mathrm{tr}_A\left(\begin{array}{c}\text{(diagram of } f \text{ with inputs } A, C \text{ and outputs } A, B)\end{array}\right)$$

which we will meet in Section 4.2.3 of the next chapter. The trace satisfies a handful of axioms to ensure that it behaves like a 'feedback loop' and that it respects the other categorical structure. Then, strict traced symmetric monoidal categories are equivalent to abstract tensor systems, in that it is possible to turn an abstract tensor system into a strict traced symmetric monoidal category and vice versa without losing any meaningful information.

We can summarise the above in the following table:

	'General' diagram	Circuit diagram
Diagrams	i.e. outputs to inputs	i.e. admits causal structure
ATS	ATS	ATS without cycles
SMC	traced SMC	SMC

3.7 Historical Notes and References

The idea of representing the composition of mathematical objects using nodes and wires dates back a very long time. A prominent example is the flow chart, which goes back (at least) to the 1920s (Gilbreth and Gilbreth, 1922). In physics, the most notable example is Feynman diagrams. Despite their name, it is now recognised that they originated in the work of Ernst Stueckelberg around 1941 and were only later independently re-derived by Richard Feynman around 1947. In a lecture at CERN after having been awarded the Nobel Prize Richard Feynman referred to this fact the moment that Stueckelberg left the room: 'He did the work and walks alone toward the sunset; and, here I am, covered in all the glory, which rightfully should be his!' (Mehra, 1994).

Penrose (1971) was the first to introduce the particular kind of diagrams we use in this book as an alternative to the abstract tensor notation (see Section* 3.6.1), which he also introduced. In fact, he already started using diagrams when he was still an undergraduate student (Penrose, 2004).

Paul Dirac introduced his new notation in Dirac (1939). The connection between Dirac notation and the corresponding use of the triangle notation for states and effects in string diagrams was observed in Coecke (2005). The idea of treating diagrams themselves as the basic structure rather than categories has recently become prominent in the quantum foundations community. It has been used, for example, in efforts towards crafting alternative formulations of quantum theory and a theory of quantum gravity (Chiribella et al., 2010; Coecke, 2011; Hardy, 2011, 2013b).

The unwillingness of many to accept diagrams as a rigorous mathematical language (alluded to in Section 3.1.3) may be largely due to the Bourbaki mindset in which mathematics has been taught for much of the previous century. The Bourbaki collective aimed to found all mathematics on a set-theoretic basis (Bourbaki, 1959–2004). Great credit should be given to John Baez, the pioneer of scientific blogging, who by means of his outstanding didactical skills managed to convince many of the amazingness of diagrams. His old weekly column 'This Week's Finds in Mathematical Physics' is still a great resource on diagrammatic reasoning (Baez, 1993–2010).

The connection between circuit diagrams and algebraic structure (in the form of symmetric monoidal categories, cf. Theorem 3.49) was established in Joyal and Street (1991), where circuit diagrams are referred to as 'progressive polarised diagrams'. Traced symmetric monoidal categories and their graphical notation were introduced in Joyal et al. (1996); however, the closely related notion of a 'compact closed category' (whose diagrams are string diagrams, which we will encounter very soon) had already been known for some time. The historical notes in the next chapter provide appropriate references.

The proof of correctness of the graphical language for traced symmetric monoidal categories has been sketched in various papers (e.g. Hasegawa et al., 2008), and a full proof is given by Kissinger (2014a). Following on from Joyal and Street's work came a plethora of variations of these diagrams and corresponding algebraic structures (see Selinger, 2011b).

Monoidal categories themselves were first introduced in Benabou (1963). Theorem 3.48 on coherence is crucial for connecting monoidal categories to diagrams and was first stated and proved by Mac Lane (1963). The 'little dance' in (3.24), a.k.a. the Eckmann–Hilton argument, is from Eckmann and Hilton (1962).

Abstract tensor systems, and their associated 'abstract index notation', were introduced by Penrose (1971) at the same time as the corresponding diagrammatic notation, in order to talk about various kinds of multilinear maps without needing to fix bases in advance. This notation is now very common in theoretical physics and especially general relativity. Penrose provides an introduction for a general audience in Penrose (2004). The equivalence between abstract tensor systems and traced symmetric monoidal categories is given by Kissinger (2014a).

If you want to read more on general category theory, some good places to start are Abramsky and Tzevelekos (2011) for the connection between categories and logic, Coecke and Paquette (2011) for the generality of the notion of symmetric monoidal categories, and Baez and Lauda (2011) for the role categorical structures have played in physics. Standard textbooks for mathematicians are Borceux (1994a,b) and Mac Lane (1998); standard textbooks for computer scientists are Barr and Wells (1990) and Pierce (1991); and standard textbooks for logicians are Lambek and Scott (1988) and Awodey (2010). In the context of category theory, the idea that morphisms represent processes emerged mostly in the context of applications in computer science, where programs are represented by morphisms in a category and data types are the objects. Similarly in logic, proofs can be represented by morphisms in a category, with propositions as the objects. This trifecta of morphisms, programs, and proofs are all connected by what's known as the *Curry–Howard–Lambek isomorphism* (Lambek and Scott, 1988).

The quote at the beginning of this chapter is taken from Bohm and Peat (1987). As discussed in Section 1.3, Bohm's views played a key role in making process ontology more prominent in physics.

4

String Diagrams

When two systems, of which we know the states by their respective representatives, enter into temporary physical interaction due to known forces between them, and when after a time of mutual influence the systems separate again, then they can no longer be described in the same way as before, viz. by endowing each of them with a representative of its own. I would not call that one but rather the characteristic trait of quantum mechanics, the one that enforces its entire departure from classical lines of thought.

– Erwin Schrödinger, 1935

By 1935, Schrödinger had already realised that the biggest gulf between quantum theory and our received classical preconceptions is that, when it comes to quantum systems, the whole is more than the sum of its parts. In classical physics, for instance, it is possible to totally describe the state of two systems – say, two objects sitting on a table – by first totally describing the state of the first system then totally describing the state of the second system. This is a fundamental property one expects of a classical, *separable* universe. However, as Schrödinger points out, there exist states predicted by quantum theory (and observed in the lab!) that do not obey this 'obvious' law about the physical world. Schrödinger called this new, totally non-classical phenomenon *Verschränkung*, which later became translated to the dominant scientific language as *entanglement*.

Quantum picturalism is all about studying the way parts compose to form a whole. In the good company of Schrödinger, we believe that the role of multiple interacting systems – especially non-separable systems – should take centre stage in the study of quantum theory.

We shall see in the next section that it is easy to say what it means for a process to be *separable* in terms of diagrams. Literally, it means that it can be broken up into pieces that are not connected to each other. On the other hand, enforcing *non*-separability requires us to refine our diagrammatic language. To this end, we introduce special states and effects called *cups* and *caps*, respectively. Intuitively, cups and caps act like pieces of wire that have been 'bent sideways'. Whereas all states in the classical world separate, these states and effects are obviously non-separable and thus witness a crucial quantum feature:

$$\frac{\psi_1 \quad \psi_2}{\bigcup} = \frac{\text{separable}}{\text{non-separable}} = \frac{\text{classical}}{\text{quantum}} \tag{4.1}$$

In addition to caps and cups, we'll add *adjoints* to the diagrammatic language, which we picture as the vertical reflection of a diagram. This reflection operation associates each state ψ to a corresponding effect that tests 'how similar to ψ' a given state is. This has a direct analogue in Dirac notation, namely reflecting a ket into a bra. It will play a key role in modelling things such as reversible state evolution (unitary processes), quantum measurements (projectors and positive processes), and the probabilities associated with quantum measurements (via the inner product).

The new diagrams we obtain by adding caps, cups, and reflection are called *string diagrams*, and we will already start to see some strikingly non-classical behaviour arising just from using the language of string diagrams.

For example, we will already encounter quantum teleportation in Section 4.4.4. Related to this feature is the fact that caps and cups confound our ability to give a fixed time-ordering to processes composed in sequence, which we will see in Section 4.4.3. We will also encounter several 'no-go' theorems for string diagrams. That is, we will show that several properties that we expect to hold for classical (deterministic) physical processes are in fact impossible in a process theory that admits string diagrams. One such no-go theorem is the *no-cloning theorem*, which says that in any process theory that admits string diagrams it is impossible to have a *cloning process*, that is, one that takes any state as input and returns two copies.

Remark 4.1 Throughout this chapter we are careful to say 'non-separable' rather than 'entangled'. For a class of quantum states called 'pure states' (introduced in Section 6.1 of Chapter 6), the notions of 'non-separable' and 'entangled' do indeed coincide. However, for more general states, identifying quantum entanglement is a bit more subtle. Thus, we will hold off on giving the most general notion of quantum entanglement until Section 8.3.5.

4.1 Cups, Caps, and String Diagrams

In this section, we first give a formal definition for *separability*, and hence *non-separability*. Next we motivate string diagrams via the principle of *process–state duality*, a correspondence between states of compound systems and processes – known in the special case of quantum theory as the *Choi–Jamiołkowski isomorphism* – that is fundamentally connected to non-separability.

We then present the definition of string diagrams in two equivalent ways. The first way is to *extend* the language of circuit diagrams with the 'caps' and 'cups' used to define process–state duality. The second, easier way is to simply allow the wires in a diagram to connect processes in arbitrary ways, including connecting inputs to inputs and outputs to outputs.

4.1.1 Separability

Definition 4.2 A *bipartite state* ψ is a state of two systems, and we call such a state $\otimes$-*separable* if there exist states ψ_1 and ψ_2 such that:

$$\text{(4.2)}$$

The following is an example of a process theory with all bipartite states $\otimes$-separable.

Proposition 4.3 All bipartite states in **functions** are $\otimes$-separable.

Proof In Example 3.35 we saw that states for **functions** are totally defined by the element that they 'point to'. If for a bipartite state this element is $(a_1, a_2) \in A_1 \times A_2$, then we denote this state as:

Now we have:

since both these states point to the same element. $\square$

On the other hand, there are many (far more interesting!) process theories that contain bipartite states that are not $\otimes$-separable, i.e. states that are $\otimes$-*non-separable*.

Proposition 4.4 There are $\otimes$-non-separable bipartite states in **relations**.

Proof Recall from Example 3.36 that states in **relations** are represented by subsets. Consider the bipartite state:

$$:: \begin{cases} * \mapsto (0,0) \\ * \mapsto (1,1) \end{cases}$$

Then:

$$C = \{(0,0), (1,1)\} \subseteq \mathbb{B} \times \mathbb{B}$$

Suppose for the sake of contradiction that:

for some $B, B' \subseteq \mathbb{B}$. Then, by (3.22):

that is:

$$(b, b') \in C \iff b \in B \text{ and } b' \in B'$$

Since $(0, 0) \in C$, it follows that $0 \in B$. Similarly, $(1, 1) \in C$ implies that $1 \in B'$. But these two facts imply $(0, 1) \in C$, which is a contradiction.   □

The crux of this proof is the fact that the (non-singleton) set:

$$\{(0, 0), (1, 1)\}$$

cannot be written as a Cartesian product of subsets of $\mathbb{B}$. If as in **functions**, every such set is a singleton, then this is always possible.

A closely related notion to ⊗-separability is the following.

Definition 4.5 We call a process:

o-*separable* if there is an effect π and a state ψ such that:

      (4.3)

Both ⊗-separability of bipartite states and o-separability of processes are examples of 'disconnectedness' of the corresponding diagrams. Whenever no confusion is possible we will simply say 'separable' in both cases.

Moreover, these two notions of separability are related to each other in the following sense.

• If the process f is o-separable, then for any bipartite state ϕ the following state is ⊗-separable:

where:

$$
\phi' := \begin{array}{c} \pi \\ \phi \end{array}
$$

- If the state ψ is $\otimes$-separable, then for any bipartite effect π the following process is o-separable:

$$
\begin{array}{c} \pi \\ \psi \end{array} \overset{(4.2)}{=} \begin{array}{c} \pi \\ \psi_1 \ \psi_2 \end{array} = \begin{array}{c} \psi_2 \\ \pi \\ \psi_1 \end{array} = \begin{array}{c} \psi_2 \\ \pi' \end{array}
$$

where:

$$
\pi' := \begin{array}{c} \pi \\ \psi_1 \end{array}
$$

We also saw earlier that in **functions** all bipartite states are $\otimes$-separable. So what kinds of process theories have all o-separable processes? The answer is, 'only the really boring ones!' When we apply a o-separable process to an arbitrary state:

$$
\begin{array}{c} f \\ \psi \end{array} \overset{(4.3)}{=} \begin{array}{c} \phi \\ \pi \\ \psi \end{array} = \begin{array}{c} \pi \\ \psi \end{array} \ \phi
$$

we cannot obtain anything other than a fixed state ϕ (ignoring the number $\pi \circ \psi$), regardless of what the input ψ was. That is, every process is essentially a *constant* process. In other words:

nothing ever happens!

Thus it shouldn't be surprising that the theory of **functions** *does* contain o-non-separable processes.

Exercise 4.6 Show that identities (i.e. plain wires) in **functions** are o-non-separable, and characterise those functions that are o-separable.

Remark 4.7 The ultimate boringness of process theories where all processes are o-separable will play an important role in several 'no-go' theorems that we develop later in this chapter. In particular, we will show that some statement P is not true by proving a theorem of the form:

If P is true, then all processes are o-separable.

As this is an absurd condition for any reasonable process theory, and in particular any theory of physics, we can take this as a proof that P is false.

4.1.2 Process–State Duality

One feature of quantum theory is that one can turn a process into a bipartite state and vice versa. This is in itself not very significant: we already demonstrated such a conversion in the previous section in order to relate $\otimes$-separability of bipartite states to $\circ$-separability of processes. What is significant about quantum theory is that this can be done in a reversible manner. That is, we have a way to turn a process into a bipartite state and then back into a process such that we obtain the original process, and vice versa. In other words, the operations that send processes to bipartite states and bipartite states to processes are inverses of each other. Therefore, in quantum theory, processes and bipartite states are in bijective correspondence.

As in the previous section, we will make use of bipartite states and effects to inter-convert processes and states. However, we will not be using just any state/effect pair but instead fix a special state and effect:

$$\text{and} \tag{4.4}$$

for each system A. We call these states and effects *cups* and *caps*, respectively. They can be used to convert processes to states and back via the following operations:

$$\tag{4.5}$$

Theorem 4.8 The operations given in (4.5) are inverses of each other, i.e. they give *process–state duality*, if and only if:

$$\tag{4.6}$$

In that case, the set of all processes with input type A and output type B is in one-to-one correspondence with the set of all bipartite states of type $A \otimes B$:

Proof First, assume equations (*a*) and (*b*). Then, if we first turn a process into a state and back:

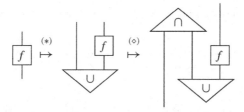

we can simplify back to the original process using (*a*):

Similarly, starting with a state:

we can use (*b*):

Conversely, if the operations given in (4.5) are mutually inverse, then:

(4.7)

for all f and ψ. Taking f to be the identity yields (*a*) immediately. Proving (*b*) using the above equations is left as an exercise. □

Exercise 4.9 Prove equation (4.6b) using both of the process–state duality equations (4.7), by setting:

Exercise 4.10 Show that process–state duality does not hold for **functions**, but that it does hold for **relations**.

Remark* 4.11 A reader familiar with quantum theory may know about a correspondence between processes and states called the *Choi–Jamiołkowski isomorphism*. Process–state duality is in fact a generalisation of the C-J isomorphism. In Chapter 6, we will introduce the process theory of **quantum maps**, in which case this generalised C-J isomorphism will actually be the familiar correspondence between bipartite (possibly mixed) quantum states and completely positive maps.

4.1.3 The Yanking Equations

Equation (4.6) is not very intuitive when viewed diagrammatically, but by means of a simple change of notation we can fix this. We write the special states and the special effects, respectively, as ∪-shaped and ∩-shaped wires:

$$\cup \;:=\; \underset{U}{\nabla} \qquad\qquad \cap \;:=\; \underset{\cap}{\triangle}$$

Equation (4.6) now becomes:

$$\cap\!\cup \;=\; | \;=\; \cup\!\cap \tag{4.8}$$

This notation expresses the fact that bent wires involving a ∪-shape and a ∩-shape can be yanked into a straight wire:

If we take this diagrammatic interpretation seriously, then there are other equations that should also hold for cups and caps. For example, we should be able to 'yank out crossings':

$$\text{⋈} = \cup \qquad\qquad \text{⋈} = \cap \qquad\qquad (4.9)$$

Or, stated with a little bit more effort:

From (4.6) and (4.9), we can also derive the equation to pull out loops:

$$\text{α} = | \qquad\qquad (4.10)$$

Exercise 4.12 Prove that equation (4.10), or equivalently:

$$\text{[diagram]} = |$$

follows from (4.6) and (4.9).

These are actually all of the equations we need to 'yank' any tangled-up piece of wire straight. In fact, we already have some redundancy, as seen in the following proposition.

Proposition 4.13 The following are equivalent:

(i) a state and an effect satisfying:

$$\text{[diagram]} = | \qquad\qquad \text{[diagram]} = \text{[diagram]}$$

$$\text{[diagram]} = \text{[diagram]}$$

(ii) a state and an effect satisfying:

Exercise 4.14 Prove Proposition 4.13.

From now on, we will refer to the three equations:

(4.11)

as the *yanking equations*.

Example* 4.15 The *Bell state* and *Bell effect* from quantum theory:

$$\frac{1}{\sqrt{2}} (|00\rangle + |11\rangle) \qquad\qquad \frac{1}{\sqrt{2}} (\langle 00| + \langle 11|)$$

provide the quintessential example of cups and caps.

In Propositions 4.3 and 4.4 we already saw that, while in **functions** all bipartite states are ⊗-separable, this is no longer the case in **relations**, as we can easily find cups and caps for every system.

Exercise 4.16 Prove that for any set A the following relations:

$$\smile \quad :: \ast \mapsto \{(a, a) \mid a \in A\} \qquad\qquad \frown \quad :: \forall a \in A : (a, a) \mapsto \ast$$

satisfy equations (4.11).

Example 4.17 Taking $A := \mathbb{B}$, we rediscover the state that we used as an example of a non-separability in the proof of Proposition 4.4:

$$\smile \quad :: \ast \mapsto \{(0, 0), (1, 1)\}$$

4.1.4 String Diagrams

Equations (4.11) do look very appealing now, but we can do even better than that. If every system-type has cups and caps satisfying (4.11), then we can introduce a more liberal notion of a diagram that has them built in.

Definition 4.18 A *string diagram* consists of boxes and wires, where we additionally allow inputs to be connected to inputs and outputs to be connected to outputs, for example:

$$(4.12)$$

Above we already indicated that we can replace the special states and the special effects of the previous section by cup-shaped wires and by cap-shaped wires, respectively. This is how we obtain a string diagram from a circuit diagram in which there are states and effects satisfying (4.11). Conversely, starting from a string diagram we can replace a wire connecting two inputs with the special cup state and a wire connecting two outputs with the special cap effect. Hence the following theorem.

Theorem 4.19 The following two notions are equivalent:

(i) string diagrams and
(ii) circuit diagrams to which we adjoin a special state and a special effect for each type, and for which (4.11) holds.

in the sense that (ii) can be unambiguously expressed as (i) and vice versa.

Bingo! We expressed non-separability in purely diagrammatic terms, and not just by adding some new boxes to (the now boring) circuit diagrams, but also by simply considering a different, more liberal kind of diagram. Thus, the most important feature of quantum theory (according to Schrödinger) is built right into the kinds of diagrams we use!

Exercise 4.20 Write diagram (4.12) in ∘ and ⊗ notation using cups and caps.

String diagrams also admit a formula-like presentation (cf. Section 3.1.3).

Definition 4.21 A *string diagram formula* is the same as a diagram formula except for the fact that matched pairs of wire names, besides consisting of an upper and a lower name, can now also consist either of two upper names or of two lower names, for example:

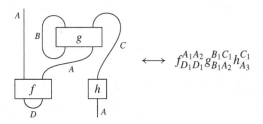

$$\longleftrightarrow \quad f^{A_1 A_2}_{D_1 D_1} g^{B_1 C_1}_{B_1 A_2} h^{C_1}_{A_3}$$

Remark 4.22 In Section 3.1.3, we introduced special box names $1_{A_1}^{A_2}$ to represent plain wires in diagram formulas. If we need explicit cups or caps in string diagram formulas, we can use two 'plain wires' with their inputs or outputs, respectively, connected together:

$$\cup^{A_1A_2} := 1_{A_3}^{A_1}1_{A_3}^{A_2} \qquad \cap_{A_1A_2} := 1_{A_1}^{A_3}1_{A_2}^{A_3}$$

This allows us to distinguish, for example:

We now explore the richness of these diagrams.

4.2 Transposition and Trace

Transposition and trace are notions you may know from linear algebra. Quite remarkably, these already arise at the very general level of string diagrams. We already know from the previous section that cups and caps let us form new processes from old ones by turning inputs into outputs and vice versa. In the case of process–state duality we have:

More generally, we can always inter-convert inputs and outputs:

This ability to inter-convert inputs and outputs for process theories that admit string diagrams means that inputs and outputs have a less profound status than in the case of other process theories that only admit circuits. In particular, any process with an input and an output has (at least) the following four inter-convertible representations:

$$\left\{\; \boxed{f} \;,\; \boxed{f} \;,\; \boxed{f} \;,\; \boxed{f} \;\right\}$$

and in the case of several inputs and outputs there are many more. We also can obtain new processes by connecting inputs and outputs together using cups and caps:

The *transpose* and the *trace* are the most prominent examples where caps and cups are used to convert processes into new processes.

4.2.1 The Transpose

In a process theory that admits string diagrams we can associate to each process another process going in the opposite direction.

Definition 4.23 The *transpose* f^T of a process f is another process:

This of course should not be confused with another well-known way to obtain a process going in the other direction:

Definition 4.24 A process f from type A to type B has an *inverse* if there exists another process f^{-1} from type B to type A such that:

$$\begin{array}{ccc} \begin{array}{c} A \\ \boxed{f^{-1}} \\ B \\ \boxed{f} \\ A \end{array} & = & A \end{array} \qquad \begin{array}{ccc} \begin{array}{c} B \\ \boxed{f} \\ A \\ \boxed{f^{-1}} \\ B \end{array} & = & B \end{array} \qquad (4.13)$$

Exercise 4.25 Show that if a process f has an inverse, it is unique.

While the transpose can be realised by a diagram involving f:

$$\boxed{f} = (1_A \otimes \cap_B) \circ (1_A \otimes f \otimes 1_B) \circ (\cup_A \otimes 1_B) \qquad (4.14)$$

this is not the case for an inverse, otherwise every process would have an inverse, which is usually not true.

Remark* 4.26 The simple fact that the transpose in linear algebra can be written using a cup and a cap as in decomposition (4.14) is surprisingly not very well known, even among specialists.

Exercise 4.27 Prove that in **relations** the transpose of a relation R is the converse relation, that is:

$$\boxed{R} :: a \mapsto b \quad \Longleftrightarrow \quad \boxed{R} :: b \mapsto a$$

A state:

$$\psi$$

has no input, so we only have to convert the output into an input:

$$\psi$$

Hence, it is no coincidence that in **relations** any system-type A has the same number of states as effects (see Example 3.36), since we have a bijective correspondence between states and effects:

$$B \quad \overset{\cong}{\longleftrightarrow} \quad B$$

This is a general phenomenon.

Proposition 4.28 For any process theory that admits string diagrams there is a bijective correspondence between states and effects of the same type. More generally, the correspondence:

$$\boxed{f} \quad \overset{\cong}{\longleftrightarrow} \quad \boxed{f}$$

yields a bijective correspondence between the processes of a fixed input and output type, and the processes with the opposite input and output type.

The transpose of a transpose yields the original process:

$$\boxed{f} \quad = \quad \boxed{f}$$

Or, symbolically:

$$(f^T)^T = f$$

An operation that 'undoes itself' like this is called an *involution*.

Example 4.29 The transpose of a cup is a cap and vice versa:

Now comes the really cool bit. We can build the definition of transpose into our diagrammatic notation. First, we deform our boxes a bit:

$$\boxed{f} \rightsquigarrow \boxed{f}$$

It doesn't matter too much <u>how</u> we deform boxes, only that we break the symmetry. For example,

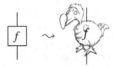

would also work, but Dave might not appreciate being skewered. Now, we express the transpose of f as a box labelled 'f', but rotated 180°:

$$\boxed{f} := \boxed{f} \qquad (4.15)$$

This notation is clearly consistent with the fact that the transpose is an involution, since, if one rotates by 180° twice, one obtains the original box again. However, the real advantage of this choice of notation comes when transposes interact with cups and caps.

Proposition 4.30 For any process f we have:

$$\boxed{f} = \boxed{f} \qquad \boxed{f} = \boxed{f}$$

Proof The first equality is established as follows:

$$\boxed{f} \overset{(4.8)}{=} \boxed{f} = \boxed{f} \overset{(4.15)}{=} \boxed{f}$$

and the proof of the second one proceeds similarly.   □

Thanks to the clever shorthand for the transpose, it now looks as if we can slide boxes across the ∪-shaped wires and ∩-shaped wires:

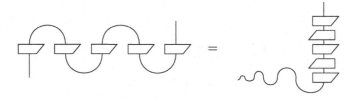

Thus we can slide boxes around on wires like beads on a necklace. For example, this is a valid equation:

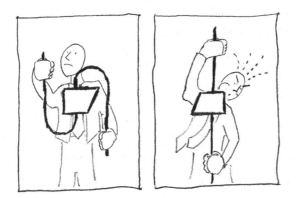

This is also very easy to remember. Just think about what would happen if we took the definition of the transpose (4.15) and yanked on the wires:

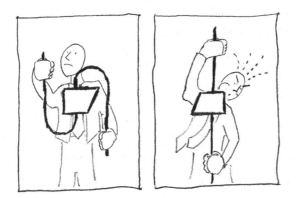

The same notational trick also applies to states and effects. To make it clear that they have been rotated, we chop off a corner:

$$ \Downarrow_\psi \quad \rightsquigarrow \quad \Downarrow_\psi \qquad \text{and we set} \qquad \triangle_\psi \; := \; \int \Downarrow_\psi \qquad (4.16) $$

It is not obvious why this is necessary at this point, since it's pretty easy to tell if a triangle as been turned upside-down. However, it will soon become very important (starting in Section 4.3), so we might as well get used to it.

Transposes also have an operational interpretation. Consider the first equality of Proposition 4.30 by interpreting the two systems as distant locations in space, each controlled by an agent, say Aleks and Bob. Then, we have:

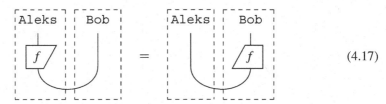

$$(4.17)$$

The equation says that, whenever two such systems are in a cup state, if Aleks applies f to his system, this will do the same thing as Bob applying the transpose of f to his system.

Now, suppose we consider (4.17) in the special case of states and effects:

Given that there is a bijective correspondence between:

- Bob's states and
- Aleks' effects ,

Aleks and Bob possess a pair of systems that are in *perfect correlation*. What does this mean? Recall that an effect can be interpreted as a successful test. Under this interpretation, we have the following property. For every state on Bob's system, there is a unique test on Aleks' system such that:

as soon as Aleks obtains *Bob's system will be in state*

A process that is equal to its own transpose is called *self-transposed*.

Example 4.31 Numbers are always self-transposed. By definition, the transpose bends all the input wires up and all the output wires down. Since a number has no wires, there's nothing to do:

$$\left(\lambda \right)^{T} = \lambda \tag{4.18}$$

4.2.2 Transposition of Composite Systems

One has to be a bit careful when dealing with joint system-types. On the one hand, to be consistent with the '180° rotation' notation for transposition, we should define the transpose this way:

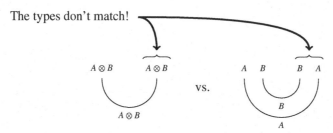

$$(4.19)$$

This would suggest that we should 'nest' the caps and cups inside of each other to define caps and cups for $A \otimes B$. However, there's a problem with defining $\cup_{A \otimes B}$ and $\cap_{A \otimes B}$ as above:

Remark* 4.32 This type mismatch vanishes if we introduce for every type A a *dual type* A^*, but this is at the cost of having two distinct types for the two systems involved in cups/caps, while in fact, these two types are often essentially the same. In Section* 4.6.2 we show how one can develop a theory of string diagrams with dual types.

We can avoid this type mismatch if we define cups/caps for $A \otimes B$ a bit differently, namely as 'criss-crossed' cups/caps:

$$(4.20)$$

$$(4.21)$$

Proposition 4.33 The cup and cap defined in equations (4.20) and (4.21) satisfy the yanking equations (4.11).

Proof For the first yanking equation of (4.11) we have:

Proofs of the other two equations are similar. □

With these 'criss-crossed' cups/caps, we obtain an alternative notion of transposition, which introduces a twist for composite systems:

$$\tag{4.22}$$

It is this twist that restores the matching of types, and since we define cups/caps on $A \otimes B$ in terms of cups/caps on A and B individually, there is no ambiguity when we apply this transpose on processes involving composite systems. But, at the same time, we lost a chunk of the elegant '180° rotation' notation for transposition.

In fact, it turns out that both versions of transposition can be useful. We will take the first version, indicated in (4.19), to be the default and continue to refer to this simply as 'the transpose'. The second version, involving the 'criss-crossed' cups and caps, will be referred to as the *algebraic transpose*, since it is in fact the one that is used in linear algebra (see Section 5.2.2). While the transpose is (still) denoted by 180° rotation, we will use the symbolic notation $(\ )^T$ to represent the algebraic transpose. So the equation:

now relates the transpose to the algebraic transpose. Of course, when a box has at most one input/output wire, these two versions coincide. Unsurprisingly, a process that is equal to its algebraic transpose is called *algebraically self-transposed*.

4.2.3 The Trace and Partial Trace

String diagrams give us a simple way of sending processes to numbers.

Definition 4.34 For a process f where the input type is the same as the output type, the *trace* is:

$$\tag{4.23}$$

and for a process g with one of its inputs having the same type as one of its outputs, the *partial trace* is:

$$\text{tr}_A \left(\boxed{\;g\;}_{A\;\;\;B}^{A\;\;\;C} \right) := A \left(\boxed{\;g\;}_{B}^{C} \right)$$

A funny property of the trace is that while in general:

$$\frac{\boxed{g}}{\boxed{f}} \neq \frac{\boxed{f}}{\boxed{g}}$$

the trace of the LHS and the RHS are in fact equal.

Proposition 4.35 (cyclicity of the trace) We have:

$$\text{tr}(f \circ g) = \text{tr}(g \circ f) \qquad \text{i.e.} \qquad \frac{\boxed{g}}{\boxed{f}} = \frac{\boxed{f}}{\boxed{g}}$$

Proof This follows from the fact that the two diagrams are equal; i.e. the LHS can be deformed into the RHS without changing how the boxes are wired together. □

Alternatively, we can give a more step-wise proof of Proposition 4.35 using Proposition 4.30. While not strictly necessary, this other proof is nice because it takes the word 'cyclicity' quite literally, yielding a ferris wheel of boxes:

$$\frac{\boxed{g}}{\boxed{f}} = \frac{\boxed{g}}{\boxed{f}} = \frac{\boxed{f}}{\boxed{g}} = \frac{\boxed{f}}{\boxed{g}}$$

Remark 4.36 In the previous section we decided to distinguish the transpose (which involves nested cups/caps) from the algebraic transpose (which involves criss-crossed cups/caps). Since equation (4.23) also involves cups and caps, we might ask whether for composite systems we need to make a similar distinction between 'nested' and 'criss-crossed' trace. Luckily, we don't:

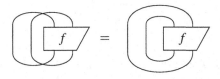

Exercise 4.37 Show that there is only one trace; i.e. any cup/cap pair satisfying the yanking equations (4.11) defines the same trace via (4.23).

4.3 Reflecting Diagrams

Having string diagrams at hand is already a substantial step towards the quantum world, since it guarantees the existence of non-separable states. They also give rise to some mathematical concepts that play a central role in quantum theory, such as transposition and trace.

We will now identify one more diagrammatic feature that makes string diagrams a lot more articulate, namely *vertical reflection* of diagrams. Vertical reflection allows us to define things like *adjoints, conjugates, inner products, unitarities*, and *positivity*, all of which are major players in any presentation of quantum theory.

Moreover, vertical reflection isn't just some extra degree of freedom that string diagrams happen to have, but has a clear operational meaning in terms of testability of states. This operational meaning will lead to a number of conditions that play an important role in quantum theory. These conditions are in fact quite standard in the literature, but they are usually just stated formally, without any conceptual justification.

4.3.1 Adjoints

In the previous section we explained how the transpose acts graphically by rotating boxes 180°, and that operationally it captures the perfect correlation between Aleks' effects on one side of a cup state and Bob's states on the other:

But what about the relationship between Aleks' effects and Aleks' states?

Recall from Section 3.4.1 that effects can be interpreted as testing a state for some property. Typically, we want to test whether a system is in a particular state. Therefore, we should have a means of relating a state ψ to the effect that tests a system for ψ. String diagrams do not yet tell us how to do this, so we extend our language by representing the effect testing for a state ψ as its vertical reflection:

$$\psi \mapsto \psi \qquad (4.24)$$

Instead of 'the effect testing for a state ψ' we'll simply say ψ's adjoint.

This reflection operation extends very naturally to all processes. If f transforms state ψ into a state ϕ:

$$\frac{f}{\psi} = \phi \qquad (4.25)$$

then f's adjoint is the process that transforms ψ's adjoint into ϕ's adjoint:

$$\frac{\psi}{f} = \phi \qquad (4.26)$$

Note that, as in the case of states, we have depicted the adjoint of an arbitrary process as its vertical reflection:

$$f \mapsto f$$

As a result, equation (4.26) is just equation (4.25) upside-down. We use † to denote the operation sending f to its adjoint $f^\dagger$. As a special case, if we take the adjoint of the effect from (4.24), we get back to ψ:

$$\psi \mapsto \psi$$

Hence, like the transpose, adjoints bijectively relate states and effects, and more generally, they bijectively relate processes from a type A to a type B to processes from B to A. Also like the transpose, this is an involution, which is plainly suggested by the diagrammatic notion:

$$f \mapsto f \mapsto f \qquad (4.27)$$

In the passage from (4.25) to (4.26), we saw one example where taking the adjoint of a process reflected its entire diagram. This actually applies to all diagrams, for example:

$$(4.28)$$

Equivalently (by Theorem 4.19), taking the adjoint preserves parallel composition and identities:

$$(4.29)$$

it reverses sequential composition and sends cups to caps:

$$(4.30)$$

and it reverses the direction of swaps:

$$(4.31)$$

Exercise 4.38 Using Exercise 3.38, prove:

$$0^\dagger = 0$$

So, we now know that adjoints are represented diagrammatically as vertical reflection. However, we haven't said how one should compute the adjoint of a process. The answer is: it depends on the theory. Since a process theory gives an interpretation for all diagrams (cf. Section 3.1.2), it must in particular give an interpretation for the vertical reflection of a diagram. This raises two questions:

1. Does there always exist such an interpretation?
2. Can there be more than one such interpretation?

The answer to the first question is yes, since one can always interpret vertical reflection as the algebraic transpose.

Exercise 4.39 Verify that the algebraic transpose provides a candidate interpretation for adjoints. In other words, show that it obeys (4.27) and (4.28). Explain why we pick the algebraic transpose instead of the transpose.

The algebraic transpose is, in some sense, a trivial interpretation for vertical reflection, since it doesn't add anything that string diagrams without adjoints didn't have already. Of course, if this were the only example, we wouldn't have bothered with adjoints in the first place. In the next chapter, we will encounter a very important non-trivial example of adjoints, which cannot be replaced by transposition. So the answer to the second question is also yes; there are multiple (trivial and non-trivial) ways to interpret the adjoint. The fact that there may be many candidate adjoints for a process theory raises yet another question:

3. What are good interpretations for the adjoint?

Any interpretation of adjoints in a process theory should be involutive and reflect diagrams in the sense of (4.28). However, to be a 'good' adjoint, it should be consistent with the idea that it sends a state to the effect that tests for that state. This has some consequences, which we will discuss now.

When we test a state ψ for ϕ, there are two extremes:

or equivalently:

since '1' is just symbolic notation for the empty diagram. In the first case, we are saying that it is impossible to get a 'yes' outcome when testing ψ for ϕ. For the process theories we'll consider in this book, the second equation has one of two interpretations: it either means that a 'yes' outcome is 'possible' (in the case of **relations**) or 'certain' (in the case of most other process theories we will encounter in this book).

In light of the interpretation of the first equation, we would expect this:

That is, it should never be impossible to get a 'yes' outcome when testing a state ψ for itself. However, there is one thing we overlooked: ψ itself could be 0, i.e. the 'impossible state'. As we saw in Section 3.4.2, 0 absorbs everything, so in particular, 0 composed with 0 will always be ... you guessed it, 0. So, we can state the condition above a bit more carefully:

$$\psi\psi = 0 \quad \Longleftrightarrow \quad \psi = 0 \qquad (4.32)$$

In words: if it is impossible to get 'yes' when testing ψ for itself, then ψ must already be impossible.

Next, we consider the other equation, i.e. when testing ψ for ϕ yields 1. If we interpret 1 as 'possible', this means that ψ and ϕ are not completely distinguishable by tests, but it doesn't tell us much beyond that. In particular, it doesn't tell us that they are equal (cf. Example 4.40).

However, when 1 means 'certain', we can say something much stronger:

$$\phi\psi = 1 \quad \Longleftrightarrow \quad \psi = \phi \qquad (4.33)$$

That is, we conclude (with certainty) that 'ψ is ϕ'.

Example 4.40 In Example 3.36 we established that in the process theory of **relations** the states for a set A correspond to subsets $B \subseteq A$. We think of these as non-deterministic states, consisting of a set of possible, 'actual' states $b \in B$. Effects for A also correspond to subsets, and we think of these as testing whether the actual state of the system is one of the elements of B. If testing B for B' yields 0, this means that none of the elements $b \in B$ is in B'. That is:

$$B'B = 0 \quad \Longleftrightarrow \quad B \cap B' = \emptyset \qquad (4.34)$$

There is in fact only one effect B' that sends a state B to 0 if and only if $B \cap B'$ is empty. That is the *relational converse* of B':

$$B' :: * \mapsto a \quad \Longleftrightarrow \quad B' :: a \mapsto * $$

By means of (4.25) $\Leftrightarrow$ (4.26), this lifts to any relation:

$$R :: a \mapsto b \quad \Longleftrightarrow \quad R :: b \mapsto a \qquad (4.35)$$

Consequently the adjoint of a relation coincides with its algebraic transpose (cf. Exercise 4.27).

Equation (4.34) arose as an instance of (4.32), so the first equation of 'goodness' uniquely fixes adjoints in the theory of **relations** to be the relational converse. But does it satisfy (4.33)? Should we expect it to? Since the number 1 in **relations** means 'possible' and not 'certain', the answer is no. In fact, there are many non-equal states where:

$$\vcenter{\hbox{$\overset{\displaystyle B'}{\underset{\displaystyle B}{\mid}}$}} = 1 \qquad\qquad (4.36)$$

because this merely means that $B \cap B' \neq \emptyset$. However, there is one situation in **relations** where the number 1 does mean certain: when comparing 'actual' (i.e. deterministic) states. That is, for $b, b' \in A$:

$$\vcenter{\hbox{$\overset{\displaystyle b'}{\underset{\displaystyle b}{\mid}}$}} = 1 \quad \Longleftrightarrow \quad b = b' \qquad\qquad (4.37)$$

The following is a prime example of adjoints, which will play a role throughout the book and are crucially different from the transpose. We give it as a 'starred' example here, as it will already be familiar to some readers. If not, don't worry, it will be introduced properly in the next chapter.

Example* 4.41 As the name suggests, adjoints in the process theory of **linear maps** are given by the *linear algebraic adjoint*, i.e. for a linear map f, the unique map $f^\dagger$ such that (in traditional notation):

$$\langle \psi | f(\phi) \rangle = \langle f^\dagger(\psi) | \phi \rangle$$

for all ψ, ϕ. In terms of matrices, this is the *conjugate-transpose*. These adjoints satisfy (4.32) and (4.33), where in the case of (4.33), we restrict to normalised states. We will see in Section 5.3.2 that, if we were to interpret adjoints using the transpose, both of these conditions will fail. So, the transpose is definitely <u>not</u> a good choice for adjoints in **linear maps**.

4.3.2 Conjugates

So, what happens when we combine adjoints with transposition? Geometry suggests that it shouldn't matter if we first reflect vertically, then rotate $180°$:

or if we first rotate 180°, then reflect vertically:

In either case, we get a horizontal reflection. Using the definition of transposition in terms of cups and caps, the transpose of the adjoint is:

and the adjoint of the transpose is:

Hence these are equal, since the following is just a deformation of diagrams:

Definition 4.42 The *conjugate* of a process is the transpose of its adjoint (or equivalently, the adjoint of its transpose). Depicted graphically:

Just as in the case of adjoints and the transpose, if we conjugate twice, we return to where we started. So all together, boxes come in quartets:

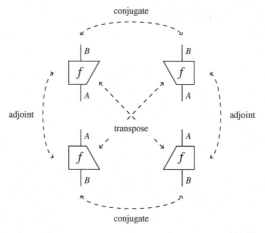

It should now also be clear why we cut out a corner for states and effects:

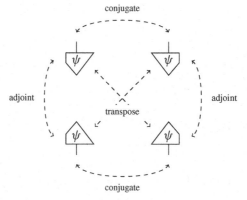

As with adjoints, conjugation mirrors entire diagrams, but horizontally rather than vertically:

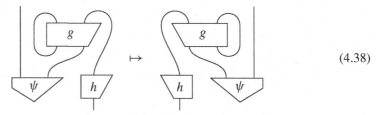

(4.38)

In particular, if a box has multiple inputs and outputs, it reverses their order:

$$\boxed{\begin{array}{c} {}^{|C} \quad {}^{|D} \\ f \\ {}_{|A} \quad {}_{|B} \end{array}} \quad \longmapsto \quad \boxed{\begin{array}{c} {}^{|D} \quad {}^{|C} \\ f \\ {}_{|B} \quad {}_{|A} \end{array}}$$

This feature of course comes from the fact that the transpose reverses input/output order, as we saw in Section 4.2.2:

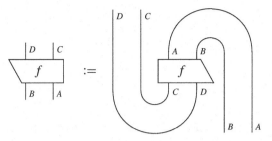

As with the transpose, we will occasionally want to avoid this. In that case, we can use the *algebraic conjugate*, which is defined in terms of the algebraic transpose as:

$$\bar{f} := (f^T)^\dagger = (f^\dagger)^T$$

It keeps the order of inputs and outputs the same by introducing a twist in the wires relative to the normal conjugate:

$$\overline{\left(\;\boxed{\begin{matrix} C & D \\ f & \\ A & B \end{matrix}}\;\right)} \;=\; \boxed{\begin{matrix} C & D \\ f \\ A & B \end{matrix}} \tag{4.39}$$

As in the case of the transpose, for single input/output wires the diagrammatic and the algebraic notions coincide.

Remark* 4.43 We will see in Chapter 5 that for **linear maps** the algebraic conjugate does exactly what one expects: it conjugates all matrix entries. So linear-algebraic adjoints, transposes, and conjugates correspond to adjoints, algebraic transposes, and algebraic conjugates, respectively.

Processes that are equal to their own conjugates will play an important role in this book. A process is said to be *self-conjugate* if:

$$\boxed{f} \;=\; \boxed{f}$$

For joint systems, this becomes:

$$\boxed{\begin{matrix} B & D \\ f \\ A & C \end{matrix}} \;=\; \boxed{\begin{matrix} D & B \\ f \\ C & A \end{matrix}} \tag{4.40}$$

which of course only makes sense if $A = C$ and $B = D$, or more generally the input and output types are palindromes such as $A \otimes B \otimes A$ or $B \otimes C \otimes C \otimes B$. On the other hand, *algebraically self-conjugate* processes on joint systems look like this:

$$\begin{array}{c} B \quad D \\ \boxed{f}\diagup \\ A \quad C \end{array} \quad = \quad \begin{array}{c} B \quad D \\ \diagup\boxed{f}\diagup \\ A \quad C \end{array} \tag{4.41}$$

Remark* 4.44 Algebraically self-conjugate matrices are those whose entries are all real numbers.

The horizontal reflection of a cup/cap is just a cup/cap again, so the following should not be surprising.

Proposition 4.45 Cups and caps are self-conjugate (and also algebraically self-conjugate).

Proof The conjugate of the cup is:

$$\bigcup = \bigcup$$

which is the same as the algebraic conjugate:

$$\bigcup = \bigcup$$

The calculation for caps is similar. ☐

Thus the 'quartet' for cups and caps is:

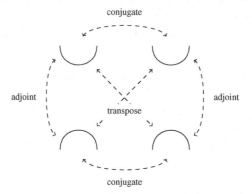

Exercise 4.46 Prove that all **relations** are algebraically self-conjugate. Of these, which relations are self-conjugate in the sense of (4.40)?

To emphasise that a state (or effect) is self-conjugate, we revert to our original notation using triangles:

$$\dot\psi := \psi\hspace{-0.5em}\diagup = \diagdown\hspace{-0.5em}\psi$$

However, we should be a bit careful with compound systems. Since conjugation reflects diagrams, states consisting of multiple, self-conjugate states are not, in general, self-conjugate:

Example 4.47 States and effects for a single system in **relations** are always self-conjugate, so we will depict them as above. In particular, we will continue to depict the states from Example 3.36 as:

For numbers, since the transpose is trivial (cf. Example 4.31), the conjugate and the adjoint coincide. Thus the 'quartet' for numbers collapses to a 'couplet' of the number and its conjugate:

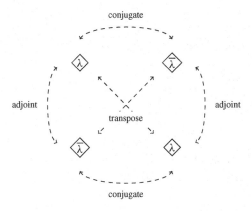

4.3.3 The Inner Product

When we first introduced Dirac notation in Section 3.4.4, we mentioned that the diagrammatic language was missing one feature: the ability to turn a ket into a bra, and vice versa. This is exactly what the adjoint does for us, so we can now amend the rules **D2** and **D3** to use adjoints:

D2: is written as $\langle\phi|$ and called a 'Dirac bra'

D3: is written as $\langle\phi|\psi\rangle$ and called a 'Dirac braket'

The composition in **D3** is important enough to give it a special name.

Definition 4.48 The *inner product* of states ψ and ϕ of the same type is:

$$\phi \, \psi$$

and these states are called *orthogonal* if:

$$\phi \, \psi \;\; = \;\; 0$$

The *squared norm* of a state ψ is the inner product of ψ with itself:

$$\psi \, \psi$$

and a state ψ is called *normalised* if:

$$\psi \, \psi \;\; = \;\; \Box$$

Remark 4.49 We say 'squared norm' because in linear algebra one typically takes the norm of ψ to be the square root of this quantity.

The meaning of the inner product immediately follows from our motivation for introducing adjoints. Since the adjoint of a state gives us the effect that tests for that state, we have:

$$\phi \, \psi \;\; := \;\; \textit{'testing state } \psi \textit{ for being state } \phi \textit{'}$$

In other words, the inner product computes 'how much commonality' states have, and orthogonality means that there is no commonality whatsoever. At the other end, we would expect to get 1 (i.e. the empty diagram) whenever we are testing a state for itself. This is true precisely when a state is normalised (which is a good reason to treat normalised states as the 'default').

Another useful intuition for 'measuring commonality' is the following. When states are distributions of some sort (e.g. probability distributions), the inner product computes how much those distributions overlap:

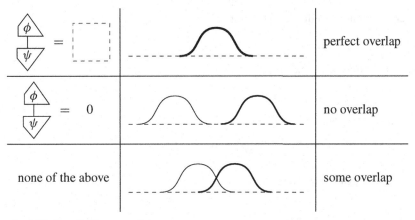

Example 4.50 In **relations** we have:

The first three inner products include cases of 'perfect overlap' as well as of 'some overlap', since both correspond to the same number. In **relations** overlap can be taken literally since:

means that the intersection of B and B' is non-empty:

B ⬭⬭ B'

On the other hand, the subsets in the last three inner products have no intersection:

B ◯ ◯ B'

The inner product and the corresponding squared norm are standard notions in linear algebra, and most of the defining properties of a linear algebraic inner product already follow from the language of string diagrams. Those readers familiar with inner products will recognise these properties more easily when denoting conjugation by $\overline{(\ )}$ and using Dirac bracket notation.

Proposition 4.51 The inner product:

1. is *conjugate symmetric*:

$$\overline{\langle\phi|\psi\rangle} = \langle\psi|\phi\rangle$$

2. preserves numbers in the second component:

$$\langle \phi | \lambda \cdot \psi \rangle = \lambda \cdot \langle \phi | \psi \rangle$$

3. conjugates numbers in the first component:

$$\langle \lambda \cdot \phi | \psi \rangle = \overline{\lambda} \cdot \langle \phi | \psi \rangle$$

4. and is *positive definite*:

$$\langle \psi | \psi \rangle = 0 \quad \Leftrightarrow \quad |\psi\rangle = 0$$

Proof   For conjugate symmetry, we have:

Next, setting:

we straightforwardly have:

as well as:

At the end of Section 4.3.1 we moreover established:

as an instance of what makes for (good) adjoints.   □

Remark 4.52 The reader may wonder what is 'positive' about 'positive definite'. This will become clear in Section 4.3.5.

Remark* 4.53 Readers familiar with inner products in linear algebra might have expected conditions 2 and 3 above to be 'linearity' and 'conjugate linearity'. The only thing missing is that the inner product should also preserve sums. This will naturally follow once we introduce sums of diagrams in Section 5.1.3.

4.3.4 Unitarity

As the inner product provides us with a measure of commonality/overlap, the natural follow-up step is to identify those processes that preserve this measure. That's what we do in this section.

Definition 4.54 A process U is an *isometry* if we have:

$$\tag{4.42}$$

In other words, $U^\dagger$ satisfies one of the two equations from Definition 4.24, needed to be an inverse of U; i.e $U^\dagger$ is a *one-sided inverse* of U.

Proposition 4.55 Isometries preserve the inner product.

Proof We have:

$$\overset{(4.42)}{=}$$

$\square$

If $U^\dagger$ satisfies both inverse equations, we obtain the following notion:

Definition 4.56 A process U is *unitary* if we have:

$$\tag{4.43}$$

We know from Exercise 4.25 that inverses are unique, so it immediately follows that, for a unitary process U:

$$\boxed{U^{-1}} \quad = \quad \boxed{U}$$

From the diagrammatic definition of unitarity we also obtain the following proposition.

Proposition 4.57 Identities and swaps are unitary, and the sequential and parallel compositions of unitary processes are again unitary.

Here are some alternative characterisations of unitarity:

Proposition 4.58 For a process f the following are equivalent:

- f is unitary.
- f is an isometry and admits an inverse.
- $f^\dagger$ is an isometry and admits an inverse.

Exercise 4.59 Prove Proposition 4.58.

In Section 5.1.5, we'll meet several more equivalent ways to recognise isometries and unitaries.

4.3.5 Positivity

In Remark 4.52 we promised to explain the 'positive' part of 'positive definite'. Its turns out that the inner product of a state with itself:

$$\begin{array}{c} \psi \\ | \\ \psi \end{array} \tag{4.44}$$

is a special case of a general notion of positivity that makes sense in any process theory.

Definition 4.60 A process f is *positive* if for some g we have:

$$\boxed{f} \quad = \quad \begin{array}{c} g \\ | \\ g \end{array} \tag{4.45}$$

So, in fact, the number given by (4.44) is positive <u>by definition</u>. In fact, for many process theories:

$$\langle \lambda \rangle \quad = \quad \begin{array}{c} \psi \\ | B \\ \psi \end{array} \qquad \text{simplifies to} \qquad \langle \lambda \rangle \quad = \quad \langle \bar{\mu} \rangle \langle \mu \rangle$$

That is, we can always take B to be 'no wire'. Thus, as we will see in the next chapter, this does in fact exactly capture the usual notion of a positive number (i.e. a real number ≥ 0) for the theory of **linear maps**. However, this notion of positivity applies not just to numbers but to any process with the same input/output system.

From (4.45), it clearly follows that positive processes are invariant under vertical reflection.

Proposition 4.61 Positive processes are *self-adjoint*, that is:

$$\begin{array}{ccc} f & = & f \end{array} \tag{4.46}$$

So in particular, positive numbers are self-adjoint, and hence, since the transpose of numbers is trivial, they are self-conjugate.

By positive definiteness, non-zero positive processes have non-zero trace. In other words, we can figure out whether a positive process is zero by computing its trace.

Proposition 4.62 For positive processes we have:

$$\begin{array}{ccc} f & = 0 & \Longrightarrow & f & = 0 \end{array}$$

Proof If f is positive, then we have:

$$\begin{array}{ccc} f & = & \begin{array}{c} g \\ g \end{array} \end{array}$$

so the trace of f is the inner product of this state:

$$\begin{array}{c} g \end{array}$$

with itself. If this inner product is zero, then so too is the above state, by positive-definiteness. Thus g itself is 0, and hence so is f. ☐

This general definition of positive process also implies a notion of positivity familiar from linear algebra. Namely, if f is positive, the number:

$$\langle \psi | f | \psi \rangle = \begin{array}{c} \psi \\ f \\ \psi \end{array}$$

is positive for all ψ. This can be easily seen by expanding Definition 4.60:

$$
\begin{array}{c}
\psi \\
f \\
\psi
\end{array}
\quad = \quad
\begin{array}{c}
\psi \\
g \\
g \\
\psi
\end{array}
$$

4.3.6 ⊗-*Positivity*

We started this chapter with the observation that, via process–state duality, we can relate ∘-separability of processes to ⊗-separability of bipartite states. Similarly, we can relate self-adjoint processes to self-conjugate states:

$$
\overset{A \quad A}{\psi} \quad = \quad \overset{A \quad A}{\psi}
$$

Proposition 4.63 A state ψ is self-conjugate if and only if the process f corresponding to it by process–state duality is self-adjoint:

self-conjugate state → ψ $\quad = \quad$ f ← self-adjoint process

Proof The conjugate of a bipartite state ψ is:

$$
\psi \quad = \quad f \quad = \quad f
$$

and this is equal to ψ itself if and only if:

$$
f \quad = \quad f
$$

that is, if and only if f is self-adjoint.   □

In the case of positivity we introduce a new name for its ⊗-counterpart.

Definition 4.64 A bipartite state ψ is $\otimes$-*positive* if for some g we have:

$$\tag{4.47}$$

Proposition 4.65 A state ψ is $\otimes$-positive if and only if the process f corresponding to it by process–state duality is positive:

$\otimes$-positive state $\longrightarrow$ $\longleftarrow$ positive process

Proof When we express a bipartite state ψ in terms of the process f, then $\otimes$-positivity becomes:

$$\overset{(4.47)}{=}$$

for some process g, which is equivalent to positivity of f. $\qquad\qquad\qquad\qquad$ $\square$

We can extend the definition of $\otimes$-positivity of states to processes.

Definition 4.66 A process f is $\otimes$-*positive* if for some g we have:

$$\tag{4.48}$$

or equivalently, for some g' we have:

$$\tag{4.49}$$

Setting:

$$:=$$

one can indeed pass from equation (4.48) to (4.49) and vice versa:

$$
\vcenter{\hbox{\includegraphics{}}}
$$

As a consequence, it is easy to see that $\otimes$-positive states are a special case of $\otimes$-positive processes, where A is the trivial system.

Exercise 4.67 Show that the sequential composition of two $\otimes$-positive processes is again a $\otimes$-positive process.

Example* 4.68 Some readers may be familiar with *density operators*, which are an example of positive 'processes'. Here we put processes in quotation marks, since in quantum theory density operators are used to represent states. Instead of density operators, in Chapter 6 we will use their $\otimes$-positive counterparts to represent quantum states. We will also use $\otimes$-positive processes, as opposed to *completely positive maps*, to represent quantum processes. The motivation for this is clear: states should be represented as states (not 'processes') and processes should be represented as processes (not 'super-processes', i.e. things that send processes to other processes).

4.3.7 Projectors

We extend our vocabulary about processes a bit more.

Definition 4.69 A *projector* is a process P that is positive and *idempotent*:

$$
\vcenter{\hbox{\includegraphics{}}} \qquad\qquad (4.50)
$$

Proposition 4.70 For a process P, the following are equivalent:

(i) It is a projector.
(ii) It is self-adjoint and idempotent.
(iii) It satisfies:

$$
\vcenter{\hbox{\includegraphics{}}} \qquad\qquad (4.51)
$$

Proof (i ⇒ ii) follows from Proposition 4.61. For (ii ⇒ iii) we have:

For (iii ⇒ i), it immediately follows from (4.51) that P is positive. Thus in particular, it is self-adjoint, so:

establishes idempotence. □

We can build projectors from any normalised state ψ as follows:

(4.52)

This is clearly positive, and for idempotence we have:

(4.53)

We will refer to these projectors as *separable projectors*. In fact, as long as $\psi \neq 0$, this recipe always yields a projector, up to a number (cf. Section 3.4.3):

Remark* 4.71 In linear algebra, separable projectors are precisely the projections onto one-dimensional subspaces.

Now recall that any process f yields the following state, via process–state duality:

If the resulting state is normalised, we obtain a separable projector:

Of course, by process–state duality, any bipartite state can be written in terms of some process f, so we have the following.

Corollary 4.72 In any process theory that admits string diagrams, every bipartite separable projector is of the form:

$$(4.54)$$

The following exercise concerns the composition of these bipartite separable projectors.

Exercise 4.73 Show that:

$$(4.55)$$

with:

$$g := f_3 \circ \bar{f}_4 \circ f_2^T \circ f_3^\dagger \circ f_1 \circ \bar{f}_2 \tag{4.56}$$

Can you generalise this particular computation into a more general statement about diagrams involving projectors of this kind?

Note in particular that the order of the processes that make up g seems at first sight to be totally unrelated to the order of projectors in the LHS of (4.55). The resulting general statement on how these bipartite projectors compose was, in the early days of diagrammatic quantum reasoning, referred to as the *logic of entanglement*.

4.4 Quantum Features from String Diagrams

While not all process theories admit string diagrams, those that do have several important features in common, which may at first sight seem weird. We already saw how process–state duality assigns to each process a state that fully captures it, that the transpose assigns a converse to each process, and more generally, that we can freely interchange inputs and outputs of processes. We also just saw that bipartite projectors exhibit some pretty funky compositional behaviour. These features have no counterpart in a process theory that doesn't admit string diagrams, such as the world of **functions** (cf. Exercise 4.10). We will discuss some other simple consequences of string diagrams, which are sometimes (perhaps prematurely) branded as 'quantum weirdness', in the remainder of this chapter.

4.4.1 A No-Go Theorem for Universal Separability

One uses the terminology 'no-go theorem' to refer to a result that establishes the impossibility of something that we might assume to be true in the light of our everyday experiences. The first no-go theorem we'll meet states that if a non-trivial theory admits string diagrams, then its bipartite states cannot all be separable. Since a $\otimes$-separable state:

$$\vcenter{\hbox{$\bigtriangledown_{\psi}$}} \;=\; \vcenter{\hbox{$\bigtriangledown_{\psi_1}$}}\;\vcenter{\hbox{$\bigtriangledown_{\psi_2}$}}$$

is described by describing the states ψ_1 and ψ_2 of the respective sub-systems, it follows that there must be states of composite systems that cannot be described merely by describing their parts.

To put this in everyday terms, suppose we have two things, like a plugstrip and a dodo. Then, it suffices to describe each thing individually to describe the whole system consisting of both things. In a $\otimes$-non-separable theory this is no longer true: the properties of the individual do not suffice to describe the properties of the whole. If this were the case for the dodo/plugstrip system, the properties of the two things would get all mixed up. For instance, the colour of the dodo's plumage could depend on whether the plugstrip has UK or EU sockets. While this situation appears to be nonsensical, there do exist some concepts in our daily world that are intrinsically $\otimes$-non-separable. One such concept is that of twins: twins are not defined by each member having a particular property, e.g. blond hair or being tall, but by the fact that whatever property one member has, the other one has that property as well. Such a concept could, for example, be modelled by the cups/caps in **relations** that we described in Exercise 4.16.

The way we establish our no-go theorem for universal separability was already alluded to in Remark 4.7. We show that if all bipartite states are ⊗-separable, then all processes are ∘-separable, so we are dealing with a process theory in which all processes are constant; i.e. nothing ever happens. As this is an absurd condition for any 'reasonable' process theory, all bipartite states cannot be ⊗-separable.

Proposition 4.74 If a theory is described by string diagrams, and all bipartite states are ⊗-separable, then all processes are ∘-separable.

Proof By assumption, the cup is ⊗-separable:

$$\smile = \psi_1 \quad \psi_2$$

So, for any process f we have:

$$f = f = f = f = \phi$$
$$\psi_2 \qquad \pi$$
$$\psi_1$$

for state $\phi := f \circ \psi_2$ and effect $\pi := \psi_1^T$.   □

If all processes are ⊗-separable, in particular, identities are ∘-separable:

$$\Big| = \frac{\phi}{\pi}$$

Hence, by looking close enough, one may discover that wires are actually not wires at all:

Conversely, we could as well have just shown that identities are o-separable, since then, any process is also o-separable:

$$
\boxed{f} \quad = \quad \Bigg| \begin{array}{c} \\ \boxed{f} \end{array} \quad = \quad \begin{array}{c} \phi \\ \pi \\ \boxed{f} \end{array}
$$

Separable states inhabit one end of the spectrum of bipartite states. At the other end of that spectrum are those states that are like cups, in the sense that they satisfy yanking-like equations for some bipartite effect:

Definition 4.75 A bipartite state ψ is called *non-degenerate*, or *cup-like*, if there exists an effect ϕ such that:

$$
\begin{array}{c} A \quad \phi \\ B \\ \psi \quad A \end{array} \stackrel{(a)}{=} \Bigg|_A \quad \text{and} \quad \begin{array}{c} \phi \quad B \\ A \\ B \quad \psi \end{array} \stackrel{(b)}{=} \Bigg|_B \qquad (4.57)
$$

Similarly, an effect ϕ is called *non-degenerate*, or *cap-like*, if there exists a state ψ satisfying the above equations.

A single cup-like state already yields a no-go theorem on separability.

Proposition 4.76 Every non-degenerate state ψ in a non-trivial process theory must be $\otimes$-non-separable.

Proof If ψ is $\otimes$-separable, then:

$$
\Bigg| \stackrel{(a)}{=} \begin{array}{c} \phi \\ \psi \end{array} = \begin{array}{c} \phi \\ \psi_1 \quad \psi_2 \end{array} = \begin{array}{c} \psi_1 \\ \phi \\ \psi_2 \end{array}
$$

i.e. a plain wire is separable. Hence ψ cannot be separable. □

Remark 4.77 In Definition 4.75 we implicitly assumed that ψ was a 'proper' bipartite state. That is, neither A nor B is the trivial, 'no wire' type. Otherwise, ψ would of course be separable in a trivial manner.

In Section 4.3.6 we saw how certain kinds of processes translate via process–state duality into certain kinds of bipartite states. In the same vein, non-degenerate bipartite states arise via process–state duality from invertible processes.

Proposition 4.78 A state ψ is non-degenerate if and only if the process f corresponding to it by process–state duality is invertible:

$$\underset{\text{invertible}}{\nearrow} \boxed{f} \;=\; \psi \quad \underset{\text{non-degenerate}}{\nwarrow}$$

Proof Setting:

$$\boxed{f^{-1}} \;:=\; \phi$$

we have:

$$\frac{\boxed{f^{-1}}}{\boxed{f}} \;=\; \phi \;\; \psi \;=\; \phi \;\; \psi$$

and:

$$\frac{\boxed{f}}{\boxed{f^{-1}}} \;=\; \phi \;\; \psi \;=\; \phi \;\; \psi$$

From these equations, it clearly follows that f and f^{-1} satisfy the inverse equations (4.13) if and only if ψ and ϕ satisfy the non-degeneracy equations (4.57). $\square$

Example* 4.79 One often encounters non-degenerate effects like ϕ in linear algebra, under the name *non-degenerate bilinear form*. In quantum entanglement theory, non-degenerate states are called *states of full (Schmidt) rank*, or *SLOCC-maximal states*, for reasons that we explain in Section 13.3.2.

Unitary processes are special kinds of invertible processes, so there also are corresponding bipartite states that further specialise cup-like states.

Definition 4.80 A state ψ is *maximally non-separable* if it corresponds to a unitary U by process–state duality, up to a number:

$$\underset{\text{unitary}}{\nearrow} \boxed{U} \;\approx\; \psi \qquad (4.58)$$
$$\underset{\text{maximally non-separable}}{}$$

Note that we use $\approx$ to allow ψ to still be normalised even when the RHS of equation (4.58) is (almost) never normalised.

Exercise 4.81 What is the squared norm of:

when U is unitary? When is this state normalised?

The following exercise indicates the utility of unitary processes in inter-converting maximally non-separable states.

Exercise 4.82 Show that if one applies a unitary to one side of a maximally non-separable state:

that one again obtains a maximally non-separable state and that this unitary can always be chosen such that the resulting state is the cup (up to a number).

Example* 4.83 In quantum theory, maximally non-separable states as defined above will be called *maximally entangled states*, or *LOCC-maximal states*, for reasons that we will explain in Section 13.3.1.

4.4.2 Two No-Go Theorems for Cloning

One feature that is characteristic of classical computation is that we can copy bits at will. For example, this book was once a PDF file on our computers, but it has since been copied and sent all over the place. While this seems like a totally obvious thing to do, 'copying' is no longer possible when we are dealing with states and processes in string diagrams! While at first this may seem really bad, this matches what goes on in the quantum world, and it also has some suprising benefits. For example, if someone wants to keep a secret, like the PIN code of a bank card, it becomes much harder in a world without copying for someone to steal the code without the card owner noticing. This principle forms the basis for quantum cryptography as well as many other quantum security protocols, as we will see in Section 9.2.6.

So far, most of the calculations we have done have been fairly trivial, and all the things that we proved were plainly obvious from the string diagram language. However, the next two theorems, and in particular the first one, will require some slightly more intricate diagram acrobatics.

By a *cloning process* for a system of type A we mean a process:

that turns an input state ψ into two copies of ψ:

$$\text{} \tag{4.59}$$

There are some obvious conditions one expects from such a process. For example, since it is producing two identical copies of a state as output, it should not matter if we interchange them:

$$\text{} \tag{4.60}$$

Furthermore, when we have two cloning processes, one for type A and one for type B, then we should be able to clone a state of type $A \otimes B$ just by cloning each system individually:

$$\text{} \tag{4.61}$$

Finally, we will assume our process theory contains at least one normalised state. That is, at least one state ψ such that:

$$\text{} \tag{4.62}$$

In essence, this a trivial assumption since otherwise there is nothing to be cloned anyway!

Theorem 4.84 Consider a process theory that admits string diagrams. If there is a cloning process of type A that satisfies (4.60), (4.61), and (4.62), then every process that has A as its input system has to be o-separable.

Proof Applying (4.61) for:

we have:

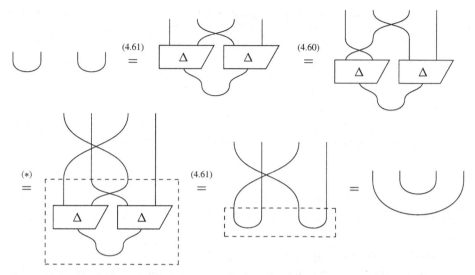

where all the wires have type A and $(*)$ is just a deformation of the diagram. Then converting the external outputs of the LHS and RHS into inputs:

we discover that an identity on a pair of systems is separable, which should already raise some eyebrows. Substituting the above equation into the dotted areas below:

so we indeed obtain that any process f that has A as its input type is $\circ$-separable. □

Thus, in the light of our conception of 'trivial' outlined in Section 4.4.1, if there is a cloning process for a certain system-type A, then the process theory is trivial with respect to type A, and if every system-type admits a cloning process, then the process theory is trivial as a whole.

The assumption that really makes things go wrong is (4.61). If one 'thinks' in the language of string diagrams, it is pretty obvious that trying to copy a non-separable state by doing something to each of the subsystems isn't going to work.

Theorem 4.84 is quite different from the one that one finds in most textbooks. First, a cloning process is usually introduced as a process with two inputs and two outputs, whose second input is in some fixed state ϕ, which gets overwritten by the copied state ψ:

$$\Delta' \circ (\psi \otimes \phi) = \psi \otimes \psi \tag{4.63}$$

Of course, a one-input, two-output cloning process arises as follows:

$$\Delta \;:=\; \Delta' \circ (- \otimes \phi) \tag{4.64}$$

and we could then just as easily use (4.64) to prove Theorem 4.84.

Second, one usually assumes that the process Δ' in equation (4.63) is unitary, which in turn means we don't need assumptions (4.60) and (4.61).

This is such a profound difference that it results in what should be considered as a different theorem, despite the fact that it aims to establish the same feature. The assumption of unitarity is motivated by quantum theory itself (cf. Section 6.2.6). In spite of the fact that this theorem relies on extra assumptions, it is still of interest because it additionally demonstrates precisely which sets of states *can* be jointly cloned by a single process, namely the *orthogonal* ones. This joint-clonility of orthogonal states is closely connected to the fact that they can be used to encode classical data within quantum systems, which we will exploit in Chapter 8 to give an elegant diagrammatic representation of classical data.

Instead of using Δ', we will use Δ defined as in equation (4.64). By unitarity of Δ', when ϕ is normalised, Δ must be an isometry:

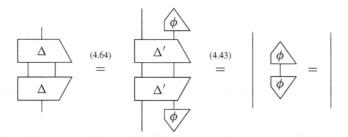

So we assume the existence of an isometry Δ satisfying the cloning equation (4.59). Of course the biggest extra assumption in the traditional no-cloning theorem is that we are dealing specifically with the theory of **quantum processes** (which of course, we haven't even defined yet!). However, this assumption is a bit overkill. All we really need is to assume a couple of things about the numbers of our theory:

(a) All non-zero numbers are *cancellable*, that is, if $\lambda \neq 0$:

$$\lambda \; \boxed{f} \; = \; \lambda \; \boxed{g} \quad \Longrightarrow \quad \boxed{f} \; = \; \boxed{g}$$

which is evidently the case for numbers representing either possibilities (cf. **relations**) or probabilities.

(b) The number 1 means 'certain', as discussed in Section 4.3.1. That is, for normalised ψ_1, ψ_2 we have:

$$\frac{\psi_2}{\psi_1} \; = \; \boxed{\;\;} \quad \Longrightarrow \quad \boxed{\psi_1} \; = \; \boxed{\psi_2}$$

Theorem 4.85 In any process theory that admits string diagrams and satisfies conditions (a) and (b) above, if two normalised states ψ_1 and ψ_2 can both be cloned by an isometry Δ, then they must either be equal or orthogonal. That is, we either have:

$$\boxed{\psi_1} \; = \; \boxed{\psi_2} \qquad \text{or} \qquad \frac{\psi_2}{\psi_1} \; = \; 0$$

Proof First, note that:

$$\frac{\psi_2}{\psi_1} \overset{(4.43)}{=} \; \frac{\psi_2 \;\; \Delta}{\Delta \;\; \psi_1} \overset{(4.59)}{=} \; \frac{\psi_2}{\psi_1} \;\; \frac{\psi_2}{\psi_1}$$

Next, consider two cases. If:

$$\frac{\psi_2}{\psi_1} \; \neq \; 0$$

by assumption (a), we can cancel this number on both sides to obtain:

$$\boxed{\;\;} \; = \; \frac{\psi_2}{\psi_1}$$

so by assumption (b) the states ψ_1 and ψ_2 are equal. On the other hand, if:

$$\frac{\overline{\psi_2}}{\underline{\psi_1}} = 0$$

then ψ_1 and ψ_2 are orthogonal. □

Theorem 4.85 immediately yields the second no-cloning theorem.

Corollary 4.86 Under the assumptions of Theorem 4.85, if a process theory has at least two states of type A that are neither equal nor orthogonal, then there is no cloning process of type A.

Remark 4.87 Note that the first no-cloning theorem applies to **relations**, whereas the second one doesn't, since it relies on condition (b). In fact, the proof of the first no-cloning theorem doesn't rely on adjoints at all, so in that sense is more general. However, the second no-cloning theorem does avoid assuming extra equations about the cloning device (notably equation (4.61) for cloning joint states), so it is not directly implied by Theorem 4.84.

Remark 4.88 There is a subtlety about the statement 'we can copy bits at will', namely, this assumes bits are *deterministic*, i.e. they have definite values. We won't discuss this any further now. A full discussion of this issue, as well as a more refined no-go theorem that avoids it, is given in Section 6.2.8.

4.4.3 As If time Flows Backwards

Consider the following equal string diagrams:

$$(4.65)$$

Suppose we interpret the LHS and the RHS as processes happening at specific points in time $t_1, \ldots, t_4$:

$$(4.66)$$

then something strange happens. When considering the LHS we have:

t_2: something happens involving g;

t_3: something happens involving f;

On the other hand, when considering the RHS:

t_2: something happens involving f;

t_3: something happens involving g.

While for the LHS the output of the g-labelled process is the input to the f-labelled process, for the RHS the output of the f-labelled process is the input to the g-labelled process. So the order in which the processes happen is reversed!

The reason that something like this can happen is that two equal diagrams may actually correspond to very different *operational scenarios*. An operational scenario is the manner in which a diagram is actually realised, e.g. by wiring together some devices in a lab. The LHS and RHS above have different operational readings, since the LHS involves three systems at times t_2 and t_3, while the RHS only involves one system at any time. The cups correspond with the creation of two systems, while caps correspond with the annihilation of two systems. As a result, we have a more complicated operational scenario in the LHS, which is then simplified in the RHS.

Despite the difference in the two operational readings, what actually happens, according to our theory, is the same. We might call this the *logical reading* of a string diagram. In this case, we can see that these diagrams have the same logical reading by tracing the logical flow of the diagram:

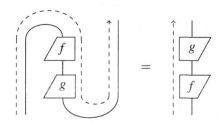

which passes first through the f-labelled box and then the g-labelled box.

In the case of the logical reading, the cup and the cap make it seem as if systems are actually travelling back in time! This has even led several researchers to propose a cup and a cap as a model for time travel:

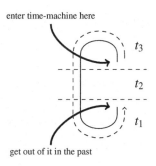

Does this mean that cups and caps provide a pathway to effectively building a time machine? Of course they don't! The main problem here is the fact that, as we shall see later, caps cannot be implemented with certainty, but only with some probability. The reason for this is what we will refer to as the *causality postulate* (see Section 6.4). In particular, if anyone proposes to you a trip in such a machine, we highly recommend you not to take it, since as we shall see in Section 6.4.4, you'd come out pretty scrambled up!

Exercise 4.89 Let $\mathbb{T} := \{0, 1, 2\}$ be a three-element set (the set of 'trits'). Define relations $f, g : \mathbb{T} \to \mathbb{T}$ as follows:

- $f :: \{0 \mapsto 1, 1 \mapsto 0, 2 \mapsto 2\}$;
- $g :: \{0 \mapsto 0, 1 \mapsto 2, 2 \mapsto 1\}$.

Note in particular that these relations do not commute:

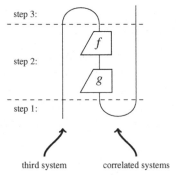

Let the cup and the cap be those of Exercise 4.16. Verify the 'time reversal' equation (4.65) by explicit composition of relations.

Exercise 4.89 demonstrates how the 'time reversal' can be realised by means of non-deterministic processes, in the following manner:

Step 1: Create non-deterministic perfect correlations for two systems.
Step 2: Apply two consecutive operations to one of the systems.
Step 3: Impose that the output matches the state of a third system.

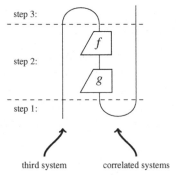

In addition to string diagrams that simply have strange logical readings, we may have diagrams that have more than one such reading, or none at all, for example:

In the left diagram, is the 'flow' from g to f or from f to g? And who knows what is even happening in the diagram on the right? However, we can still make sense of such diagrams by thinking about wires less as 'flow', which has a definite direction, and more as 'forcing the value at both ends to be the same', which is a symmetric concept.

4.4.4 Teleportation

We'll now meet one of the 'killer apps' of quantum theory for the first time: quantum teleportation. Perhaps surprisingly, the real 'meat' of quantum teleportation can already be described using just the diagrammatic concepts we've met so far. In Chapter 6, we will revisit this protocol and fill in the details, such as careful definitions of *quantum states* and *measurements*.

Here is the challenge. Aleks and Bob are far apart, Aleks possesses a system in a state ψ, and Bob needs this state. Suppose they also share another state: the cup state. So, we have this situation:

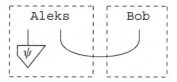

Starting from this arrangement, is there something Aleks and Bob can do in the regions marked '?' below, that results in Bob obtaining ψ?

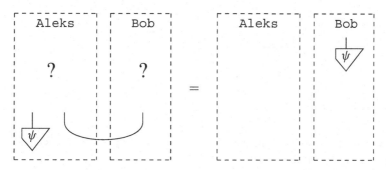

Here is a simple solution:

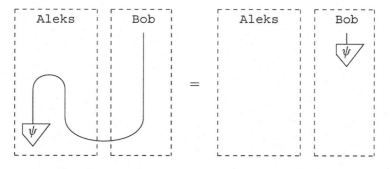

But is the cap (i.e. an <u>effect</u>) really a process that Aleks can 'do'? Well, not exactly. We'll see in Chapter 6 that effects arise as the result of quantum measurements. Now, a tricky issue about measurement is that Aleks might not get the effect he wants (i.e. the cap), but rather the cap with some (non-deterministic) error, which we can represent as a box:

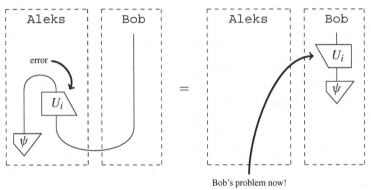

Bob's problem now!

Aleks doesn't know in advance which error he will get, just that it will be in some set $\{U_0, U_1, \ldots, U_{n-1}\}$. It could be the case that U_0 is the identity process, i.e. no error. If he's feeling lucky, he'll just hope this happens, and if it doesn't, he'll try the whole thing again with a new ψ and a new cup state. This is called *post-selection*. Thinking about it a bit more, Aleks realises this seems a bit wasteful. Moreover, if he only possesses one system in the state ψ, that state will be lost forever. Instead, Aleks decides to call up Bob and tell him to fix the error. Bob can achieve that as follows, in the particular case that U_i is a unitary process:

$$\text{fix} \rightarrow \boxed{U_i} \quad \text{error} \rightarrow \boxed{U_i} \quad \boxed{\psi} \quad = \quad \boxed{\psi}$$

So, all Aleks needs to tell Bob is the value of $i \in \{0, \ldots, n-1\}$ so he knows which correction to perform. And poof! Bob now has ψ :

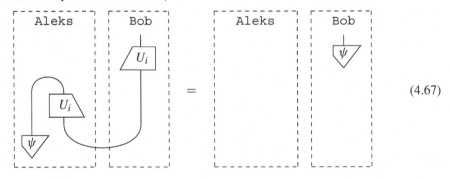

(4.67)

So this is it: quantum teleportation. In summary:

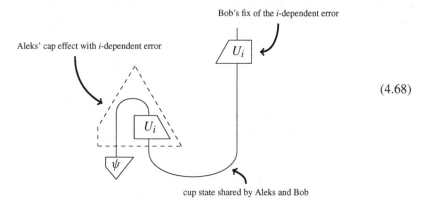

$$(4.68)$$

One important thing to note is that it is crucial for Aleks to send Bob the value of i, otherwise Bob can't fix the error. In that case, Bob will only get noise, as we shall see in Section 6.4.4. This is why, for example, quantum teleportation is compatible with relativity theory, which forbids sending any signal faster than the speed of light. As a consequence, when we use the term 'teleportation', we don't really mean magically beaming something through space. Instead, the teleportation protocol uses one kind of information to send another kind of information. In this protocol, Aleks communicates some *classical data* to Bob, and as a result, Bob obtains *quantum data* (i.e. a quantum state). This is genuinely surprising, since, as we shall see in Chapter 7, the space of possible quantum states is infinitely larger than the number of possible classical values.

Exercise 4.90 Write the teleportation protocol of (4.67):

1. as a diagram formula, and
2. algebraically, using $\otimes$ and $\circ$.

We're not quite ready to fully describe quantum teleportation. However, there exists a totally classical analogue to quantum teleportation that can be described using **relations**.

Example 4.91 ('classical' teleportation) Suppose Aleks and Bob both have envelopes with a card inside that says either '0' or '1'. They don't know which it is, but they do know that they both have the <u>same</u> card. We can represent this non-deterministic state as the cup relation:

$$\cup :: \begin{cases} * \mapsto (0,0) \\ * \mapsto (1,1) \end{cases}$$

Now, suppose Aleks has a bit ψ that he wants to send to Bob. As before, they have a telephone, but unlike before, Aleks has a classical bit, so he could just look at it, ring up Bob, and tell him what it is. On the other hand, Aleks is a bit paranoid and doesn't want any potential eavesdroppers to get a hold of ψ. So instead of just telling ψ to Bob, he computes the parity of ψ together with the bit stored in his envelope. That is, he looks at both bits and tells Bob over the phone whether they are the same, which corresponds to the effect M_0, or different, which corresponds to the effect M_1:

$$M_0 :: \begin{cases} (0,0) \mapsto * \\ (1,1) \mapsto * \end{cases} \qquad M_1 :: \begin{cases} (0,1) \mapsto * \\ (1,0) \mapsto * \end{cases}$$

If they are the same, Bob knows that the card in his envelope is the bit ψ. If they are different, he knows that ψ is the negation of the bit in his envelope. In other words, based on the outcome of Aleks' parity measurement, Bob chooses a correction to apply to the bit in his envelope:

$$U_0 :: \begin{cases} 0 \mapsto 0 \\ 1 \mapsto 1 \end{cases} \qquad U_1 :: \begin{cases} 0 \mapsto 1 \\ 1 \mapsto 0 \end{cases}$$

Noting that the effect M_i can be rewritten in terms of a cap and U_i, we can stick this all together, giving us this picture:

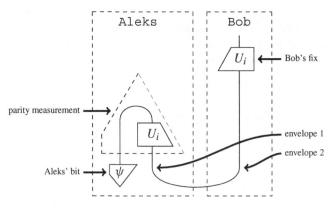

which, of course, is just teleportation!

Remark 4.92 In the world of computer security, Example 4.91 is called *one-time pad encryption*. The bits in the envelopes correspond to the shared key, or 'pad'; the parity measurement *encrypts* Aleks' bit; and Bob's correction *decrypts* it. Above we noted that a teleportation protocol uses one kind of information to send another. This interpretation applies here as well. In this case, Aleks sends *public* (i.e. encrypted) data to Bob, and Bob receives *private* (i.e. unencrypted) data at the end. So, the analogy between classical and quantum teleportation goes like this:

	Aleks sends	Bob receives	Using a shared
One-time pad encryption:	public data	private data	encryption key
Quantum teleportation:	classical data	quantum data	quantum state

Bob's fix of the error (complements white box)

Aleks' cap effect with error (cf. white box)

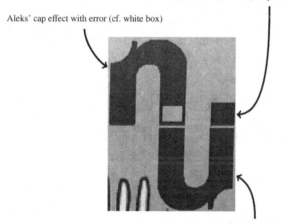

cup state shared by Aleks and Bob

Figure 4.1 Teleportation is everywhere when you know how to recognise it. Here it is in the highly recommended Cafe Cantine in Istanbul's Taksim square area.

4.5 Summary: What to Remember

1. *String diagrams* realise what Schrödinger singled out as the characteristic trait of quantum theory: non-separability. They have the following two equivalent characterisations:

Diagrams in which wires may connect inputs to inputs and outputs to outputs, resulting in cup- and cap-shaped wires:

Circuits for which each type has a special state and a special effect, the *cup state* and *cap effect*, which satisfy equations:

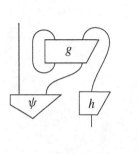

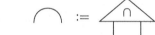

These two characterisations are related by setting:

and then the defining equations of the cup state and the cap effect become the *yanking equations*:

2. For string diagrams there is a bijective correspondence between processes and bipartite states, called *process–state duality*, which is realised as follows:

$$\boxed{f} \;\mapsto\; \boxed{f} \qquad \boxed{\psi} \;\mapsto\; \boxed{\psi}$$

3. The string diagram language enables us to define the *transpose*, which we represent the transpose as a 180° rotation:

$$\boxed{f} \;:=\; \boxed{f}$$

As a consequence, boxes can slide along cups and caps:

4. It also enables us to define the *trace* (see below), which gives us a way to assign numbers to processes and obeys *cyclicity*:

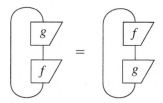

5. We also introduce vertical reflection of diagrams. *Adjoints* give an interpretation for vertical reflection within a process theory and associate to each state the effect that tests for that state. They also enable us to define an *inner product*, as well as *isometries*, *unitary* processes, *positive* processes, and *projectors* (see below for each of these).

6. Combining vertical reflection with transpose-rotation, we get a second, horizontal reflection, called the *conjugate*:

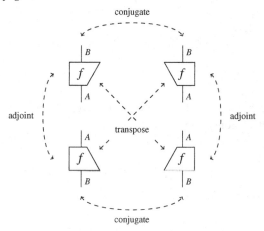

Conjugates enable us to define ⊗-*positive* bipartite states (see below).

7. We identified several 'physical' features of string diagrams:

- In any non-trivial theory there <u>always exist</u> non-separable states, most notably the cup-state. This state establishes perfect correlations and can create the illusion of systems travelling back in time:

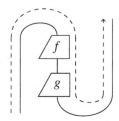

- There <u>does not exist</u> an operation:

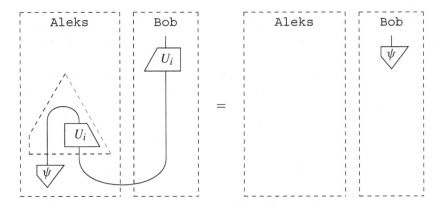

that clones all states ψ. In particular, the only states that can jointly be cloned by an isometry must be orthogonal.

- We can describe *teleportation*, which boils down to this equation:

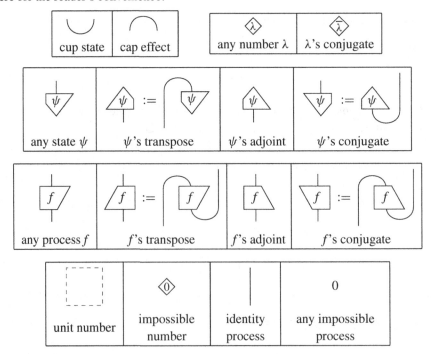

8. We introduced a plethora of diagrammatically defined concepts, which we summarise here for the reader's convenience:

cup state	cap effect

any number λ	λ's conjugate

any state ψ	ψ's transpose	ψ's adjoint	ψ's conjugate

any process f	f's transpose	f's adjoint	f's conjugate

unit number	impossible number	identity process	any impossible process

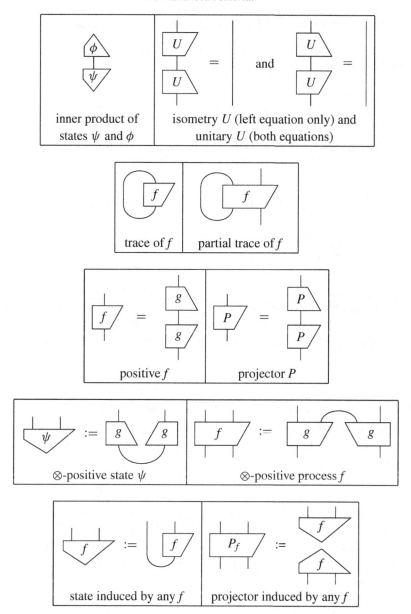

4.6 Advanced Material*

Following on from the advanced material in the previous chapter, we will now look at how the additional structure of string diagrams is defined in terms of abstract tensor systems, on the one hand, and symmetric monoidal categories, on the other. In the case of the latter, before we can do so we will need to refine our treatment of cups and caps a bit more.

4.6.1 String Diagrams in Abstract Tensor Systems*

String diagrams arise in abstract tensor systems just by adjoining special tensors $\cup^{AB}$ and $\cap_{AB}$ satisfying:

$$\cap_{AB}\cup^{BC} = \delta_A^C$$

$$\cap_{BA} = \cap_{AB}$$

$$\cup^{BA} = \cup^{AB}$$

Drawing these equations as pictures, these are just the normal identities we expect from caps and cups:

In tensorial language, the cups and caps are sometimes called *index-raising* and *index-lowering* operations, because composing other tensors with them raises or lowers an index (i.e. changes an input to an output or vice versa). This trick should already be familiar from Section 4.2:

$$f'^{BC}_A = f^B_{AC'}\cup^{C'C} \qquad\qquad f'^B_{AC} = f^{BC'}_A \cap_{C'C}$$

The tensor $\cap_{AB}$ plays a key role in *differential geometry*, which is a branch of mathematics that studies geometrical properties from inside curved surfaces. The cap is called the *metric* of a space. In that context, a more familiar notation for cups and caps is the following:

$$g^{\mu\nu} := \cup^{AB} \qquad\qquad g_{\mu\nu} := \cap_{AB}$$

where $\mu := A$ and $\nu := B$. Much like the inner product can be used to calculate lengths of vectors in plain old Cartesian space, the metric can be used to calculate lengths of paths in more general spaces, which can be curved and distorted. These spaces play a central role in general relativity, where the effects of gravity amount to distortions in spacetime (hence the use of 'g').

4.6.2 Dual Types and Self-Duality*

In order to simplify the presentation of caps and cups, we assume throughout this book that caps and cups satisfy the property of *self-duality*:

 same type

or, in other words, that all of our types are *self-dual*. We can relax the requirement that these types are the same. Rather than taking cups and caps as the main actors in the definition, the focus now moves to the 'other type' involved in cups and caps, and 'having cups and caps' becomes 'having duals'.

Definition 4.93 For any type A, the type A^* is called the *dual type* of A if there exists a *cup state* and a *cap effect*:

satisfying the *yanking equations*:

(4.69)

A type A is called *self-dual* if $A = A^*$.

Recalling Proposition 4.13 where we gave two equivalent presentations of the yanking equations, one can see that the equations above generalise version (ii). If A is self-dual and additionally satisfies:

(4.70)

we say it is *coherently self-dual*. If $A \neq A^*$, this isn't even a meaningful equation anymore, since the types on the LHS and RHS don't match. However, just by deforming equations (4.69), we can see that the LHS above gives a cup for A^*. Pairing with the appropriate cap yields:

So, A is in fact a dual type for A^*. Thus we can let:

$$(A^*)^* := A$$

in which case (4.70) becomes:

$$
\text{} \tag{4.71}
$$

When $A = A^*$, we have two ways to build a cup with the same type (either as $\cup_A$ or $\cup_{A^*}$), so (4.70) says that they should be equal.

For many examples in mathematics, it may indeed be more natural to treat a type as different from its dual type. An important example comes from **linear maps**, which are presented in Chapter 5. For those unfamiliar with vector spaces, linear maps, or the tensor product, it might be worth reading Chapter 5 before looking at the next example.

Example 4.94 For a finite-dimensional vector space A, let A^* be the *dual space* of A. That is, the elements $\xi \in A^*$ are themselves linear maps from A into $\mathbb{C}$. These form a vector space by letting addition and scalar multiplication act 'point-wise':

$$
(\xi + \eta)(v) := \xi(v) + \eta(v) \qquad\qquad (\lambda \cdot \xi)(v) := \lambda\xi(v)
$$

Furthermore, any basis $\{\phi_i\}_i$ in A fixes a *dual basis* $\{\widetilde{\phi}_i\}_i$ via:

$$
\widetilde{\phi}_i(\phi_j) := \delta_i^j
$$

Now, we can show that A^* is the *dual type* of A, by defining a cup state and cap effect. The choice of cap effect is very natural; we just take the effect that evaluates $\xi \in A^*$ at the vector $v \in A$:

$$
\text{} \;::\; (v \otimes \xi) \mapsto \xi(v)
$$

The cup state is given by summation over a basis:

$$
\text{} \;::\; 1 \mapsto \sum_i \widetilde{\phi}_i \otimes \phi_i
$$

It is possible to check that this cup and cap satisfy (4.69), and one can also show that, unlike their self-dual variations, these cups and caps don't depend on the choice of basis $\{\phi_i\}_i$.

If we transpose $f : A \to B$ with respect to these new caps and cups, we get a new map $f^* : B^* \to A^*$:

$$
\text{}
$$

which is sometimes called the *linear operator transpose*. Concretely, f^* sends an element of B^* to an element of A^* by precomposing with f:

$$f^*(\xi) := \xi \circ f$$

which again doesn't depend on a choice of basis (unlike the normal transposition).

Even when it is possible to choose A to be self-dual, having the freedom to chose A^* can sometimes be helpful. For instance, consider the 'nested caps' problem from Section 4.2.2:

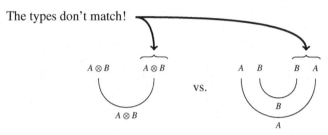

This problem goes away if we let:

$$(A \otimes B)^* := B^* \otimes A^*$$

But we seem to have lost some notational niceness when $A \neq A^*$. Rather than representing caps and cups as a piece of wire, we seem forced to use explicit boxes for $\cap$ and $\cup$, otherwise we'll end up with wires that say one type at one end and another type at the other:

That doesn't look right! However, there is a very elegant way to deal with duals graphically. We just introduce a *direction* to the wires:

$$A\!\uparrow \;\; := \;\; \Big|\, A \qquad\qquad A\!\downarrow \;\; := \;\; \Big|\, A^*$$

Wires of 'normal' types A are depicted as wires directed upwards (i.e. as 'stuff flowing forwards in time'), whereas wires of dual types A^* are depicted as wires directed downwards (i.e. as 'stuff flowing backwards in time'). In either case, we label the wire simply as A and use the direction to tell us whether the wire is A or A^*. This little tweak to the notation allows us to once again represent caps and cups as pieces of wire, but using directed wires this time:

Equations (4.69) become:

and (4.71) becomes:

Maps to and from duals are depicted as boxes connected to wires with the directions reversed. For example, a box with input type $A \otimes B^*$ and output type $C^* \otimes D$ looks like this:

Just as we were able to define string diagrams without referring to caps and cups (cf. Definition 4.18), we can define *directed string diagrams* as diagrams where we are allowed to connect any two wires, provided that the types and directions are compatible:

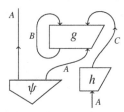

That is, if we connect an input to an output, the types should be the same. If we connect an input to an input (or an output to an output), the types should be dual to each other. For instance, the output of h with type C^* is connected to the output of g with type C.

4.6.3 Dagger Compact Closed Categories*

A compact closed category is a symmetric monoidal category where every object has a dual. We can say this in categorical language as follows.

Definition 4.95 A *compact closed category* is a symmetric monoidal category $\mathcal{C}$ such that for every object $A \in \mathrm{ob}(\mathcal{C})$, there exists another object $A^* \in \mathrm{ob}(\mathcal{C})$ and morphisms:

$$\epsilon_A : A \otimes A^* \to I \qquad\qquad \eta_A : I \to A^* \otimes A$$

such that:

$$(\epsilon_A \otimes 1_A) \circ (1_A \otimes \eta_A) = 1_A \qquad (1_{A^*} \otimes \epsilon_A) \circ (\eta_A \otimes 1_{A^*}) = 1_{A^*} \qquad (4.72)$$

Remark 4.96 The adjective 'closed' means that for every two objects A and B there is a special object $[A \to B]$ whose states $\psi : I \to [A \to B]$ encode morphisms in $\mathcal{C}(A, B)$. For example, in the category that has sets as objects and functions as morphisms, $[A \to B]$ is the set of all functions from A to B, whereas in the category that has vector spaces as objects and linear maps as morphisms, $[A \to B]$ is the vector space of linear maps from A to B. 'Compact closed' means that these special objects take the form:

$$[A \to B] := A^* \otimes B$$

i.e. that the category has process–state duality.

We can also account for adjoints in category-theoretic terms.

Definition 4.97 A *dagger functor* for a symmetric monoidal category is an operation $\dagger$ that doesn't alter objects:

$$A^\dagger := A$$

reverses morphisms:

$$(f : A \to B)^\dagger := f^\dagger : B \to A$$

is involutive:

$$(f^\dagger)^\dagger = f$$

and respects the symmetric monoidal category structure:

$$(g \circ f)^\dagger = f^\dagger \circ g^\dagger \qquad (f \otimes g)^\dagger = f^\dagger \otimes g^\dagger \qquad \sigma_{A,B}^\dagger = \sigma_{B,A}$$

A *dagger compact closed category* is a compact closed category $\mathcal{C}$ with a dagger functor satisfying:

$$\epsilon_A^\dagger = \eta_{A^*}$$

Just as Theorem 3.49 gave the soundness/completeness for circuit diagrams with respect to symmetric monoidal categories, it is possible to show the same for string diagrams and dagger compact closed categories.

Theorem 4.98 String diagrams are sound and complete for dagger compact closed categories. That is, two morphisms f and g are provably equal using the equations of a dagger compact closed category if and only if they can be expressed as the same string diagram.

We can now extend our table from Section* 3.6.3.

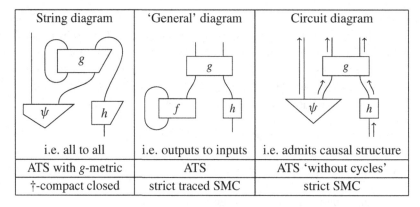

String diagram	'General' diagram	Circuit diagram
i.e. all to all	i.e. outputs to inputs	i.e. admits causal structure
ATS with g-metric	ATS	ATS 'without cycles'
†-compact closed	strict traced SMC	strict SMC

4.7 Historical Notes and References

The manner in which bipartite projectors compose, discussed in Section 4.3.7, is taken from Coecke (2003) and was the formal basis for the first formal graphical presentation of quantum teleportation. So while it is therefore fair to say that this paper kicked off the diagrammatic approach for depicting quantum processes, it was refused by a *Physical Review Letters* editor (without even consulting referees) on the basis of being too 'speculative'. The idea that these bipartite projectors should be interpreted in terms of cups, caps, and boxes as depicted in (4.54) first appeared in Abramsky and Coecke (2005, 2014a). Independently, a topology-based analysis of quantum teleportation was given in Kauffman (2005).

But in fact, cups and caps appeared much earlier, in Penrose's diagrammatic calculus for abstract tensor systems (Penrose, 1971), where they represented the metric tensors of spacetime geometry. However, the further development of string diagrams and obtaining an understanding of their meaning in terms of mathematical models mainly took place within the category theory community. Around the same time as Penrose's paper, string diagrams also appeared in Kelly's paper, which introduced compact closed categories (Kelly, 1972), which as we explained in Section* 4.6.3 are the category-theoretic version of process theories that admit string diagrams. The use of additional boxes besides cups and caps in the categorical context is due to Yetter (see e.g. Freyd and Yetter, 1989), who used the name 'coloured tangles'.

Process–state duality is known, in the context of quantum theory, as the Choi–Jamiołkowski isomorphism (Jamiołkowski, 1972; Choi, 1975). In fact, the isomorphisms presented respectively by Jamiołkowski and Choi were not equivalent. Choi's is the one that we will use throughout most of this book and is basis dependent. Jamiołkowski's is the basis-independent counterpart, resulting from the cups and caps of Example* 4.94.

Dagger compact closed categories were defined in Abramsky and Coecke (2004, 2005) in order to axiomatise the manner in which bipartite projectors compose, resulting in a presentation of basic quantum theory in category-theoretic terms. Around the same time, similar ideas were proposed by Baez (2006). Asymmetric boxes for representing transposes and adjoints are taken from Selinger (2007). As mathematical entities, dagger compact closed categories had already appeared in Baez and Dolan (1995) as a special case of a more general construct in *n*-categories. Theorem 4.98 is widely regarded as folklore, with a formal statement in Selinger (2007), citing an 'implicit proof' first occurring in (Kelly and Laplaza, 1980).

The notion of ⊗-positivity was introduced in Selinger (2007) as (abstract) complete positivity. The version of the no-cloning theorem presented in Theorem 4.85, which preceded the emergence of quantum information, was independently proved by Dieks (1982) and Wootters and Zurek (1982). The version presented in Theorem 4.84 that relies on the existence of cups rather exploiting unitarity is taken from Abramsky (2010).

The word 'teleportation' was first coined in 1931 by anomalistics writer Charles Fort (1931) in a book entitled *Lo!* consisting of two parts, one of which is entirely devoted to teleportation. However, the concept really took off with the role of the transporter in the adventures of the starship *Enterprise* in Gene Roddenberry's (1966) *Star Trek*. Quantum teleportation was first proposed by Bennett et al. (1993), and its first experimental realisation was in Bouwmeester et al. (1997). A diagrammatic presentation of classical teleportation was in Coecke et al. (2008b), and a more detailed account is in Stay and Vicary (2013).

The observation related to time-reversal of Section 4.4.3 is taken from Coecke (2003, 2014a), and the results of this paper were experimentally simulated in Laforest et al. (2006). The inspiration for the fact that cups and caps can be used to simulate time travel also came from this paper and was first stated as such in Svetlichny (2009). This was later picked up again in Lloyd et al. (2011), which received a lot of media attention, with some headlines claiming that time travel had been realised. Much earlier, a different theory of quantum time travel had appeared in Deutsch (1991).

The best overview for a wide variety of monoidal categories and their associated notions of diagram is Selinger (2011b). If you want to read more about string diagrams in the context of physics, logic, and topology, we suggest Baez and Stay (2011), and in the context of physics and linguistics, Heunen et al. (2012a). To learn about a whole variety of (mostly) diagrammatic constructions used in higher-dimensional category theory, see Leinster (2004).

The quote at the beginning of this chapter is taken from Schrödinger (1935).

5

Hilbert Space from Diagrams

> I would like to make a confession which may seem immoral: I do not
> believe absolutely in Hilbert space any more.
> — *John von Neumann, letter to Garrett Birkhoff, 1935*

We have now seen how processes described by string diagrams already exhibit some quantum-like features. It is natural to ask how much extra work is needed to go from string diagrams to Hilbert spaces and linear maps, the mathematical tools von Neumann used to formulate quantum theory in the late 1920s. The answer is: not that much.

We start by considering what it takes for two processes to be equal. In many process theories, it suffices for them to agree on a relatively small number of states. This feature leads very naturally to the notion of *basis*, and we can use adjoints to identify a particularly handy type of basis, called an *orthonormal basis* (ONB). When all types admit a basis, any process can be completely described by a collection of numbers called its *matrix*.

Now, such a matrix identifies a particular process uniquely, but for any matrix to represent a process we need to add a bit more structure. Therefore, we allow processes with the same input/output wires to be combined into one, or *summed* together. If a process theory admits string diagrams, has ONBs for every type, and has sums of processes, we can describe sums, sequential composition, parallel composition, transpose, conjugate, and adjoint all in terms of operations on matrices. We call this the *matrix calculus* of a process theory.

Thus, by adding ONBs and sums, we have very nearly recovered the full power of *linear algebra*, but with the added generality that the numbers λ are still very unrestricted (in particular, they need not be the elements of some field like the real or complex numbers). In fact, a matrix calculus for relations makes perfect sense, where the numbers are booleans.

The final step towards Hilbert spaces and linear maps consists of requiring the numbers of the process theory to be the *complex numbers*. Thus, we define **linear maps** as the process theory admitting string diagrams where:

1. every type has a finite ONB;
2. for any $n \in \mathbb{N}$ there is a type with an ONB of size n;
3. processes of the same type can be summed; and
4. the numbers are the complex numbers.

The system-types in this process theory are then called *Hilbert spaces*. For those familiar with the Hilbert space formalism of quantum theory, the most notable thing is the absence of any reference to the *tensor product* of Hilbert spaces, nor to *(multi-)linearity* of maps. The reason for this is that the language of string diagrams gives us these for free!

Our presentation of Hilbert spaces and linear maps is quite different from the 'bottom-up' presentation one usually encounters. There, one typically starts with small things, namely vectors, then defines special sets of vectors called vector spaces, and specialises these to Hilbert spaces. Then, to turn this into a process theory, one puts in an awful lot of work defining linear maps, bilinear maps, composition and tensor product using a whole bag of tools from set theory and algebra. These set-theoretic definitions are intrinsically *reductionist* in spirit: they are all about understanding bigger things in terms of smaller things. This is much like the particle physicist's dream to have a 'theory of everything', which explains the world entirely in terms of its smallest parts. By contrast, in our presentation a thing is understood in terms of how it interacts with other things. Thus, it makes sense to adopt a 'top-down' approach, where the whole process theory of **linear maps** is defined by first stating how things compose, then filling in the remaining blanks.

Now, this book is about using diagrams for reasoning about quantum processes, so why do we even bother to introduce Hilbert spaces? In fact, in Chapter 8 things like the existence of ONBs and sums will be accounted for in terms of a new diagrammatic primitive, and in Chapter 9 something similar will be done to account for the fact that there are multiple ONBs for a single system. Nonetheless, there are at least three good reasons to introduce Hilbert spaces:

1. Readers who are familiar with Hilbert space quantum theory may find it useful to see the diagrammatic concepts represented in the language of Hilbert spaces, because of their familiarity with that language.
2. Readers who are unfamiliar with Hilbert space quantum theory can translate what they learn in this book into the language used in most other texts on quantum theory.
3. Sometimes the hybrid approach we introduce in this section, combining diagrammatic reasoning with sums, is convenient for calculation. Such a blend has already proved to be very useful in some areas of pure mathematics such as knot theory, where one encounters equations like this one:

$$\text{\huge$\diagup\kern-0.8em\diagdown$} \;=\; \lambda \;\Big|\;\Big| \;+\; \lambda^{-1} \;\text{\huge$\smile$}_{\text{\huge$\frown$}}$$

And there is more. Even if one doesn't care at all about the Hilbert space formulation of quantum theory, one may still ask the question: 'What can one actually prove using string diagrams?' When it comes to process equations, the answer to this question is, surprisingly, 'Exactly what one can prove in Hilbert spaces'!

Using logical terminology, this means there exists a *completeness theorem* for Hilbert spaces with respect to string diagrams. We carefully explain what this means in Section 5.4.1.

On the other hand, string diagrams come with substantially less baggage than Hilbert spaces, so they leave enough room to seek out an alternative to Hilbert space for the formalisation of quantum theory. So, two seemingly unrelated historical developments: abandoning the Hilbert-space formalism for quantum theory (as von Neumann desired), versus embracing non-separability as the crucial feature of quantum theory (as Schrödinger insists), go hand in hand in our tale:

5.1 Bases and Matrices

We now show how some standard notions from linear algebra, most notably bases and matrices, emerge for certain process theories, and how the special processes and operations that we identified in the previous chapters for diagrams then transform into standard linear-algebraic concepts.

Throughout this section we will assume that process theories admit string diagrams and have zero processes for every system-type.

5.1.1 Basis for a Type

Many of the processes that we have studied so far have this property:

$$\left(\text{for all } \psi : \quad \frac{f}{\psi} = \frac{g}{\psi} \right) \quad \Longrightarrow \quad f = g \tag{5.1}$$

that is, if two processes do the same thing to all states, then they are equal. In other words, a process is <u>uniquely fixed</u> by what it does to states.

Examples 5.1 Both **functions** and **relations** satisfy (5.1).

For **functions**, in Example 3.35 we saw that the states of A:

$$\boxed{a} \ :: \ * \mapsto a$$

are in a bijective correspondence with the elements $a \in A$. By this correspondence we have:

Thus, we can translate (5.1) as:

$$(\text{for all } a \in A : f(a) = g(a)) \implies f = g$$

which is of course true for any functions f and g.

For **relations**, we saw in Example 3.36 that the states of type A:

are in a bijective correspondence with the subsets $A' \subseteq A$. However, in the light of (5.1), considering <u>all</u> states $A' \subseteq A$ is overkill. If we just restrict to singletons, we already have:

$$(\text{for all } a \in A : R(a) = S(a)) \implies R = S$$

That is, a relation is fixed by what it does to singletons.

As we just saw for **relations**, it is not always necessary to know what a process does to every state to fix it uniquely. Sometimes it suffices to know how it acts only on a special subset of the states.

Definition 5.2 A *basis* for a type A is a minimal set of states:

such that for all processes f and g:

(5.2)

where 'minimal' means that no state can be removed from $\mathcal{B}$ without sacrificing (5.2). The *dimension* $\dim(A)$ of the type A is the minimum size of a basis for A.

Remark* 5.3 For a well-behaved process theory like **relations**, all bases for a particular system will be the same size. In that case, we can just as well define $\dim(A)$ to be the size of any basis. For the theory of **linear maps**, this result is commonly known as the *dimension theorem*.

Exercise 5.4 Show that for a set A in **relations** the singletons:

$$\mathcal{B}_A := \left\{ \; \bigtriangledown_a \; \middle| \; a \in A \right\}$$

form a basis, that is, that no element can be removed from $\mathcal{B}_A$ without losing the property of being a basis. Also show that all bases are of this form and, consequently, that the dimension of a set A in **relations** is its number of elements.

Henceforth, we will use:

$$\left\{ \; \bigtriangledown_i \; \right\}_i$$

as a shorthand for:

$$\left\{ \; \bigtriangledown_1 \;, \ldots, \; \bigtriangledown_n \; \right\}$$

The best kinds of bases are those whose states are perfectly distinguishable by testing. That is, if we test the i-th state for being the j-th state, we should get a 'yes' outcome with certainty if and only if $i = j$. We give these sets of states (and in particular, bases) a special name.

Definition 5.5 A set of states:

$$\mathcal{A} := \left\{ \; \bigtriangledown_i \; \right\}_i$$

is *orthonormal* if for all i, j we have:

$$\begin{array}{c} \bigtriangleup_j \\ \bigtriangledown_i \end{array} = \delta_i^j \tag{5.3}$$

where δ_i^j is the *Kronecker delta*:

$$\delta_i^j = \begin{cases} \; \fbox{ } & \text{if } i = j \\ \\ \; 0 & \text{otherwise} \end{cases}$$

If $\mathcal{A}$ forms a basis, it is called an *orthonormal basis (ONB)*.

Recall from Section 4.3.3 that we can think of the inner product in (5.3) as measuring the 'overlap' between states. In that case, we can think of an ONB as a basis whose elements don't overlap, as in the following example.

Example 5.6 We saw in Example 4.50 that the inner product of two states in **relations** is 1 if and only if they intersect. Since the intersection of any two (different) singletons is empty, the unique bases in **relations** from Exercise 5.4 are ONBs:

$$
\begin{array}{c}
\triangle\!\!\!\!\!\!\bigtriangledown \\[-2pt] {}^{b}_{a}
\end{array}
=
\begin{cases}
\;\fbox{} & \text{if } a = b \\[8pt]
\;\emptyset & \text{if } a \neq b
\end{cases}
$$

Some ONBs aren't unique.

Example 5.7 In Section 5.3, we will see that **linear maps** admit many different ONBs for a single system-type, and there is no unique choice of 'preferred' ONB. This fact is very important for many quantum phenomena.

Some ONBs are even invisible.

Example 5.8 Since for all states ψ and ϕ of a system A, we have:

$$
\frac{\triangledown_{\psi}}{\fbox{}} = \frac{\triangledown_{\phi}}{\fbox{}} \implies \triangledown_{\psi} = \triangledown_{\phi}
$$

the empty diagram forms an ONB for the 'no wire'-type, since in this case orthonormality just means:

$$
\frac{\Box}{\Box} = \Box
$$

Other ONBs are full of colour.

Example 5.9 Consider a process theory of **lamps&detectors**, where states are lamps producing light of a certain colour, and effects are detectors that click when they detect light of a certain colour. Numbers arise when composing a lamp with a detector. The number 0 means 'no click', i.e. nothing detected, and 1 means 'loudest click', i.e. maximal intensity detected. Our interpretation of adjoints dictates that the adjoint of a lamp is the detector for light of the same colour. Then red, green, and blue lamps are 'orthonormal', because they will never detect each other's light:

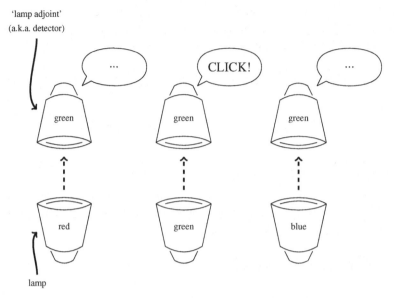

and they also form a basis. For example, suppose we have some unknown detector, and this happens:

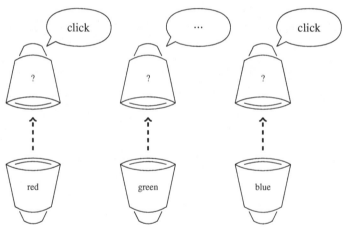

then we can conclude that:

The minimality condition in Definition 5.2 can be tricky to verify for a generic basis. Fortunately, in the case of an ONB, this condition is automatic.

Proposition 5.10 If an orthonormal set of states of type A satisfies (5.2) for all pairs of processes, then it must be minimal, and hence an ONB.

Proof Let:

$$\mathcal{A} = \left\{ \begin{array}{c} \bigtriangledown \\ i \end{array} \right\}_i$$

be an orthonormal set of states satisfying (5.2). Suppose it is not minimal; that is, there is some state *i* such that:

$$\mathcal{A}' := \mathcal{A} - \left\{ \begin{array}{c} \bigtriangledown \\ i \end{array} \right\}$$

still satisfies (5.2). First, note that effect *i* cannot be equal to the zero effect (depicted here as $\boxed{0}$ to avoid confusion) because:

$$\begin{array}{c} i \\ i \end{array} = 1 \neq 0 = \begin{array}{c} 0 \\ i \end{array}$$

However, since the states in $\mathcal{A}$ are orthonormal, for all $j \neq i$ we have:

$$\begin{array}{c} i \\ j \end{array} = 0 = \begin{array}{c} 0 \\ j \end{array}$$

But then by (5.2) it follows that

$$\begin{array}{c} i \end{array} = \boxed{0}$$

which is a contradiction. $\qquad\qquad\square$

Now is probably a good time for a quick word of warning about ONBs.

⚠**Warning 5.11** In Section 3.4.3 we introduced the notion of processes being equal 'up to a number' and showed that the ≈-relation plays well with diagrams. On the other hand, it does <u>not</u> play well with ONBs. Just because we prove that for all *i*:

$$\begin{array}{c} f \\ i \end{array} \approx \begin{array}{c} g \\ i \end{array} \tag{5.4}$$

it does <u>not</u> follow that:

$$\begin{array}{c} f \end{array} \approx \begin{array}{c} g \end{array}$$

In order for this to be true, each of the instances of equation (5.4) should hold up to the <u>same</u> number, otherwise we may not be able to find a single pair of numbers λ, μ such that

$\lambda f = \mu g$. For example, if f and g are effects, this is just saying f and g are non-zero on the same set of ONB states.

For most uses of bases in this book we will use ONBs, with one notable exception: *tomography* (cf. Section 7.4). To account for this, we will prove a number of results in the following sections for the general case of non-orthonormal bases as well. It will also be convenient (or sometimes even necessary) to choose a basis consisting of self-conjugate states. If a basis is self-conjugate, then the corresponding effects are also self-conjugate. We denote these self-conjugate states and effects as follows:

$$ \bigtriangledown_{i} := \bigtriangledown_{\overline{i}} = \overline{\bigtriangledown}_{i} \quad \text{and} \quad \bigtriangleup^{i} := \bigtriangleup_{\overline{i}} = \overline{\bigtriangleup}^{i} $$

Example* 5.12 In linear algebra, self-conjugate ONBs are those whose matrices consist only of real numbers. The canonical example is:

$$ \begin{pmatrix} 1 \\ 0 \\ \vdots \\ 0 \end{pmatrix} , \dots , \begin{pmatrix} 0 \\ \vdots \\ 0 \\ 1 \end{pmatrix} $$

We will see in Section 5.2.3 that we can turn any ONB into a self-conjugate one just by choosing the correct cups/caps (which are non-unique in general). An example where it becomes necessary to deal with a non-self-conjugate basis is when studying multiple distinct bases for the same system. It may not, in general, be possible to make a single choice of cap/cup to make all bases simultaneously self-conjugate.

Example* 5.13 In quantum computing, the X-basis, Y-basis, and Z-basis for qubits have the property that only two out of three can be made simultaneously self-conjugate.

5.1.2 Matrix of a Process

When we have basis states around, we can often prove things about processes by proving things about states. However, because we have adjoints, we also have the associated basis effects, so we can actually do better. To prove equality of processes, it suffices to just look at numbers.

Theorem 5.14 Suppose that $\mathcal{B}$ is a basis for A and that $\mathcal{B}'$ is a basis for B. Then for all f and g with input type A and output type B:

$$ \left(\text{for all } \bigtriangledown_{i} \in \mathcal{B}, \ \blacktriangledown \in \mathcal{B}' : \quad \begin{matrix} j \\ f \\ i \end{matrix} = \begin{matrix} j \\ g \\ i \end{matrix} \right) \implies \boxed{f} = \boxed{g} \tag{5.5} $$

Proof We can prove this using one basis at a time. First, take any state j in $\mathcal{B}'$. Then, for all states i in $\mathcal{B}$, if we have:

$$\hat{j}\,\Big[f\Big]\,\hat{i} \;=\; \hat{j}\,\Big[g\Big]\,\hat{i}$$

then, since $\mathcal{B}$ is a basis, it follows that:

$$\hat{j}\,\Big[f\Big] \;=\; \hat{j}\,\Big[g\Big]$$

and applying the adjoint to both sides, we have:

$$\Big[f\Big]\,\check{j} \;=\; \Big[g\Big]\,\check{j}$$

The above equation holds for all states $j \in \mathcal{B}'$, so since $\mathcal{B}'$ is a basis:

$$\Big[f\Big] \;=\; \Big[g\Big]$$

Applying the adjoint to both sides again, we conclude that $f = g$. $\qquad\square$

Exercise 5.15 Show that the converse to Theorem 5.14 is also true if in addition we require $\mathcal{B}$ and $\mathcal{B}'$ to be minimal, that is, whenever condition (5.5) holds, then (5.2) holds both for $\mathcal{B}$ and $\mathcal{B}'$.

When the bases in Theorem 5.14 are orthonormal, we give a familiar name to the numbers that uniquely identify a process.

Definition 5.16 The numbers:

$$\mathbf{f} := \left(f_i^j \;\Big|\; \check{i} \in \mathcal{B}, \; \check{j} \in \mathcal{B}' \right)$$

where $\mathcal{B}$ and $\mathcal{B}'$ are ONBs and:

$$f_i^j := \hat{j}\,\Big[f\Big]\,\check{i} \tag{5.6}$$

is called the *matrix* of f. The numbers f_i^j are called the *matrix entries*.

Usually we will use the same notation for a process f and its matrix, but when we wish to distinguish the two, as above, we will use a boldface notation $\mathbf{f}$ for the matrix.

In school, you may have seen a matrix written this way:

$$\begin{pmatrix} f_1^1 & f_2^1 & \cdots & f_m^1 \\ f_1^2 & f_2^2 & \cdots & f_m^2 \\ \vdots & \vdots & \ddots & \vdots \\ f_1^n & f_2^n & \cdots & f_m^n \end{pmatrix}$$

Note how for each matrix entry we give the row index (which arises from an output basis element) as a superscript and the column index (which arises from an input basis element) as a subscript. This 'tensor-style' notation (cf. Section* 3.6.1) will come in handy when we have multiple input/output wires.

The matrix of a process with input A and output B will have $\dim(A)$ columns and $\dim(B)$ rows. In Example 5.8 we saw that the 'no wire' type has a one-element basis. So, states give $n \times 1$ matrices, called *column vectors*, and effects give $1 \times m$ matrices, called *row vectors*:

Numbers, of course, give 1×1 matrices:

$$(\lambda)$$

but we typically don't bother to write them that way.

Though they are typically associated with linear maps, matrices are actually more general. For example, they provide a convenient alternative representation for relations.

Example 5.17 We saw earlier that each type in **relations** has a unique ONB given by the singletons, and in Example 3.36, we showed that there are only two numbers:

$$0 := \emptyset \text{ (a.k.a. 'impossible')} \qquad 1 := \boxed{} \text{ (a.k.a. 'possible')}$$

Then we have:

$$= \begin{cases} 1 & \text{if } R :: a \mapsto b \\ 0 & \text{otherwise} \end{cases}$$

Clearly these numbers fully characterise R since they identify precisely the pairs (a, b) such that $R :: a \mapsto b$ by assigning the number 1 to them. We can label the columns of R's matrix by the elements of A, and the rows by the elements of B. Then, we see a 1 whenever the elements of the given row and column are related, and 0 everywhere else:

$$R :: \begin{cases} a_1 \mapsto b_4 \\ a_2 \mapsto b_2 \\ a_2 \mapsto b_3 \\ a_3 \mapsto b_4 \end{cases} \quad \leftrightarrow \quad \begin{array}{c} \begin{array}{ccc} a_1 & a_2 & a_3 \end{array} \\ \begin{array}{c} b_1 \\ b_2 \\ b_3 \\ b_4 \end{array} \left(\begin{array}{ccc} 0 & 0 & 0 \\ 0 & 1 & 0 \\ 0 & 1 & 0 \\ 1 & 0 & 1 \end{array} \right) \end{array}$$

This matrix is sometimes called an *adjacency matrix* of R.

Not only can we represent processes by matrices, but the diagrammatic operations of transpose, conjugate, and adjoint also correspond to familiar operations on matrices.

Theorem 5.18 Let f be a process with associated matrix $\mathbf{f}$. The matrix of $f^\dagger$ is the *adjoint matrix* $\mathbf{f}^\dagger$, which is defined as:

$$(\mathbf{f}^\dagger)^j_i := \overline{(\mathbf{f}^i_j)}$$

Proof The matrix entries of $f^\dagger$ are computed as:

where (4.18) is the fact that numbers are self-transposed. □

Note that the above theorem works for any ONB, not just the self-conjugate ones. In contrast, for transposition and conjugation to work correctly on matrices, we need to assume that the ONBs are self-conjugate.

Theorem 5.19 Let f be a process with associated matrix $\mathbf{f}$ for self-conjugate ONBs $\mathcal{B}$ and $\mathcal{B}'$. The matrix of f^T is the *transposed matrix* $\mathbf{f}^T$, which is defined as:

$$(\mathbf{f}^T)^j_i := \mathbf{f}^i_j$$

The matrix of $\overline{f}$ is the *conjugate matrix* $\overline{\mathbf{f}}$, which is defined as:

$$\overline{\mathbf{f}}^j_i := \overline{(\mathbf{f}^j_i)}$$

Proof The matrix entries of the transpose of f are computed as:

$$(5.7)$$

and for the matrix entries of the conjugate we have:

$$\overline{\begin{array}{c} j \\ \bigtriangledown \\ f \\ \bigtriangledown \\ i \end{array}} \overset{(*)}{=} \left(\overline{\begin{array}{c} j \\ \bigtriangleup \\ f \\ \bigtriangledown \\ i \end{array}} \right)$$

where the equations marked $(*)$ rely on $\mathcal{B}$ and $\mathcal{B}'$ being self-conjugate. If this were not the case, the resulting matrices would not be in terms of $\mathcal{B}$ and $\mathcal{B}'$, but rather in terms of their conjugate bases. □

So the transpose $\mathbf{f}^T$ of a matrix $\mathbf{f}$ is obtained by interchanging the roles of the row and column indices; the conjugate $\overline{\mathbf{f}}$ is obtained by conjugating each of the entries; and the adjoint $\mathbf{f}^\dagger$ consists of applying both of these operations. For that reason, the adjoint matrix is also sometimes called the conjugate-transpose. As in the case of diagrams, the order doesn't matter:

$$\mathbf{f}^\dagger = \overline{(\mathbf{f}^T)} = (\overline{\mathbf{f}})^T$$

Exercise 5.20 Characterise the matrix of a self-adjoint process.

So, we can already treat several operations on processes as operations on matrices. However, we haven't yet reached the full power of *matrix calculus*. Theorem 5.14 says that $f = g$ if and only if f and g have the same matrix. In other words, when there exists an f with a particular matrix, it is unique. But then, given some arbitrary matrix, nothing guarantees (yet) that there will always exist an f that has that matrix.

Suppose we fix a bunch of numbers, and write them all in an $n \times m$ matrix:

$$\begin{pmatrix} g_1^1 & g_2^1 & \cdots & g_m^1 \\ g_1^2 & g_2^2 & \cdots & g_m^2 \\ \vdots & \vdots & \ddots & \vdots \\ g_1^n & g_2^n & \cdots & g_m^n \end{pmatrix} \tag{5.8}$$

How might we go about obtaining some process g that has this matrix? First, fix ONBs $\mathcal{B}$ and $\mathcal{B}'$. Then, for any i, j, it is possible to build a process $\widetilde{g}_{ij}$ that agrees with g on the i-th input element of $\mathcal{B}$ and the j-th output element of $\mathcal{B}'$ and is zero everywhere else:

$$\boxed{\widetilde{g}_i^{\,j}} = \langle g_i^j \rangle \,\, \begin{array}{c} \bigtriangledown \\ j \\ \bigtriangleup \\ i \end{array}$$

If we compute the matrix of $\widetilde{g}_i^j$, it indeed has precisely one non-zero entry, g_i^j, at the (i,j)-th position:

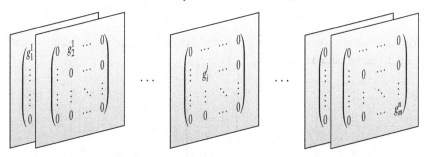

We can define a whole stack of these $\widetilde{g}_i^j$-processes, for all i,j:

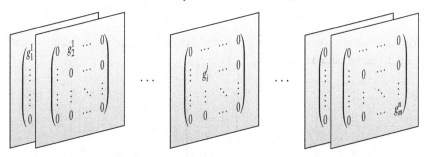

Then if we could only 'overlay' them somehow, we would get g.

It turns out, for certain process theories, 'overlaying' processes makes perfect sense. If we denote this 'overlaying' procedure as a *sum* of processes, g can be expressed as:

$$g := \sum_{ij} \left\langle g_i^j \right\rangle \quad \tag{5.9}$$

Thus, it only remains to make precise what we mean by 'sums of processes'.

5.1.3 Sums of Processes

Sums are not something you can do for any old processes; for example, what is the sum of two babies? However, for many process theories, it has a perfectly well-defined mathematical meaning. This meaning directly emerges from the intuition behind overlaying diagrams:

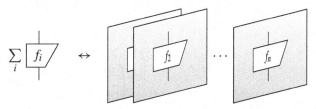

Definition 5.21 We say a process theory *has sums* if the following three conditions are satisfied:

- **Condition 1:** For any two processes f, g with the same input and output types, $f + g$ is a process. We always assume that '$+$' is associative, is commutative, and has a unit given by the zero process:

$$(f + g) + h = f + (g + h) \qquad f + g = g + f \qquad f + 0 = f = 0 + f$$

and for a set $\{f_i\}_i$ of processes, we write:

$$\sum_i \boxed{f_i} := \boxed{f_1} + \boxed{f_2} + \cdots + \boxed{f_N}$$

- **Condition 2:** Sums *distribute over diagrams*, that is, any time a summation occurs in a diagram, it can be pulled outside:

- **Condition 3:** Sums preserve adjoints:

$$\left(\sum_i \boxed{f_i} \right)^\dagger = \sum_i \boxed{f_i}$$

Note that **Condition 2** subsumes distributivity of sums with respect to parallel and sequential composition, e.g.:

$$\left(\sum_i \boxed{f_i} \right) \boxed{g} = \sum_i \left(\boxed{f_i} \ \boxed{g} \right)$$

and:

$$\left(\sum_i \boxed{f_i} \right) = \sum_i \left(\boxed{g} \atop \boxed{f_i} \right)$$

An important example of this is *linearity* of maps with respect to states:

$$\left(\sum_i \langle\!\langle \lambda_i \rangle\!\rangle\, \psi_i\right) f \;=\; \sum_i \left(\langle\!\langle \lambda_i \rangle\!\rangle\, \frac{f}{\psi_i}\right)$$

Another one is the inner product. We have full-fledged linearity for states:

$$\left(\sum_i \langle\!\langle \lambda_i \rangle\!\rangle\, \psi_i\right) \phi \;=\; \sum_i \left(\langle\!\langle \lambda_i \rangle\!\rangle\, \frac{\phi}{\psi_i}\right) \tag{5.10}$$

and conjugate-linearity for effects:

$$\left(\sum_i \langle\!\langle \overline{\lambda_i} \rangle\!\rangle\, \phi_i\right)_\psi \;=\; \sum_i \left(\langle\!\langle \overline{\lambda_i} \rangle\!\rangle\, \frac{\phi_i}{\psi}\right) \tag{5.11}$$

Distributivity also helps us derive the matricial counterpart to sums, which unsurprisingly results in something that looks like the linear-algebraic sum of matrices.

Theorem 5.22 Let $\{f_k\}_k$ be processes with associated matrices $\{\mathbf{f}_k\}_k$. The matrix of the process $\sum\limits_k f_k$ is the sum of the matrices $\sum\limits_k \mathbf{f}_k$, where:

$$\left(\sum_k \mathbf{f}_k\right)_i^j := \sum_k (\mathbf{f}_k)_i^j$$

Proof We can use **Condition 2** to compute the matrix of $\sum\limits_k f_k$ as follows:

$$\left(\sum_k \frac{j}{f_k}{i}\right) \;=\; \sum_k \left(\frac{j}{f_k}{i}\right) \;=\; \sum_k (\mathbf{f}_k)_i^j$$

$\square$

If there are multiple summations in a diagram, we can pull them all outside, though we may need to do some re-indexing (as in $(*)$ below):

$$\frac{\left(\sum_i g_i\right)}{\left(\sum_i f_i\right)} \;\overset{(*)}{=}\; \frac{\left(\sum_j g_j\right)}{\left(\sum_i f_i\right)} \;=\; \sum_i \left(\frac{\left(\sum_j g_j\right)}{f_i}\right) \;=\; \sum_{ij} \left(\frac{g_j}{f_i}\right)$$

where we used:

$$\sum_{ij} \boxed{f_{ij}} \qquad \text{as shorthand for} \qquad \sum_{i}\sum_{j} \boxed{f_{ij}}$$

The order we pull out the sums doesn't matter, so if we always use distinct letters for indexes, we can forget about brackets and write the summation symbols anywhere in the picture:

$$\frac{\sum_{j}\boxed{g_j}}{\sum_{i}\boxed{f_i}} = \sum_i \frac{\sum_j \boxed{g_j}}{\boxed{f_i}} = \frac{\boxed{g_j}\,\sum_i}{\boxed{f_i}\,\sum_j} = \frac{\boxed{g_j}}{\boxed{f_i}}\,\sum_{ij}$$

However, we tend be boring and stick to writing sums on the left.

Remark 5.23 One can remember all the above rules concerning sums simply by thinking of the sum-symbol as a 'number':

$$\left\langle\!\! \sum_i \!\!\right\rangle$$

Of course it is not a number in the usual sense, but it can freely move around the diagram, just like a number.

As with ONBs, a quick word of warning about sums is warranted.

⚠**Warning 5.24** What we said in Warning 5.11 about the $\approx$-relation <u>not</u> playing well with ONBs also applies to sums. Just because we prove that for all i:

$$\boxed{f_i} \approx \boxed{g_i} \tag{5.12}$$

it does <u>not</u> follow that:

$$\sum_i \boxed{f_i} \approx \sum_i \boxed{g_i}$$

In order for this to be true, again each of the instances of equation (5.12) should hold up to the <u>same</u> number.

While the existence of sums rules out babies, relations are still hopping along.

Example 5.25 In **relations** sums are unions. Setting:

$$\sum_i R_i := \bigcup_i R_i$$

it can be shown straightforwardly that **Conditions 1–3** are satisfied. Applying it to the two numbers 0 and 1 we obtain:

$$0 + 0 = 0 \qquad 0 + 1 = 1 \qquad 1 + 0 = 1 \qquad 1 + 1 = 1$$

So '+' for numbers in **relations** is the boolean 'or' operation. Writing '·' for (parallel/sequential) composition of numbers, we also have:

$$0 \cdot 0 = 0 \qquad 0 \cdot 1 = 0 \qquad 1 \cdot 0 = 0 \qquad 1 \cdot 1 = 1$$

That is, we obtain the boolean 'and' operation, and **Condition 2** now yields the usual *distributive law* for 'or' and 'and':

$$x \cdot (y + z) = (x \cdot y) + (x \cdot z)$$

So, in a process theory with sums, numbers always have a 'plus' operation as well as a 'times' (i.e. composition), with a distributive law between them. Hence they are starting to look a lot more like actual numbers.

Example 5.26 We can regard any natural number n as a number in our process theory. We just take an n-fold sum of 1s (a.k.a. empty diagrams):

$$\langle\!\langle n \rangle\!\rangle := \underbrace{1 + \cdots + 1}_{n \text{ times}} \tag{5.13}$$

If the numbers of the process theory are the real or complex numbers, then these correspond precisely to the natural numbers (seen as a subset of $\mathbb{R}$ or $\mathbb{C}$). However, this need not be the case. For instance, if the numbers are booleans as in Example 5.25, all ns are the same:

$$\underbrace{1 + \cdots + 1}_{n \text{ times}} = 1$$

One may even want to consider process theories with *subtraction*, where for every process f, there exists another process $-f$ such that $f + (-f) = 0$. As usual, we can abbreviate $f + (-g)$ as $f - g$. In terms of overlays this can be thought of as a layer that 'neutralises' another layer. Note that, by distributivity, this is equivalent to assuming a special number '-1' such that $1 - 1 = 0$. If we include subtraction, the numbers of the process theory form a *ring*. Without subtraction, they form a weaker kind of structure, sometimes called a *unital semiring* or *rig* (because it is a ring without *negative* numbers).

Remark 5.27 The assumption of the existence of additive inverses is actually quite strong. In particular, it rules out the theory of **relations**, because the booleans contain no number that behaves as the additive inverse of 1. Our only two options are 0 and 1, and neither works:

$$1 + 0 = 1 \neq 0 \qquad\qquad 1 + 1 = 1 \neq 0$$

In the following sections, we will see how every matrix now corresponds to a process, and how composition of processes corresponds to composing the corresponding matrices. All together, this is what we're aiming for.

Definition 5.28 For a process theory

- that admits string diagrams,
- has a (fixed, self-conjugate) ONB for every system-type, and
- has sums satisfying the conditions in Definition 5.21,

the *matrix calculus* of that process theory refers to the matrices associated with its processes, along with operations for summation, sequential composition, parallel composition, transposition, conjugation, and adjoints of those matrices.

As explained in the proof of Theorem 5.19, we only rely on self-conjugate ONBs to have matricial counterparts to transposition and conjugation.

5.1.4 Processes from Matrices

With sums in hand, we are now able to build processes from matrices, as we discussed at the end of Section 5.1.1.

Theorem 5.29 Fix an ONB $\mathcal{B}$ with m elements and an ONB $\mathcal{B}'$ with n elements. Then, for a collection of numbers g_i^j where i ranges from 1 to m and j ranges from 1 to n, the process:

$$\boxed{g} := \sum_{ij} \langle g_i^j \rangle \qquad (5.14)$$

has the following matrix:

$$\begin{pmatrix} g_1^1 & g_2^1 & \cdots & g_m^1 \\ g_1^2 & g_2^2 & \cdots & g_m^2 \\ \vdots & \vdots & \ddots & \vdots \\ g_1^n & g_2^n & \cdots & g_m^n \end{pmatrix}$$

Proof It suffices to show that the following numbers are equal, for all i, j:

$$\boxed{g} = \langle g_i^j \rangle \qquad (5.15)$$

This is easily seen by substituting (5.14) in (5.15):

$$
\begin{array}{ccccccc}
\raisebox{-1em}{g} & = & \displaystyle\sum_{kl} \left\langle g_k^l \right\rangle & = & \displaystyle\sum_{kl} \delta_i^k\, \delta_l^j\, \left\langle g_k^l \right\rangle & = & \left\langle g_i^j \right\rangle
\end{array}
$$

$\square$

A process written as in the RHS of equation (5.14) is said to be in *matrix form*, which means it is written as a sum, with each summand consisting of:

(1) a number,
(2) an ONB state, and
(3) an ONB effect.

It is called 'matrix form' because the sum explicitly refers to all of the elements of the matrix of g. For the special case of states, the matrix form is just a state written as a sum of basis states with coefficients:

$$
\psi = \sum_i \left\langle \psi^i \right\rangle\, i
$$

We can turn this around to give another characterisation of ONBs.

Theorem 5.30 Suppose there exists a basis $\mathcal{B}$ for a type A. Then, another orthonormal set of states:

$$
\mathcal{A} := \left\{\, i \,\right\}_i
$$

forms an ONB for A if and only if it *spans* A; that is, if any state ψ of type A can be written in the following form, for some numbers λ_i :

$$
\psi = \sum_i \left\langle \lambda_i \right\rangle\, i \tag{5.16}
$$

Proof For ($\Rightarrow$), assume $\mathcal{A}$ is an ONB. Then equation (5.16) is given by the matrix form of the state ψ. For ($\Leftarrow$), assume $\mathcal{A}$ spans A, and:

$$
\text{for all } i \in \mathcal{A} : \quad \frac{f}{i} = \frac{g}{i}
$$

Since $\mathcal{A}$ spans A, we can express any state ψ as:

$$\psi = \sum_i \langle \lambda_i \rangle \; i$$

So:

$$\frac{f}{\psi} = \sum_i \langle \lambda_i \rangle \frac{f}{i} = \sum_i \langle \lambda_i \rangle \frac{g}{i} = \frac{g}{\psi}$$

In particular, f and g agree on all the states in the basis $\mathcal{B}$, so $f = g$.   □

Another particularly useful special case is the matrix form of an identity process. Given an ONB, the matrix entries of the identity are:

$$\frac{j}{i} = \delta_i^j$$

If we write these in a matrix, we get 1s down the diagonal (where $i = j$), and 0s everywhere else:

$$\left| \quad \leftrightarrow \quad \begin{pmatrix} 1 & 0 & \cdots & 0 \\ 0 & 1 & \cdots & 0 \\ \vdots & \vdots & \ddots & \vdots \\ 0 & 0 & \cdots & 1 \end{pmatrix} \right.$$

i.e. we get the *identity matrix*. Translating this to matrix form yields:

$$\left| \; = \sum_i \delta_i^j \; \frac{j}{i} = \sum_i \frac{i}{i} \right.$$

Hence we have the following theorem.

Theorem 5.31 For any ONB we have:

$$\left| \; = \sum_i \frac{i}{i} \right. \tag{5.17}$$

We refer to such a decomposition as a *resolution of the identity*.

The matrix form for the identity process gives us a handy way to compute the matrix form for an arbitrary process:

The converse of Theorem 5.31 is also true and provides a second, very succinct alternative characterisation of ONBs.

Theorem 5.32 A set of states:

$$\mathcal{A} := \left\{ \; \bigtriangledown_i \; \right\}_i$$

is an ONB if and only if:

$$= \delta_i^j \qquad\qquad = \sum_i \qquad\qquad (5.18)$$

Proof From Theorem 5.31, any ONB satisfies (5.18). Conversely, let $\mathcal{A}$ satisfy (5.18), and suppose that for processes f and g we have:

Then:

so $\mathcal{A}$ indeed forms a basis, and hence an ONB. □

Another interesting corollary to Theorem 5.31 concerns the number given by a 'circle'. We have:

$$
\bigcirc \;=\; \sum_i \; \Yup{i}{i} \;=\; \sum_i \; \downtriangle{i}{i} \;=\; \downtriangle{1}{1} + \cdots + \downtriangle{D}{D} \;=\; \langle\!\langle D \rangle\!\rangle
$$

where, following Example 5.26:

$$
\langle\!\langle D \rangle\!\rangle \;:=\; \underbrace{1 + \cdots + 1}_{D \text{ times}} \tag{5.19}
$$

So the circle counts the number of basis vectors. In other words, it gives the dimension! Well, at least most of the time. Recall from Example 5.26 that $\langle\!\langle D \rangle\!\rangle$ may sometimes not be the actual natural number D; for example, in **relations** it will be either 0 (for zero basis vectors) or 1 (for one or more basis vectors). However, in the case of **linear maps** $\langle\!\langle D \rangle\!\rangle$ will always be dim(A).

Corollary 5.33 For D the dimension of a system-type A, we have:

$$
{}_A\bigcirc \;=\; \langle\!\langle D \rangle\!\rangle
$$

where $\langle\!\langle D \rangle\!\rangle$ is defined as in (5.19).

Remark 5.34 Actually, we don't like sums! The reason why is obvious:

$$
\text{wire} \longrightarrow \Big| \;=\; \sum_i \; \Yup{i}{i} \longleftarrow \text{no wire}
$$

Sums completely mess up the fact that diagrams are all about 'what is connected to what', and it is this fact that makes diagrams so appealing. Hence, we will always get rid of sums whenever we can find a better, diagrammatic alternative.

We end this section with a definition of (classical) probability distributions using matrices. A probability distribution assigns to each member of a collection of mutually exclusive events a real number between 0 and 1, such that all these numbers together add up to 1. Probability distributions, and their matrix representation in particular, will become increasingly important in this book.

Definition 5.35 Suppose the numbers of a process theory contain the positive real numbers. A *probability distribution* is a matrix of the form:

$$\begin{pmatrix} p^1 \\ \vdots \\ p^n \end{pmatrix} \tag{5.20}$$

with positive real matrix entries p^i summing to 1:

$$\sum_i p^i = \boxed{}$$

Equivalently, probability distributions can also be represented as states of the following form:

$$\tikz{p} := \sum_i p^i \tikz{i}$$

with p^i as above. The probability distributions corresponding to basis states:

$$\tikz{1} \leftrightarrow \begin{pmatrix} 1 \\ 0 \\ 0 \\ \vdots \\ \vdots \\ 0 \\ 0 \end{pmatrix} \quad \cdots \quad \tikz{i} \leftrightarrow \begin{pmatrix} 0 \\ \vdots \\ 0 \\ 1 \\ 0 \\ \vdots \\ 0 \end{pmatrix} \quad \cdots \quad \tikz{n} \leftrightarrow \begin{pmatrix} 0 \\ 0 \\ \vdots \\ \vdots \\ 0 \\ 0 \\ 1 \end{pmatrix}$$

are called *point distributions*.

Since a probability distribution is a matrix (or a state in matrix form), it always comes with a choice of ONB. We take this ONB to be self-conjugate for the simple reason that conjugation has no part in probability theory.

5.1.5 Matrices of Isometries and Unitaries

In this and the following section we characterise isometries, unitaries, positive processes, projectors, and $\otimes$-positive states in terms of their matrices.

In the case of the first two, which are the ones that we consider in this section, it will be useful to look at the rows and columns of these matrices.

Definition 5.36 Given ONBs for the input and output types of a process f, the *columns* of f are the matrices of the following set of states:

$$\left\{ \begin{matrix} f \\ 1 \end{matrix} \;, \ldots, \; \begin{matrix} f \\ m \end{matrix} \right\}$$

whereas the *rows* are the matrices of the following set of effects:

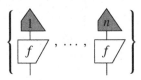

As the name suggests, the columns of f embed as columns in the overall matrix of f:

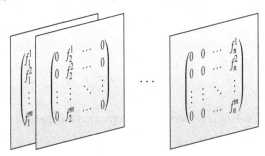

and similarly for the rows:

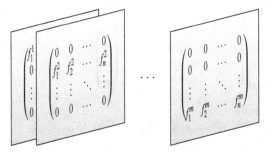

We say a set of column vectors forms an ONB if the associated states do. Similarly, we say row vectors form an ONB if the (adjoints of) the associated effects do. This now gives us a way to recognise matrices of isometries.

Proposition 5.37 For a process f, the following are equivalent:

(1) f is an isometry;
(2) f sends orthonormal sets of states to orthonormal sets of states;
(3) the columns of f are orthonormal; and
(4) the rows of $f^\dagger$ are orthonormal.

Proof For $(1 \Rightarrow 2)$, given any orthonormal set of states

$$\mathcal{A} := \left\{ \begin{matrix} \scriptstyle\bigtriangledown \\ \scriptstyle i \end{matrix} \right\}_i$$

we need to show that the following set is also orthonormal:

This follows from the fact that f is an isometry:

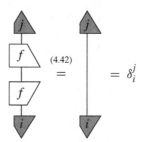

$(2 \Rightarrow 3)$ is immediate, since the columns of f are obtained by applying f to a particular orthonormal set, namely the fixed ONB on its input system. $(3 \Rightarrow 4)$ follows from the fact that the rows of $f^\dagger$ are just the adjoints of the columns of f. For $(4 \Rightarrow 1)$, assume the rows of $f^\dagger$ are orthonormal. Then, we note by orthonormality that $f^\dagger \circ f$ has the same matrix as the identity process:

$$= \delta_i^j =$$

Thus, f is an isometry.   □

Using this proposition, it's now straightforward to recognise unitary processes in terms of their matrices. Essentially, we just need to replace orthonormal sets with orthonormal bases.

Proposition 5.38 The following are equivalent:

(1) f is unitary;
(2) f sends ONBs to ONBs;
(3) the columns of f form an ONB; and
(4) the rows of f form an ONB.

Proof (1 ⇒ 2) If f is unitary, then it is an isometry. Thus, for any ONB:

$$\mathcal{B} := \left\{ \begin{array}{c} \rule{0pt}{1em} \\ i \end{array} \right\}_i$$

we know from Proposition 5.37 that the states:

are orthonormal. It only remains to show that these states form a basis. For this, we show that $g = h$ precisely when g and h agree on all of these states. Assume that, for all i:

$$\begin{array}{ccc} g \\ f \\ i \end{array} = \begin{array}{c} h \\ f \\ i \end{array}$$

Since $\mathcal{B}$ is a basis, it follows that:

$$\begin{array}{cc} g \\ f \end{array} = \begin{array}{c} h \\ f \end{array} \hspace{3cm} (5.21)$$

Combining this with unitarity of f yields:

$$\begin{array}{c} g \end{array} \overset{(4.43)}{=} \begin{array}{c} g \\ f \\ f \end{array} \overset{(5.21)}{=} \begin{array}{c} h \\ f \\ f \end{array} \overset{(4.43)}{=} \begin{array}{c} h \end{array}$$

(2 ⇒ 3) is immediate. For (3 ⇒ 1), we know from the previous proposition that f is an isometry, so it suffices to show that $f^\dagger$ is one too. Using (5.17):

Since the columns of f form an ONB, by Theorem 5.31, the RHS above is also a resolution of the identity:

Finally, $(3 \Leftrightarrow 4)$ follows from noting that f is unitary if and only if $f^\dagger$ is unitary. $\qquad\square$

The theorem above gives us a characterisation of unitaries as precisely those maps that send ONBs to ONBs. The next corollary immediately follows.

Corollary 5.39 Unitaries can equivalently be represented as:

for some pair of ONBs (with the same number of elements).

Remark 5.40 Proposition 5.37 tells us that isometries send orthonormal sets of states to orthonormal sets of sets. In particular, it sends an ONB to some *subset* of another ONB. Thus, we can still use the representation (5.39) for isometries, provided we relax the requirement on:

from being an ONB to an orthonormal set. In other words, we can think of isometries as unitaries that have been 'extended' to a larger output system.

5.1.6 Matrices of Self-Adjoint and Positive Processes

In Theorem 5.18 we characterised the matrix for the adjoint of a process:

$$(f^\dagger)^j_i = \overline{(f^i_j)}$$

From this, it directly follows that the matrix of a self-adjoint process is:

$$
\begin{pmatrix}
f^1_1 & f^1_2 & \cdots & f^1_n \\
\overline{f^1_2} & f^2_2 & \cdots & f^2_n \\
\vdots & \vdots & \ddots & \vdots \\
\overline{f^1_n} & \overline{f^2_n} & \cdots & f^n_n
\end{pmatrix}
$$

where, in particular, the elements on the diagonal are self-conjugate:

$$f^i_i \;\overset{(5.25)}{=}\; \cdots \;\overset{(4.46)}{=}\; \cdots \;\overset{(5.25)}{=}\; (f^i_i)^\dagger = \overline{f^i_i}$$

Unfortunately, matrices of positive processes cannot be recognised quite so easily, but we do at least know what the numbers on the diagonal look like:

$$f^i_i \;\overset{(5.25)}{=}\; \cdots \;\overset{(4.45)}{=}\; \cdots \;=\; \cdots \tag{5.22}$$

i.e. they are positive numbers. Then, by process–state duality we have the following.

Corollary 5.41 For a $\otimes$-positive state ψ the numbers:

$$\psi^{ii} :=$$

are positive.

Fortunately, for certain processes called *diagonalisable* processes, characterising the numbers on the diagonal is good enough.

Definition 5.42 An *eigenstate* of a process f is a non-zero state ψ such that for some number λ we have:

$$\vcenter{\hbox{[diagram: f applied to ψ]}} = \lambda \vcenter{\hbox{[diagram: ψ]}}$$

and f is called *diagonalisable* if there exists an ONB $\mathcal{B}$ such that all of the basis states are eigenstates of f, that is:

$$\text{for all } \vcenter{\hbox{[i]}} \in \mathcal{B}, \text{ there exists } \lambda_i : \quad \vcenter{\hbox{[diagram: f on i]}} = \lambda_i \vcenter{\hbox{[i]}} \tag{5.23}$$

The following is the reason for the terminology.

Proposition 5.43 If a process f is diagonalisable, then its matrix in the basis of eigenstates is a *diagonal matrix* :

$$\vcenter{\hbox{[diagram: f]}} = \sum_i \lambda_i \vcenter{\hbox{[diagram: i over i]}} \quad\leftrightarrow\quad \begin{pmatrix} \lambda_1 & 0 & \cdots & 0 \\ 0 & \lambda_2 & \cdots & 0 \\ \vdots & \vdots & \ddots & \vdots \\ 0 & 0 & \cdots & \lambda_n \end{pmatrix} \tag{5.24}$$

Proof For the matrix entries of a diagonalisable process f we have:

$$\vcenter{\hbox{[diagram: j, f, i]}} \overset{(5.23)}{=} \lambda_i \vcenter{\hbox{[diagram: j, i]}} = \lambda_i \, \delta^j_i \tag{5.25}$$

from which (5.24) follows. $\qquad\qquad\square$

If a process is diagonalisable, we can characterise self-adjointness, positivity, and projectors in terms of the numbers on the diagonal.

Proposition 5.44 For a diagonalisable process f:

(i) f is self-adjoint if and only if all λ_i are self-conjugate;
(ii) f is positive if and only if all λ_i are positive;
(iii) f is a projector if and only if all λ_i are positive and satisfy $(\lambda_i)^2 = \lambda_i$.

Proof Part (i) follows from the characterisation of self-adjoint matrices at the beginning of this section. For (ii), first assume f is positive. Then, by (5.22) the λ_i are all positive. Conversely, assume each of the λ_i is positive. Then, for some μ_i, $\lambda_i = \mu_i^\dagger \circ \mu_i$. For all the process theories in this book, it suffices to assume each μ_i is also a number (as opposed to a more general state), so we first consider that case. Letting:

$$
\boxed{g} \;:=\; \sum_i \mu_i \;\; \bigtriangledown_{\!\!i}^{} \bigtriangleup_{\!\!i}
$$

one can easily verify that $f = g^\dagger \circ g$. For a more general process theory, μ_i need not be a number, in which case g needs to be constructed more carefully (cf. Exercise* 5.45). For (iii), let f be a projector. By (i) we know the λ_i must all be positive. Moreover we have:

$$
\lambda_i \;\overset{(5.25)}{=}\; \boxed{f} \;\overset{(4.50)}{=}\; \boxed{f} \;\overset{(5.23)}{=}\; \boxed{f} \;\overset{(5.25)}{=}\; (\lambda_i)^2
$$

Conversely, if the λ_i are positive, so is f, and f is clearly a projector precisely when $(\lambda_i)^2 = \lambda_i$. $\qquad\qquad\qquad\qquad\qquad\qquad\qquad\qquad\qquad\qquad\qquad\qquad\qquad\square$

Put equivalently, a diagonalisable process f is self-adjoint, positive, or a projector if and only if the numbers λ_i are respectively self-adjoint, positive, or projectors themselves.

Exercise* 5.45 Prove (ii) from Proposition 5.44 in the case where the processes μ_i are each states of type A_i. That is, where:

$$
\lambda_i \;=\; \boxed{\begin{array}{c} \mu_i \\ A_i \\ \mu_i \end{array}}
$$

In all of the process theories we will consider in this book, the only numbers satisfying $\lambda^2 = \lambda$ are 0 and 1, so diagonalisable projectors have only zeroes and ones on the diagonal.

Why all this fuss about diagonalisable processes? One might think this is a very restrictive condition. For instance, diagonalisability in some process theories is totally uninteresting.

Example 5.46 For **relations**, the only ONBs available are the singleton bases. As a consequence, the only diagonalisable processes are those whose matrices are already diagonal.

However, as we will see in Section 5.3.3.1, all self-adjoint processes in the theory of **linear maps** are diagonalisable. Since positive processes and projectors are also self-adjoint, Proposition 5.44 gives a complete characterisation in that context.

On the other hand, there are many non-self-adjoint processes that cannot be diagonalised in (virtually) any process theory.

Exercise 5.47 Show that in any process theory in which non-zero numbers are cancellable (see Section 4.4.2), for any non-zero states ψ, ϕ where:

$$= 0$$

the process:

is not diagonalisable in any ONB.

5.1.7 Traces of Matrices

In Section 4.2.3, we defined the trace of a process:

$$\mathrm{tr}(f) :=$$

We can derive the matrix form of the trace of f using the decomposition of the identity:

$$\tag{5.26}$$

In terms of the matrix of f, this is the sum over all entries of the form f_i^i. In other words, the trace of a matrix is the sum of its diagonal entries:

$$\mathrm{tr} \begin{pmatrix} f_1^1 & \cdots & \cdots & \cdot \\ \vdots & f_2^2 & \cdots & \cdot \\ \vdots & \vdots & \ddots & \vdots \\ \cdot & \cdot & \cdots & f_D^D \end{pmatrix} = \sum_i f_i^i$$

Note that the diagonal entries of a matrix depend on the choice of basis in which the matrix is written. One might be tempted to assume the trace is therefore basis-dependent. However, this is evidently not the case, since we initially defined the trace without any reference to an ONB. The crucial point is in (5.26): we can decompose the identity using any ONB, and the result will be the same.

One can characterise the partial trace similarly. First note that we can decompose the trace as a sum, similar to before:

However, now rather than *numbers* being summed over, we are summing over processes from B to C. If we compare the matrices of each of these smaller processes with the big matrix of f, we will see them occurring as block matrices along the main diagonal:

where:

is the matrix of the process

So, the matrix of the partial trace of f is computed by summing up these block matrices:

where the size and the shape of the blocks depends on the dimensions of the system-types B and C.

5.2 Matrix Calculus

We can now represent all processes by matrices, and conversely, each matrix represents a process. We can identify special processes easily in terms of their matrices, and also have matricial counterparts for transposition, conjugation, and adjoints. However, rather than processes in isolation, what we truly care about is forming diagrams with processes. So we need to figure out how diagrams translate into compositions of matrices.

We have essentially two ways forward here. One way, suggested by Theorem 4.19, is to treat string diagrams as circuit diagrams to which we adjoin cups and caps for each type and provide matricial counterparts to sequential composition, parallel composition, and cups/caps. We will do this over the next three sections. In Section 5.2.4 we will then show a more direct way to compute the matrix of a string diagram, going via the string diagram formulas introduced in Definition 4.21.

Once all of this is in place, we show that matrices provide not only a manner for representing process theories, but also an easy way to construct 'abstract' process theories where all that needs to be specified are the numbers that serve as matrix entries.

5.2.1 Sequential Composition of Matrices

First we investigate how sequential composition of processes can be seen as an operation on the matrices of those processes.

Theorem 5.48 Let f and g be processes with associated matrices $\mathbf{f}$ and $\mathbf{g}$. The matrix of $g \circ f$ is the *matrix product* $\mathbf{g}\mathbf{f}$, which is defined by:

$$(\mathbf{g}\mathbf{f})_i^k := \sum_j \mathbf{g}_j^k \, \mathbf{f}_i^j \tag{5.27}$$

Proof The matrix entries for $g \circ f$ are:

$$(g \circ f)_i^k = \overset{(5.17)}{=} \sum_j \quad = \sum_j \quad = \sum_j \mathbf{g}_j^k \, \mathbf{f}_i^j = (\mathbf{g}\mathbf{f})_i^k$$

□

There is an easy way to compute the matrix product (5.27) using rows and columns. First note that the matrix of a state consists of a single column, while the matrix of an effect consists of a single row. Applying the composition formula (5.27) to the case of a state ψ and an effect ϕ:

we obtain:

$$\begin{pmatrix} \phi_1 & \phi_2 & \cdots & \phi_n \end{pmatrix} \begin{pmatrix} \psi^1 \\ \psi^2 \\ \vdots \\ \psi^n \end{pmatrix} = \phi_1 \psi^1 + \cdots + \phi_n \psi^n$$

This is sometimes called the *dot product* of ψ and ϕ, though really it is just sequential composition. Examining the resulting matrix entries in the composition formula (5.27) for general processes f and g:

$$(g \circ f)^j_i = g^j_1 f^1_i + \cdots + g^j_n f^n_i$$

we see that we can compute the entry in the i-th column and the j-th row of the matrix of $g \circ f$ by taking the dot product of the i-th column of f with the j-th row of g:

resulting in:

$$\begin{pmatrix} & \vdots & \\ g^j_1 & \cdots & g^j_n \\ & \vdots & \end{pmatrix} \begin{pmatrix} & f^1_i & \\ \cdots & \vdots & \cdots \\ & f^m_i & \end{pmatrix} = \begin{pmatrix} & \vdots & \\ \cdots & g^j_1 f^1_i + \cdots + g^j_n f^m_i & \cdots \\ & \vdots & \end{pmatrix}$$

which is of course how most readers will have learned to do matrix composition in school.

5.2.2 Parallel Composition of Matrices

Next up is parallel composition. We will compute the matrix corresponding to $f \otimes g$, given that we have the matrices of f and g. To do this, we must first establish how we obtain an ONB for a joint system-type:

given that we have ONBs for A and B.

Theorem 5.49 Let:

$$\mathcal{B} := \left\{ \begin{array}{c} i \end{array} \right\}_i \qquad \text{and} \qquad \mathcal{B}' := \left\{ \begin{array}{c} j \end{array} \right\}_j$$

be ONBs for A and B, respectively. Then the set of states:

$$\left\{ \begin{array}{c} i \quad j \end{array} \right\}_{ij} \tag{5.28}$$

forms an ONB for $A \otimes B$ called the *product basis*.

Proof We can show that the basis condition (5.2) holds using one basis at a time, much as we did in the proof of Theorem 5.14. Assume any pair of processes f, g with input type $A \otimes B$ agrees on all of the states in (5.28):

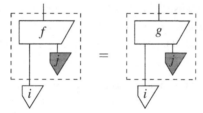

which we can rewrite as:

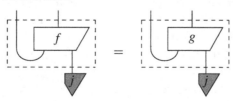

Since $\mathcal{B}$ is a basis, it follows that:

so for all states j in $\mathcal{B}'$ we also have:

and since $\mathcal{B}'$ is a basis, it follows that:

and hence:

Orthonormality of states (5.28) means:

which follows from orthonormality of $\mathcal{B}$ and $\mathcal{B}'$. Hence, by Proposition 5.10, we have an ONB.   □

If A has a basis of size D, then $A \otimes A$ has D^2 basis states. By extension, a system consisting of n copies of A has D^n basis states. So, the dimension grows exponentially with the number of systems!

Remark* 5.50 Note how the proof of Theorem 5.49 relies crucially on cups and caps. So, string diagrams force the *tensor product* on us for describing composite systems. There simply is no other option available. This is interesting in light of the fact that physicists have asked for a long time why we use the tensor product for describing composite systems, and not the *direct sum*. We leave it as an exercise to the interested reader to try to construct a process theory where linear maps are combined by means of the direct sum. What goes wrong?

Suppose now we wish to write the matrix for a process of the form:

Something we have implicitly assumed so far when writing down matrices of processes is that we have given some ordering to basis states. Otherwise, how would we know where to position each of the numbers in the matrix? When they are labelled $i \in \{1, \ldots, D\}$, this is clearly the case. So, in order to write the matrix of f in terms of product bases, we should fix an ordering of product basis states. First, we can make our lives a bit easier by numbering basis elements from 0. Then, for bases:

and

we can rely on the fact that each integer k where $0 \leq k < DD'$ can be decomposed as $iD' + j$ for unique i and j. We can therefore number elements in the product basis as:

$$\underset{iD'+j}{\bigtriangledown} := \underset{i}{\bigtriangledown}\,\underset{j}{\bigtriangledown} \tag{5.29}$$

This kind of convention extends naturally to any number of systems, and if all systems have the same dimension D, we can think of (5.29) as a representation of some number in base-D. For example, when all system-types have dimension $D = 2$ we have:

$$\underset{117}{\bigtriangledown} := \underset{1}{\bigtriangledown}\,\underset{1}{\bigtriangledown}\,\underset{1}{\bigtriangledown}\,\underset{0}{\bigtriangledown}\,\underset{1}{\bigtriangledown}\,\underset{0}{\bigtriangledown}\,\underset{1}{\bigtriangledown} \tag{5.30}$$

since the number 117 can be written in base-2 as 1110101.

Remark 5.51 The fact that we can encode bit strings as basis states of N systems is very important for quantum computation. We already encountered a similar encoding in Example 3.27 when representing bit strings as N-fold Cartesian products of the set $\mathbb{B}$.

When working with product bases, it will almost always be more convenient to index a matrix entry by individual digits, thus avoiding the extra number-juggling from equation (5.29). Given a string of individual basis states as in the RHS of (5.30), we call the state on the far left the *most significant* basis state, because changing it will change the 'encoded' number the most, and the state on the far right the *least significant* basis state. This follows computer science terminology, where one refers to the 'most significant bit' or 'least significant bit' in a bit string. This is of course purely a matter of convention. The important thing is to pick a convention and stick to it.

For Q a two-dimensional system the dimension of $Q \otimes Q$ is four, so the matrix of a state Ψ of type $Q \otimes Q$ is a column vector with four elements:

$$\underset{\psi}{\bigtriangledown} \quad\leftrightarrow\quad \begin{pmatrix} \psi^{00} \\ \psi^{01} \\ \psi^{10} \\ \psi^{11} \end{pmatrix} \qquad \text{where} \qquad \psi^{ij} := \underset{\psi}{\overset{\overset{i}{\triangle}\,\overset{j}{\triangle}}{\bigtriangledown}}$$

The use of subscripts and superscripts keeps things tidy, even with multiple indices around:

$$\boxed{g} \quad\leftrightarrow\quad \begin{pmatrix} g^0_{00} & g^0_{01} & g^0_{10} & g^0_{11} \\ g^1_{00} & g^1_{01} & g^1_{10} & g^1_{11} \end{pmatrix} \qquad \text{where} \qquad g^k_{ij} := \overset{\overset{k}{\triangle}}{\underset{\underset{i}{\bigtriangledown}\,\underset{j}{\bigtriangledown}}{\boxed{g}}}$$

So in general, we will encounter matrices with many upper and lower indices:

$$\left(f^{j_1 j_2 \cdots j_N}_{i_1 i_2 \cdots i_M} \ \middle|\ 0 \leq i_k < D_k\,,\, 0 \leq j_k < D'_k \right) \tag{5.31}$$

Remark* 5.52 This is sometimes called *tensor notation*, where a tensor is just a matrix with lots of indices. It is a precursor to *abstract tensor notation* (cf. Section* 3.6.1), and hence also to diagram formulas, neither of which depend on fixing bases.

Now, suppose we consider a separable state $\psi \otimes \phi$ of type $Q \otimes Q$. Since this state is formed from states ψ and ϕ, we would expect there to be a relationship between its matrix and the matrices of ψ and ϕ. It can be easily verified that this is indeed the case:

$$
\psi \;\leftrightarrow\; \begin{pmatrix} \psi^0 \\ \psi^1 \end{pmatrix}
\qquad
\phi \;\leftrightarrow\; \begin{pmatrix} \phi^0 \\ \phi^1 \end{pmatrix}
\qquad
\psi\,\phi \;\leftrightarrow\; \begin{pmatrix} \psi^0\phi^0 \\ \psi^0\phi^1 \\ \psi^1\phi^0 \\ \psi^1\phi^1 \end{pmatrix}
$$

The final matrix consists of all the ways to form a product of one number from the first matrix and one number from the second matrix. This is called the *Kronecker product* of matrices. Let $\boldsymbol{\phi}$ be the matrix for ϕ. Then we can write this more succinctly using a *block matrix*:

$$
\psi\,\phi \;\leftrightarrow\; \begin{pmatrix} \psi^0\boldsymbol{\phi} \\ \psi^1\boldsymbol{\phi} \end{pmatrix}
$$

where $\psi^i\boldsymbol{\phi}$ means 'multiply all of the elements in $\boldsymbol{\phi}$ by ψ^i'. The Kronecker product works not only for matrices of states, but also for matrices of any processes of the form:

$$
f \quad g
$$

Theorem 5.53 Let f and g be processes with matrices $\mathbf{f}$ and $\mathbf{g}$. The matrix of $f \otimes g$ is the *Kronecker product* $\mathbf{f} \otimes \mathbf{g}$, which is defined by:

$$
(\mathbf{f} \otimes \mathbf{g})^{kl}_{ij} := \mathbf{f}^k_i \mathbf{g}^l_j \tag{5.32}
$$

Proof The matrix entries for $f \otimes g$ are given by:

$$
\begin{matrix} k & l \\ f & g \\ i & j \end{matrix}
$$

which exactly matches Definition (5.32). $\square$

In terms of block matrices, for:

$$
\mathbf{f} := \begin{pmatrix} f^1_1 & \cdots & f^1_m \\ \vdots & \ddots & \vdots \\ f^n_1 & \cdots & f^n_m \end{pmatrix}
\qquad \text{and} \qquad
\mathbf{g} := \begin{pmatrix} g^1_1 & \cdots & g^1_{m'} \\ \vdots & \ddots & \vdots \\ g^{n'}_1 & \cdots & g^{n'}_{m'} \end{pmatrix}
$$

the Kronecker product is the $(nn') \times (mm')$ matrix:

$$f \middle/ g \quad \leftrightarrow \quad \begin{pmatrix} f_1^1 \mathbf{g} & \cdots & f_m^1 \mathbf{g} \\ \vdots & \ddots & \vdots \\ f_1^n \mathbf{g} & \cdots & f_m^n \mathbf{g} \end{pmatrix}$$

This applies to any dimension of matrix, so we can compute Kronecker products of matrices of states, effects, and more general processes. For example, the Kronecker product of the 2×1 matrix of a state ψ and the 2×2 matrix of a process f is computed as follows:

$$\begin{pmatrix} \psi^0 \\ \psi^1 \end{pmatrix} \otimes \begin{pmatrix} f_0^0 & f_1^0 \\ f_0^1 & f_1^1 \end{pmatrix} = \begin{pmatrix} \psi^0 \mathbf{f} \\ \psi^1 \mathbf{f} \end{pmatrix} = \begin{pmatrix} \psi^0 f_0^0 & \psi^0 f_1^0 \\ \psi^0 f_0^1 & \psi^0 f_1^1 \\ \psi^1 f_0^0 & \psi^1 f_1^0 \\ \psi^1 f_0^1 & \psi^1 f_1^1 \end{pmatrix}$$

Exercise 5.54 Give the matrices of these processes:

assuming all of the systems are two-dimensional.

We conclude this section by looking at transposition of matrices over compound systems. Just like when we first computed transposes of matrices in Theorem 5.19, we will assume for the remainder of this section that all ONBs are self-conjugate.

In Section 4.2.2 we pointed out that there are two choices of transposition for processes on joint systems: the usual transpose, which is just a rotation, and the algebraic transpose, which uses the 'criss-crossed' cups and caps. For the transpose we have:

$$\tag{5.33}$$

and for the algebraic transpose we have:

$$\tag{5.34}$$

So, we see that the transpose interchanges superscripts and subscripts and reverses the order, whereas the algebraic transpose just interchanges superscripts and subscripts:

$$f_{ij}^{kl} \overset{(5.33)}{\rightsquigarrow} f_{lk}^{ji} \qquad\qquad \text{vs.} \qquad\qquad f_{ij}^{kl} \overset{(5.34)}{\rightsquigarrow} f_{kl}^{ij}$$

Therefore, it is the <u>algebraic</u> transpose that performs the usual transpose of a matrix from linear <u>algebra</u>:

$$\begin{pmatrix} f_{00}^{00} & f_{01}^{00} & f_{10}^{00} & f_{11}^{00} \\ f_{00}^{01} & f_{01}^{01} & f_{10}^{01} & f_{11}^{01} \\ f_{00}^{10} & f_{01}^{10} & f_{10}^{10} & f_{11}^{10} \\ f_{00}^{11} & f_{01}^{11} & f_{10}^{11} & f_{11}^{11} \end{pmatrix} \overset{(5.34)}{\rightsquigarrow} \begin{pmatrix} f_{00}^{00} & f_{00}^{01} & f_{00}^{10} & f_{00}^{11} \\ f_{01}^{00} & f_{01}^{01} & f_{01}^{10} & f_{01}^{11} \\ f_{10}^{00} & f_{10}^{01} & f_{10}^{10} & f_{10}^{11} \\ f_{11}^{00} & f_{11}^{01} & f_{11}^{10} & f_{11}^{11} \end{pmatrix}$$

The algebraic transpose also keeps the most significant basis state on the left, and the least significant on the right:

Contrast this with the usual, 'rotational' transpose. While handy from a diagrammatic point of view, it's not very good at counting:

Since conjugation is defined in terms of transposition, this distinction appears there as well.

Exercise 5.55 Show that conjugation reverses the order of input/output indices, whereas algebraic conjugation does not:

$$f_{ij}^{kl} \rightsquigarrow \overline{f_{ji}^{lk}} \qquad\qquad \text{vs.} \qquad\qquad f_{ij}^{kl} \rightsquigarrow \overline{f_{ij}^{kl}}$$

5.2.3 Matrix Form of Cups and Caps

We now have matricial counterparts to sequential composition and parallel composition, i.e. a matricial representation for circuits. In order to obtain a matricial representation for string diagrams, we additionally need matricial counterparts for cups and caps. We begin by computing their matrix form.

Proposition 5.56 For any ONB we have:

$$\smile = \sum_i \qquad\qquad \frown = \sum_i \qquad\qquad (5.35)$$

Proof   The ONB decomposition of the cup follows immediately from the matrix form of
the identity process, as in equation (5.17):

$$\cup = \left| \quad \sum_i \bigvee_i^{\,i} \right| = \sum_i \vee_i \vee_i$$

The ONB decomposition of the cap is obtained by taking the adjoint.   □

So (5.35) gives us the matrix form for cups and caps in a product basis. But rather than
two copies of the same ONB, this product basis consists of an ONB and its conjugated
counterpart. At first it may even seem that this violates the yanking laws, e.g.:

$$\aleph = \cup$$

But it doesn't. The key point is that Proposition 5.56 holds for any ONB. In particular, we
can apply the proposition to the conjugate of our original ONB and obtain the following
equivalent characterisation of cups and caps:

$$\cup = \sum_i \overline{\vee_i}\, \vee_i \qquad\qquad \cap = \sum_i \triangle_i \overline{\triangle_i} \tag{5.36}$$

Using the two equivalent forms, the yanking laws follow straightforwardly. However, we
can avoid this complication by using self-conjugate ONBs, in which case (5.35) becomes:

$$\cup = \sum_i \vee_i \vee_i \qquad\qquad \cap = \sum_i \triangle_i \triangle_i \tag{5.37}$$

Remark 5.57   Translating equations (5.37) to Dirac notation, we recover the usual definition of the Bell state and the Bell effect that one encounters in the quantum computing
literature:

$$\sum_i \vee_i \vee_i = \sum_i |ii\rangle \qquad\qquad \sum_i \triangle_i \triangle_i = \sum_i \langle ii|$$

The matrices for cups and caps in two dimensions are:

$$\cup \;\leftrightarrow\; \begin{pmatrix} 1 \\ 0 \\ 0 \\ 1 \end{pmatrix} \qquad\qquad \cap \;\leftrightarrow\; \begin{pmatrix} 1 & 0 & 0 & 1 \end{pmatrix}$$

Exercise 5.58   Give the matrix of cups and caps in three and four dimensions. What is the
relationship between the matrix of a cup/cap and the identity matrix?

In fact, we could just as well take (5.37) to be the definition of the caps and cups for a system of type A. We can then verify the yanking equations directly, e.g.:

Exercise 5.59 Prove the yanking equation above without using a self-conjugate basis.

Near to the end of Section 5.1.1 we stated that we can always choose (unique) cups/caps to make a given ONB self-conjugate. We are now ready to show which cups/caps to choose for this purpose.

Proposition 5.60 An ONB:

$$\mathcal{B} := \left\{ \; \overset{|}{\underset{i}{\triangledown}} \; \right\}_i$$

is self-conjugate if and only if we have:

$$\smile \; = \; \sum_i \; \overset{|}{\underset{i}{\triangledown}} \; \overset{|}{\underset{i}{\triangledown}}$$

Proof Assuming the cup and cap are as given above, we compute conjugates of basis states in $\mathcal{B}$ as the transpose of the adjoint:

So each of the basis states is indeed self-conjugate. Conversely, suppose caps and cups are chosen such that $\mathcal{B}$ is self-conjugate, that is, for all i:

$$\overset{|}{\underset{i}{\triangledown}} \; = \; \overset{|}{\underset{i}{\triangledown}} \tag{5.38}$$

Then, using Proposition 5.56 we obtain:

$$\smile \; = \; \sum_i \; \overset{|}{\underset{i}{\triangledown}} \overset{|}{\underset{i}{\triangledown}} \; \overset{(5.38)}{=} \; \sum_i \; \overset{|}{\underset{i}{\triangledown}} \overset{|}{\underset{i}{\triangledown}}$$

$\square$

5.2.4 String Diagrams of Matrices

In principle, the matricial counterparts of parallel/sequential composition, as well as the matrices of cups and caps, give us everything we need to compute the overall matrix of a string diagram. There is, however, a much more direct method, which we already encountered in Section 3.3.3 for the particular case of **relations**. True, we didn't know then what a string diagram was, but that turns out to not matter that much.

Recall that for a process:

$$
\begin{array}{c}
B_1 \quad \cdots \quad B_N \\
\boxed{f} \\
A_1 \quad \cdots \quad A_M
\end{array}
$$

we can write the matrix as:

$$
\mathbf{f} = \left(\mathbf{f}_{i_1 \ldots i_M}^{j_1 \ldots j_N} \,\middle|\, 0 \le i_k < D_k,\, 0 \le j_k < D'_k \right)
$$

where D_k is the dimension of A_k, D'_k is the dimension of B_k, and:

$$
\mathbf{f}_{i_1 \ldots i_M}^{j_1 \ldots j_N} :=
\begin{array}{c}
\widehat{j_1} \quad \cdots \quad \widehat{j_N} \\
\boxed{f} \\
\widecheck{i_1} \quad \cdots \quad \widecheck{i_M}
\end{array}
$$

Theorem 5.61 The matrix of a diagram is the corresponding diagram formula subject to the substitutions:

- box names f become corresponding matrices $\mathbf{f}$;
- wire names A_k become indices $0 \le i_k < D_k$; and
- all repeated indices are summed over.

For example, the matrix $\mathbf{m}$ of:

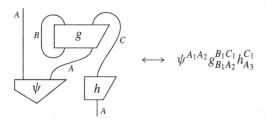

$$
\longleftrightarrow \quad \psi^{A_1 A_2}\, g_{B_1 A_2}^{B_1 C_1}\, h_{A_3}^{C_1}
$$

has entries:

$$
\mathbf{m}_{i_3}^{i_1} = \sum_{i_2 j_1 k_1} \psi^{i_1 i_2}\, \mathbf{g}_{j_1 i_2}^{j_1 k_1}\, \mathbf{h}_{i_3}^{k_1}
$$

If a diagram includes transposes, conjugates, or adjoints, then update the matrix entries of those processes accordingly (cf. Theorems 5.18 and 5.19).

Exercise 5.62 Compute the matrix of:

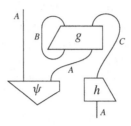

5.2.5 Matrices as Process Theories

Matrices are not just a convenient manner of describing processes of many process theories, but they also allow us to construct a wide range of process theories, which will always admit string diagrams.

Consider a set X, whose elements we will write as $\langle\!\langle x \rangle\!\rangle$, $\langle\!\langle y \rangle\!\rangle$, ... with:

(i) a multiplication satisfying the properties of parallel (or equivalently sequential) composition of numbers, that is:

$$(\langle\!\langle x \rangle\!\rangle \langle\!\langle y \rangle\!\rangle)\langle\!\langle z \rangle\!\rangle = \langle\!\langle x \rangle\!\rangle(\langle\!\langle y \rangle\!\rangle \langle\!\langle z \rangle\!\rangle) \qquad\qquad \boxed{\ } \ \langle\!\langle x \rangle\!\rangle = \langle\!\langle x \rangle\!\rangle$$

$$\langle\!\langle x \rangle\!\rangle \langle\!\langle y \rangle\!\rangle = \langle\!\langle y \rangle\!\rangle \langle\!\langle x \rangle\!\rangle \qquad\qquad\qquad 0\langle\!\langle x \rangle\!\rangle = 0$$

(ii) a sum obeying the conditions of Definition 5.21:

$$\langle\!\langle x \rangle\!\rangle + 0 = \langle\!\langle x \rangle\!\rangle \qquad\qquad \left(\sum_i \langle\!\langle x_i \rangle\!\rangle\right)\langle\!\langle y \rangle\!\rangle = \sum_i \langle\!\langle x_i \rangle\!\rangle \langle\!\langle y \rangle\!\rangle$$

(iii) a (possibly trivial) conjugation operation satisfying the properties of an adjoint (or equivalently a conjugate) of numbers:

$$\overline{\left(\overline{\langle\!\langle x \rangle\!\rangle}\right)} = \langle\!\langle x \rangle\!\rangle \qquad \overline{\left(\langle\!\langle x \rangle\!\rangle \langle\!\langle y \rangle\!\rangle\right)} = \langle\!\langle x \rangle\!\rangle \langle\!\langle y \rangle\!\rangle \qquad \overline{\left(\sum_i \langle\!\langle x_i \rangle\!\rangle\right)} = \sum_i \overline{\langle\!\langle x_i \rangle\!\rangle}$$

Then, we can define a process theory whose processes are matrices with entries taken from X. In fact, all of the work of defining this process theory is done already, since we now know how to combine matrices into string diagrams and take adjoints, transposes, etc. Thus, just giving the set X of numbers tells us all about the processes of our new theory.

But what are the system-types? They should be the stuff that mediates when we can compose processes. For matrices, this is just the number of rows and columns. Hence, we can take the types simply to be the natural numbers $\mathbb{N}$. Since numbers are 1×1 matrices, the 'no wire' type is '1'.

We can spell this out a bit more precisely.

Definition 5.63 Let X be a set of numbers with composition and sums satisfying the conditions spelled out above. We construct the process theory **matrices**(X) as follows:

(1) The systems are the natural numbers $\mathbb{N}$.
(2) The processes with input type $m \in \mathbb{N}$ and output type $n \in \mathbb{N}$ are all $n \times m$ matrices with entries in X.
(3) Diagrams are computed as in Theorem 5.61.

Of course, instead of the single rule (3) one could take:

(3a) The cups and caps are the matrices obtained in Exercise 5.58.
(3b) Adjoints are given by the conjugate-transpose of matrices.
(3c) Parallel composition is given by the Kronecker product.
(3d) Sequential composition is given by the matrix product.

Moreover, we have seen in this chapter that matrices can faithfully represent processes in a process theory where every system-type has a finite ONB. The following result should then be unsurprising.

Theorem 5.64 If for a process theory we have:

(1) each system-type has a finite ONB;
(2) there is at least one system-type of every dimension $D \in \mathbb{N}$; and
(3) processes of the same type admit sums

and if the numbers of this process theory are X, then this process theory is *equivalent* to the process theory **matrices**(X).

So what does it mean for two process theories to be equivalent? This is actually a bit more subtle than one may think at first. The naïve approach is to simply require that:

- there is a bijection between system-types $A, B, \ldots$ of theory 1 and system-types $A', B', \ldots$ of theory 2;
- for all system-types A, B (and corresponding system-types A', B') there is a bijection between processes from A to B in theory 1 and processes from A' to B' in theory 2; and
- these bijections preserve string diagrams, for example, if ψ, g, h are in correspondence with ψ', g', h', then so too are:

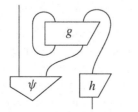 and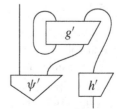

However, this notion of equivalence is often too strict. For example, if a process theory has many types with n-element ONBs, then in **matrices**(X) these all get smooshed onto the same type 'n'. Still, for all practical purposes **matrices**(X) is 'equivalent' to the process theory we started with. Intuitively, equivalent process theories are defined as above, but with the additional ability to 'smoosh together' types that are essentially the same (in this case, types with the same dimension). This description of equivalence will suffice for our purposes, but readers interested in the full details are referred to Section* 5.6.4.

Remark* 5.65 As detailed in Section* 5.6.4, this more subtle notion of equivalence is known in category theory as an *equivalence of categories*, whereas the more naïve notion is called an *isomorphism of categories*.

In the case of the theory of **relations**, we already saw in Example 5.17 that the numbers to consider are the booleans $\mathbb{B}$. Let **finrelations** be the sub-theory of **relations** that is obtained by restricting the types to finite sets. In order to establish equivalence between the process theory **finrelations** and **matrices**$(\mathbb{B})$ we need to identify all sets with the same number of elements. Doing so we can conclude the following.

Corollary 5.66 The theory **finrelations** is equivalent to **matrices**$(\mathbb{B})$.

But we can pick many other sets of numbers X in order to form process theories admitting string diagrams, for example, natural numbers $\mathbb{N}$, integers $\mathbb{Z}$, rational numbers $\mathbb{Q}$, real numbers $\mathbb{R}$, or some totally wacky 'numbers' like the open sets of a topological space.

Exercise* 5.67 Consider the process theory **matrices**$(\mathbb{Z}_2)$, i.e. matrices over the two-element field, which differs from $\mathbb{B}$ in that $1 + 1 = 0$ rather than $1 + 1 = 1$. What are its properties? How does it relate to **relations**? Note that the 'complete' answer to this question is still a topic of active research.

5.3 Hilbert Spaces

We are now ready to define the process theory of Hilbert spaces and linear maps, which is an important stepping stone towards the theory of quantum processes. In fact, almost all of the pieces of the puzzle are in place. The only remaining piece is to tell you what the numbers are.

5.3.1 Linear Maps and Hilbert Spaces from Diagrams

We saw in Example 3.35 that in **functions**, numbers are trivial. That is, there is only one number. In Example 3.36 we saw that in **relations** there are two numbers 0 and 1, corresponding to 'impossible' and 'possible'. We just saw that given any kind of numbers, we can construct a process theory for these, for example all the real numbers. For Hilbert spaces and linear maps even real numbers are not enough. The new numbers we need will (for the first time in this book) have non-trivial conjugation, and hence will yield adjoints that do not coincide with the transpose. In other words, finally we will be able to capture the full richness of string diagrams.

Definition 5.68 The process theory of **linear maps** is defined to be the set of all processes described by string diagrams where:

(1) Each type has a finite ONB.
(2) There is at least one system-type of every dimension $D \in \mathbb{N}$.
(3) Processes of the same type admit sums.
(4) The numbers are the complex numbers $\mathbb{C}$.

A *Hilbert space* is a system-type in **linear maps**, and we'll denote the D-dimensional systems assumed in (2) as $\mathbb{C}^D$.

First note that Definition 5.68 doesn't <u>really</u> say what a Hilbert space is. As far as we're concerned, it's just some type that tells us which linear maps can be plugged together. In fact, we know from Theorem 5.64 that we could define **linear maps** equivalently as **matrices**($\mathbb{C}$), so for our purposes, the system-types might as well just be natural numbers (i.e. dimensions). This is in line with our usual attitude that processes are more important than systems. For those really worried about this omission, we give the usual set-theoretic definition of Hilbert spaces in Section 5.4.2 and show how it connects to Definition 5.68.

We also haven't recalled yet what a *complex number* is. Before doing so, we want to stress that much of this book can be understood without developing a profound familiarity with complex numbers since, as we mentioned in the introduction, our ultimate goal is to replace matrices of complex numbers with diagrams. A complex number consists of a pair of real numbers $a, b \in \mathbb{R}$, which we write as:

$$a + ib \tag{5.39}$$

One can pretty much compute with complex numbers as one does with real numbers, provided one adopts the following additional equation:

$$i^2 = -1$$

From condition (4) of Definition 5.68, it should be understood that composition of numbers is the usual *multiplication of complex numbers*:

$$\langle\lambda_1\rangle \langle\lambda_2\rangle \quad \rightsquigarrow \quad (a_1 + ib_1)(a_2 + ib_2) = (a_1 a_2 - b_1 b_2) + i(a_1 b_2 + b_1 a_2)$$

sums are given by *addition of complex numbers*:

$$\langle\lambda_1\rangle + \langle\lambda_2\rangle \quad \rightsquigarrow \quad (a_1 + ib_1) + (a_2 + ib_2) = (a_1 + a_2) + i(b_1 + b_2)$$

and, hence, 0 (the 'absorb everything' diagram) and 1 (the empty diagram) are the actual numbers 0 and 1 in $\mathbb{C}$. Finally, conjugation of numbers corresponds to *complex conjugatation*:

$$\langle\lambda\rangle \mapsto \langle\overline{\lambda}\rangle \quad \rightsquigarrow \quad a + ib \mapsto a - ib$$

Crucially, this conjugation is non-trivial!

Remark* 5.69 In fact, if we want an involution that preserves products and sums, we have no choice but to choose the conjugation to be trivial for the real numbers. Furthermore, complex conjugation is the unique (non-trivial) product- and sum-preserving involution for complex numbers that fixes the subset $\mathbb{R}$.

In light of Example 5.26, complex numbers form a *field*, which means that, along with subtraction, we can have *division by non-zero numbers*. That is, for all $\lambda \neq 0$, there exists a number $\frac{1}{\lambda}$ such that:

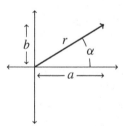

Complex numbers can also be written in *polar form*:

$$re^{i\alpha} \tag{5.40}$$

for r a positive real number and α an angle, called the *complex phase*. The two forms (5.39) and (5.40) are related as follows:

$$\begin{cases} a = r\cos(\alpha) \\ b = r\sin(\alpha) \end{cases} \qquad\qquad \begin{cases} r = \sqrt{a^2 + b^2} \\ \alpha = \arctan\left(\frac{b}{a}\right) \end{cases}$$

which can be visualised in the *complex plane* as follows:

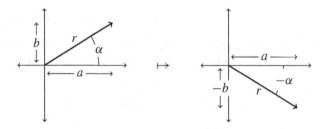

From this, it's fairly easy to see that conjugation flips the complex phase:

$$re^{i\alpha} \quad \mapsto \quad re^{-i\alpha}$$

Indeed, conjugation simply reflects the complex plane:

Our choice of 'reflection' to represent conjugation in diagrams is inspired by this fact:

$$\langle \lambda \rangle = \begin{matrix} \phi \\ \psi \end{matrix} \quad \mapsto \quad \begin{matrix} \phi \\ \psi \end{matrix} = \langle \bar{\lambda} \rangle$$

5.3.2 Positivity from Conjugation

Using standard terminology, a complex number is 'positive' if it is both real and ≥ 0. Since numbers are, in particular, processes, this definition of 'positive' should agree with our prior notion of positivity for processes. This is indeed the case.

Proposition 5.70 For a complex number λ the following are equivalent:

1. It is real and ≥ 0.
2. There exists a complex number μ such that:

$$\langle \lambda \rangle = \langle \bar{\mu} \rangle \langle \mu \rangle \tag{5.41}$$

3. It is positive in the sense of Definition 4.60; i.e. there exists ψ such that:

$$\langle \lambda \rangle = \begin{matrix} \psi \\ \psi \end{matrix} \tag{5.42}$$

Proof For $(1 \Rightarrow 2)$, suppose λ is real and ≥ 0. Then letting $\mu = \bar{\mu} = \sqrt{\lambda}$, we have equation (5.41). $(2 \Rightarrow 3)$ is immediate, since (5.41) is a special case of (5.42). For $(3 \Rightarrow 1)$, first note that any complex number multiplied by its conjugate is real and ≥ 0:

$$\bar{\mu}\mu = (a+ib)(a-ib) = a^2 + b^2$$

Computing λ in (5.42) using the matrix of ψ, we have:

$$\lambda = \begin{pmatrix} \overline{\psi^1} & \overline{\psi^2} & \cdots & \overline{\psi^n} \end{pmatrix} \begin{pmatrix} \psi^1 \\ \psi^2 \\ \vdots \\ \psi^n \end{pmatrix} = \sum_i \overline{\psi_i}\psi_i$$

Since this is a sum of numbers of the form $\bar{\mu}\mu$, it must be real and ≥ 0. □

In Example* 4.41 we claimed that the transpose in **linear maps** does not provide 'good' adjoints. In particular, positive-definiteness:

$$\begin{matrix} \psi \\ \psi \end{matrix} = 0 \quad \Longleftrightarrow \quad \begin{matrix} \psi \end{matrix} = 0 \tag{5.43}$$

fails to hold for the transpose. We are now ready to substantiate that claim, then show how complex conjugation fixes the problem.

First, consider this (incorrect) expression of positive definiteness, involving the transpose:

$$
\begin{pmatrix} \psi^1 & \psi^2 & \cdots & \psi^n \end{pmatrix}
\begin{pmatrix} \psi^1 \\ \psi^2 \\ \vdots \\ \psi^n \end{pmatrix} = 0
\qquad \Longleftrightarrow \qquad
\begin{pmatrix} \psi^1 \\ \psi^2 \\ \vdots \\ \psi^n \end{pmatrix} =
\begin{pmatrix} 0 \\ 0 \\ \vdots \\ 0 \end{pmatrix}
$$

While this works for real numbers, it fails for complex numbers. For example:

$$
\begin{pmatrix} 1 & i \end{pmatrix} \begin{pmatrix} 1 \\ i \end{pmatrix} = 1 + (-1) = 0
\qquad \text{while} \qquad
\begin{pmatrix} 1 \\ i \end{pmatrix} \neq \begin{pmatrix} 0 \\ 0 \end{pmatrix}
$$

The problem is, for complex numbers, unlike real numbers, $(\psi^i)^2$ might be less than zero (in this case -1). This is why we need the conjugate-transpose, i.e. the adjoint, for (5.43). Translating this into matrix form, we get:

$$
\begin{pmatrix} \overline{\psi^1} & \overline{\psi^2} & \cdots & \overline{\psi^n} \end{pmatrix}
\begin{pmatrix} \psi^1 \\ \psi^2 \\ \vdots \\ \psi^n \end{pmatrix} = 0
\qquad \Longleftrightarrow \qquad
\begin{pmatrix} \psi^1 \\ \psi^2 \\ \vdots \\ \psi^n \end{pmatrix} =
\begin{pmatrix} 0 \\ 0 \\ \vdots \\ 0 \end{pmatrix}
$$

or, equivalently:

$$
\sum_i \overline{\psi^i}\, \psi^i = 0 \qquad \Longleftrightarrow \qquad \text{for all } i: \ \psi^i = 0 \tag{5.44}
$$

By Proposition 5.70 all of the numbers $\overline{\psi^i}\, \psi^i$ are real and greater than 0. So, if any of the ψ^i are non-zero, the sum in (5.44) will be greater than zero.

5.3.3 *Why Mathematicians Love Complex Numbers*

In this section, we will look at a number of ways in which passing to complex numbers makes mathematicians' lives a lot easier. While many of the results in this section rely on the explicit matrix forms of linear maps, we will later be able to give their diagrammatic counterparts.

5.3.3.1 *The Spectral Theorem*

As promised in Section 5.1.6, we present the following theorem.

Theorem 5.71 All self-adjoint linear maps f are *diagonalisable*; that is, there exists some ONB such that:

$$\boxed{f} \;=\; \sum_i r_i \;\;\overset{\displaystyle \overset{|}{\boxed{i}}}{\underset{\underset{|}{\boxed{i}}}{\vee\wedge}} \tag{5.45}$$

where all r_i are real numbers. Moreover, if f is positive, then $r_i \geq 0$ for all i, and if f is a projector, then $r_i \in \{0, 1\}$ for all i. Consequently, every projector P can be written in the following form:

$$\boxed{P} \;=\; \sum_i \;\;\overset{\displaystyle \overset{|}{\boxed{i}}}{\underset{\underset{|}{\boxed{i}}}{\vee\wedge}}$$

for some orthonormal <u>set</u> (i.e. not necessarily a full ONB).

Proof We merely sketch this proof, as the full proof can be found in any book on linear algebra. There are two key ingredients. The first is a standard result that every linear map from a system to itself has at least one eigenstate. This is a consequence of the fact that the numbers in **linear maps** are complex numbers and of the 'fundamental theorem of algebra'. The second is that self-adjoint linear maps preserve orthogonality with eigenstates. That is, for any eigenstate ψ, if ϕ is orthogonal to ψ, then so too is $f \circ \phi$:

$$\overset{\boxed{\psi}}{\underset{\boxed{\phi}}{\boxed{f}}} \;=\; \overset{\boxed{\psi}}{\underset{\boxed{\phi}}{\boxed{f}}} \;=\; \bar{\lambda}\,\overset{\boxed{\psi}}{\underset{\boxed{\phi}}{}} \;=\; 0$$

This enables us to keep choosing new eigenstates for f orthogonal to all the previous ones until we build an ONB. Once we have established that f is indeed diagonalisable, the fact that each of the r_i in Theorem 5.71 are real, positive, or $\in \{0, 1\}$ follows from Proposition 5.44. $\qquad\square$

Theorem 5.71 is known as the *spectral theorem*. The following equivalent presentation is particularly relevant for us. First we feed (5.45) into process–state duality:

$$\boxed{f} \;=\; \sum_i r_i \;\;\overset{\displaystyle \overset{|}{\boxed{i}}}{\underset{\underset{|}{\boxed{i}}}{\vee\wedge}}$$

and then we exploit the correspondences established in Section 4.3.6:

Corollary 5.72 For all self-conjugate bipartite states ψ in **linear maps** there exists some ONB such that:

$$\widehat{\psi} = \sum_i r_i \bigvee_i \bigvee_i \tag{5.46}$$

where all r_i are real numbers. If ψ is $\otimes$-positive, then $r_i \geq 0$ for all i.

The origin of the word 'spectral' in 'spectral theorem' is the following.

Definition 5.73 For a self-adjoint linear map f we call the list $(r_i)_i$ from Theorem 5.71 the *spectrum* of f, and we denote it by $\mathsf{spec}(f)$. Similarly, for a self-conjugate bipartite state ψ, we also call the numbers from Corollary 5.72 the *spectrum* of ψ, denoted $\mathsf{spec}(\psi)$.

In Section 8.2.5 we will provide a diagrammatic counterpart to Theorem 5.71 and hence also to Corollary 5.72.

Now we will use the spectral theorem to prove a fact about adjoints of **linear maps** that comes from taking the idea that an adjoint really is a reflection seriously. Suppose we have a ∘-non-separable process. One could then imagine that it has some internal structure, say a collection of tubes or machines connecting some inputs to outputs:

If we now compose this process with its adjoint, i.e. its vertical reflection, then these internal connections match up:

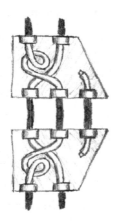

so one expects the resulting process also to be o-non-separable. Consequently, if the composite of a process with its adjoint is o-separable:

$$
\frac{\boxed{f}}{\boxed{f}} = \frac{\phi}{\psi}
$$

then also that process should be o-separable:

$$
\boxed{f} = \frac{\phi'}{\psi'}
$$

The spectral theorem indeed guarantees that $f^\dagger \circ f$ detects separability of f.

Proposition 5.74 For any linear map f we have:

$$
\left(\exists \psi, \phi : \quad \frac{\boxed{f}}{\boxed{f}} = \frac{\phi}{\psi} \right) \iff \left(\exists \psi', \phi' : \quad \boxed{f} = \frac{\phi'}{\psi'} \right) \tag{5.47}
$$

Proof The direction ($\Leftarrow$) is trivial. For ($\Rightarrow$), assume $f^\dagger \circ f$ is o-separable. Then, since it is positive, we can diagonalise $f^\dagger \circ f$:

$$
\frac{\boxed{f}}{\boxed{f}} = \sum_i r_i \; \frac{i}{i}
$$

but then, since $f^\dagger \circ f$ is o-separable, it must be the case that:

$$
\frac{\boxed{f}}{\boxed{f}} = r_j \; \frac{j}{j}
$$

for some j. Otherwise, $f^\dagger \circ f$ could produce (non-zero) orthogonal states as output, which can never happen if it is $\circ$-separable. Thus, for all $i \neq j$:

by positive-definiteness. A resolution of the identity completes the proof:

□

Remark 5.75 Note that there is an analogy between condition (5.47) and positive-definiteness. While positive-definiteness tells us that the number $\psi^\dagger \circ \psi$ allows one to detect whether the state ψ is zero, condition (5.47) tells us that $f^\dagger \circ f$ allows one to detect whether f is $\circ$-separable.

And now comes the really interesting bit. While in Example 4.40 we established that **relations** obey the positive-definiteness condition, **relations** do not satisfy condition (5.47).

Exercise 5.76 Verify that the relation with matrix:

$$\begin{pmatrix} 1 & 1 \\ 0 & 1 \end{pmatrix}$$

does not satisfy (5.47).

So in **relations** the intuition explained above that a gap in $f^\dagger \circ f$ causes a gap in f fails to hold, and this can be traced back to the failure of the spectral theorem for **relations**. The main reason for this is that in **relations** there simply aren't enough numbers to guarantee a diagonal representation for every relation.

Exercise 5.77 Along the same vein as Proposition 5.74, show using the spectral theorem that $f^\dagger \circ f$ detects whether f is invariant under a projector; that is, show that for any projector P:

$$\left(\quad = \quad \right) \iff \left(\quad = \quad \right)$$

5.3.3.2 The Dimension Theorem

We (somewhat perversely) defined $\dim(A)$ in Section 5.1.1 as the size of the <u>smallest</u> basis of a type A. This is quite annoying, since if we want to figure out the dimension of A we need to look at all bases of A, just to make sure that we have the smallest one. Fortunately, in **linear maps** we don't have to do that, because the complex numbers form a field. We can therefore rely on the following theorem, called the *dimension theorem*.

Theorem 5.78 If a process theory admits a matrix calculus and its numbers form a field, then all bases for a given type A are the same size, namely $\dim(A)$.

Thanks to the dimension theorem, in **linear maps** $\dim(A)$ is the size of <u>any</u> basis for A. As a consequence, any orthonormal set of size $\dim(A)$ is an ONB. To see this, we only need the following proposition.

Proposition 5.79 Any orthonormal set:

$$\mathcal{A} = \left\{ \quad \right\}$$

extends to an ONB containing $\mathcal{A}$.

Proof It is straightforward to show that:

$$P := \sum_i \qquad \text{and} \qquad Q := \quad - \quad P$$

are projectors. Using the spectral theorem we can diagonalise Q in terms of a second orthonormal set of states $\mathcal{A}'$:

$$Q = \sum_i$$

Since P is a projector, it easily follows that $P \circ Q = 0$, and from this it then also easily follows that all of states in $\mathcal{A}'$ must be orthogonal to those in $\mathcal{A}$. Moreover, since:

$$\sum_i \; \overset{i}{\underset{i}{\diamondsuit}} \;+\; \sum_i \; \overset{i}{\underset{i}{\blacklozenge}} \;=\; \boxed{P} \;+\; \boxed{Q} \;=\; \Big|$$

$\mathcal{A} \cup \mathcal{A}'$ gives a resolution of the identity, and thus, by Theorem 5.32, $\mathcal{A} \cup \mathcal{A}'$ is an ONB, which, evidently, contains $\mathcal{A}$. $\qquad\square$

Thus, if we have an orthonormal set $\mathcal{A}$ of size dim(A), there is an ONB $\mathcal{B}$ containing $\mathcal{A}$. But, by Theorem 5.78, $\mathcal{B}$ must be of size dim(A), so $\mathcal{A} = \mathcal{B}$.

Corollary 5.80 Any orthonormal set of size dim(A) is an ONB.

Let's now have a look at how the dimension theorem genuinely can make life easier (rather than just 'less annoying'). Recall that in Propositions 5.37 and 5.38, we gave characterisations of isometries and unitaries in terms of their behaviour on orthonormal sets. Combining this with the dimension theorem yields the following.

Proposition 5.81 For any isometry:

$$\overset{\displaystyle B}{\underset{\displaystyle A}{\boxed{U}}}$$

dim(A) $\leq$ dim(B), and U is furthermore unitary if dim(A) = dim(B). So in particular, any isometry from a Hilbert space to itself must be unitary.

Proof Proposition 5.79 implies that any orthonormal set on B must be of size $\leq$ dim(B). Then, by Proposition 5.37, U sends any ONB $\mathcal{B}$ on A to an orthonormal set $\mathcal{B}'$ of size dim(A). Thus, dim(A) $\leq$ dim(B). If dim(A) = dim(B), then by Corollary 5.80, $\mathcal{B}'$ is an ONB for B. Thus by Proposition 5.38, U is a unitary. $\qquad\square$

5.3.4 Classical Logic Gates as Linear Maps

We assumed the existence of Hilbert spaces $\mathbb{C}^n$ for every n. Of particular interest is $\mathbb{C}^2$, because of its role in quantum computation, namely in the description of quantum bits, or *qubits*. The Hilbert space $\mathbb{C}^2$ has a two-element basis:

$$\left\{ \underset{0}{\triangledown} , \underset{1}{\triangledown} \right\}$$

which is referred to as the *computational basis* or *Z-basis*. By analogy with (classical) bits, we choose to count the basis elements of $\mathbb{C}^2$ from zero.

Remark 5.82 In Dirac notation, these basis states are denoted as:

$$\{|0\rangle, |1\rangle\}$$

Classical *logic gates* are functions:

which form the basic building blocks in digital circuits. Thanks to the computational basis, we can now write any logic gate as a linear map, by means of the following translation, for $a, b \in \{0, 1\}$:

or more generally, for $a_1, \ldots, a_m, b_1, \ldots, b_n \in \{0, 1\}$:

Each such mapping of a particular input contributes a term:

and the linear map is obtained as the sum of all the terms of this kind. For instance, the AND *gate*:

$$\text{AND} :: \begin{cases} (0, 0) \mapsto 0 \\ (0, 1) \mapsto 0 \\ (1, 0) \mapsto 0 \\ (1, 1) \mapsto 1 \end{cases}$$

induces the linear map:

Similarly, for an *exclusive*-OR, or XOR *gate* :

$$\text{XOR} :: \begin{cases} (0,0) \mapsto 0 \\ (0,1) \mapsto 1 \\ (1,0) \mapsto 1 \\ (1,1) \mapsto 0 \end{cases}$$

the induced linear map is:

We will see in Chapter 12 that this process of turning logic gates into linear maps plays a key role in the *circuit model of quantum computing*. However, in that context one must restrict to logic gates that yield unitary linear maps, for reasons that will become clear in the following chapter. The AND linear map defined above is not a unitary map. In fact, it doesn't even have an inverse.

Exercise 5.83 Show that a logic gate will define a unitary linear map if and only if it has an inverse (see Definition 4.24).

A simple example of a logic gate with an inverse is the NOT *gate*:

$$\text{NOT} :: \begin{cases} 0 \mapsto 1 \\ 1 \mapsto 0 \end{cases}$$

which induces the linear map:

Another important example is the *controlled*-NOT, or CNOT, *gate*:

$$\text{CNOT} :: \begin{cases} (0,0) \mapsto (0,0) \\ (0,1) \mapsto (0,1) \\ (1,0) \mapsto (1,1) \\ (1,1) \mapsto (1,0) \end{cases}$$

The reason it's called a controlled-NOT gate is that the first bit 'controls' whether we apply a NOT to the second bit. Its induced linear map is:

$$
\text{CNOT} :=
\langle 0|\langle 0| + \langle 0|\langle 1| + \langle 1|\langle 1| + \langle 1|\langle 0|
$$

the matrix of which is:

$$
\begin{pmatrix}
1 & 0 & 0 & 0 \\
0 & 1 & 0 & 0 \\
0 & 0 & 0 & 1 \\
0 & 0 & 1 & 0
\end{pmatrix}
$$

Exercise 5.84 Use Proposition 5.38 to prove that the NOT and CNOT linear maps are both unitary.

Note how the matrix of the CNOT linear map has an identity matrix in the top left corner and a NOT matrix in the bottom right corner. This is a common shape for matrices of 'controlled-(...)' gates, as the first basis element is 'selecting' which of these two smaller matrices to apply to the second:

$$
\text{CNOT} \quad\leftrightarrow\quad \begin{pmatrix} 1 & 0 \\ 0 & 1 \end{pmatrix}
\qquad\qquad
\text{CNOT} \quad\leftrightarrow\quad \begin{pmatrix} 0 & 1 \\ 1 & 0 \end{pmatrix}
$$

Note also that the CNOT linear map is self-adjoint, which justifies a diagrammatic representation that is invariant under vertical reflection, which we will encounter in the next section. On the other hand, it is not invariant under horizontal reflection.

Exercise 5.85 Prove that the CNOT linear map is not invariant under horizontal reflection:

$$
\text{CNOT} \quad\neq\quad \text{CNOT}
$$

Hence, its diagrammatic representation shouldn't be, either.

5.3.5 The X-Basis and the Hadamard Linear Map

An alternative perspective on the CNOT gate involves the COPY *gate*:

$$
\text{COPY} :: \begin{cases} 0 \mapsto (0,0) \\ 1 \mapsto (1,1) \end{cases}
$$

whose induced linear map is:

One way to perform a CNOT is to COPY the first bit, then XOR a copy of the first bit with the second bit. That is:

$$(5.48)$$

Exercise 5.86 Prove equation (5.48).

In most textbooks, one encounters the CNOT gate written this way:

This evokes the same 'COPY+XOR' interpretation of CNOT, given that copying in classical circuits is usually expressed as a small black dot, and the XOR operation as the symbol $\oplus$:

However, we will adopt a more symmetric-looking notation:

This is because these XOR and COPY linear maps are very closely related. They actually both arise as COPY maps, but for different bases. To see this, let's consider another basis for $\mathbb{C}^2$. First, note that since $-1 \in \mathbb{C}$, we can subtract states:

Let the *X-basis* be:

It is easy to check that:

$$\delta_i^j$$

and since both of these states take real-valued coefficients in the Z-basis, they are self-conjugate. Thus we now have two self-conjugate ONBs for $\mathbb{C}^2$:

and

Remark 5.87 In Dirac notation, the X-basis is usually denoted as:

$$\{|+\rangle, |-\rangle\}$$

referring to the fact that we take the sum and the difference of the states of the Z-basis. We will use these bases in Section 6.1.2 to label the Z- and X-axes on a sphere of states for a qubit, which justifies the seemingly ad hoc names 'Z-basis' and 'X-basis'.

Now we can define two versions of all of the maps before, one for the Z-basis and one for the X-basis. For instance, the COPY linear map now has two versions:

Evidently, such COPY maps make sense for Hilbert spaces of all dimensions n. Later on, we use these maps so much that it's convenient to write them just as dots of the appropriate colour:

By writing the adjoints of these maps by vertical reflection as usual:

we can write the XOR-part of a CNOT gate as a grey dot, up to a number.

Proposition 5.88

Proof   We show this by computing matrix entries for the Z-basis. Using:

$$\text{(diagram)} = \text{(diagram)} = \frac{1}{\sqrt{2}} \qquad \text{(diagram)} = \text{(diagram)} = (-1)^i \frac{1}{\sqrt{2}}$$

for the RHS we have:

$$\text{(diagram)} = \left(\text{(diagram)} + \text{(diagram)} \right) = \frac{1}{\sqrt{2}} \cdot \frac{1}{2} \left(1 + (-1)^{i+j+k} \right)$$

This will be equal to $\frac{1}{\sqrt{2}}$ if $i+j+k$ is even, and 0 otherwise. But then, $i+j+k$ is even if and only if $i \oplus j = k$. If we ignore the $\frac{1}{\sqrt{2}}$-factor, this gives precisely the matrix of XOR.   □

Hence, we let:

$$\text{(diagram)} := \text{(diagram)} \approx \boxed{\text{CNOT}}$$

If we don't ignore numbers we have:

$$\sqrt{2}\;\;\text{(diagram)} = \boxed{\text{XOR}}$$

and hence:

$$\sqrt{2}\;\;\text{(diagram)} = \boxed{\text{CNOT}}$$

Example 5.89 Since it depends on subtraction, the X-basis doesn't seem to make sense for the theory of **relations** (see Remark 5.27). However, since the COPY process of the X-basis is the same as XOR, this process does live in **relations**. Thus the 'shadow' of this basis, and crucially its interaction with the Z-basis, occurs there.

Exercise 5.90 Prove that CNOT obeys:

$$\text{(diagram)} \approx \text{(diagram)} \qquad \text{(diagram)} \approx \text{(diagram)}$$

We will see in Chapter 9 that these are not just random diagrammatic equations, but rather capture the fundamental concept of *complementarity* in quantum theory.

We can also define a unitary linear map that translates the *X*-basis into the *Z*-basis by following the recipe of Corollary 5.39.

Definition 5.91 The *Hadamard map* is:

$$\boxed{H} := \sum_i \overset{i}{\underset{i}{\bigtriangledown \!\!\bigtriangleup}}$$

The reason why we don't draw the Hadamard map asymmetrically, like most boxes, is the following.

Proposition 5.92 The Hadamard linear map *H* is self-conjugate and self-adjoint and, hence, also self-transposed:

$$\boxed{H} = \boxed{H} = \boxed{H}$$

Proof We can see that *H* is self-adjoint and self-conjugate if we expand the *X*-basis in terms of the *Z*-basis:

$$\boxed{H} = \overset{0}{\underset{0}{\bigtriangledown\!\!\bigtriangleup}} + \overset{1}{\underset{1}{\bigtriangledown\!\!\bigtriangleup}} = \frac{1}{\sqrt{2}}\left(\overset{0}{\underset{0}{\bigtriangledown\!\!\bigtriangleup}} + \overset{1}{\underset{0}{\bigtriangledown\!\!\bigtriangleup}} + \overset{0}{\underset{1}{\bigtriangledown\!\!\bigtriangleup}} - \overset{1}{\underset{1}{\bigtriangledown\!\!\bigtriangleup}} \right) \tag{5.49}$$

$\square$

From (5.49), we see that the matrix of *H* in the *Z*-basis is:

$$\boxed{H} \leftrightarrow \frac{1}{\sqrt{2}}\begin{pmatrix} 1 & 1 \\ 1 & -1 \end{pmatrix}$$

Exercise 5.93 What is the matrix of *H* in the *X*-basis?

Since the adjoint of a unitary map is its inverse, we also have the following.

Corollary 5.94 The Hadamard linear map is self-inverse:

$$\boxed{H}\ \boxed{H} = \ |$$

Remark 5.95 Just as we did for the CNOT gate, later we will also introduce an alternative notation for the NOT gate, namely:

The reason for it being grey is that it, like XOR, can be expressed naturally in terms of the X-basis, which will be explained later. So, here are all of the operations we met in this section:

We will encounter the copy maps again in Section 8.2.2 when expressing *classical data* in diagrammatic terms. Decorations such as the π in the case of the NOT gate will be explained in Section 9.1, and the interaction of the two colours will be explained in Section 9.2. All of these pieces together will play an important role in a very powerful diagrammatic language called the *ZX-calculus*, described in Section 9.4.

5.3.6 Bell Basis and Bell Maps

We will now give a basis for $\mathbb{C}^2 \otimes \mathbb{C}^2$ that is not simply a product of two bases for $\mathbb{C}^2$. For the following four states:

$$
\left\{
\begin{array}{l}
\begin{array}{c}\boxed{B_0}\end{array} := \tfrac{1}{\sqrt{2}}\left(\begin{array}{c}0\;\;0\end{array} + \begin{array}{c}1\;\;1\end{array} \right) \\[2em]
\begin{array}{c}\boxed{B_1}\end{array} := \tfrac{1}{\sqrt{2}}\left(\begin{array}{c}0\;\;1\end{array} + \begin{array}{c}1\;\;0\end{array} \right) \\[2em]
\begin{array}{c}\boxed{B_2}\end{array} := \tfrac{1}{\sqrt{2}}\left(\begin{array}{c}0\;\;0\end{array} - \begin{array}{c}1\;\;1\end{array} \right) \\[2em]
\begin{array}{c}\boxed{B_3}\end{array} := \tfrac{1}{\sqrt{2}}\left(\begin{array}{c}0\;\;1\end{array} - \begin{array}{c}1\;\;0\end{array} \right)
\end{array}
\right.
\tag{5.50}
$$

it is straightforward to show that they form an orthonormal set, e.g.:

$$
\begin{array}{l}
\begin{array}{c}\boxed{B_0}\\\boxed{B_0}\end{array} = \tfrac{1}{2}\left(\begin{array}{c}0\;0\\0\;0\end{array} + \begin{array}{c}1\;1\\0\;0\end{array} + \begin{array}{c}0\;0\\1\;1\end{array} + \begin{array}{c}1\;1\\1\;1\end{array} \right) \\[2em]
= \tfrac{1}{2}(1+0+0+1) = 1
\end{array}
$$

$$
\begin{array}{l}
\begin{array}{c}\boxed{B_1}\\\boxed{B_0}\end{array} = \tfrac{1}{2}\left(\begin{array}{c}0\;1\\0\;0\end{array} + \begin{array}{c}1\;0\\0\;0\end{array} + \begin{array}{c}0\;1\\1\;1\end{array} + \begin{array}{c}1\;0\\1\;1\end{array} \right) \\[2em]
= \tfrac{1}{2}(0+0+0+0) = 0
\end{array}
$$

$$\text{(diagram)} = \tfrac{1}{2}\left(\text{(diagram)} - \text{(diagram)} + \text{(diagram)} - \text{(diagram)} \right)$$

$$= \tfrac{1}{2}(1 - 0 + 0 - 1) = 0$$

and so on. So the B_i are an orthonormal set of size $\dim(\mathbb{C}^2 \otimes \mathbb{C}^2) = 4$, hence by Corollary 5.80, they form an ONB. Indeed, we can write each of the elements of the product basis for $\mathbb{C}^2 \otimes \mathbb{C}^2$ in terms of the states (5.50):

$$\text{(diagram)} = \tfrac{1}{\sqrt{2}}\left(\text{(diagram } B_0 \text{)} + \text{(diagram } B_2 \text{)} \right)$$

$$\text{(diagram)} = \tfrac{1}{\sqrt{2}}\left(\text{(diagram } B_1 \text{)} + \text{(diagram } B_3 \text{)} \right)$$

$$\text{(diagram)} = \tfrac{1}{\sqrt{2}}\left(\text{(diagram } B_1 \text{)} - \text{(diagram } B_3 \text{)} \right)$$

$$\text{(diagram)} = \tfrac{1}{\sqrt{2}}\left(\text{(diagram } B_0 \text{)} - \text{(diagram } B_2 \text{)} \right)$$

Definition 5.96 The ONB (5.50) is called the *Bell basis*.

By Proposition 5.38 we know that unitaries always send ONBs to ONBs. So, another way to show that the states (5.50) form an ONB is to show that they arise from applying some unitary to another ONB.

Exercise 5.97 Show that the Bell basis arises from applying the following unitary linear map:

$$\sqrt{2}\ \text{(diagram with } H \text{)}$$

to the product basis:

$$\left\{ \text{(diagram)} \right\}_{ij}$$

In other words, show that the states (5.50) can be expressed as:

$$\left\{ \sqrt{2}\ \text{(diagram with } H \text{)} \right\}_{ij}$$

and hence that they form an ONB.

The first of the Bell states is just a cup multiplied by $\frac{1}{\sqrt{2}}$, while the other three are variations obtained by flipping bits or signs. From this, it is easy to see that the Bell states are all non-separable. Moreover, they are all maximally non-separable in the sense of Definition 4.80, a fact that plays a crucial role in quantum teleportation. We can show maximal non-separability by relating certain unitary linear maps, called the *Bell maps*, to the Bell states as follows:

$$(5.51)$$

where:

Exercise 5.98 Show that the matrix forms of the Bell maps are:

$$(5.52)$$

and hence their corresponding matrices are:

$$\boxed{B_0} \leftrightarrow \begin{pmatrix} 1 & 0 \\ 0 & 1 \end{pmatrix} \qquad \boxed{B_1} \leftrightarrow \begin{pmatrix} 0 & 1 \\ 1 & 0 \end{pmatrix}$$

$$(5.53)$$

$$\boxed{B_2} \leftrightarrow \begin{pmatrix} 1 & 0 \\ 0 & -1 \end{pmatrix} \qquad \boxed{B_3} \leftrightarrow \begin{pmatrix} 0 & -1 \\ 1 & 0 \end{pmatrix}$$

Also show that these linear maps are unitary.

Unsurprisingly, we call the matrices associated with the Bell maps the *Bell matrices*. However, in most textbooks, one doesn't tend to encounter the Bell matrices, but rather their close cousins.

Remark 5.99 The following matrices are known as the *Pauli matrices*:

$$\sigma_0 = 1 := \begin{pmatrix} 1 & 0 \\ 0 & 1 \end{pmatrix} \qquad \sigma_1 = \sigma_X := \begin{pmatrix} 0 & 1 \\ 1 & 0 \end{pmatrix}$$

$$\sigma_2 = \sigma_Y := \begin{pmatrix} 0 & -i \\ i & 0 \end{pmatrix} \qquad \sigma_3 = \sigma_Z := \begin{pmatrix} 1 & 0 \\ 0 & -1 \end{pmatrix}$$

They are almost exactly the same as the Bell matrices:

$$\sigma_0 = B_0 \qquad\qquad \sigma_1 = B_1 \qquad\qquad \sigma_3 = B_2$$

where the only difference is that:

$$\sigma_2 = iB_3 \tag{5.54}$$

We shall see in the next chapter that multiplying by i has no effect on the physical meaning of this process. So why should one care about these Pauli matrices? We don't! But others do. The reason for (5.54) is to make all of the Pauli matrices self-adjoint. In the usual presentation of quantum theory, self-adjoint matrices are used to define measurements (see Remark 7.25). However, our definition of measurement doesn't depend on self-adjoint linear maps, so we'll stick to Bell maps.

Note from equation (5.51) that one obtains all Bell states by applying certain processes to a fixed state. This fact is directly related to how the Bell states and Bell maps play a role in quantum teleportation. We will explain this in detail in the next chapter, but the following is a preview.

Exercise 5.100 Compute the state:

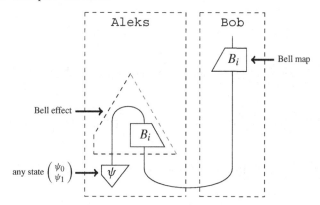

using the matrix representations of all of its ingredients and conclude that, up to a number, we indeed obtain ψ, for every i.

The relationship that the Bell basis and the Bell maps enjoy is a particular case of a general relationship between ONBs and certain families of linear maps induced by processes-state duality.

Definition 5.101 A set of linear maps:

$$
\left\{\ \vcenter{\hbox{$\begin{array}{c} B \\ \boxed{i} \\ A \end{array}$}}\ \right\}_i
\tag{5.55}
$$

is called a *Hilbert–Schmidt ONB* if the associated bipartite states:

$$
\vcenter{\hbox{$\smash{\overbrace{}}$}}\; i = \frac{1}{\sqrt{D}}\; \vcenter{\hbox{i}}
\tag{5.56}
$$

form an ONB, for $D = \dim(A)$.

Taking the inner product of states (5.56) yields:

$$
\vcenter{\hbox{$\begin{array}{c} j \\ i \end{array}$}} \;=\; \frac{1}{D}\;\vcenter{\hbox{$\begin{array}{c} j \\ i \end{array}$}} \;=\; \frac{1}{D}\,\mathrm{tr}\left(\vcenter{\hbox{$\begin{array}{c} j \\ i \end{array}$}}\right)
$$

The rightmost expression (without the $\frac{1}{D}$) is sometimes known as the *Hilbert–Schmidt inner product* of linear maps. Evidently, we have an alternative expression for orthonormality of (5.55) in terms of this inner product:

$$
\mathrm{tr}\left(\vcenter{\hbox{$\begin{array}{c} j \\ i \end{array}$}}\right) \;=\; D\,\delta_i^j
$$

Recalling Definition 4.80 of maximally non-separable states, the following are moreover equivalent:

- the basis states $\vcenter{\hbox{$i$}}$ are maximally non-separable, and

- the corresponding basis maps $\vcenter{\hbox{$\boxed{i}$}}$ are unitary.

5.4 Hilbert Spaces versus Diagrams

So what do string diagrams really have to do with Hilbert spaces and linear maps? We defined Hilbert spaces and linear maps as a special kind of string diagrams. However, in order to get there, we needed to assume bases, sums, and the fact that the numbers are complex numbers, which all has a bit of a brute-force feel to it. Moreover, many other things are also described by string diagrams, for example, sets and relations, so the relationship with Hilbert spaces and linear maps may not be that special.

However, we will present a surprising result, tightly relating diagrammatic reasoning to Hilbert spaces and linear maps. Given that the additional structure of Hilbert spaces allows us to prove lots and lots of stuff about processes, one might be tempted to think that we can prove more equations between diagrams of linear maps than we could for just plain old diagrams, i.e. without using bases, sums, etc. This turns out not to be the case. The equations that are true between diagrams of linear maps are precisely those that can be proven just by deforming diagrams. So, the first thing we will see in this section is a precise statement of that fact.

The second thing we'll do is derive the usual, set-theoretic definition of Hilbert space from the one based on string diagrams. We'll see in Section 5.4.2 that this definition, along with linear maps and the tensor product, falls out automatically by regarding the system-types for **linear maps** not simply as labels, but instead as the sets of all their states.

Note that material in the next two sections is optional and won't be used explicitly later in this book. However, many readers may find them useful for understanding the power of diagrammatic proofs and connecting our notion of Hilbert space to the usual one encountered in the literature.

5.4.1 String Diagrams Are Complete for Linear Maps

In this section we aim to characterise which <u>new</u> equations between diagrams appear once we assume that the diagrams actually represent linear maps, with the extra structure of sums, bases, and complex numbers around. And as already indicated above, the quite surprising answer is: none! That is, the language of string diagrams is already powerful enough to capture all of the equations that hold for generic diagrams of linear maps. Using terminology borrowed from logic:

> String diagrams are *complete* for **linear maps**.

Another technical way to state this, which emphasises the fact that adding ONBs, sums, and complex numbers is 'innocent', is as follows:

> **linear maps** are a *conservative extension* of string diagrams.

Recall from Section 3.1.2 that the data making up a string diagram are the boxes in it, including specification of the input and output wires, and how these boxes are wired together. That is, by a 'string diagram' we mean the 'drawing' of boxes and wires without interpretation within a process theory. In the same vein, string diagrams are equal if they can be deformed into each other. In order to distinguish wires and boxes, we used labels:

Diagrams also come with a notion of vertical reflection, which is crucial to the completeness result we will soon present. This result is about comparing these 'free', or 'interpretation-free', diagrams, with diagrams interpreted in the process theory of **linear maps**. Let us make the idea of interpretation a bit more formal.

Definition 5.102 An *interpretation in linear maps* $[\![\ ]\!]$ of a string diagram D consists of a choice of Hilbert space $[\![A]\!]$ for every type A occurring in D and a choice of linear map $[\![f]\!]$ (respecting input/output types) for every box f occurring in D.

In particular, an interpretation defines a linear map $[\![D]\!]$ obtained by plugging all the linear maps $[\![f]\!]$, $[\![g]\!]$, ...into the diagram D and evaluating them as linear maps (using e.g. Theorem 5.61).

Example 5.103 Consider this diagram:

The diagram D has two types and two boxes, so here is an interpretation:

$$\left\{ [\![A]\!] := \mathbb{C}^3, \ [\![B]\!] := \mathbb{C}^2, \ [\![f]\!] := \begin{pmatrix} 1 & 2 & 0 \\ 3 & 1 & 4 \\ 5 & 5 & 5 \\ 0 & 1 & 0 \end{pmatrix}, \ [\![g]\!] := \begin{pmatrix} 1 & 0 \\ 1 & 1 \end{pmatrix} \right\}$$

Then, under this interpretation:

$$[\![D]\!] = \begin{pmatrix} 1 & 2 & 0 \\ 4 & 3 & 4 \\ 5 & 5 & 5 \\ 5 & 6 & 5 \end{pmatrix}$$

Similarly, we can give an interpretation for a collection of diagrams $D, E, F, \ldots$ by interpreting all of the boxes and wires occurring in any of the diagrams, which in turn yields linear maps $[\![D]\!], [\![E]\!], [\![F]\!], \ldots$

Now consider the following two equations, which we encountered earlier:

Both equations are valid in the process theory of **linear maps**; however, in the first one f and g can be any linear maps (with appropriately matching types), while the second one only holds in the specific case of CNOT. In other words, the second equation involves only a single interpretation of the diagrams in **linear maps**, while the first one ranges over all possible interpretations in **linear maps**. The equations we are concerned with here are those of first kind, involving all possible interpretations in **linear maps**.

We are now ready to state the completeness theorem.

Theorem 5.104 String diagrams are complete for **linear maps**. That is, for any two string diagrams D, E, the following are equivalent:

- $D = E$
- For all interpretations of D, E into **linear maps**, $[\![D]\!] = [\![E]\!]$.

Proof (sketch) The proof of this theorem goes beyond the scope of this book. Roughly speaking, it involves taking some diagram D and defining an interpretation that is 'characteristic' of D. That is, the interpretation yields some fixed linear map L for D, and for any other $D' \neq D$, it is possible to show that $[\![D']\!]$ cannot possibly be equal to L. □

So the bottom line is:

> *An equation between string diagrams holds for all Hilbert spaces and linear maps if and only if the string diagrams are the same.*

This provides compelling evidence for the fact that string diagrams capture the compositional content of linear maps.

The condition that $[\![D]\!] = [\![E]\!]$ for <u>all</u> interpretations is fairly strong, so one still might ask if Theorem 5.104 holds for any process theory that admits string diagrams. This is not the case.

Theorem 5.105 String diagrams are <u>not</u> complete for **relations**. That is, there exist non-equal string diagrams $D \neq E$, where for all interpretations as **relations**, $[\![D]\!] = [\![E]\!]$.

Proof We just need to find two distinct string diagrams that are equal for all interpretations as **relations**. Recall that the only numbers in **relations** are 0 and 1. In both cases $\lambda^2 = \lambda$, so the following two diagrams:

 and

are equal for all interpretations as **relations**. However, they are clearly not equal as string diagrams. □

In other words, there are equations between string diagrams that hold for (all) sets and relations that are not string diagram equalities. In light of the fact that the only real difference between **linear maps** and **relations** is the choice of numbers, it might come as a surprise that something very simple like the booleans breaks completeness, whereas $\mathbb{C}$, with all of its extra structure, retains it.

5.4.2 The Set-Theoretic Definition of Hilbert Spaces

This section can safely be skipped by the reader who is perfectly happy with our definition of the process theory of **linear maps** (and hence Hilbert spaces). The main goal here is to connect our definition to the traditional one. In other words, we will define a Hilbert space H as a set with some extra structure (sums, scalar multiplication, inner product) and linear maps as functions that preserve (some of) this structure. Compound systems are formed by constructing the *tensor product* of Hilbert spaces, which we will also define. We provide this alternative definition for two reasons:

- for comprehensiveness, so the reader has some exposure to the traditional definition of Hilbert space, and
- for contrast, to emphasise the different spirit of the two definitions.

Recall from Section 3.4.1 that for **functions**, the states of type A are in one-to-one correspondence with the elements of the set A. We will define a set-theoretic counterpart $\tilde{H}$ to 'our Hilbert spaces' H (i.e. system-types in **linear maps**) by mimicking this one-to-one correspondence:

$$\tilde{H} := \left\{ \begin{array}{c} |H \\ \psi \end{array} \; \middle| \; \psi \text{ is a state of type } H \right\}$$

That is, given a type H in **linear maps** we define a *Hilbert space* $\tilde{H}$ to be its set of states. Instead of states, we call the elements of $\tilde{H}$ *vectors*. In what follows we show that we also get all of the set-theoretic Hilbert space operations (along with the tensor product!) 'for free' by expressing them in terms of states. We begin with sums and scalar multiplication.

Since **linear maps** has sums, we have the following:

- $\boxed{0} \in \tilde{H}$

- if $\boxed{\psi} \in \tilde{H}$, $\boxed{\phi} \in \tilde{H}$ then $\boxed{\psi} + \boxed{\phi} \in \tilde{H}$

- if $\langle \lambda \rangle \in \mathbb{C}$, $\boxed{\psi} \in \tilde{H}$ then $\langle \lambda \rangle \boxed{\psi} \in \tilde{H}$

In other words, the set of states in H forms a *complex vector space*.

Definition 5.106 A *complex vector space* is a set $\tilde{V}$ with a distinguished element $0 \in \tilde{V}$ and two operations:

- $\psi + \phi$ for all $\psi, \phi \in \tilde{V}$
- $\lambda \cdot \psi$ for all $\lambda \in \mathbb{C}$ and $\psi \in \tilde{V}$

called *sum* and *scalar multiplication*, respectively, satisfying:

1. $0 + \psi = \psi$
2. $(\psi + \phi) + \xi = \psi + (\phi + \xi)$

3. $\psi + \phi = \phi + \psi$

4. $0 \cdot \psi = 0$

5. $\lambda \cdot (\lambda' \cdot \psi) = (\lambda\lambda') \cdot \psi$

6. $\lambda \cdot (\psi + \phi) = \lambda \cdot \psi + \lambda \cdot \phi$

7. $(\lambda + \lambda') \cdot \psi = \lambda \cdot \psi + \lambda' \cdot \psi$

for all $\psi, \phi, \xi \in \widetilde{V}$ and $\lambda, \lambda' \in \mathbb{C}$.

We typically write the scalar product $\lambda \cdot \psi$ simply as $\lambda\psi$. As in Section 5.1.3, once it is established that $(- + -)$ is associative and commutative, it is more convenient to use summation notation, and replace equations 6 and 7 above with their equivalents using the summation symbol:

6'. $\lambda \left(\sum_i \psi_i \right) = \sum_i (\lambda\psi_i)$

7'. $\left(\sum_i \lambda_i \right) \psi = \sum_i (\lambda_i \psi)$

Exercise 5.107 Use **Conditions 1–3** in Definition 5.21 to show that $\widetilde{H}$ satisfies the seven equations of a complex vector space.

In **linear maps** every system-type has a finite basis:

$$\mathcal{B} = \left\{ \; \underset{i}{\nabla} \; \right\}_i$$

From Theorem 5.30 it follows that any $\psi \in \widetilde{H}$ can be written in this basis:

$$\underset{\psi}{\nabla} = \sum_i \langle \psi_i \rangle \; \underset{i}{\nabla}$$

This means that our complex vector space is *finite dimensional.*

Definition 5.108 A *basis* for a vector space $\widetilde{V}$ is a minimal set of vectors

$$\{\phi_i\}_i \subseteq \widetilde{V}$$

that *spans* $\widetilde{V}$; that is, for all $\psi \in \widetilde{V}$ there exist $\{\phi_i\}_i \subseteq \mathbb{C}$ such that:

$$\psi = \sum_i \lambda_i \phi_i \tag{5.57}$$

If $\{\phi_i\}_i$ is finite, then $\widetilde{V}$ is *finite-dimensional* and the number of elements in $\{\phi_i\}_i$ is the *dimension* of $\widetilde{V}$.

The theory of **linear maps** has adjoints, so we can form:

$$\begin{array}{c} \triangle\psi \\ | \\ \nabla\phi \end{array}$$

for all $\psi, \phi \in \widetilde{H}$. This gives us all the structure we need to make the following definition.

Definition 5.109 A *finite-dimensional Hilbert space* is a finite-dimensional vector space $\widetilde{H}$, with an additional operation:

$$\langle - | - \rangle : \widetilde{H} \times \widetilde{H} \to \mathbb{C}$$

called the *inner product*, satisfying:

1. $\langle \psi | \lambda \phi + \xi \rangle = \lambda \langle \psi | \phi \rangle + \langle \psi | \xi \rangle$
2. $\langle \lambda \psi + \phi | \xi \rangle = \overline{\lambda} \langle \psi | \xi \rangle + \langle \phi | \xi \rangle$
3. $\langle \psi | \phi \rangle = \overline{\langle \phi | \psi \rangle}$
4. $\langle \psi | \psi \rangle > 0$ for all $\psi \neq 0$

where $\overline{(-)}$ is complex conjugation.

Note that 1 and 3 imply 2. We already encountered most of these properties in Proposition 4.51, with the exception of the sum-preservation part of condition 1, which as we saw in Section 5.1.3, amounts to:

and similarly for condition 2.

Remark* 5.110 We only define finite-dimensional Hilbert spaces here, because that is all we need for this book. The definition of an arbitrary Hilbert space is more complicated, as the set of vectors is required to be topologically closed under taking limits of certain sequences of vectors. This condition is automatically satisfied in finite dimensions.

The usual notion of ONB for a Hilbert space matches ours.

Definition 5.111 A basis $\{\phi_i\}_i$ for a finite-dimensional Hilbert space $\widetilde{H}$ is *orthonormal* if we have:

$$\langle \phi_i | \phi_j \rangle = \delta_i^j$$

The next thing to do is to introduce linear maps as certain functions between sets. General *multilinear* maps (i.e. maps with many inputs/outputs) require a bit of work to define using this definition of Hilbert space. However, maps with precisely one input and one output are straightforward. Given Hilbert spaces $\widetilde{H}$ and $\widetilde{K}$, by considering functions of the form:

we indeed recover the usual properties of linear maps.

Definition 5.112 For vector spaces (or Hilbert spaces) $\widetilde{H}$ and $\widetilde{K}$, a *linear map* $f : \widetilde{H} \to \widetilde{K}$ is a function from the set $\widetilde{H}$ to the set $\widetilde{K}$ satisfying:

$$f(\lambda\psi) = \lambda f(\psi) \quad \text{and} \quad f(\psi + \phi) = f(\psi) + f(\phi)$$

These two conditions are referred to as *linearity*.

We can express both conditions more compactly using summation:

$$f\left(\sum_i \lambda_i \psi_i\right) = \sum_i \lambda_i f(\psi_i) \tag{5.58}$$

As we saw in Section 5.1.3, this amounts to:

Remark 5.113 We initially defined the set-theoretic counterpart to Hilbert spaces by mimicking the one-to-one correspondence between sets A and functions from the set representing the 'no wire'–type to the set A. In the case of Hilbert spaces, it's linearity that makes this one-to-one correspondence work. First, note that $\mathbb{C}$ forms a particularly simple vector space, for which the number 1 by itself forms a one-element ONB, since all the other numbers $\lambda \in \mathbb{C}$ can be decomposed as follows (cf. equation (5.57)):

$$\lambda = \lambda \cdot 1$$

and the orthonormality condition is trivially satisfied. Addition of 'vectors' in $\mathbb{C}$ is just addition of complex numbers, by rule 7 from Definition 5.106:

$$\lambda + \lambda' = \lambda \cdot 1 + \lambda' \cdot 1 = (\lambda + \lambda') \cdot 1 = \lambda + \lambda'.$$

Now, a vector $\psi \in \tilde{H}$ defines a unique linear map:

$$\tilde{\psi} : \mathbb{C} \to \tilde{H} :: 1 \mapsto \psi$$

since by linearity we now have, for all $\lambda \in \mathbb{C}$:

$$\tilde{\psi} :: \lambda = \lambda \cdot 1 \mapsto \lambda \cdot \psi$$

Taking $\tilde{H}$ to be $\mathbb{C}$ itself, each $\lambda \in \mathbb{C}$ defines a unique linear map:

$$\tilde{\lambda} : \mathbb{C} \to \mathbb{C} :: 1 \mapsto \lambda$$

Hence there also is a one-to-one correspondence between the elements of $\mathbb{C}$ and the linear maps from $\mathbb{C}$ to itself. Thus linear maps, vectors, and numbers in $\mathbb{C}$ are all instances of the same thing, which you knew already, because they are all certain kinds of processes.

For string diagrams, and hence **linear maps**, the adjoint is:

For traditional linear maps we have the following standard result.

Proposition 5.114 For Hilbert spaces $\tilde{H}, \tilde{K}$ and a linear map $f : \tilde{H} \to \tilde{K}$ there exists a unique linear map $f^\dagger$ such that for all $\psi \in \tilde{H}, \phi \in \tilde{K}$:

$$\langle \phi | f(\psi) \rangle = \langle f^\dagger(\phi) | \psi \rangle \qquad (5.59)$$

The unique linear map $f^\dagger$ in the proposition is also called the *adjoint*. With string diagrams, equation (5.59) is a tautology. Indeed, since:

- $|f(\psi)\rangle$ translates as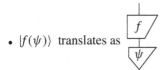

it follows that:

- $\langle f^\dagger(\phi)|$ translates as

and so both the LHS $\langle \phi | f(\psi) \rangle$ and RHS $\langle f^\dagger(\phi) | \psi \rangle$ of (5.59) are:

The notions of unitary and positive maps are identical for both the traditional presentation of Hilbert spaces and ours. That is, they are maps that take the form of Definition 4.56 and Definition 4.60, respectively.

The final, and trickiest, piece of the set-theoretic counterpart to **linear maps** is defining the parallel composition of Hilbert spaces. We need to define an operation $\tilde{\otimes}$ that mimics at the level of Hilbert spaces $\tilde{H}$ and $\tilde{K}$ the result of composing the types H and K in parallel. That is:

$$\tilde{H} \,\tilde{\otimes}\, \tilde{K} \quad \text{should match} \quad \widetilde{H \otimes K}$$

For two Hilbert spaces $\tilde{H}$ and $\tilde{K}$, let's look at some of the properties of $\widetilde{H \otimes K}$. Any states $\psi \in \tilde{H}$ and $\phi \in \tilde{K}$ yield a state:

in $\widetilde{H \otimes K}$. So, at first glance, one might be tempted to define $\tilde{H} \,\tilde{\otimes}\, \tilde{K}$ as the Cartesian product $H \times K$, which is what we did for **functions**. However, this quickly goes wrong, since $\widetilde{H \otimes K}$ contains lots of states that are non-separable. Recall from Section 5.2.2 that we can build a basis for $H \otimes K$ in terms of the individual bases of H and K. In particular, given ONBs:

$$\left\{ \; i \; \right\}_i \qquad \text{and} \qquad \left\{ \; j \; \right\}_j$$

for H and K respectively, we can write all states $\Phi \in \widetilde{H \otimes K}$ as follows:

$$\mathord{\Phi} \;=\; \sum_i \langle \Phi_{i,j} \rangle\; \overline{i}\, \overline{j}$$

So given Hilbert spaces $\widetilde{H}$ and $\widetilde{K}$, we want to form a Hilbert space for which:

$$\left\{\, \overline{i}\; \overline{j}\, \right\}_{ij}$$

plays the role of an ONB. First we need to form the set of vectors, and one way to do this is to simply consider the set of 'formal sums':

$$\widetilde{H} \,\widetilde{\otimes}\, \widetilde{K} := \left\{ \sum_{ij} \lambda_{ij} \phi_i \,\widetilde{\otimes}\, \phi'_j \;\middle|\; \lambda_{ij} \in \mathbb{C} \right\}$$

where $\{\phi_i\}_i$ and $\{\phi'_j\}_j$ are ONBs of $\widetilde{H}$ and $\widetilde{K}$, respectively. We say 'formal sums' since at the moment the sum-symbol really doesn't have any meaning. Nor does the $\widetilde{\otimes}$-symbol between ϕ_i and ϕ'_j. We could just as well have represented these elements as lists of numbers indexed by (i,j), i.e. matrices of the λ_{ij} coefficients. Next, we need to make this set into a vector space, by letting these formal sums act like 'actual' sums in our definition of vector addition and scalar multiplication:

$$\left(\sum_{ij} \lambda_{ij} \phi_i \,\widetilde{\otimes}\, \phi'_j \right) + \left(\sum_{ij} \lambda'_{ij} \phi_i \,\widetilde{\otimes}\, \phi'_j \right) := \sum_{ij} (\lambda_{ij} + \lambda'_{ij}) \phi_i \,\widetilde{\otimes}\, \phi'_j$$

$$\lambda \cdot \left(\sum_{ij} \lambda_{ij} \phi_i \,\widetilde{\otimes}\, \phi'_j \right) := \sum_{ij} \lambda \lambda_{ij} \phi_i \,\widetilde{\otimes}\, \phi'_j$$

Also, in this form, we can clearly relate the elements of $\widetilde{H}$ and $\widetilde{K}$ to the corresponding ones in $\widetilde{H} \,\widetilde{\otimes}\, \widetilde{K}$. More specifically, to

$$\sum_i \lambda_i \phi_i \in \widetilde{H} \qquad \text{and} \qquad \sum_i \lambda'_i \phi'_i \in \widetilde{K}$$

we can associate the element:

$$\sum_{ij} \lambda_i \lambda'_j \phi_i \,\widetilde{\otimes}\, \phi'_j \in \widetilde{H} \,\widetilde{\otimes}\, \widetilde{K}$$

We denote this mapping also by the symbol $\widetilde{\otimes}$, and hence obtain:

$$\left(\sum_i \lambda_i \phi_i \right) \widetilde{\otimes} \left(\sum_j \lambda'_j \phi'_j \right) = \sum_{ij} \lambda_i \lambda'_j \phi_i \,\widetilde{\otimes}\, \phi'_j$$

This equation is called $\widetilde{\otimes}$-*bilinearity*. Also as the case of linearity, this follows diagrammatically just from shuffling numbers and summations around:

$$\sum_i \langle \lambda_i \rangle \phi_i \quad \sum_j \langle \lambda_j' \rangle \phi_j' \quad = \quad \sum_{ij} \langle \lambda_i \rangle \langle \lambda_j' \rangle \phi_i \quad \phi_j'$$

Horizontal composition of linear maps on separable vectors is defined as:

$$(f \mathbin{\tilde{\otimes}} g)(\psi \mathbin{\tilde{\otimes}} \phi) := f(\psi) \mathbin{\tilde{\otimes}} g(\phi)$$

The definition for non-separable vectors then follows from linearity:

$$(f \mathbin{\tilde{\otimes}} g)(\Psi) = (f \mathbin{\tilde{\otimes}} g)\left(\sum_i \psi_i \mathbin{\tilde{\otimes}} \phi_i\right) = \sum_i f(\psi_i) \mathbin{\tilde{\otimes}} g(\phi_i)$$

Diagrammatically this is nothing but:

Finally, it remains to show that $\tilde{H} \mathbin{\tilde{\otimes}} \tilde{K}$ is a Hilbert space, i.e. that it has an inner product. For separable vectors in $\tilde{H} \mathbin{\tilde{\otimes}} \tilde{K}$, we can define it as:

$$\langle \phi \mathbin{\tilde{\otimes}} \phi' | \psi \mathbin{\tilde{\otimes}} \psi' \rangle_{\tilde{H} \mathbin{\tilde{\otimes}} \tilde{K}} := \langle \phi | \psi \rangle_{\tilde{H}} \, \langle \phi' | \psi' \rangle_{\tilde{K}}$$

The definition for non-separable vectors again follows from linearity:

$$\langle \Phi | \Psi \rangle_{\tilde{H} \mathbin{\tilde{\otimes}} \tilde{K}} = \left\langle \sum_i \phi_i \mathbin{\tilde{\otimes}} \phi_i' \,\middle|\, \sum_j \psi_j \mathbin{\tilde{\otimes}} \psi_j' \right\rangle_{\tilde{H} \mathbin{\tilde{\otimes}} \tilde{K}} = \sum_{ij} \langle \phi_i | \psi_j \rangle_{\tilde{H}} \langle \phi_i' | \psi_j' \rangle_{\tilde{K}}$$

As with the rest, this equation falls right out of the picture:

Definition 5.115 The operation $\tilde{\otimes}$ on Hilbert spaces and linear maps that we constructed above is called the *tensor product*.

Remark 5.116 Above we explicitly relied on ONBs of Hilbert spaces in order to form their tensor product. This can be avoided, which then also shows that the tensor doesn't depend on the choices of ONBs. However, this ONB-independent construction is more involved.

Now we have all the ingredients to define a new process theory:

- Systems types are set-theoretic Hilbert spaces $\tilde{H}$.
- Processes are set-theoretic linear maps $\tilde{f}$.
- Horizontal composition is set-theoretic composition.

- Vertical composition is the tensor product $\widetilde{\otimes}$.
- The no-wire type is the one-dimensional Hilbert space $\mathbb{C}$.

Let's call it $\widetilde{\textbf{linear maps}}$.

Theorem 5.117 **linear maps** and $\widetilde{\textbf{linear maps}}$ are equivalent process theories in the sense explained in Section 5.2.5.

5.5 Summary: What to Remember

1. **Linear maps** is a process theory admitting string diagrams such that:

    (1) Each type has a finite *orthonormal basis* (ONB).
    (2) There is at least one system-type of every dimension $D \in \mathbb{N}$.
    (3) Processes of the same type admit sums.
    (4) The numbers are the complex numbers $\mathbb{C}$.

Here, an ONB is a set of states satisfying:

$$\includegraphics{} = \delta_i^j$$

and:

$$\left(\text{for all } i : \quad \frac{f}{i} = \frac{g}{i} \right) \implies f = g$$

Admitting sums means that for any set $\{f_i\}_i$ of processes of the same type there exists a process of that type:

$$\sum_i f_i$$

such that these sums can always be pulled out of diagrams:

$$\left(\sum_i h_i \right) f \;\; g \;\; = \;\; \sum_i \left(h_i \;\; f \;\; g \right)$$

and preserve adjoints:

$$\left(\sum_i \boxed{f_i} \right)^{\dagger} = \sum_i \boxed{f_i}$$

A *Hilbert space* is a type in **linear maps**, and the traditional set-theoretic notion of Hilbert space arises as the set of all the states for a fixed type:

$$\tilde{H} := \left\{ \left. \begin{array}{c} H \\ \boxed{\psi} \end{array} \right| \psi \text{ is a state of type } H \right\}$$

2. In **linear maps** each process admits a *matrix*:

$$\boxed{f} \leftrightarrow \begin{pmatrix} f_1^1 & f_2^1 & \cdots & f_m^1 \\ f_1^2 & f_2^2 & \cdots & f_m^2 \\ \vdots & \vdots & \ddots & \vdots \\ f_1^n & f_2^n & \cdots & f_m^n \end{pmatrix} \quad \text{where} \quad \left\langle f_i^j \right\rangle := \boxed{f}$$

which allows them to be written in *matrix form*:

$$\boxed{f} = \sum_{ij} \left\langle f_i^j \right\rangle$$

The matricial counterparts to sums, sequential composition, parallel composition, transposes, and adjoints of processes are usual matrix operations from linear algebra. The matrix **m** of a diagram:

$$\longleftrightarrow \quad \psi^{A_1 A_2} g_{B_1 A_2}^{B_1 C_1} h_{A_3}^{C_1}$$

is computed via its diagram formula:

$$\mathbf{m}_{i_3}^{i_1} = \sum_{i_2 j_1 k_1} \psi^{i_1 i_2} \mathbf{g}_{j_1 i_2}^{j_1 k_1} \mathbf{h}_{i_3}^{k_1}$$

3. The process theory of **linear maps** is equivalent to **matrices**($\mathbb{C}$):

   (1) The systems are the natural numbers $\mathbb{N}$.
   (2) The processes with input type $m \in \mathbb{N}$ and output type $n \in \mathbb{N}$ are all $n \times m$ matrices with entries in $\mathbb{C}$.
   (3) Diagrams are computed as explained above.

Moreover, for any set of numbers X admitting sums, the matrix calculus **matrices**(X) is a process theory admitting string diagrams. The process theory **relations** (restricted to finite sets) also arises in this manner, as **matrices**$(\mathbb{B})$.

4. We introduced a handful of linear maps that will come in handy later:

- The *Bell basis* is:

$$
B_0 := \frac{1}{\sqrt{2}}\left(\quad_0 \quad_0 + \quad_1 \quad_1 \right)
$$

$$
B_1 := \frac{1}{\sqrt{2}}\left(\quad_0 \quad_1 + \quad_1 \quad_0 \right)
$$

$$
B_2 := \frac{1}{\sqrt{2}}\left(\quad_0 \quad_0 - \quad_1 \quad_1 \right)
$$

$$
B_3 := \frac{1}{\sqrt{2}}\left(\quad_0 \quad_1 - \quad_1 \quad_0 \right)
$$

- The corresponding *Bell maps* are:

$$
B_0 \leftrightarrow \begin{pmatrix} 1 & 0 \\ 0 & 1 \end{pmatrix} \qquad\qquad B_2 \leftrightarrow \begin{pmatrix} 1 & 0 \\ 0 & -1 \end{pmatrix}
$$

$$
B_1 \leftrightarrow \begin{pmatrix} 0 & 1 \\ 1 & 0 \end{pmatrix} \qquad\qquad B_3 \leftrightarrow \begin{pmatrix} 0 & -1 \\ 1 & 0 \end{pmatrix}
$$

- The *copy* maps for the Z-basis (white) and the X-basis (grey) are:

- The *Hadamard* and *CNOT* maps are:

$$
H := \quad_0 + \quad_1 \qquad\qquad \sqrt{2}\ \leftrightarrow \begin{pmatrix} 1 & 0 & 0 & 0 \\ 0 & 1 & 0 & 0 \\ 0 & 0 & 0 & 1 \\ 0 & 0 & 1 & 0 \end{pmatrix}
$$

5. The *spectral theorem* states that self-adjoint linear maps *diagonalise*:

$$
\boxed{f} \;=\; \sum_i r_i \; \overset{i}{\underset{i}{\diamond}}
$$

for some ONB and real numbers r_i. This has many consequences, an important one being that $f^\dagger \circ f$ detects separability of f:

$$
\left(\exists \psi, \phi : \; \boxed{\begin{smallmatrix}f\\f\end{smallmatrix}} \;=\; \begin{smallmatrix}\phi\\\psi\end{smallmatrix} \right) \;\Longleftrightarrow\; \left(\exists \psi', \phi' : \; \boxed{f} \;=\; \begin{smallmatrix}\phi'\\\psi'\end{smallmatrix} \right)
$$

Interpretationally, this fact naturally follows from taking seriously the idea that an adjoint really is a reflection:

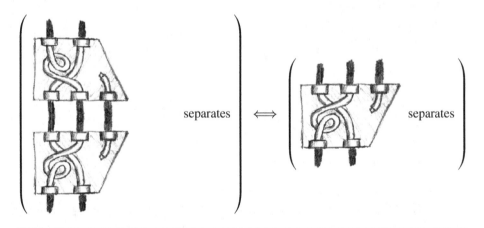

$$
\left(\text{separates} \right) \;\Longleftrightarrow\; \left(\text{separates} \right)
$$

⚠️ The statement of the spectral theorem involving sums is only a temporary one. In Chapter 8, we will picturalise this theorem.

6. String diagrams are *complete* for **linear maps**: the equations that hold between diagrams of linear maps are exactly all string diagram equations.

7. Many diagrammatic concepts have a corresponding representation using a (self-conjugate) ONB. Here is a summary:

Concept	Diagram	In terms of basis
process		$\sum_{ij} \left\langle f_i^j \right\rangle$ with $\left\langle f_i^j \right\rangle =$
state		$\sum_{i} \left\langle \psi^i \right\rangle$ with $\left\langle \psi^i \right\rangle =$
identity		$\sum_{i}$
cups		$\sum_{i}$
caps		$\sum_{i}$
dimension		$\sum_{i} \boxed{} = \dim(H)$
trace		$\sum_{i}$
transpose		$\sum_{ij} \left\langle f_j^i \right\rangle$ with $\left\langle f_i^j \right\rangle$ as above
conjugate		$\sum_{ij} \left\langle \overline{f_i^j} \right\rangle$ with $\left\langle f_i^j \right\rangle$ as above
adjoint		$\sum_{ij} \left\langle \overline{f_j^i} \right\rangle$ with $\left\langle f_i^j \right\rangle$ as above

5.6 Advanced Material*

In this section, we look at the definition of Hilbert space beyond finite dimensions and highlight some interesting examples, as well as some of the difficulties they cause. After that, we continue our venture into category theory a bit more, by looking at how category theoreticians represent sums and ONBs. We also explain how knot theoreticians use a similar blend of diagrammatic reasoning and sums in their work. Last but not least, we explain how the pioneers of category theory effectively formulated what it means for two process theories to be equivalent.

5.6.1 Beyond Finite Dimensions*

Recall that the inner product gives us a norm: $\|v\| := \sqrt{\langle v|v \rangle}$. This plays a much bigger role in the infinite-dimensional case, because it allows one to talk about the convergence of sequences, which is the missing piece of the definition of a Hilbert space, which we can safely ignore in the finite-dimensional case. The full definition of a Hilbert space goes like this.

Definition 5.118 A (possibly infinite-dimensional) *Hilbert space* is a complex inner product space that is additionally *Cauchy complete*. That is, for any (Cauchy) convergent sequence $(v_i)_i$ of vectors in H, there exists $v \in H$ such that $\|v_i - v\| \to 0$.

This means we can take limits of sequences, so we can hit Hilbert spaces (and hence quantum mechanics) with a whole big bag of tools from *functional analysis*. For one thing, having limits around allows one to define infinite sums:

$$\sum_{i=0}^{\infty} \psi_i := \lim_{n \to \infty} \sum_{i=0}^{n} \psi_i$$

There are two examples of Hilbert spaces that have historically played a major role in quantum mechanics as (equivalent) presentations of the quantum mechanical state space.

Example 5.119 L^2 is the set of all functions $\psi : \mathbb{R}^n \to \mathbb{C}$ whose 'squared-integrals' are finite:

$$\int \overline{\psi(x)} \psi(x) dx < \infty$$

This forms a Hilbert space by letting:

$$(\lambda_1 \psi + \lambda_2 \phi)(x) := \lambda_1 \psi(x) + \lambda_2 \phi(x) \quad \text{and} \quad \langle \psi | \phi \rangle := \int \overline{\psi(x)} \phi(x) dx$$

Example 5.120 ℓ^2 is the set of all (countably infinite) sequences of complex numbers, whose 'squared sums' are finite:

$$\sum_{i=0}^{\infty} \overline{a_i} a_i < \infty$$

This forms a Hilbert space by letting:

$$\lambda_1 (a_i)_i + \lambda_2 (b_i)_i := (\lambda_1 a_i + \lambda_2 b_i)_i \qquad \text{and} \qquad \langle (a_i)_i | (b_i)_i \rangle := \sum_{i=0}^{\infty} \overline{a_i} b_i$$

Both L^2 and ℓ^2 allow one to express the state of a quantum particle and compute its position or momentum. What is surprising is not only that these are both the same kind of mathematical object (an infinite-dimensional Hilbert space), but that they are in fact *isomorphic* Hilbert spaces (see Section 5.7 for more details).

So, this brings us back to **Q3** from Section 2.3:

Why don't infinite dimensional Hilbert spaces play a role in this book?

Or in the majority of quantum computing, for that matter? The short answer is that many things just don't seem to work as well. Suppose, for example, that we take the caps and cups that we know (and love) by now, and try to run them in infinite dimensions. Naïvely, things look good at first:

(assuming for the moment that the 'infinite sums' above are well-defined). However, if we introduce a circle, things really go wrong:

(5.60)

This problem comes from the fact that caps and cups are not *bounded* linear maps. That is, they do not satisfy the property that $\|M(v)\| < \infty$ whenever $\|v\| < \infty$. Thus, infinite-dimensional Hilbert spaces and bounded maps form a process theory, but since we have no caps and cups, only circuit diagrams are well-defined.

One might argue that this is a problem with string diagrams. On the other hand, situation (5.60) occurs not just for caps and cups, but for any perfect correlations. Thus:

The Hilbert space formalism explicitly rules out perfect correlations between infinite-dimensional systems.

It's far from clear whether this is a 'property of nature' or simply an artifact of the mathematics of Hilbert spaces. This suggests two interesting questions:

1. How much of the process-theoretic development of quantum theory can be extended to infinite dimensions (i.e. without recourse to string diagrams)?
2. Can the theory of Hilbert spaces and bounded linear maps be modified to accommodate string diagrams and/or perfect correlations in infinite dimensions?

Some progress has been made towards both of these questions (see Section 5.7), but there is still much to do.

5.6.2 Categories with Sums and Bases*

A large subdiscipline of category theory (sometimes called Abelian category theory) studies categories whose morphisms can be added. A category that has some extra structure on its sets of morphisms is called an *enriched category*. The extra structure we are interested in is that of a *commutative monoid*, i.e. a set M with an associative, commutative, and unital sum. For example, by Definition 5.21, **Condition 1**, processes of the same type in a process theory with sums form such a commutative monoid. **Condition 2** moreover guarantees that the categories corresponding to process theories with sums also have the following structure.

Definition 5.121 A category C is said to be *enriched in commutative monoids* if for every pair of objects A and B, the set $C(A, B)$ forms a commutative monoid and the monoid structure is compatible with $\circ$-composition:

$$\begin{cases} 0 \circ f = 0 = g \circ 0 \\ h \circ (f + g) = (h \circ f) + (h \circ g) \\ (g + h) \circ f = (g \circ f) + (h \circ f) \end{cases}$$

A related notion to the *sums* of morphisms is the *direct sum* of objects (e.g. vector spaces). In categorical terms, this is called a *biproduct*.

Definition 5.122 Let C be a category that is enriched in commutative monoids. Then, C has *biproducts* if for every pair of objects A_1, A_2, there exists a third object $A_1 \oplus A_2$, along with a pair of maps:

$$\iota_j : A_j \to A_1 \oplus A_2$$

called *injections*, and a pair of maps:

$$\pi_j : A_1 \oplus A_2 \to A_j$$

called *projections*, satisfying:

$$\iota_1 \circ \pi_1 + \iota_2 \circ \pi_2 = 1_{A_1 \oplus A_2} \qquad \pi_k \circ \iota_j = \begin{cases} 1_{A_j} & \text{if } j = k \\ 0 & \text{otherwise} \end{cases}$$

Look familiar yet? If not, let's get the dagger involved.

Definition 5.123 Let C be a dagger compact closed category that is enriched in commutative monoids. Then, C has *dagger biproducts* if for every pair of objects A_1, A_2, there exists a third object:

$$A_1 \oplus A_2$$

along with a pair of maps:

$$\iota_j : A_j \to A_1 \oplus A_2$$

satisfying:

$$\iota_1 \circ \iota_1^\dagger + \iota_2 \circ \iota_2^\dagger = 1_{A_1 \oplus A_2} \qquad \iota_k^\dagger \circ \iota_j = \begin{cases} 1_{A_j} & \text{if } j = k \\ 0 & \text{otherwise} \end{cases}$$

If $A_1 = A_2 := I$, the trivial object, these conditions look like this:

That's right:

forms a two-dimensional ONB! Using the equations in Definition 5.123, it is possible to show that $\oplus$ is associative (up to isomorphism), so we can define biproducts of three objects:

$$A \oplus B \oplus C := (A \oplus B) \oplus C \cong A \oplus (B \oplus C)$$

or, similarly, N objects. Objects that are N-fold biproducts of I:

$$A \cong I \oplus I \oplus \cdots \oplus I \tag{5.61}$$

then have N-dimensional ONBs. To see if an object admits something like an ONB, we can first decompose it into its *irreducible* components.

Definition 5.124 An object is called *irreducible* if it cannot be written as a biproduct of two non-zero objects.

The reason **relations** and **linear maps** have ONBs for every type is that the only irreducible object is the trivial system. In other words, we can decompose any object as in (5.61). However, there are many interesting algebraic categories (rings, modules, (representations of) algebras, etc.) that have a much richer collection of irreducibles. Interestingly, we still have a useful matrix calculus, but now instead of matrices of numbers we get matrices of processes. Suppose we look at some complicated map like this:

$$g : A_1 \oplus \cdots \oplus A_m \to B_1 \oplus \cdots \oplus B_n$$

Then g can be decomposed into a 'matrix form' that generalises the matrix form given in equation (5.14):

$$
\boxed{g} \; = \; \sum_{jk} \; \boxed{\iota_k \, / \, \boxed{g_j^k} \, / \, \iota_j \, \backslash}
\qquad \leftrightarrow \qquad
\begin{pmatrix}
g_1^1 & g_2^1 & \cdots & g_m^1 \\
g_1^2 & g_2^2 & \cdots & g_m^2 \\
\vdots & \vdots & \ddots & \vdots \\
g_1^n & g_2^n & \cdots & g_m^n
\end{pmatrix}
$$

Rather than numbers, the components in the matrices are all morphisms

$$
g_j^k : A_j \to B_k
$$

If we start to compose matrices, everything looks just like normal matrix composition, where sums are sums and 'multiplication' is composition of morphisms. So, we can pretend we are doing normal matrix calculus, but actually, we're doing something much more general (and powerful!).

5.6.3 Sums in Knot Theory*

In this book wires are like electrical wires, in the sense that only connections matter. In contrast, if they were ropes, then one may care if they are knotted up or braided together. In particular, one may wish to distinguish these two diagrams:

$$
\times \quad \neq \quad \times
$$

There is an area of mathematics called knot theory that is all about ropes either being knotted/braided or not. In fact, figuring out if a rope is knotted/braided at all turns out to be extremely difficult.

The study of knots and braids not only is of interest for its own sake but is also the foundation of a particular model of quantum computation called *topological quantum computation*, where the structure of braids is used to encode quantum gates. In the historical material at the end of this section we provide some pointers to the existing literature.

In the introduction we already gave an example of an equation that one encounters in any modern knot theory textbook:

$$
\times \; = \; \lambda \; \big| \; \big| \; + \; \lambda^{-1} \; \smile\!\!\frown \tag{5.62}
$$

We will now explain its use, as an example of a hybrid calculus that uses both diagrams and traditional mathematical notions. The *Kauffman bracket* (5.62) allows one to associate a polynomial to any given knot.

Consider the *trefoil knot*:

Applying (5.62) to the lowermost crossing we obtain:

$$\lambda \quad + \quad \lambda^{-1}$$

Deforming the diagrams this becomes:

$$\lambda \quad + \quad \lambda^{-1}$$

Then, a second application of (5.62) yields:

$$\lambda \quad + \quad + \quad \lambda^{-2} \quad = \left(\lambda + 1 + \lambda^{-2}\right)$$

Besides (5.62) a second equation is:

$$= 1$$

so we are left with a polynomial in λ:

$$\lambda + 1 + \lambda^{-2}$$

This polynomial is called the *bracket polynomial*, and a suitably normalised version of it is called the *Jones polynomial*, which is a *knot invariant*. That is, it is the same for all knots that can be deformed into each other. This makes it it an extremely valuable tool for the classification of knots.

5.6.4 Equivalence of Symmetric Monoidal Categories*

In Section 5.2.5, we gave an informal definition of equivalence for process theories. In this section, we will make this definition formal, using the language of category theory.

In order to talk about equivalence of categories, we first must say what it means to interpret the morphisms (i.e. processes) in one symmetric monoidal category as morphisms in another. We do this using *functors*, which are the standard kind of 'map' one considers between categories.

Definition 5.125 A (strict) *symmetric monoidal functor F* from SMC $\mathcal{C}$ to SMC $\mathcal{D}$ consists of a function mapping objects of $\mathcal{C}$ to objects of $\mathcal{D}$:

$$F : \mathrm{ob}(\mathcal{C}) \to \mathrm{ob}(\mathcal{D})$$

and for every pair of objects A, B, a function mapping morphisms of $\mathcal{C}$ to morphisms of $\mathcal{D}$ (also written as 'F'):

$$F : \mathcal{C}(A, B) \to \mathcal{D}(F(A), F(B))$$

such that 'F preserves diagrams', i.e.:

1. F preserves parallel composition and unit for objects:

$$F(A \otimes B) = F(A) \otimes F(B) \qquad\qquad F(I) = I$$

2. F preserves parallel and sequential composition for morphisms:

$$F(f \otimes g) = F(f) \otimes F(g) \qquad\qquad F(g \circ f) = F(g) \circ F(f)$$

3. F preserves identity morphisms and swaps:

$$F(1_A) = 1_{F(A)} \qquad\qquad F(\sigma_{A,B}) = \sigma_{F(A),F(B)}$$

As in the definition of monoidal category, the adjective 'strict' means equations involving parallel composition of objects are 'real equations'; i.e. they hold on the nose, rather than only up to isomorphism. If we dropped it, condition 1 above would become:

$$F(A \otimes B) \cong F(A) \otimes F(B) \qquad\qquad F(I) \cong I$$

and we would need to require that these isomorphisms obey certain coherence equations like those for a non-strict SMC.

The simplest functor is the identity functor:

$$1_{\mathcal{C}} : \mathcal{C} \to \mathcal{C}$$

which just sends every object and morphism to itself. We can define sequential composition of functors in the obvious way:

$$(G \circ F)(X) = G(F(X)) \qquad\qquad (G \circ F)(f) = G(F(f))$$

in which case it clearly follows that:

$$F \circ 1_{\mathcal{C}} = F = 1_{\mathcal{D}} \circ F$$

The stricter sense in which two categories can be 'the same' is isomorphism, which is defined just like isomorphism between other kinds of objects.

Definition 5.126 Two symmetric monoidal categories are *isomorphic* if there exist symmetric monoidal functors:

$$F : \mathcal{C} \to \mathcal{D} \qquad\qquad G : \mathcal{D} \to \mathcal{C}$$

such that:

$$G \circ F = 1_{\mathcal{C}} \qquad\qquad F \circ G = 1_{\mathcal{D}}$$

In other words, if we take a round trip through F and G, we get back to <u>exactly</u> where we started. However, in many cases, such as the equivalence between **linear maps** and **matrices**($\mathbb{C}$), we don't get back where we started. However, we end up somewhere that 'looks exactly the same' as where we started. In other words, for every object X, the object $G(F(X))$ is isomorphic to X. Let's give this isomorphism a name:

$$\eta_X : X \to G(F(X))$$

Of course, we don't get just a single such isomorphism but a whole family of them $\eta_X, \eta_Y, \ldots$ for every object in $\mathcal{C}$.

That tells us what it means for objects to look exactly the same after a round trip, but what does it mean for morphisms to look the same? In other words, what does it mean for:

$$f : X \to Y \qquad \text{and} \qquad G(F(f)) : G(F(X)) \to G(F(Y))$$

to do 'the same thing'? It means if we 'encode' X as $G(F(X))$ via the isomorphism η_X, then do $G(F(f))$, then 'decode' the result, this should be the same thing as just doing f:

$$\eta_Y^{-1} \circ G(F(f)) \circ \eta_X = f \tag{5.63}$$

This means that η is not just any old family of isomorphisms, but is what's called a *natural isomorphism*. By moving η_Y to the right, we have:

$$G(F(f)) \circ \eta_X = \eta_Y \circ f \tag{5.64}$$

a condition that is usually expressed using a *commutative diagram*:

$$
\begin{array}{ccc}
X & \xrightarrow{\;\;f\;\;} & Y \\
{\scriptstyle \eta_X}\big\downarrow & & \big\downarrow{\scriptstyle \eta_Y} \\
G(F(X)) & \xrightarrow{G(F(f))} & G(F(Y))
\end{array}
\tag{5.65}
$$

A commutative diagram is a visual way to express equations between morphisms. It asserts that any two paths with the same start and end point yield the same morphism by sequential composition. So, the diagram above says exactly the same thing as equation (5.64).

Remark 5.127 The definition of natural isomorphism actually makes sense for any pair of functors:

$$D, E : C \to D$$

in which case it becomes a family of isomorphisms:

$$\{\iota_X : D(X) \to E(X)\}_{X \in C}$$

satisfying:

$$
\begin{array}{ccc}
D(X) & \xrightarrow{\;D(f)\;} & D(Y) \\
{\scriptstyle \iota_X}\big\downarrow & & \big\downarrow{\scriptstyle \iota_Y} \\
E(X) & \xrightarrow{\;E(f)\;} & E(Y)
\end{array}
$$

Then, (5.65) is the special case where $D := 1_C$ and $E := G \circ F$. We can also drop the requirement that each ι_X be an isomorphism, in which case such a family is called a *natural transformation*.

Example 5.128 Let $C \times C$ be the category whose objects are pairs (A, B) of objects in C and whose morphisms are pairs (f, g) of morphisms in C. Parallel composition in an SMC then induces two functors from the *product category* $C \times C$ to the category C:

$$P_1 :: \begin{cases} (A, B) \mapsto A \otimes B \\ (f, g) \mapsto f \otimes g \end{cases} \qquad\qquad P_2 :: \begin{cases} (A, B) \mapsto B \otimes A \\ (g, f) \mapsto g \otimes f \end{cases}$$

The swap morphisms in C then give a natural isomorphism from P_1 to P_2. Naturality in this case gives the following commutative diagram:

$$
\begin{array}{ccc}
A \otimes B & \xrightarrow{\;f \otimes g\;} & A' \otimes B' \\
{\scriptstyle \sigma_{A,B}}\big\downarrow & & \big\downarrow{\scriptstyle \sigma_{A',B'}} \\
B \otimes A & \xrightarrow{\;g \otimes f\;} & B' \otimes A'
\end{array}
$$

which, as an equation between diagrams, should look very familiar:

Similarly, the associativity and unit isomorphisms for non-strict SMCs mentioned in Section* 3.6.2, which respectively are:

$$\alpha_{A,B,C} : (A \otimes B) \otimes C \to A \otimes (B \otimes C)$$

$$\lambda_A : I \otimes A \to A \qquad\qquad \rho_A : A \otimes I \to A$$

also give natural isomorphisms for suitably defined functors.

The final ingredient we need for capturing 'sameness' of SMCs is to define a natural isomorphism that respects parallel composition of objects:

$$\eta_{X \otimes Y} = \eta_X \otimes \eta_Y$$

which is called a *monoidal natural isomorphism*. Then, an equivalence of SMCs is a pair of functors where each round trip ($G \circ F$ and $F \circ G$) yields something (monoidally, naturally) isomorphic to what we started with.

Definition 5.129 Two symmetric monoidal categories are *equivalent* if there exist symmetric monoidal functors $F : C \to D$ and $G : D \to C$ such that there exist monoidal natural isomorphisms:

$$\{\eta_X : X \to G(F(X))\}_{X \in \mathrm{ob}C} \qquad \text{and} \qquad \{\epsilon_Y : Y \to F(G(Y))\}_{Y \in \mathrm{ob}D}$$

So, we should be able to see the relationship between **linear maps** and **matrices**($\mathbb{C}$) in these terms. As SMCs, these are usually called FHilb (for finite-dimensional Hilbert spaces) and Mat($\mathbb{C}$), respectively. For Hilbert spaces $A, B, \ldots$, fix bases:

$$\left\{\left. \overset{\mid}{\underset{\phi_i}{\bigtriangledown}} \right\}_i \right. , \quad \left\{\left. \overset{\mid}{\underset{\phi_j}{\blacktriangledown}} \right\}_j \right. , \quad \cdots$$

Then, let:

$$F : \mathrm{FHilb} \to \mathrm{Mat}(\mathbb{C})$$

be the functor that sends each Hilbert space to its dimension and each linear map to the matrix with elements:

$$f_i^j = \begin{array}{c} \blacktriangle \\ \phi_j \\ \mid \\ \boxed{f} \\ \mid \\ \bigtriangledown \\ \phi_i \end{array}$$

In the other direction, let:

$$G : \mathrm{Mat}(\mathbb{C}) \to \mathrm{FHilb}$$

be the functor that sends each natural number $d \in \mathrm{ob}(\mathrm{Mat}(\mathbb{C}))$ to the Hilbert space $\mathbb{C}^d$ and each matrix with entries f_i^j to the linear map:

So, we have:

$$A \;\overset{F}{\mapsto}\; d = \dim(A) \;\overset{G}{\mapsto}\; \mathbb{C}^d$$

Now, $A \neq \mathbb{C}^d$, but since these two Hilbert spaces have the same dimension, there is a unitary:

which is, of course, also an isomorphism. If we now look at the round trip of a linear map, we have:

which is not exactly the same linear map. But it can be 'decoded' as we did in equation (5.63) using the natural isomorphisms:

The other round trip is even easier, because it just sends a matrix to itself:

$$\mathbf{f} \;\overset{G}{\mapsto}\; \sum_{ij} \mathbf{f}^{j}_{i} \;\; \overset{F}{\mapsto}\; \mathbf{f}$$

So the second natural isomorphism is just given by identity morphisms:

$$\boxed{\;\epsilon_d\;} \;:=\; \Big|$$

for all $d \in \mathrm{ob}(\mathrm{Mat}(\mathbb{C}))$.

5.7 Historical Notes and References

The name 'Hilbert space' was coined by John von Neumann in his mathematical formulation of quantum theory, ultimately culminating in his famous book on the subject (von Neumann, 1932). Hilbert spaces united two seemingly different mathematical formalisations of quantum mechanics: the wave-function picture due to Schrödinger (1926) and matrix mechanics originating with Heisenberg (1925) and formalised by Born and Jordan (1925). In mathematics, many examples of Hilbert spaces had already been studied extensively, including by David Hilbert when studying integral equations, but it was von Neumann who was the first one to provide an axiomatic treatment, so they may as well have been coined 'von Neumann spaces'. A detailed account on the history of Hilbert spaces can be found in Bourbaki (1981, 1987).

The name 'matrix' was coined by Sylvester in 1848. The origins of the complex numbers trace back way over 1000 years, to Arabic algebra. Euler introduced the notation i and Hamilton was the first to treat complex numbers as a pair of real numbers, which constituted the first algebraic definition of the complex numbers. Gauss coined the name 'complex number'.

As mentioned in Remark 5.99, the Bell maps are typically introduced as the (marginally different) Pauli matrices, which first occurred in the Pauli equation, a variation on Schrödinger's equation that takes into account the interaction of the spin of a particle with an external electromagnetic field.

Unfortunately, in the literature, the terminology concerning Bell states and the Bell basis is a bit all over the place, mainly for historical reasons. First, the 'Bell state' typically refers to the first element of the Bell basis (i.e. the cup), whereas the 'Bell states' refer to the whole basis. Sometimes the Bell state is also called the 'EPR-state', but more commonly, *EPR-state* is understood to be B_3 from (5.50). The reason for singling out this state is that it is the only *antisymmetric* state of the Bell states, and in fact, the only antisymmetric state for a pair of two-dimensional quantum systems. Here, *antisymmetry* means that if we swap the two systems, we get the same state up to (-1):

Systems that obey this antisymmetry are called *fermions*. The three other Bell states are *bosons*, which means that if we swap the two systems, we get the same state on the nose.

The use of sums and matrix structure in addition to string diagrams for modelling quantum systems was initiated by Abramsky and Coecke (2004), in the form of categorical biproducts, the category-theoretic counterpart to having a matrix calculus, as explained in Section* 5.6.2. On a personal note, the second author of that paper never really liked biproducts. If one were to stick with biproducts, this chapter would have been where this book ends. Aiming beyond biproducts has produced the remaining 66.6%.

Biproducts arose from the study of Abelian categories, or categories that resemble the category of Abelian groups in several important ways. These first occurred in Mac Lane (1950), following discussions with Eilenberg about the appropriate categories for treating homology and cohomology of topological spaces. Following further developments by Buchsbaum (1955) and Grothendieck (1957), such categories grew to play a role in mainstream algebraic topology and geometry. A standard text is Freyd (1964).

Category theory was invented by Eilenberg and Mac Lane (1945) precisely to say what it means for process theories to be equivalent. However, the concept that a morphism should be thought of as a process didn't become widespread until much later (see Baez and Lauda, 2011). The process theories Eilenberg and Mac Lane originally had in mind only had sequential composition, not parallel composition, so their notion of equivalence only had to preserve ∘. However, with the advent of parallel composition, i.e. monoidal categories (Benabou, 1963; Mac Lane, 1963), came the notion of monoidal equivalence of categories introduced in Section* 5.6.4.

Selinger (2011a) proved completeness of string diagrams for **linear maps**, building further on a similar result for vector spaces (i.e. without adjoints) due to Hasegawa, Hofmann, and Plotkin (2008).

Discussions of quantum protocols in terms of non-self-dual cups and caps, including the upshot of doing so, are in Coecke et al. (2008a), Paquette (2008), and Kissinger (2012a). Much earlier they had been extensively studied in mathematics, for example in the area of quantum groups (Kassel, 1995). The Jones polynomial, which first appeared in Jones (1985), earned Vaughan Jones the Fields Medal, sometimes also called the Nobel Prize for mathematics. A few years later Louis Kauffman introduced his bracket as an easy manner to produce the Jones polynomial, in Kauffman (1987). Kauffman (1991) surveys the interplay between knots and physics. Panangaden and Paquette (2011) survey topological quantum computing, and the interplay between crossings and the kind of diagrams this book is all about is surveyed in Fauser (2013).

6

Quantum Processes

The art of progress is to preserve order amid change, and to preserve change amid order.

– Alfred North Whitehead, Process and Reality, 1929

From now on, no more babies, plug-strips, or cooking: we now focus exclusively on quantum processes. Naturally, our first goal is to construct the theory of **quantum processes**. Fortunately, the work that we have done so far brings us fairly close to that goal.

For one thing, the process theory of **linear maps** is not too far away from full-blown **quantum processes**. In particular, the theory of **quantum processes** will inherit its string diagram description from **linear maps**. This provides us with a high-level language that will make reasoning about quantum processes very easy, for example in the context of quantum computation and quantum protocol design. This also means that we already know several features of quantum processes from Section 4.4, namely those that happen to hold in all process theories admitting string diagrams: the existence of non-separable states, the no-cloning theorem, teleportation, and so on.

Note that we say **linear maps** are <u>not too far away</u> from **quantum processes**, not that they <u>are</u> **quantum processes**. In this chapter, we will proceed from **linear maps** to **quantum processes** in a few steps (Fig. 6.1), which correct the shortcomings of the former. A first issue is that, as a model of quantum processes, **linear maps** contain some redundant data, namely 'global phases'. These will never be detectable by quantum measurements (which we'll meet in the next chapter) and hence have no discernable effect on which process actually happened.

But so what? Who cares? We could just carry on using linear maps, and ignore global phases whenever necessary. In fact, many textbooks on quantum theory would do just that. On the other hand, the generalised Born rule:

$$\left. \begin{array}{l} \text{test} \left\{ \begin{array}{c} \phi \end{array} \right. \\ \text{state} \left\{ \begin{array}{c} \psi \end{array} \right. \end{array} \right\} \text{probability} \tag{6.1}$$

wouldn't work, because the numbers generated are not positive real numbers (which we can interpret as probabilities), but complex numbers.

We address this issue by performing a simple construction on the process theory of **linear maps** to produce another theory called **quantum maps**, which truly describes the processes that may take place in quantum theory.

First, in Section 6.1 we turn **linear maps** into **pure quantum maps** by *doubling* every process. This magically solves the problem of global phases and probabilities in one fell swoop.

After this, we show in Section 6.2.1 that pure quantum maps fail to include one very important process: the process of *discarding* a system. Often we wish to ignore some part of a larger system when it is out of our control (e.g. a potential eavesdropper in a security protocol) or simply irrelevant (e.g. some electrons flying around on Mars). By adding this discarding process to pure quantum maps, we obtain **quantum maps**. Many new *impure* (or *mixed*) quantum maps arise by composing pure ones with the discarding process.

An alternative interpretation of the impurity of certain quantum maps is *probabilistic mixing*. In classical physics, probabilities can be seen as a way of accounting for our lack of knowledge about the state of a system. For example a branch of physics called *statistical mechanics* describes the states of a system using probability distributions, since it's nearly impossible to know what each little particle is up to. This is also what probabilistic mixing means in quantum theory: having a lack of knowledge about which process is actually happening.

Note that we said quantum maps describe processes that '<u>may</u> take place'. Unlike classical physics, quantum theory has processes that are <u>irreducibly</u> non-deterministic. That is, there exist non-deterministic processes that cannot be accounted for solely by a lack of knowledge about the quantum system. Regardless of how perfectly we know the state of a system, such processes will not have a fixed outcome until after they occur. This is the feature of quantum theory that Einstein found deeply upsetting, as he famously said, 'God does not play dice.' To account for this 'quantum dice-throwing', the third and final step in this chapter is to define **quantum processes** as collections of **quantum maps** that together make up the alternatives of what may happen. The rule that tells us which **quantum maps** together make up valid **quantum processes** is called the *causality postulate*.

These irreducibly non-deterministic processes are absolutely essential to quantum theory, for at least two reasons:

1. *Quantum measurement*, which is our only means of interacting with quantum systems, is a non-deterministic quantum process. (Recall Dave's non-deterministic travels narrated in the introductory chapter.)
2. The causality postulate places a strong restriction on which quantum maps can occur as deterministic quantum processes. For example, the only deterministic quantum effect is discarding. On the other hand, every quantum map can be realised as part of a non-deterministic quantum process. This fact is crucial to realising quantum teleportation, among many other things.

The causality postulate is tightly connected to the concept of causality in physics. In particular, we will see in Section 6.3.2 that it forbids 'faster-than-light signalling', hence

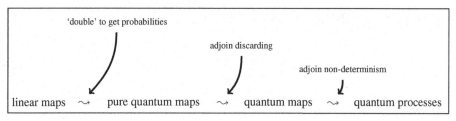

Figure 6.1 The passage from **linear maps** to **quantum processes**.

guaranteeing that quantum theory is not in conflict with that other funky physical theory, *relativity theory*. We don't want to make Einstein even more upset than he already is!

6.1 Pure Quantum Maps from Doubling

In the previous chapter we put a lot work into defining and studying **linear maps**, so one may think that quantum theory is all about these. This is almost true. Recall from Section 3.4.1 that effects can be interpreted as 'tests', and from Section 4.3.1 that the adjoint of a state ψ corresponds to testing whether a system is in the state ψ. When such a test is composed with a state, the number produced should be the probability of that test returning 'yes'.

However, in the case of **linear maps**, if we pick any old states ψ and ϕ, it's quite likely that their inner product won't even be a real number, much less a probability (i.e. a real number between zero and one). Taking this into account, the process theory of **linear maps** is not an appropriate candidate for describing quantum processes. However, we can turn it into one by means of what we call *doubling*, and call the resulting process theory **pure quantum maps**.

Moreover, doubling has two other nice consequences:

1. It automatically eliminates redundant *global phases* (Section 6.1.2).
2. It makes space for two new ingredients that didn't exist in **linear maps**: the *discarding map* (Section 6.2.1.1) and *classical wires* (Chapter 8).

6.1.1 Doubling Generates Probabilities

Recall from Proposition 5.70 that if we multiply a number by its conjugate we automatically get a positive number, so:

$$0 \leq \begin{array}{c}\phi \quad \phi \\ \psi \quad \psi\end{array} \tag{6.2}$$

Furthermore, if ψ and ϕ are both normalised, this will be a real number between 0 and 1, i.e. a probability. Rather than proving this fact directly, we prove a more general fact that will be useful later.

Lemma 6.1 For any ONB and any normalised state ψ :

$$\sum_i \;\; = 1 \qquad\qquad (6.3)$$

Proof We have:

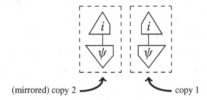

where the last step uses the fact that numbers are self-transposed. □

This theorem says that any normalised state along with any ONB, considering 'doubled inner products':

(mirrored) copy 2   copy 1

yields a probability distribution (cf. Definition 5.35), that is, a list of positive real numbers that sums up to one.

Remark 6.2 In the next chapter we will see that ONBs represent certain quantum measurements, and then Lemma 6.1 will guarantee that the probabilities for all of the possible outcomes add up to 1.

We showed in Proposition 5.79 that any orthonormal set extends to an ONB. In particular, a single normalised state ϕ extends to an ONB:

$$\left\{ \boxed{1} := \phi, \boxed{2}, \ldots, \boxed{n} \right\}$$

For this ONB, the only way (6.3) can hold is if:

$$\phi\,\phi \;\; \leq 1$$

so, we get the following.

Corollary 6.3 For normalised states and effects we have:

$$
0 \leq \quad \vcenter{\hbox{(doubled inner-product diagram)}} \quad \leq 1 \tag{6.4}
$$

These doubled inner-product diagrams constitute the main mechanism for computing probabilities in quantum theory, and they are what is called the 'Born rule' in standard textbooks. At first, this new thing doesn't look like the generalised Born rule we met back in Section 3.4. However, it will soon!

Remark 6.4 One typically encounters the second and/or the third of the following three equivalent forms of the Born rule:

$$
\vcenter{\hbox{(diagram)}} \;=\; \vcenter{\hbox{(diagram)}} \;=\; \vcenter{\hbox{(diagram)}}
$$

The first equation follows from the fact that numbers are self-transposed, and the second is just a diagram deformation. In more traditional notation, these alternative forms become, respectively:

$$
\langle\psi|\phi\rangle\langle\phi|\psi\rangle = |\langle\psi|\phi\rangle|^2 \qquad\qquad \mathrm{tr}\big(P_{|\phi\rangle}\,\rho_{|\psi\rangle}\big)
$$

where

$$
P_{|\psi\rangle} = \rho_{|\psi\rangle} := |\psi\rangle\langle\psi|
$$

The upshot of our expression of the Born rule is what much of the remainder of this section is about: by transforming states and effects into a 'doubled' form, the Born rule reduces to the simplest way we can produce a number from a state and an effect, namely composing them.

A key point is that the probabilities in the doubled inner product (6.4) do not depend precisely on ψ but rather on 'ψ-doubled':

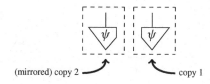

(mirrored) copy 2   copy 1

So in order to realise them as an instance of the generalised Born rule (6.1), we can simply treat ψ-doubled as a first-class citizen. That is, we treat it as a state:

$$\widehat{\psi} := \psi \,\, \psi$$

in a new 'doubled-process theory'. In other words, for any state ψ of a Hilbert space A we define a new state $\widehat{\psi}$ that is the doubled version of ψ, and to $\widehat{\psi}$ we attribute a new type $\widehat{A}$, which is secretly just two copies of A. Diagrammatically:

$$| := ||$$

Similarly, we define new effects for this new type $\widehat{A}$:

$$\widehat{\phi} := \phi \,\, \phi$$

Together the new state and effect yield:

$$\left. \begin{array}{l} \text{test} \left\{ \widehat{\phi} \right. \\ \\ \text{state} \left\{ \widehat{\psi} \right. \end{array} := \begin{array}{c} \phi \,\, \phi \\ \psi \,\, \psi \end{array} \right\} \text{probability}$$

Bingo! We now see the Born rule (6.4)) from quantum theory arising as a special case of the generalised Born rule (6.1).

We call these doubled states and effects *pure quantum states* and *pure quantum effects*, respectively. We sometimes will use the following notation:

$$\mathrm{double}\left(\psi \right) := \widehat{\psi} = \psi \,\, \psi$$

$$\mathrm{double}\left(\phi \right) := \widehat{\phi} = \phi \,\, \phi$$

Remark 6.5 This 'doubling trick' is closely related to a construction that is familiar in quantum theory, namely, the passage from a pure state vector $|\psi\rangle$ to its associated *density operator*:

$$\widetilde{\psi} := |\psi\rangle\langle\psi|$$

This has the same data as a doubled state, which can be seen just by transposing the effect ψ into the conjugate of the state ψ:

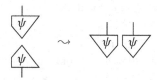

If ψ is normalised, then by (4.53) the density operator $\widetilde{\psi}$ is a projector. These projectors play the role of quantum states in the traditional literature. However, the process-theoretic paradigm takes states to be processes with no inputs. Clearly, density operators break this convention, which becomes a bit of a pain later on (cf. Remark 6.50). On the other hand, our doubled states (conveniently) retain it.

6.1.2 Doubling Eliminates Global Phases

As we already mentioned, the probabilities produced by the Born rule do not depend precisely on ψ, but rather on 'ψ-doubled'. This distinction might seem trivial, until one realises that the correspondence between states and doubled states is not one-to-one, but rather many-to-one. This phenomenon already occurs for numbers. There are many complex numbers λ such that:

$$\overline{\lambda}\lambda = 1 \tag{6.5}$$

for example, $1, -1, i, -i$.

In Section 5.3.1, we saw that it is possible to write any complex number as:

$$re^{i\alpha}$$

Equation (6.5) just means $r = 1$, so any such λ is a number of the form:

$$e^{i\alpha}$$

By definition, these numbers vanish when they are multiplied by their conjugate, so they have no effect on a doubled state. In fact, this is the only data that gets lost in the doubled state.

Proposition 6.6 Two states ψ and ϕ become the same state when doubled if and only if they are equal up to some number $e^{i\alpha}$, i.e.:

$$\Yleft{\psi}\,\Yleft{\psi} = \Yleft{\phi}\,\Yleft{\phi} \qquad \Longleftrightarrow \qquad \Yleft{\psi} = e^{i\alpha}\,\Yleft{\phi}$$

for $\alpha \in [0, 2\pi)$. The number $e^{i\alpha}$ is called a *global phase*.

The proof is an instance of a more general one, given below as the proof of Theorem 6.17.

In fact, from the very start of quantum theory, as formulated by von Neumann, global phases were declared meaningless but remained an explicit part of the formalism. Most textbooks on quantum theory deal with this fact by reserving the term 'quantum state' for an 'equivalence class' of states, namely those that are equal up to a global phase. However, doubling gives us a more elegant and simpler way to deal with this problem.

The usual justification for declaring global phases physically meaningless is that they are not empirically accessible; i.e. they cannot be discovered by means of quantum measurements, which we'll study in Chapter 7. This is already apparent from the fact that all of the probabilities produced by quantum theory come from the Born rule, which only makes use of doubled states and effects. Thus there is really no point in distinguishing two states that differ only by a global phase.

Ignoring global phases has a useful practical consequence as well: it allows our puny human brains to actually picture quantum systems in a geometric way. This is something quite handy for physics, and we do it all the time.

The states of the simplest non-trivial quantum system, *qubits*, live in $\widehat{\mathbb{C}^2}$. Since we can represent such a state with two complex numbers, we can do it with four real numbers. So, naïvely, one may think we need four-dimensional space to write down the state of a qubit. But that's one too many dimensions for (most) humans to picture! The job becomes much easier by just looking at normalised states. Suppose:

$$\Psi = a\,|0\rangle + b\,|1\rangle$$

Then, if ψ is normalised, it must be the case that $|a|^2 + |b|^2 = 1$. Thus, if we want to know $|a|$ and $|b|$, we can ask Pythagoras:

If you remember your trigonometry, this means $|a| = \cos\theta$ and $|b| = \sin\theta$. If you flunked trig, just take our word for it.

As a matter of convention, it's slightly more convenient to use $\frac{\theta}{2}$ instead of θ, so let $|a| = \cos\frac{\theta}{2}$ and $|b| = \sin\frac{\theta}{2}$. Then, we can then drop the absolute values by introducing complex phases, which gives us:

$$\Psi = \cos\frac{\theta}{2}e^{i\beta}\,|0\rangle + \sin\frac{\theta}{2}e^{i\gamma}\,|1\rangle$$

So, we've replaced four real parameters with three angles. This is where doubling comes in. Since doubling kills global phases, the angle β is actually redundant, because we can multiply the whole thing by $e^{-i\beta}$. For some α (namely $\alpha := \gamma - \beta$), we can therefore write the quantum state $\widehat{\psi}$ conveniently as:

$$\widehat{\psi} := \text{double}\left(\cos\frac{\theta}{2}\,\psset{0} + \sin\frac{\theta}{2}e^{i\alpha}\,\pset{1}\right)$$

Since the quantum state is now totally described by two angles, we can plot it on a sphere, called the *Bloch sphere*:

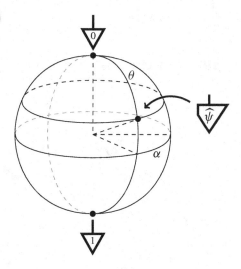

This picture is useful for our intuition. For example, the more 'similar' two states are, that is, the higher the value of their inner product, the closer they are on the Bloch sphere. In particular, orthogonal states are always antipodes. Remember Dave's travels from Section 1.1? We now know what sort of sphere he was hopping around on:

Exercise 6.7 Show that the following points:

$$\vcenter{\hbox{$\blacktriangledown_0$}} \ :=\ \text{double}\left(\tfrac{1}{\sqrt{2}}\left(\vcenter{\hbox{$\triangledown_0$}} + \vcenter{\hbox{$\triangledown_1$}}\right)\right)$$

$$\vcenter{\hbox{$\triangledown_1$}} \ :=\ \text{double}\left(\tfrac{1}{\sqrt{2}}\left(\vcenter{\hbox{$\triangledown_0$}} - \vcenter{\hbox{$\triangledown_1$}}\right)\right)$$

$$\vcenter{\hbox{$\blacktriangledown_0$}} \ :=\ \text{double}\left(\tfrac{1}{\sqrt{2}}\left(\vcenter{\hbox{$\triangledown_0$}} + i\,\vcenter{\hbox{$\triangledown_1$}}\right)\right)$$

$$\vcenter{\hbox{$\blacktriangledown_1$}} \ :=\ \text{double}\left(\tfrac{1}{\sqrt{2}}\left(\vcenter{\hbox{$\triangledown_0$}} - i\,\vcenter{\hbox{$\triangledown_1$}}\right)\right)$$

are located on the Bloch sphere as follows:

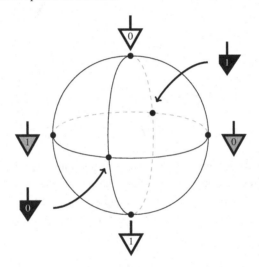

6.1.3 The Process Theory of Pure Quantum Maps

We will now construct a full-blown theory of doubled processes. Besides states and effects, arbitrary linear maps can also be doubled:

$$\text{double}\left(\boxed{f}\right) \ :=\ \boxed{\widehat{f}} \ =\ \boxed{f \quad f} \tag{6.6}$$

In fact, this extends to processes with any number of inputs and outputs. However, we should be a bit careful with which pairs of thin wires should be taken together to form a thick wire. We should always pair the first input/output of f with the first input/output of f's conjugate, the second with the second, and so on:

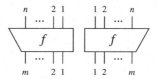

Note that since f's conjugate is the mirror image of f, we count inputs and outputs from right to left rather than left to right. As a result, pairing up inputs and outputs introduces a 'twist' in the wires connected to the conjugate process:

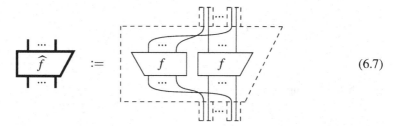 (6.7)

One of the benefits of using the new thick boxes and wires is that the extra complexity of all this twisting remains hidden within the notation.

Doubling all processes yields the following new process theory.

Definition 6.8 The process theory of **pure quantum maps** has as types $\widehat{A}$ for all Hilbert spaces A in **linear maps**, and has as processes doubled linear maps $\widehat{f}$ for all processes f in **linear maps**.

The processes in **pure quantum maps** aren't really all that new; they are just special kinds of linear maps. In other words, **pure quantum maps** is a subtheory of **linear maps**:

$$\textbf{pure quantum maps} \subseteq \textbf{linear maps}$$

Notably, the fact that **linear maps** admits string diagrams is inherited by **pure quantum maps**. Applying equation (6.7) to the cup yields a 'twisted' double cup:

$$\smile := \text{double}\left(\smile\right) = \boxed{} = $$

and a corresponding double cap:

$$\frown := $$

When we compose these two, the 'twists' cancel out, yielding the first yanking equation:

$$\text{(diagram)}$$

Concerning the other two yanking equations, first applying equation (6.7) to the swap map yields a double swap:

$$\text{(diagram)}$$

and when we compose it with a cap or a cup, we obtain:

$$\text{(diagram)}$$

Since there are cups and caps, there also is a notion of transposition for the doubled theory:

$$\text{(diagram)}$$

which turns out to coincide with transposition in the undoubled theory.

Proposition 6.9 Doubling preserves transposition:

$$\text{(diagram)}$$

Proof In the single input/output case we have:

$$\text{(diagram)}$$

The many input/output case from equation (6.7) can be shown similarly. □

The fact that doubling preserves transposes is an instance of a more general fact that doubling preserves diagrams. This can be best seen by decomposing a string diagram into its constituent processes. We already know that doubling sends cups/caps to cups/caps. Evidently, doubling preserves sequential composition:

$$
\text{double}\left(\vcenter{\hbox{\includegraphics{}}} \right) = \vcenter{\hbox{\includegraphics{}}}
$$

Exercise 6.10 Show that doubling preserves parallel composition:

$$
\text{double}\left(\vcenter{\hbox{\includegraphics{}}} \right) = \vcenter{\hbox{\includegraphics{}}}
$$

The doubled theory also inherits its adjoints from the non-doubled theory:

$$
\vcenter{\hbox{\includegraphics{}}} \; := \; \vcenter{\hbox{\includegraphics{}}} \tag{6.8}
$$

If we put all of these pieces together, we can conclude the following.

Corollary 6.11 Doubling preserves string diagrams:

$$
\text{double}\left(\vcenter{\hbox{\includegraphics{}}} \right) = \vcenter{\hbox{\includegraphics{}}}
$$

As a result, any of the calculations we have previously done for diagrams of **linear maps** lifts straightforwardly to **pure quantum maps** just by doubling all of the diagrams.

Example 6.12 In Section 5.3.6, we realised teleportation in the theory of **linear maps** using the Bell maps. To pass to quantum maps, we simply double everything (except Aleks and Bob of course):

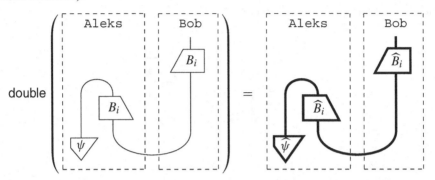

Example 6.13 In Sections 5.3.4 and 5.3.5 we showed how to turn classical logic gates into linear maps. Now, by relying on doubling, we can turn them into *quantum (logic) gates*. For example, the *quantum NOT gate* is:

$$\pi \;\; := \;\; \text{double}\left(\pi \right)$$

the *quantum CNOT gate* is:

$$\circ\!\!-\!\!\bullet \;\; := \;\; \text{double}\left(\circ\!\!-\!\!\bullet \right)$$

and the *Hadamard gate* is:

$$\boxed{H} \;\; := \;\; \text{double}\left(\boxed{H} \right)$$

which doesn't even have a classical counterpart. *Quantum circuits* constitute the application of quantum gates to a fixed number of qubits, for example:

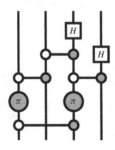

Aside from the Hadamards, we could already do everything above using classical logic gates. Things will start to get much more interesting when we introduce *phase gates* in Chapter 9.

Example 6.14 When doubling the Bell matrices of Section 5.3.6, then:

$$\boxed{B_3} \;\leftrightarrow\; \begin{pmatrix} 0 & -1 \\ 1 & 0 \end{pmatrix}$$

becomes self-transposed, since doubling eliminates global phases:

For the corresponding Bell state we then also have:

Generally, we will carry on using the same terminology for **pure quantum maps** that we use for processes in any other theory. However, certain concepts are important enough to get dedicated 'quantum' names.

Definition 6.15 A $\otimes$-non-separable pure quantum state is called an *entangled state*.

So, an entangled pure state is any bipartite state $\widehat{\psi}$ in **pure quantum maps** that does not factor into single-system states $\widehat{\psi}_1$ and $\widehat{\psi}_2$:

An example is the (doubled) Bell state:

$$\mathrm{double}\left(\frac{1}{\sqrt{D}} \;\cup\; \right) = \frac{1}{D} \cup$$

As mentioned in Remark 4.1, we will need to refine this definition of entanglement once we pass from pure states to more general quantum states. We do this in Section 8.3.5.

6.1.4 Things Preserved by Doubling

By Corollary 6.11, doubling preserves string diagrams, so if two diagrams are equal, then their doubled counterparts will also be equal.

Corollary 6.16 Any equation that holds between string diagrams of 'single' processes also holds for its doubled version:

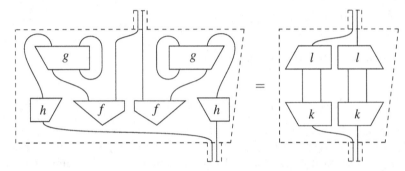

We can see this directly by unfolding both sides of the doubled equation:

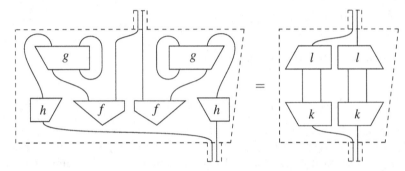

Then clearly the equality between diagrams of 'single' processes together with its conjugated version yields the doubled equation.

The converse of Corollary 6.16 is almost true. Since doubling eliminates global phases, we need to reintroduce them to get 'if and only if'.

Theorem 6.17 Let D and D' be arbitrary diagrams in **linear maps**, and $\widehat{D}$ and $\widehat{D}'$ be their doubled versions in **pure quantum maps**; then:

$$\left(\exists e^{i\alpha} : \boxed{D} = e^{i\alpha} \boxed{D'} \right) \iff \boxed{\widehat{D}} = \boxed{\widehat{D}'}$$

Proof ($\Rightarrow$) directly follows from Corollary 6.16. For ($\Leftarrow$), we can use Corollary 6.11 to replace a diagram of doubled maps with one big, doubled map. So, it suffices to show that for any linear maps f and g such that:

$$\boxed{\widehat{f}} = \boxed{\widehat{g}} \tag{6.9}$$

there exists some α such that:

$$\boxed{f} = e^{i\alpha} \boxed{g}$$

Let λ and μ be defined as:

Then:

$$\lambda\bar\lambda = \overset{(6.9)}{=} = \mu\bar\mu$$

where the dotted lines indicate where we use (6.9). There are two cases: either $\lambda \neq 0$ or $\lambda = 0$. In the first case, divide both sides of the equation above by $\lambda\bar\lambda$:

$$1 = \frac{\mu\,\overline{\mu}}{\lambda\,\overline{\lambda}} = \left(\frac{\mu}{\lambda}\right)\overline{\left(\frac{\mu}{\lambda}\right)}$$

So $\frac{\mu}{\lambda}$ is a global phase, i.e.:

$$\frac{\mu}{\lambda} = e^{i\alpha}$$

for some α, and:

$$\lambda \boxed{f} = \overset{(6.9)}{=} = \mu \boxed{g}$$

so we indeed obtain $f = e^{i\alpha} g$.

In the second case:

$$\boxed{f}\boxed{f} = \boxed{f}\boxed{f} = \lambda = 0 \tag{6.10}$$

Thus by positive definiteness (see Section 5.3.2):

$$\boxed{f} = 0$$

so $f = 0$, hence $\widehat{f} = 0$, and hence by assumption, $\widehat{g} = 0$. This can only be the case if g itself is zero. To see this, attach cups/caps to $\widehat{g}$:

$$\begin{array}{c} g \quad g \end{array} = 0$$

and using positive definiteness just as we did in (6.10) we obtain $g = 0$. Hence $f = e^{i\alpha}g = 0$ where now α can be any angle.   □

Of course, if in Theorem 6.17 we replace equality with equality up to a number (cf. Section 3.4.3), the phase vanishes.

Corollary 6.18 For D and D' arbitrary diagrams in **linear maps**:

$$\boxed{D} \approx \boxed{D'} \quad \Longleftrightarrow \quad \boxed{\widehat{D}} \approx \boxed{\widehat{D'}}$$

Remark 6.19 Besides what is actually being proven in Theorem 6.17, there is another surprising lesson to be learned from this proof. The crux of this proof is purely diagrammatic! In particular, we define the numbers λ and μ as diagrams:

$$\lambda := \begin{array}{c} f \quad f \end{array} \qquad \mu := \begin{array}{c} g \quad f \end{array}$$

and we show by means of diagram substitutions that:

$$\lambda \boxed{f} = \mu \boxed{g} \quad \Longleftrightarrow \quad \boxed{\widehat{f}} = \boxed{\widehat{g}}$$

where $\lambda\,\bar{\lambda} = \mu\,\bar{\mu}$, and the only place we use some specific structure of **linear maps**, namely the complex numbers, is to put this equation in a form involving $e^{i\alpha}$, which is also what requires the case-distinction. But this can be entirely avoided (see references in Section 6.7). This implies that not just for **linear maps**, but also for any other process theory, doubling identifies precisely those processes that differ only by these 'generalised global phases'. The lesson to take from this is to not judge a book by its cover: what at first looks not at all diagrammatic (e.g. eliminating global phases) may turn out to be fundamentally diagrammatic.

We now show how a number of properties of processes coincide in the doubled world and the undoubled world:

Proposition 6.20 A pure quantum map $\widehat{f}$ is an isometry (respectively unitary) if and only if f is an isometry (respectively unitary).

Proof If f is an isometry, $\widehat{f}$ is an isometry by Corollary 6.16. Conversely, we can use Theorem 6.17 to convert doubled equations into single equations up to a phase:

That is, for $\widehat{f}$ an isometry, f is an isometry up to a global phase $e^{i\alpha}$. We can furthermore show that $\alpha = 0$. First, for any normalised state ψ, we have:

The LHS of the above equation is the inner product of a state with itself, so it must be positive. Since the only global phase that is also a positive real number is 1, $e^{i\alpha} = 1$, so f is an isometry. Unitarity is proven similarly. □

We now prove a result similar to Proposition 6.20 for positive maps and projectors, but with a small caveat. If f is an isometry/unitary, then $e^{i\alpha}f$ is also an isometry/unitary, so we can choose any 'undoubled representative' of $\widehat{f}$. However, this is not the case for positive maps and projectors, where we need to choose a particular representative to get the desired property.

Proposition 6.21 A pure quantum map $\widehat{f}$ is positive (respectively a projector) if and only if there exists a positive linear map (respectively a projector) f' with $\widehat{f'} = \widehat{f}$.

Proof Again, one direction follows from Corollary 6.16. For the other direction, suppose $\widehat{f}$ is positive and apply Theorem 6.17 (moving the global phase to the LHS):

$$(6.11)$$

From this, we see that $f' := e^{-i\alpha} f$ is positive. Since doubling removes global phases, $\widehat{f} = \widehat{f'}$. If $\widehat{f}$ is a projector, then (6.11) holds for $g := f$ (cf. Proposition 4.70). Hence f' is also a projector:

$\square$

The following is one more variation on the same theme.

Exercise 6.22 Show that a pure quantum state $\widehat{\psi}$ is normalised if and only if ψ is normalised and that two pure quantum states $\widehat{\psi}$ and $\widehat{\phi}$ are orthogonal if and only if ψ and ϕ are orthogonal.

A consequence of all of the results above is that we can (mostly) work with pure quantum maps as if we were working with plain old linear maps, but with the added benefit that the Born rule is now nothing but a state-effect encounter and that the redundant global phases are eliminated.

Of course there are some exceptions; otherwise, there wouldn't have been any point in doubling everything in the first place.

6.1.5 Things Not Preserved by Doubling

The definition of **linear maps** included three requirements:

- the numbers are the complex numbers;
- there exist sums for processes; and
- there exists an ONB for each type.

Now, we will see that none of these defining features of **linear maps** is preserved under doubling! More specifically:

- the doubled numbers are not the complex numbers;
- sums in the doubled theory are not doubled sums; and
- doubled ONBs are not ONBs in the doubled theory.

However, the things we obtain instead all play a key role in quantum theory.

6.1.5.1 The Doubled Numbers Are 'Probabilities'

This was of course our initial motivation to do doubling.

Proposition 6.23 The numbers in **pure quantum maps** are the positive real numbers.

Proof This follows from Proposition 5.70 characterising positive numbers:

$$p \;=\; \langle\overline{\mu}\rangle\,\langle\mu\rangle \;=\; \langle\widehat{\mu}\rangle$$

In particular, every positive number is a pure quantum map. $\qquad\square$

6.1.5.2 Sums in the Doubled Theory Represent 'Mixing'

Sums of doubled processes:

$$\sum_i \boxed{\widehat{f_i}}$$

are in general <u>not</u> the double of the summed processes:

$$\mathrm{double}\left(\sum_i \boxed{f_i}\right)$$

In fact, non-trivial sums of doubled maps are not even **pure quantum maps**, so they can't be obtained by doubling at all.

The crucial point is that unfolding the doubling operation yields two <u>independent</u> summations (i.e. over different indices):

$$\sum_i \boxed{f_i}\ \ \sum_j \boxed{f_j}$$

We can split this sum into the parts where $i = j$ and where $i \neq j$:

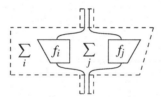

$$\sum_i \left[\ \boxed{f_i}\ \ \boxed{f_i}\ \right] \;+\; \sum_{i\neq j} \left[\ \boxed{f_i}\ \ \boxed{f_j}\ \right]$$

Thus the LHS equals the RHS, plus the 'off-diagonal' terms:

$$\mathrm{double}\left(\sum_i \boxed{f_i}\right) \;=\; \sum_i \boxed{\widehat{f_i}} \;+\; \sum_{i\neq j}\left[\ \boxed{f_i}\ \ \boxed{f_j}\ \right]$$

These off-diagonal terms will not go to zero in general. We can already see this for the case of numbers. Letting $\lambda_0 = \lambda_1 = 1$, we have:

$$\mathsf{double}\left(\sum_i \langle\!\langle \lambda_i \rangle\!\rangle\right) = \mathsf{double}(1+1) = 4 \neq 2 = 1 + 1 = \sum_i \langle\!\langle \lambda_i \rangle\!\rangle$$

This is a feature, not a bug. Summing pure quantum maps has an important conceptual meaning, which has no counterpart when summing the underlying linear maps. It can be given a clear physical interpretation as introducing some uncertainty in which process happened. This is called 'mixing' and will be discussed in detail in Section 6.2.7.

The other kind of sums, that is, those that are made in **linear maps** before doubling, gives rise to *quantum superpositions*. We discuss these in Section 7.1.2. So in the end we have two kinds of sums around, each meaning a different thing.

On the other hand, the fact that doubling doesn't preserve sums confirms what we already pointed out in Remark 5.34, namely, that sums are out of place in diagrammatic reasoning, and they may easily cause mistakes. Thus it is prudent to avoid them as much as possible. Over the next two chapters, many concepts will be initially introduced with sums, but gradually replaced by their purely diagrammatic counterparts.

6.1.5.3 Doubled ONB Effects Are 'Quantum Measurements'

If an ONB in **linear maps**:

$$\mathcal{B} = \left\{ \; \bigtriangledown_{\!i} \; \right\}_i$$

contains at least two states, then:

$$\mathsf{double}\,(\mathcal{B}) := \left\{ \; \bigtriangledown_{\!i} \; \right\}_i$$

is <u>not</u> a basis in **pure quantum maps**. To see this, consider two states:

$$\bigtriangledown_{\!\psi} := \sum_j \bigtriangledown_{\!j} \qquad\qquad \bigtriangledown_{\!\phi} := \sum_j e^{i\alpha_j} \bigtriangledown_{\!j}$$

where the α_j are all distinct. Since ϕ has at least two terms with non-equal coefficients $e^{i\alpha_j}$, ψ and ϕ are not within a global phase of each other, so:

$$\bigtriangledown_{\!\psi} \neq \bigtriangledown_{\!\phi}$$

However, one can easily verify that for all i:

$$\genfrac{}{}{0pt}{}{\bigtriangleup_{\!i}}{\bigtriangledown_{\!\psi}} = \genfrac{}{}{0pt}{}{\bigtriangleup_{\!i}}{\bigtriangledown_{\!\phi}}$$

So $\mathsf{double}(\mathcal{B})$ cannot possibly be a basis.

The reason why double ($\mathcal{B}$) is not a basis in **pure quantum maps** is that it misses out on all of the 'local' phase information (i.e. the numbers $e^{i\alpha_j}$ above) from ϕ, which (unlike global phases) are highly relevant to the quantum state $\widehat{\phi}$. An important physical consequence of this fact is that *quantum measurements* can extract only a fraction of the information about the state of a system. In the next chapter, we will define quantum measurements and show that measurements defined in terms of ONB effects are in fact the best we can do (though still pretty poor!) when it comes to extracting information from a single quantum state.

At this point, you might start to wonder whether the theory of **pure quantum maps** has bases at all. We saw from Theorem 5.14 that bases allow us to completely characterise processes by a finite set of numbers (namely, its matrix). This is the basis of *quantum tomography*, which as we'll see in the next chapter, is all about identifying a state or a process by means of quantum measurements. The only way this would be possible is if there are still bases around. Thankfully, this is the case.

Theorem 6.24 For any ONB:

$$\mathcal{B} = \left\{ \; \vat{j} \; \right\}_j$$

on a Hilbert space A, the set of states double($\mathcal{B}$) can be extended to a (non-orthogonal) basis in **linear maps** for the type $A \otimes A$, consisting entirely of pure quantum states. Hence, in particular this is also a basis in **pure quantum maps** for the doubled system-type $\widehat{A}$.

Proof There are many ways to extend a basis for pure states. For example, let $\mathcal{A}$ be the set of all states of the form:

$$\psi_{jk} := \begin{cases} \frac{1}{2}\left(\vat{j} + \vat{k} \right) & \text{if } j \le k \\[2em] \frac{1}{2}\left(\vat{j} + i\,\vat{k} \right) & \text{if } j > k \end{cases}$$

Each basis state $j \in \mathcal{B}$ is then given as ψ_{jj}, so $\mathcal{B} \subseteq \mathcal{A}$. To show:

$$\text{double}(\mathcal{A}) = \left\{ \; \doublepsi{jk} \; \right\}_{jk}$$

is a basis for $A \otimes A$, it suffices to show that any element of a product basis:

$$\left\{ \; \vat{j}\,\vat{k} \; \right\}_{jk} \qquad\qquad (6.12)$$

can be obtained using sums of states of the form $\lambda\psi_{jk}$ for $\lambda \in \mathbb{C}$, which is left as an exercise to the reader. $\qquad\square$

So, we can still distinguish processes by how they act on states, but it's not nearly as easy as it used to be! In particular, there is no way to build an ONB of pure quantum states.

Convention 6.25 Even though Theorem 6.24 implies that the 'dimension' (in the sense of Definition 5.2) of a quantum system $\widehat{A}$ is D^2 when the dimension of A is D, when we talk about the dimension of a quantum system, we are always referring to the dimension of the (undoubled) Hilbert space. For instance, we still refer to $\widetilde{\mathbb{C}^2}$ as a two-dimensional quantum system.

Remark 6.26 In the statement of the previous theorem we distinguished between double(A) being a basis in **linear maps** and being a basis in **pure quantum maps**. These are indeed two different things. A basis in **linear maps** needs to uniquely fix all linear maps via:

whereas a basis for **pure quantum maps** only needs to uniquely fix linear maps of the form $\widehat{f}, \widehat{g}, \ldots$ via:

which is a weaker requirement. We will use the full strength of Theorem 6.24 in Section 6.2.1.2.

6.2 Quantum Maps from Discarding

In the previous section we transformed **linear maps** into **pure quantum maps** via doubling. So far, doubling was merely a way of recasting an existing thing into something derived, which we depicted by showing it in boldface. However, the new types are made from two wires rather than one, and since two is more than one, there is actually some extra space for genuinely new stuff, which has no counterpart in the underlying process theory of **linear maps**. We will now show that we are missing out on a very important process, namely the process of *discarding* a system, which is hiding in this extra space. In fact, this is the only extra thing we need to throw in to get all **quantum maps**.

6.2.1 Discarding

When we introduced effects back in Section 3.4.1, the first example we gave was 'discarding a system'. Here, 'discarding a system' may mean ignoring it, destroying it, or maybe firing it off into space, never to be seen again. Alternatively, it could be the case that there is some piece of a bigger system that we simply do not have access to, so we can think of it as being 'discarded' for all intents and purposes.

In this section, we will home in on discarding, by looking at its behaviour and showing that this behaviour forces us to make one particular choice. We begin with the following realisation.

6.2.1.1 Discarding Is Not a Pure Quantum Map

To see this, it suffices to consider how a discarding process should behave on the state of a single system. Discarding should do nothing but remove that state from the picture:

$$
\begin{array}{c}\raisebox{0pt}{$\widehat{?}$} \\ \widehat{\psi}\end{array} \;=\; \square
$$

or put another way: it is a test that succeeds with certainty, but otherwise tells us nothing about the state. In particular, the discarding process cannot depend on the state of the system that gets discarded.

Proposition 6.27 For non-trivial Hilbert spaces, i.e. with dimension > 1, there exists no pure quantum effect $\widehat{\phi}$ such that for all normalised pure quantum states $\widehat{\psi}$ we have:

$$
\begin{array}{c}\widehat{\phi} \\ \widehat{\psi}\end{array} \;=\; \square \tag{6.13}
$$

Proof Suppose that $\widehat{\phi}$ is a pure discarding effect. Clearly $\widehat{\phi}$ cannot be zero. Thus there exists some λ such that $\lambda\phi$ is normalised. By Theorem 5.79, $\lambda\phi$ extends in an ONB. Since the dimension is at least two, let ϕ' be a distinct state in that ONB. It must therefore be orthogonal to $\lambda\phi$, and hence also orthogonal to ϕ. Consequently:

$$
\begin{array}{c}\phi \\ \phi'\end{array} \;=\; 0
$$

and hence:

$$
\begin{array}{c}\widehat{\phi} \\ \widehat{\phi'}\end{array} \;=\; 0 \;\neq\; \square
$$

which is a contradiction. $\qquad\square$

Remark 6.28 In Section 4.3.3, we gave a good reason for considering the normalised states by default, namely that these are precisely the states that return 'yes' with certainty when tested for themselves. To even get a discarding map in the first place, this restriction is essential. If for some $\widehat{\psi}$ we have (6.13), then for $2\widehat{\psi}$ this would no longer be the case:

$$2\;\overline{\underline{\bigtriangledown_{\psi}}} \;=\; 2\;\begin{bmatrix} \\ \end{bmatrix} \;\neq\; \begin{bmatrix} \\ \end{bmatrix}$$

so (6.13) would simply be impossible to satisfy.

Since we cannot use any pure effect, we propose the following instead.

Definition 6.29 We define *discarding* to be the effect:

$$\overline{\overline{\top}} \;:=\; \begin{bmatrix} \cap \\ \end{bmatrix} \tag{6.14}$$

This effect indeed behaves as required.

Proposition 6.30 For any normalised pure quantum state $\widehat{\psi}$ we have:

$$\overline{\underline{\bigtriangledown_{\psi}}} \;=\; \begin{bmatrix} \\ \end{bmatrix} \tag{6.15}$$

Proof Since $\widehat{\psi}$ is normalised, so is ψ (cf. Exercise 6.22), so we have:

$$\overline{\underline{\bigtriangledown_{\psi}}} \;=\; \begin{bmatrix} \cap \\ \bigtriangledown_{\psi}\;\bigtriangledown_{\psi} \end{bmatrix} \;=\; \begin{matrix} \psi \\ \psi \end{matrix} \;=\; \begin{bmatrix} \\ \end{bmatrix}$$

$\square$

Discarding as defined in (6.14) certainly doesn't look like a pure quantum effect:

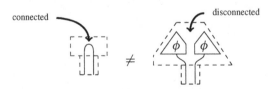

which is consistent with Proposition 6.27. It also behaves as expected on states, so it looks like we're done. But before we hang up our hats, it's worth asking whether this is our only choice for a discarding effect (spoiler alert: yes!).

6.2.1.2 There Is Only One Linear Map Fit for Purpose

Theorem 6.31 The discarding map defined in Definition 6.29 is the unique linear map sending all normalised pure quantum states to 1.

Proof Suppose there exists some other effect:

that sends all normalised pure states to 1. From Theorem 6.24, we saw that there exists a basis $\mathsf{double}(\mathcal{B}')$ of pure quantum states for $A \otimes A$ in **linear maps**. Let $\mathsf{double}(\mathcal{B}'')$ be the basis formed by normalising each of the states in $\mathsf{double}(\mathcal{B}')$ (which is clearly still a basis). Then, for all $\widehat{\phi}_{jk} \in \mathsf{double}(\mathcal{B}'')$:

$$
\begin{array}{ccccc}
\raisebox{-1em}{\includegraphics{disc}} & = & \raisebox{-1em}{\includegraphics{box}} & = & \raisebox{-1em}{\includegraphics{d}}
\end{array}
$$

Since **d** and discarding agree on a basis, they must therefore be equal. Hence discarding is unique. □

Now that we know that discarding is uniquely defined by its intended behaviour, we can derive what it should be in certain special cases.

Exercise 6.32 Show that:

$$
\widehat{H}_1 \otimes \cdots \otimes \widehat{H}_n \; \overline{\overline{\top}} \quad := \quad \overline{\overline{\top}}_{\widehat{H}_1} \; \overline{\overline{\top}}_{\widehat{H}_2} \; \cdots \; \overline{\overline{\top}}_{\widehat{H}_n} \tag{6.16}
$$

and that:

$$
\widehat{\mathbb{C}} \; \overline{\overline{\top}} \quad := \quad \raisebox{-0.8em}{\includegraphics{emptybox}} \tag{6.17}
$$

(noting that $\widehat{\mathbb{C}}$ is the 'no wire' system for **pure quantum maps**).

6.2.1.3 Discarding Does Not Preserve Pure Quantum States

If we start with a pure quantum state on two systems and discard one system, the resulting state typically won't be a pure quantum state. In fact, the only case where it will be a pure state is when we start with something that is $\otimes$-separable.

Proposition 6.33 For any pure quantum state $\widehat{\psi}$, the *reduced state*:

is a pure state if and only if $\widehat{\psi}$ is $\otimes$-separable:

$$\text{\includegraphics{}} \quad = \quad \text{\includegraphics{}}$$

Proof For ($\Leftarrow$), we have:

$$\text{\includegraphics{}} \quad = \quad \text{\includegraphics{}} \qquad\qquad (6.18)$$

In the proof of Proposition 6.30 we already saw that:

$$\text{\includegraphics{}} \quad = \quad \text{\includegraphics{}}$$

which is a positive number. Since every positive number is pure (cf. Proposition 6.23), the state (6.18) is a pure quantum state. For ($\Rightarrow$), assume there exists some pure state $\widehat{\phi}$ such that:

$$\text{\includegraphics{}} \quad = \quad \text{\includegraphics{}} \qquad\qquad (6.19)$$

We will rely on Proposition 5.74, which states that $f^{\dagger} \circ f$ is $\circ$-separable if and only if f is. Unfolding the doubled maps in (6.19):

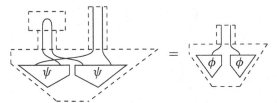

we obtain:

$$\text{\includegraphics{}} \quad = \quad \text{\includegraphics{}}$$

By process–state duality and using transposition we get:

$$\text{\includegraphics{}} \quad = \quad \text{\includegraphics{}}$$

Then, by Proposition 5.74, there exist ψ_1 and ψ_2 such that:

(Note that there is no loss of generality by depicting the effect ψ_2 in conjugate form.) Thus, again using process–state duality, ψ is $\otimes$-separable:

Doubling the equation above yields the required condition. □

As reduced states are not, in general, pure quantum states, we need to introduce a more general family of states to account for the fact that parts of systems may be discarded.

6.2.2 Impurity

When we compose arbitrary pure quantum maps with the discarding map, lots of new stuff emerges. For example, consider the transpose (or equivalently, the adjoint) of the discarding map:

This state is so important that its normalised version gets a special name.

Definition 6.34 The *maximally mixed state* is:

This maximally mixed state is an example of a reduced state (cf. Proposition 6.33), namely what's left after discarding half of a Bell state:

As we mentioned in the introduction, the term 'mixed' has to do with a lack of knowledge about the actual state of the system. In Section 6.2.7 we'll see that the maximally mixed state means we have no knowledge whatsoever about the state of a system.

We can now generate other new quantum states either by applying a pure quantum map to a system in a maximally mixed state:

$$\widehat{f} \;=\; f \quad f \;=\; f \quad f$$

or, equivalently, by discarding one system of a bipartite pure state (represented here by means of process–state duality):

$$\widehat{f} \;=\; \widehat{f}$$

The resulting form is moreover totally generic.

Theorem 6.35 Any state obtained by composing pure quantum maps and discarding is a *quantum state*, i.e. a state of the form:

$$\rho \;:=\; f \quad f \tag{6.20}$$

Proof For a diagram consisting of some pure quantum maps and discarding:

we can always pull all of the discarding maps (or maximally mixed states) down to the bottom, using caps as necessary:

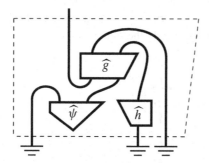

We can then combine all of the maximally mixed states into a single state (cf. equation (6.16) upside-down). Thus, we obtain a pure quantum map, applied to a maximally mixed state. □

Recalling the definition of ⊗-positivity from Section 4.3.6, we obtain the following.

Corollary 6.36 Quantum states are ⊗-positive states in **linear maps**.

The form of quantum states generalises that of pure quantum states, where the map f in (6.20) has a trivial input wire. In particular, not all quantum states are pure states. When they fail to be pure states, we call them *impure*. Unfolding the form of an impure quantum state ρ and a pure one $\widehat{\psi}$ the difference comes down to the presence of a wire connecting the left half to the right half:

So *purity* itself is a diagrammatic notion.

Remark 6.37 Note that we should take the absence of a wire as evidence for purity. Conversely, the presence of a wire only indicates the possibility of being impure. For example, suppose f itself is disconnected:

$$\widehat{f}\ \widehat{f} \;=\; \widehat{\psi}\ \widehat{\psi}\ \pi\ \pi \;=\; \widehat{\psi}\ \diamond\lambda\ \diamond\lambda\ \widehat{\psi}$$

with $\lambda := \sqrt{\pi^\dagger \circ \pi}$. Then the resulting state is pure.

Example 6.38 In quantum teleportation it doesn't matter if the state that we teleport is impure, since the picture stays exactly the same:

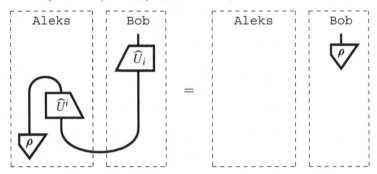

While this is a proper generalisation, the reason that it doesn't require any additional work is the compositional nature of string diagrams. The key to the above equality being true is the fact that have:

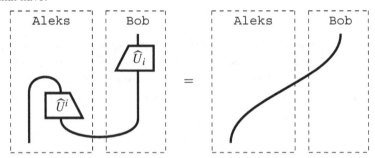

Then, clearly it doesn't matter whether we plug in $\widehat{\psi}$ or ρ at the input.

6.2.3 Weight and Causality for Quantum States

Now, the cautious reader may have noticed something sneaky about Definition 6.34, namely, that the maximally mixed state is not normalised in the sense of Definition 4.48:

$$\frac{1}{D} \quad \frac{1}{D} \quad = \quad \frac{1}{D^2} \quad = \quad \frac{1}{D} \tag{6.21}$$

since the circle is equal to the dimension D, as we saw in Corollary 5.33. On the other hand, if we discard the maximally mixed state we do get:

$$\frac{1}{D} \quad = \quad$$

So unlike in the case of pure quantum states, where in Proposition 6.30 we showed that the squared norm and the number arising from discarding coincide, this ceases to be the case for the maximally mixed state and, more generally, will no longer hold for impure quantum states. This justifies the introduction of a new name for the numbers obtained when discarding.

Definition 6.39 The *weight* of a quantum state ρ is:

$$\overline{\underline{\underline{\Ydown{\rho}}}}$$

and ρ is *causal* if its weight is 1, i.e.:

$$\overline{\underline{\underline{\Ydown{\rho}}}} \;=\; \boxed{\;\;}\;\; \tag{6.22}$$

How should we interpret this quantity? The Born rule tells us that:

$$\text{effect}\left\{\;\overline{\underline{\underline{\;}}}\atop \text{state}\left\{\;\Ydown{\rho}\right.\right\}\text{probability}$$

So, the weight is the result of performing a trivial test on the state (i.e. 'is this a state?'). Normally, we would expect this test to return 'yes' with probability 1. However, we will see later in this chapter that a state may be the result of a non-deterministic process. In this case, the weight then tells us what the probability is to end up in this state. Causal states are then those that occur with certainty. In other words, what we call a quantum state is really a combination of two things:

$$\Ydown{\rho'} \;=\; p\,\Ydown{\rho}$$

an 'actual' state of a system (the causal state ρ) and some probability p that it occurred (i.e. the weight of ρ'). We continue this discussion in Section 6.4.1 when we introduce non-deterministic quantum processes properly.

Ignoring non-determinism for the moment, the only 'actual' states are the causal ones. So, we can more fundamentally interpret the causality equation as follows:

> *If a state is discarded, it may as well never have existed.*

This is obviously a reasonable, and furthermore necessary, assumption to make. There are many systems out there (e.g. on Mars) that we have no control over and know nothing about. So, we ignore (i.e. discard) them in our calculations. If we weren't allowed to do so, we pretty much couldn't do any science. There is also a hint of relativity theory here: if something is sufficiently far away (i.e. or 'space-like separated', to use relativity lingo), it should not be able to affect what is happening locally, because light simply cannot travel fast enough. This feature, called 'non-signalling', is fundamental to relativity theory. In Section 6.3, we will see how non-signalling and causality are closely connected.

For states, the constraint imposed by causality is fairly weak: it restricts the weight to 1. Hence, any non-zero state can be turned into a causal one simply by composing it with the appropriate number. However, we shall see in Section 6.2.6 that for arbitrary processes the constraint is more drastic, and, as we will show in Theorem 6.54, for effects it is pretty extreme!

For pure states, we already know that normalisation and causality coincide. More generally, we have the following.

Proposition 6.40 For any pure state $\widehat{\psi}$:

$$
\raisebox{-1.2em}{\includegraphics{eq1}} = \left(\raisebox{-1.2em}{\includegraphics{eq2}}\right)^{2}
$$

Proof   This can be shown by unfolding, then transposing the effects:

$$
\raisebox{-1.2em}{\includegraphics{eq3}} = \raisebox{-1.2em}{\includegraphics{eq4}} = \raisebox{-1.2em}{\includegraphics{eq5}} = \raisebox{-1.2em}{\includegraphics{eq6}}
$$

$\square$

The fact that normalisation and causality coincide for pure states is then just the special case where the squared norm and the (squared) weight are both equal to 1.

Example 6.41 Since the circle is equal to D, the Bell state is causal:

$$
\raisebox{-1.2em}{\includegraphics{eq7}} = \frac{1}{D}\raisebox{-1.2em}{\includegraphics{eq8}} = \raisebox{-1.2em}{\includegraphics{eq9}}
$$

Given that the Bell state is moreover pure, it is also normalised:

$$
\raisebox{-1.2em}{\includegraphics{eq10}} = \frac{1}{D^{2}}\raisebox{-1.2em}{\includegraphics{eq11}} = \raisebox{-1.2em}{\includegraphics{eq12}}
$$

However, as we already saw in (6.21), the squared norm can be smaller than the squared weight for impure states. The fact that these two quantities do not always coincide is actually useful, since it gives us an effective way to figure out when a state is pure.

Proposition 6.42 For any quantum state ρ we have:

$$\vcenter{\hbox{\includegraphics{rho_rho_discard}}} \leq \left(\vcenter{\hbox{\includegraphics{rho_ground}}}\right)^2 \tag{6.23}$$

and we have equality if and only if ρ is pure.

Proof Since ρ is a $\otimes$-positive state, by the spectral theorem (and Corollary 5.72 in particular) there exists some ONB and positive numbers r_i such that:

$$\vcenter{\hbox{\includegraphics{rho}}} := \sum_i r_i \vcenter{\hbox{\includegraphics{ii}}}$$

It is then straightforward to compute the squared norm and the weight:

$$\vcenter{\hbox{\includegraphics{norm}}} = \sum_{ij} r_i r_j \vcenter{\hbox{\includegraphics{jjii}}} = \sum_i r_i^2$$

$$\vcenter{\hbox{\includegraphics{weight}}} = \sum_i r_i \vcenter{\hbox{\includegraphics{cap}}} = \sum_i r_i$$

Thus:

$$\left(\vcenter{\hbox{\includegraphics{weight}}}\right)^2 = \left(\sum_i r_i\right)^2 = \sum_i r_i^2 + \sum_{i \neq j} r_i r_j$$

Since all $r_i r_j \geq 0$ the first claim follows. If ρ is pure, then by Proposition 6.21, the squared norm and the squared weight coincide. Conversely, assume the squared norm and the squared weight are equal. Then:

$$\sum_{i \neq j} r_i r_j = 0$$

which is only true if, for all $i \neq j$, we have $r_i r_j = 0$. In that case, at most one r_i is non-zero, so:

$$\vcenter{\hbox{\includegraphics{rho}}} = r_i \vcenter{\hbox{\includegraphics{ii}}}$$

which is a pure quantum state. $\qquad\square$

Proposition 6.42 does much more than provide a means of detecting whether a state is pure. As a causal state becomes more and more impure, the squared norm will go lower and lower, until it hits $1/D$ for the maximally mixed state. Thus, it is a quantity that actually measures the impurity of a state. We will explore this and other measures of impurity in Section 13.2.

We can also visualise the difference between the quantities in (6.23). For the weight we have:

$$ \tag{6.24} $$

and for the squared norm we have:

$$ \tag{6.25} $$

In the case of a pure state it is the absence of the following wires that results in the squared weight and the squared norm becoming the same:

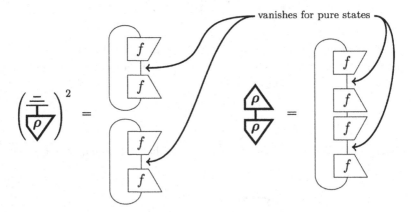

Otherwise, it is the difference in wiring that causes the difference in numbers.

Remark 6.43 Recall from Remark 6.5 that we could translate between a pure state and its representation as a density operator, which in that case was a projector. This generalises to (causal) mixed states, whose associated density operator is a positive map that has trace 1.

The translation between our doubled representation and the density operator representation is provided by process–state duality:

$$\tilde{\rho} \quad = \quad \overset{\text{[diagram]}}{f \quad f} \qquad \mapsto \qquad \tilde{\rho} := \overset{\text{[diagram]}}{f \quad f} \quad = \quad \overset{f}{f}$$

In the density operator representation, from (6.24) it follows that discarding a state means taking its trace:

$$\overset{\rho}{=} \quad = \quad \overset{f}{f} \quad = \quad \mathrm{tr}(\tilde{\rho})$$

so a density operator with trace 1 corresponds to a causal state. Similarly, from (6.25) it follows that:

$$\overset{\rho}{\underset{\rho}{\diamond}} \quad = \quad \begin{matrix} f \\ f \\ f \\ f \end{matrix} \quad = \quad \mathrm{tr}\left(\tilde{\rho}^2\right)$$

and so (6.23) becomes, in terms of density operators:

$$\mathrm{tr}\left(\tilde{\rho}^2\right) \;\leq\; (\mathrm{tr}\,(\tilde{\rho}))^2$$

6.2.4 The Process Theory of Quantum Maps

We passed from pure quantum states to quantum states by adjoining discarding. In fact, the entire process theory of quantum maps is also obtained in that manner.

Definition 6.44 The process theory of **quantum maps** has as types doubled Hilbert spaces $\widehat{A}$ and as processes all diagrams made from pure quantum maps and discarding:

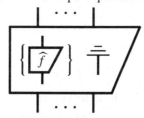

and the following should come as no surprise.

Theorem 6.45 The theory of **quantum maps** admits string diagrams.

Proof **Quantum maps** inherit their caps and cups from **pure quantum maps**, so it suffices to show that **quantum maps** have adjoints. The adjoint of a pure quantum map is another pure quantum map, and the adjoint of discarding is just its transpose:

$$\left(\stackrel{=}{\top}\right)^{\dagger} = \stackrel{=}{\cup}$$

which is the composition of a pure quantum map (the cup) with discarding. Thus it is again a quantum map. Since adjoints need to preserve diagrams and all diagrams in **quantum maps** are made up of pure quantum map and discarding, every quantum map has an adjoint.   □

Just as we saw with quantum states in the previous section, we can put any quantum map in a 'normal form' by grouping all of the discarding maps together into a single effect:

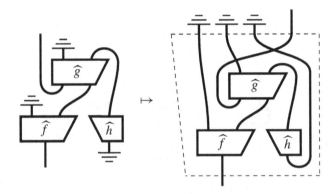

Hence we have the following.

Proposition 6.46 All quantum maps are of the form:

for some linear map f.

So, quantum maps are precisely those linear maps that are $\otimes$-positive processes (cf. Definition 4.66). Proposition 6.46 also shows that arbitrary quantum maps arise by ignoring part of the output of a process. In other words, for any quantum map Φ, we know that there must exist a pure quantum map $\widehat{f}$ such that:

$$\Phi = \widehat{f} \tag{6.26}$$

Definition 6.47 We refer to the pure quantum map $\widehat{f}$ in (6.26) as a *purification* of the quantum map Φ.

It might be tempting to think that quantum maps are precisely those maps that send quantum states to quantum states, but this is not the case. For example, the swap linear map, when conceived as a linear map from type $\widehat{A}$ to $\widehat{A}$:

$$\tag{6.27}$$

clearly sends quantum states to quantum states:

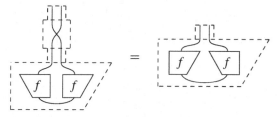

(it actually conjugates them!) However ...

Proposition 6.48 The linear map (6.27) is <u>not</u> a quantum map.

Proof For any quantum map Φ and any linear map f the following number:

will also be a quantum map, namely a positive number. However, if instead of Φ we take (6.27), we obtain:

This number is not of the form of a linear map composed with its adjoint, so it cannot be positive for all f. And indeed, picking:

$$\Box f \quad \leftrightarrow \quad \begin{pmatrix} 0 & -1 \\ 1 & 0 \end{pmatrix}$$

yields -2, which is a contradiction. $\qquad\qquad\square$

Being a quantum map is actually stronger than just preserving states of a single system. The reason is that quantum maps should not only be well behaved in that special case but should be well behaved when they are included in any diagram. For example, applying a quantum map to just one part of a state on multiple systems, as in (6.28) below, should again yield a quantum state. In fact, this (seemingly more specific) condition is actually equivalent to being a quantum map.

Theorem 6.49 A linear map:

is a quantum map if and only if for all quantum states ρ we have that:

$\qquad\qquad$ (6.28)

is again a quantum state.

Proof If Φ is a quantum map, (6.28) must be a quantum state simply because **quantum maps** is a process theory. For the other direction, let ρ in (6.28) be the doubled cup state. Then by assumption, this is a quantum state, so by Proposition 6.46, there exists a pure quantum state $\widehat{\psi}$ such that:

So by process–state duality it follows that:

and, hence, Φ is indeed a quantum map. $\qquad\qquad\square$

Remark 6.50 We pick things up where we left off in Remark 6.43. First note that we can equivalently represent a general quantum map as follows:

$$\tag{6.29}$$

We can then reshape it a bit:

Here, the small dashed box represents a hole where we can insert a density operator $\tilde{\rho}$ and obtain $\tilde{\Phi}(\tilde{\rho})$. The 'super-operator' (i.e. a map taking operators to operators):

$$\tilde{\Phi}(\ ) := \tag{6.30}$$

is usually called a *completely positive map* (CP-map). They are also commonly seen in an equivalent form, involving the trace:

$$\tilde{\Phi}(\ ) :=$$

The 'complete' part refers to the fact that a CP-map preserves positivity (i.e. 'being a quantum state') even when applied to just part of a system, as in (6.28). The super-operator analogue to preserving positivity only when applied to entire systems is called simply *positive*. In the case of (6.27) the positive super-operator looks like this:

which, by Proposition 6.48, is not of the form (6.30).

When specialising Proposition 6.42 to the case of a bipartite state:

we now obtain a means of determining if a quantum map is pure, as well as a quantity that represents the degree of impurity for general quantum maps.

Corollary 6.51 For any quantum map Φ we have:

$$\begin{array}{|c|}\hline \Phi \\\hline \Phi \\\hline\end{array} \;\leq\; \left(\begin{array}{|c|}\hline \Phi \\\hline\end{array}\right)^{2} \tag{6.31}$$

and we have equality if and only if Φ is pure.

6.2.5 Causality for Quantum Maps

We now generalise the definition of causality from states to maps.

Definition 6.52 We call a quantum map Φ *causal* if we have:

$$\begin{array}{|c|}\hline \Phi \\\hline\end{array} \;=\; \top \tag{6.32}$$

In fact, causality for quantum maps means preservation of causal states.

Proposition 6.53 A quantum map is causal if and only if it sends causal quantum states to causal quantum states.

Proof For a causal quantum map Φ and a causal quantum state ρ we have:

$$\begin{array}{ccccc} \overline{\underline{\overline{}}} \\ \boxed{\Phi} \\ \rho \end{array} \quad = \quad \begin{array}{c} \overline{\underline{\overline{}}} \\ \rho \end{array} \quad = \quad \boxed{}$$

so ($\Rightarrow$) indeed holds. For ($\Leftarrow$), suppose Φ sends any causal state ρ to another causal state ρ'. Then we have:

$$\begin{array}{ccccc} \overline{\underline{\overline{}}} \\ \boxed{\Phi} \\ \rho \end{array} \quad = \quad \begin{array}{c} \overline{\underline{\overline{}}} \\ \rho' \end{array} \quad = \quad \boxed{}$$

Thus the following effect:

$$\begin{array}{c} \overline{\underline{\overline{}}} \\ \boxed{\Phi} \end{array}$$

sends all causal states (and in particular, all normalised pure states) to 1. Then, by Theorem 6.31 on uniqueness of discarding:

$$\begin{array}{c} \overline{\underline{\overline{}}} \\ \boxed{\Phi} \end{array} \quad = \quad \overline{\underline{\overline{}}}$$

so the quantum map Φ is indeed causal. $\square$

We can also interpret (6.32) directly:

> *If the output of a process is discarded, it may as well have never happened.*

This is a straight generalisation of the interpretation we gave for causal states in Section 6.2.3.

Despite seeming innocent at first, causality has a somewhat shocking consequence when applied to effects. By equation (6.17), discarding the 'no wire' system is the same as doing nothing. Since effects have no outputs, causality reduces to this equation:

$$\begin{array}{c} \rho \end{array} \quad = \quad \overline{\underline{\overline{}}}$$

which forces any causal effect to be equal to discarding!

Theorem 6.54 There is a unique causal quantum effect: discarding.

So causal effects in **quantum maps** are utterly uninteresting. If causality is going to play such an important role, why did we even bother to introduce effects at all? Don't worry, your time hasn't been wasted! Once we consider non-deterministic quantum processes, we will be able to realise all quantum effects non-deterministically, and this will be vital for applications such as quantum teleportation.

Remark 6.55 The analogue to causality for CP-maps (cf. Remark 6.50) is that they are trace preserving:

$$\mathrm{tr}\,(\widetilde{\Phi}(\widetilde{\rho})) \;=\; \mathrm{tr}\,(\widetilde{\rho})$$

since trace-preserving CP-maps send density operators to density operators, just as causal quantum maps send causal quantum states to causal quantum states, as we saw in Proposition 6.53.

6.2.6 Isometry and Unitarity from Causality

So what is the constraint imposed by causality for pure quantum maps? We already saw in Section 6.2.3 that for pure states causality means normalisation. Realising that a normalised state is just a special case of an isometry, when the input system is trivial, we can generalise this statement.

Theorem 6.56 For pure quantum maps, the following are equivalent:

1. $\widehat{U}$ is causal:

2. U is an isometry:

3. $\widehat{U}$ is an isometry:

Proof Unfolding the causality equation, we have:

$$
\raisebox{-1em}{\includegraphics{placeholder}} = \raisebox{-1em}{\includegraphics{placeholder}} \tag{6.33}
$$

so we obtain the isometry equation by bending the left input wire up:

$$
\raisebox{-1em}{\includegraphics{placeholder}} = \raisebox{-1em}{\includegraphics{placeholder}} = \raisebox{-1em}{\includegraphics{placeholder}} = \raisebox{-1em}{\includegraphics{placeholder}}
$$

and hence we obtain 1 ⇔ 2. By Theorem 6.20 we obtain 2 ⇔ 3. □

We already saw above in Theorem 6.54 that discarding, which of course is not pure, is the only causal quantum effect. Hence, there are no pure causal quantum effects. More generally, Theorem 6.56 implies that there exist no pure causal quantum maps from $\widehat{A}$ to $\widehat{B}$ if $\dim(A) > \dim(B)$, because in that case there exist no isometries from A to B (cf. Proposition 5.81).

Now recall from Proposition 5.81 that any isometry from a Hilbert space to itself must be a unitary. This yields an easy corollary to Theorem 6.56.

Corollary 6.57 A pure quantum map from a system $\widehat{A}$ to itself is causal if and only if it is a unitary.

Of course, this fact depends on the dimension theorem, which we have only established for **linear maps** in particular. On the other hand, Theorem 6.56 does not rely on any special properties of linear maps. A similarly general consequence is the following. By Proposition 4.58, invertible isometries are unitary. Combining this with Theorem 6.56 thus yields the following.

Corollary 6.58 For a pure quantum map, the following are equivalent:

1. It is causal and invertible.
2. It is unitary.

Remark 6.59 In many textbooks, unitarity is assumed from the start, without (much) justification. However, causality, with its simple physical interpretation, is much easier to justify. Thus, a pleasant consequence of doubling is that isometry (and hence unitarity) fall out so easily, as in equation (6.33).

We already made implicit use of the unitarity requirement.

Example 6.60 Recall that in Section 4.4.4 where we presented quantum teleportation we assumed that $\widehat{U}_i$ needed to be unitary in the RHS of:

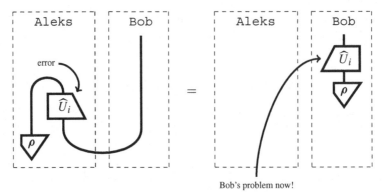

Bob's problem now!

in order for Bob to be able to fix the error. Now we know why: in order for Bob's correction to be causal, it needs to be an isometry, and in order to undo $\widehat{U}_i$ it must furthermore be unitary.

So what about impure processes? If we combine the fact that all causal pure quantum maps are isometries with the fact that all quantum maps can be purified, we can immediately see that every causal quantum map can be represented as an isometry with one of its outputs discarded.

Theorem 6.61 (Stinespring dilation I) For every causal quantum map Φ there exists an isometry $\widehat{U}$ such that:

$$\raisebox{-1em}{\includegraphics{phi}} \quad = \quad \raisebox{-1em}{\includegraphics{U}} \qquad\qquad (6.34)$$

Proof By Proposition 6.46 we know that there always exists a pure quantum map $\widehat{U}$ such that 6.34 holds. By causality of Φ, it follows that $\widehat{U}$ must also be causal:

$$\raisebox{-1em}{\includegraphics{U2}} \quad = \quad \raisebox{-1em}{\includegraphics{phi2}} \quad = \quad \raisebox{-1em}{\includegraphics{ground}}$$

which, by Theorem 6.56 implies that $\widehat{U}$ is an isometry.   □

In fact, we can boost this result from isometries to unitaries, but this requires a bit more work. First, note that we can replace any isometry with a unitary and a pure state.

Lemma 6.62 For any isometry U:

$$\raisebox{-1.5em}{\includegraphics{Uhat}}$$

there exists a unitary U' and a pure quantum state $\widehat{\psi}$ such that:

$$
\widehat{U} \;=\; \widehat{U'}\;\widehat{\psi}
$$

Proof Fix ONBs for A and B such that:

$$
\left\{\; \overset{U}{\underset{i}{\square}} \;\right\} \;\subseteq\; \left\{\; \blacktriangledown \;\right\}_j
$$

Then, fix a linear map U' from $A \otimes B$ to $B \otimes A$, which sends:

$$
\mathcal{B} := \left\{\; \overset{i}{\triangledown}\;\blacktriangledown \;\right\}_{ij} \qquad \text{to} \qquad \mathcal{B}' := \left\{\; \blacktriangledown\;\overset{i}{\triangledown} \;\right\}_{ji}
$$

such that:

$$
\overset{U'}{\underset{i\;\;0}{\triangledown}} \;=\; \overset{U\;\;0}{\underset{i}{\triangledown}} \tag{6.35}
$$

and the remaining basis states of $\mathcal{B}$ are mapped (injectively) to the remaining basis states of $\mathcal{B}'$. This is always possible since $\dim(A \otimes B) = \dim(B \otimes A)$, and yields a bijection on the ONBs $\mathcal{B}$ and $\mathcal{B}'$. Thus by Proposition 5.38, U' is unitary. Since equation (6.35) holds for all white ONB-states, we have:

$$
\overset{U'}{\underset{0}{\square}} \;=\; \overset{U\;\;0}{\square}
$$

Doubling this equation and discarding the first output yields:

$$
\overset{\widehat{U'}}{\underset{0}{\triangledown}} \;=\; \widehat{U}\;\overset{\overline{\overline{\;}}}{\underset{0}{\triangledown}} \;=\; \widehat{U}
$$

which completes the proof, for:

$$
\overset{}{\underset{\psi}{\triangledown}} \;:=\; \overset{}{\underset{0}{\triangledown}}
$$

$\square$

Now we can conclude the following.

Corollary 6.63 (Stinespring dilation II) Every causal quantum map Φ arises from some unitary $\widehat{U}$ by plugging some pure causal quantum state $\widehat{\psi}$ into one of its inputs and discarding one of its outputs:

$$\begin{array}{c}\Phi\end{array} = \begin{array}{c}\widehat{U}\\\widehat{\psi}\end{array} \tag{6.36}$$

Remark 6.64 Stinespring dilation is used by some people to justify a point of view where the only real quantum processes are just the pure, unitary ones, of which we only have access to some small part. This belief is sometimes referred to as the 'Church of the larger Hilbert space'. While this point of view is perfectly consistent with quantum theory, it is not very convenient for thinking process theoretically. For example, one cannot build a process theory that includes both unitaries and pure quantum states without also considering general isometries. Indeed, unitary quantum maps U and pure quantum states ψ yield (non-unitary) isometries when composed in parallel:

We will return to this point in Section 7.3.2 when we discuss von Neumann's formulation of quantum theory.

6.2.7 Kraus Decomposition and Mixing

One way to understand impure quantum maps, by their very definition, is in terms of discarding parts of a larger system. We now give an alternative interpretation in terms of *mixing*. For now, this will involve explicit sums. However, in Section 8.3.4 mixing will be given a purely diagrammatic treatment, as part of our general strategy for eliminating sums.

A first step towards mixing is to replace the discarding map by a sum over an ONB. We can do this by decomposing the cap with equation (5.37):

$$\overline{\overline{\top}} = \overline{\underline{\bigcap}} = \sum_i \overline{\bigwedge}^i \overline{\bigwedge}^i = \sum_i \overline{\bigwedge}^i \tag{6.37}$$

Combining this with purification, we can write any quantum map as a sum of pure quantum maps:

$$
\boxed{\Phi} \; = \; \overline{\widehat{f}} \;\; \overset{(6.37)}{=} \;\; \overset{\sum_i \triangle_i}{\widehat{f}} \; = \; \sum_i \boxed{\widehat{f_i}}
$$

where:

$$
\boxed{\widehat{f_i}} \; := \; \overset{\triangle_i}{\widehat{f}}
$$

Conversely, the sum of any finite set of pure quantum maps is a quantum map (recall that sums only exist for processes of the same type):

$$
\sum_i \boxed{\widehat{f_i}} \; = \; \overline{\widehat{f}}
$$

where:

$$
\boxed{f} \; := \; \sum_i \bigvee^i \boxed{f_i} \tag{6.38}
$$

So, in summary, we have the following.

Theorem 6.65 The sum of pure quantum maps is a quantum map, and any quantum map Φ can be written as a sum of pure quantum maps:

$$
\boxed{\Phi} \; = \; \sum_i \boxed{\widehat{f_i}} \tag{6.39}
$$

Such a representation is known as a *Kraus decomposition*.

Remark 6.66 For completely positive maps (cf. Remark 6.50), a Kraus decomposition becomes:

$$
\begin{array}{c} g \\ \hline g \end{array} \; = \; \sum_i \begin{array}{c} \widehat{f_i} \\ \hline f_i \end{array}
$$

which one usually encounters in the following form:

$$\tilde{\Phi}(\tilde{\rho}) := \sum_i f_i \, \tilde{\rho} \, f_i^\dagger$$

The sum of pure quantum maps is almost never pure, so the theory of **pure quantum maps** isn't closed under sums. However, we have the following theorem.

Theorem 6.67 The sum of any finite set of quantum maps:

$$\sum_i \boxed{\Phi_i}$$

is a quantum map; i.e. the theory of **quantum maps** is *closed under sums*.

Proof Since all of the Φ_i have Kraus decompositions:

$$\boxed{\Phi_i} \;=\; \sum_j \boxed{\hat{f}_{ij}}$$

then we can expand their sum as:

$$\sum_i \boxed{\Phi_i} \;=\; \sum_i \sum_j \boxed{\hat{f}_{ij}} \;=\; \sum_{ij} \boxed{\hat{f}_{ij}}$$

which by Theorem 6.65 is a quantum map.   □

In Theorem 6.65 we said 'a' Kraus decomposition since Kraus decompositions are not unique. There are, however, some special ones. For Kraus decompositions for quantum states, by the spectral theorem (and Corollary 5.72 in particular), we have the following.

Corollary 6.68 Every quantum state ρ has a Kraus decomposition of the following form:

$$\boxed{\rho} \;=\; \sum_i r_i \, \bigtriangledown{i} \tag{6.40}$$

for some ONB and positive real numbers r_i.

If we apply Corollary 6.68 to:

$$\boxed{\Phi} \tag{6.41}$$

we can decompose (6.41) over an ONB of biparite states. Equivalently, we can decompose Φ itself over a Hilbert–Schmidt ONB (cf. Definition 5.101).

Corollary 6.69 Every quantum map Φ has a Kraus decomposition of the following form:

$$\boxed{\Phi} = \sum_i r_i \boxed{i} \tag{6.42}$$

for some Hilbert–Schmidt ONB and positive real numbers r_i.

We showed in Theorem 6.67 that any sum of quantum maps is again a quantum map. Of course, if we add together <u>causal</u> quantum maps, what we get will no longer be causal:

$$\boxed{\Psi} + \boxed{\Phi} = \;\bar{\bar{\top}}\; + \;\bar{\bar{\top}}\; = 2\;\bar{\bar{\top}}\;$$

However, instead of ordinary sums, we can consider *convex combinations* of causal quantum maps.

Definition 6.70 A *convex combination* or *mixture* of a family of causal quantum maps $\{\Phi_i\}_i$ is a sum of the form:

$$\sum_i p^i \boxed{\Phi_i} \tag{6.43}$$

where the numbers p^i sum to 1.

In this case causality is preserved.

Theorem 6.71 Every convex combination of causal quantum maps is again a causal quantum map.

Proof By Theorem 6.67, the map (6.43) is a quantum map, and by causality of each of the quantum maps Φ_i we have:

$$\sum_i p^i \boxed{\Phi_i} = \sum_i p^i \;\bar{\bar{\top}}\; = \;\bar{\bar{\top}}\;$$

$\square$

We refer to the operation that produces (6.43) out of $\{p^i\}_i$ and $\{\Phi_i\}_i$ as *mixing*. For a mixture we can interpret each p^i as the probability that the process Φ_i happens. In other words, there is a lack of knowledge – represented by the probability distribution $\{p^i\}_i$ – about which pure process Φ_i out of the $\{\Phi_i\}_i$ is happening. For example, the causal quantum state:

$$\boxed{\rho} = \sum_i p^i \boxed{\psi_i} \tag{6.44}$$

can be interpreted as a system that is in one of the pure states $\widehat{\psi_i}$, but we don't know which one. We only know the probability p^i that it is in the i-th state. We can in fact write any causal quantum state in the following form.

Theorem 6.72 Every causal quantum state can be regarded as a mixture of <u>pure</u> causal quantum states. Moreover, these pure causal quantum states can always be chosen to form an ONB.

Proof Since ONB states are always causal, this follows immediately from Corollary 6.68. □

So, it is tempting to believe that all impurity can be reduced to this kind of situation. However, we quickly hit a snag if we try to do this for other more general maps than states.

Theorem 6.73 Not every causal quantum map can be regarded as a mixture of <u>pure</u> causal quantum maps.

Proof Discarding cannot be decomposed in pure causal quantum effects, since there aren't any (cf. Theorem 6.54). □

The idea of mixing is not just important as a conceptual interpretation of impure quantum states (and some impure quantum maps), but it also provides a geometric picture for those states. In (6.44), ρ is a convex combination of pure states. The natural way to picture this is to think of ρ as lying somewhere between the pure states, where each p^i determines just how close it is to the i-th pure state ($1 :=$ 'at the same place', $0 :=$ 'as far away as possible'). Recall from Section 6.1.2 that we can picture causal (i.e. normalised) pure quantum states in $\mathbb{C}^2$ as living on the surface of a sphere called the Bloch sphere. If we include arbitrary causal quantum states, we will include all convex combinations of states on the surface of the sphere. Therefore, we get not just a sphere, but a whole ball, called the *Bloch ball*. In this ball, a mixed state:

$$\rho := p\,\psi_1 + (1-p)\,\psi_2$$

is pictured as a point inside the sphere:

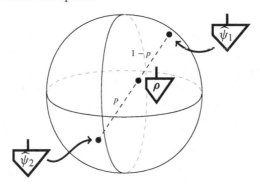

A special case of mixing is mixing over an ONB, which enables us to encode any probability distribution.

Proposition 6.74 Given a fixed ONB, probability distributions can be equivalently represented as causal quantum states of the form:

$$\bigtriangledown_P := \sum_i p^i \; \bigtriangledown_i \tag{6.45}$$

Proof We need to prove that the numbers p^i are all positive, which follows from Corollary 5.72, and that causality (cf. (∗) below) forces the probabilities to sum up to one:

$$\sum_i p^i = \sum_i p^i \; \overset{(6.45)}{=} \; \overset{(*)}{=} 1$$

□

In the two-dimensional case probability distributions then become:

$$\bigtriangledown_P := p \; \bigtriangledown_0 + (1-p) \; \bigtriangledown_1$$

that is, they depend only on the number $p \in [0, 1]$, and we can visualise them within the Bloch ball as a line connecting the two doubled basis states:

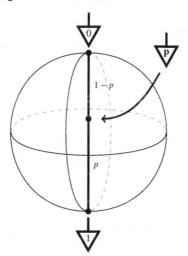

If this were classical probability theory, that would be the end of the story: any probabilistic state decomposes uniquely into a probability distribution over the pure states (i.e. those corresponding to point distributions) spanning the space of probability distributions. However, for quantum states, the situation is quite different, in that this decomposition into pure states is usually <u>not</u> unique: a quantum state ρ may decompose as many different mixtures of pure states. The most extreme example of this phenomenon is the maximally mixed state, which can be seen as an equal mixture of the pure states corresponding to any orthonormal basis. Indeed, using Proposition 5.56, we can decompose the cup across any $\overline{\text{ONB}}$:

$$\frac{1}{D}\; \underset{\underline{}}{\perp} \;=\; \frac{1}{D}\; \bigcup \;=\; \frac{1}{D}\sum_i \; \bigvee\!\!\bigvee \;=\; \sum_i \frac{1}{D}\; \bigvee \tag{6.46}$$

This also means that the maximally mixed state is 'equally distant' from any pure state, which explains the name 'maximally mixed': it has no bias towards any of the pure states.

So, given a mixed state, there is no unique interpretation in terms of a set of pure states. In fact, there is no particular reason to decompose ρ as a mixture over a basis or as a mixture of just two pure states. Why not three, or four:

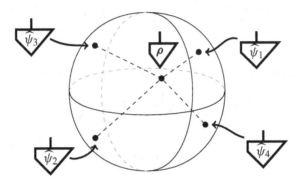

or a billion? In fact, mixed states are often used to describe enormous numbers of quantum systems, called *ensembles*, in a variety of pure states. Such an ensemble could represent, for example, all the photons in a laser beam.

Example 6.75 (noise) One can think of the maximally mixed state as total noise, in that it doesn't have any bias towards any meaningful data, i.e. any pure state. Consider the following process:

$$\frac{1}{D}\; \begin{array}{l} \rule{0pt}{0pt} \\ \underset{\text{\fontsize{6pt}{6pt}\selectfont =}}{\perp} \;\;\leftarrow \text{output pure noise} \\ \underset{\top}{\underline{}} \;\;\leftarrow \text{discard any input} \end{array}$$

This process converts any input into noise. The fact that nothing from the input remains at the output is clear from the diagram in that it is ∘-separable. We can now mix this *noise map* with any other process:

$$(1-p)\; \boxed{\Phi} \;+\; p\,\frac{1}{D}\; \underset{\top}{\overset{\perp}{\rule{0pt}{0pt}}}$$

to make it a bit noisy. The bigger p is, the more noisy it is.

This is all we say about mixing for now. Section 8.3.4, where we picturalise mixing, contains diagrammatic counterparts of some of the results we have seen here, as well as some new results. In Chapter 13 we will introduce a *resource theory* that will let us compare and quantify mixedness.

6.2.8 The No-Broadcasting Theorem

We already saw in Section 4.4.2 that any process theory that admits string diagrams satisfies a no-cloning theorem. Thus, in **quantum maps** there exists no quantum map Δ that clones all input states ρ:

Even if we restrict ourselves to pure states $\widehat{\psi}$, then there still do not exist cloning maps, since pure quantum maps already admit string diagrams. Restricting to causal states doesn't help either, since non-causal states can always be made causal by multiplying by a number.

Section 4.4, where we first introduced no-cloning, was entitled 'Quantum Features from String Diagrams'. But how justified was this title? When one says 'quantum feature' one typically doesn't just mean something that <u>happens</u> to be true for quantum theory, but moreover, something that fails to hold classically.

In the previous section, we saw that we can represent a probability distribution as a quantum state as follows:

$$
\vcenter{\hbox{\includegraphics{p}}} := \sum_i p^i \vcenter{\hbox{\includegraphics{i}}}
$$

The pure states are then given by point distributions:

$$
\vcenter{\hbox{\includegraphics{j}}}
$$

while all other probability distributions are mixtures of point distributions. This results in the following analogy:

	pure	mixed
probability distributions	$\vcenter{\hbox{$j$}}$	$\vcenter{\hbox{$p$}} := \sum_i p^i \vcenter{\hbox{i}}$
causal quantum states	$\vcenter{\hbox{$\widehat{\psi}$}}$	$\vcenter{\hbox{$\rho$}} := \sum_i p^i \vcenter{\hbox{$\widehat{\phi_i}$}}$

If no-cloning is a truly quantum feature, then it should be the case that we can clone classical states, i.e. probability distributions.

The good news is that there does exist a cloning map for point distributions, which we can represent as a quantum map:

$$
\begin{array}{c} \tilde{\Delta} \end{array} := \sum_i \begin{array}{c} i \quad i \\ \Delta \\ i \end{array} \quad :: \quad j \mapsto j \; j \tag{6.47}
$$

so while we cannot clone pure quantum states, we can clone pure probability distributions. However, the bad news is that this does not extend to general (i.e. mixed) probability distributions:

$$
\sum_i p^i \quad \overset{\tilde{\Delta}}{\mapsto} \quad \sum_i p^i \quad \neq \quad \left(\sum_i p^i \right)\left(\sum_i p^i \right) \tag{6.48}
$$

Just as in the quantum case, there is no map that clones all probability distributions. So 'no-cloning' only separates <u>pure</u> classical and quantum theories. What we really want is a weaker criterion, which still holds in the (mixed) classical world but fails for quantum maps.

This can be realised as follows. Rather than asking for the existence of a cloning map, we can ask for the existence of a *broadcasting map*, that is, a map that takes in a state ρ and outputs two systems with the property that if we discard <u>either</u> system, the overall state will be ρ:

$$
\begin{array}{c} \Delta \\ \rho \end{array} \quad = \quad \rho \quad = \quad \begin{array}{c} \Delta \\ \rho \end{array} \tag{6.49}
$$

Here the term 'broadcasting' is used in the same sense as a television broadcast. When you receive a TV show, you only really care that it comes to you correctly, and not about what's going on everywhere else.

First we show that broadcasting is indeed weaker than cloning.

Proposition 6.76 Any cloning map is also a broadcasting map.

Proof For any cloning map Δ we have:

$$
\begin{array}{c} \Delta \\ \rho \end{array} \quad = \quad \rho \; \rho \quad = \quad \rho
$$

and the second equation also holds by symmetry.   □

The converse is not true. In particular, we can broadcast probability distributions.

Exercise 6.77 Show that $\widetilde{\Delta}$ is a broadcasting map for probability distributions.

However, despite the fact that broadcasting is weaker than cloning, we still cannot broadcast quantum states. Thus no-broadcasting is a truly quantum feature, which doesn't depend on any extra assumptions such as purity. To prove no-broadcasting for quantum states, we first need to generalise the bipartite state in Proposition 6.33 from a pure state to any quantum state.

Proposition 6.78 If the reduced state of any quantum state is pure:

$$\text{(diagram)} \qquad\qquad (6.50)$$

then the state ρ separates as follows:

$$\text{(diagram)}$$

for some (causal) quantum state ρ'.

Proof Assuming (6.50), we first purify ρ (cf. Definition 6.47):

$$\text{(diagram)}$$

and substitute this into (6.50):

$$\text{(diagram)}$$

By Proposition 6.33, $\widehat{\phi}$ is $\otimes$-separable, in which case:

$$\text{(diagram)}$$

Just by multiplying $\widehat{\psi}_2$ by some number, we can always choose ρ' to be causal. In that case, equation (6.50) implies that $\widehat{\psi}_2 = \widehat{\psi}$. The proof of the converse is straightforward. $\qquad\square$

... and then we generalise to processes.

Proposition 6.79 If the *reduced map* of any quantum map Φ is pure:

$$\text{(diagram)} \qquad\qquad (6.51)$$

then Φ separates as follows:

$$\text{[diagram]} \qquad (6.52)$$

for some (causal) quantum state ρ.

Proof   Bend the wire in (6.51):

$$\text{[diagram]}$$

Since the reduced state of the tripartite quantum state is pure, by Proposition 6.78 it separates as follows:

$$\text{[diagram]}$$

Unbend the wire and we're done.   □

   Now we are ready to prove the no-broadcasting theorem.

Theorem 6.80   Quantum states cannot be broadcast.

Proof   By Theorem 6.24 there exists a basis consisting of quantum states for any type, so equations (6.49) are equivalent to:

$$\text{[diagram]} \qquad (6.53)$$

We now show there exists no such Δ. By equation (6.53) (l) the reduced state of Δ is pure, so by Proposition 6.79 we have:

$$\text{[diagram]} \qquad (6.54)$$

for some state ρ. Hence it follows that:

$$\text{[diagram]}$$

Since the identity is ∘-separable, so is every other process involving that type, and hence the system must be trivial for Δ to exist. ☐

Remark 6.81 Even though we posed Theorem 6.80 specifically for **quantum maps**, we can derive this result for many process theories, much as we did for the no-cloning theorems in Section 4.4.2. Indeed, defining broadcasting directly via equation (6.53), no-broadcasting holds for the doubled version of any (non-trivial) process theory satisfying (5.47). Alternatively, if we are working in a process theory that didn't arise from doubling, we can define 'purity' of quantum processes as equation (6.51), implying equation (6.52). Then, no-broadcasting holds for any system whose identity process is 'pure' in the sense of (6.51). There is, however, a caveat with this approach, which we reveal in Section 6.7. Can you guess what it is?

6.3 Relativity in Process Theories

By now it must be clear that causality plays a central role in quantum theory. For example, it forces all pure quantum maps to be isometries and general quantum maps to arise as a purification of an isometry via Stinespring dilation. We motivated causality with this motto:

> *If the output of a process is discarded, it may as well have never happened.*

This is embodied in the equation:

$$\stackrel{\doublestackbar}{\boxed{\Phi}} \ = \ \stackrel{\doublestackbar}{\top} \tag{6.55}$$

This also means that if a process is happening somewhere else, and its output never reaches us, we don't need to care about it. As we already noted, this is crucial to being able to even do science, in that it allows us to safely ignore parts of the universe that won't affect us.

The goal of this section is to show that the causality postulate guarantees compliance of quantum theory with the theory of relativity. In particular, causality guarantees that:

> *Nothing can attain speeds that are faster than light.*

This is perhaps the most unexpected, and compelling, of its interpretations. In fact, we shall also see that quantum picturalism already has relativistic aspects built in and that causality also has the fact that the laws of physics should be observer independent built in.

6.3.1 Causal Structure

The theory of relativity teaches us about the structure of space and time, or in short, *spacetime*. Rather than there being a single theatre in which things take place with a single clock ticking, Aleks and Bob each have their own private screening of reality as well as their own personal perception of how fast things progress. Relativity in its simplest form, called *special relativity*, is extracted from two principles:

1. all observers experience the same laws of physics, and
2. the speed of light through empty space is constant.

From these principles and a healthy dose of creativity one can then derive the entire theory of relativity, including the following features:

- *Relativity of simultaneity*, that is, two processes that appear to Aleks as happening at the same time, might appear to Bob as happening at very different times, and vice versa.
- *No faster-than-light travel*, that is, no object can ever attain speeds beyond the speed of light, and can only achieve that speed if it has no mass, since otherwise this would require an infinite amount of energy.

The first of these features implies that this scenario:

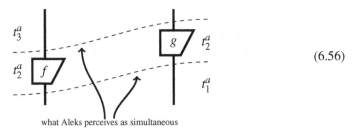

$$(6.56)$$

what Aleks perceives as simultaneous

is equal to this scenario:

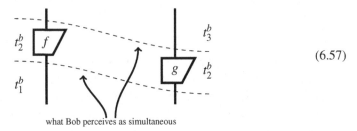

$$(6.57)$$

what Bob perceives as simultaneous

because in each case, for at least one observer these processes appear as simultaneous. Of course, we have assumed this right from the start, since those diagrams are equal! So by using the diagrammatic language we account for this important aspect of the theory of relativity.

The notion of simultaneity, which has lost its absolute (i.e. objective) meaning, makes way for the weaker (but still objective) notion of *spatial separation*, which allows for processes to appear as simultaneous to some, but not to others. We can depict this situation as two points without a connection between them:

Since no signal from Aleks to Bob can exceed the speed of light, when they are very far apart it takes some time before anything sent by Aleks will reach Bob, and vice versa. To visualise this, we can extend the picture above:

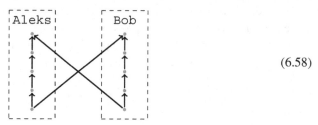

(6.58)

The points represent processes, performed at a certain location in spacetime, and the arrows indicate the possibility of one process to effect another.

We see in (6.58) that Aleks and Bob are free to perform a number of processes locally, but Aleks' processes can't have an effect on Bob's (and vice versa) until some time has passed after they are performed. Note how we have omitted such details as 'how far' Aleks and Bob are apart, or 'how long' each of the processes take. Such a drawing of points and arrows is what is called a *causal structure*.

The set of all points reachable by following the arrows from a single point is called the *causal future* of that point. One should think of these as all of the spacetime points one could reach from a given point by travelling less than or equal to the speed of light. In relativity parlance, this is called the *future light cone*. Similarly, there is a *past light cone*, and together these form the *light cone* for that point.

There are of course infinitely many causal structures one could imagine to describe processes interacting across space and time. Essentially, the only limit on what makes a 'valid' causal structure is that one (usually) assumes it does not include any directed cycles, much as we did in the case of circuit diagrams in Section 3.2.3.

Remark* 6.82 One can also define a causal structure as a *partially ordered set*. That is, a set with a relation $\leq$ where, for all a, b, c:

- $a \leq a$,
- $(a \leq b$ and $b \leq c) \implies a \leq c$, and
- $(a \leq b$ and $b \leq a) \implies a = b$

There is a tight relationship between the directed acyclic graphs derived from circuits (cf. Remark* 3.21) and causal structures, in that each directed acyclic graph can be turned into a partially ordered set by taking the transitive closure. Hence, in Remark* 6.82 we were right to characterise circuits as those diagrams that admit a causal structure.

Examples of causal structures are the V-shape and pitchfork-shape:

We will encounter the pitchfork-shape in Section 11.1 when we prove that quantum theory is *non-local*, and the V-shape will help us establish the connection between the causality postulate and the theory of relativity.

Assume now that we have some fixed causal structure that represents spacetime and a process theory that tells us which processes can take place (e.g. causal quantum maps). We can now think of those processes as happening at points in spacetime by laying them out on top of the causal structure:

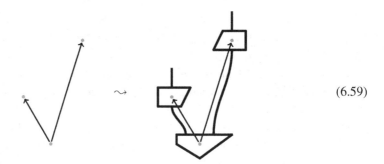

$$(6.59)$$

Notably, the arrows in the causal structure tell us where wires are allowed and where they are forbidden:

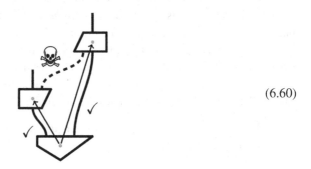

$$(6.60)$$

A process theory is said to be *non-signalling* if each process Ψ in a diagram with a fixed causal structure can have an influence only on processes in the causal future of Ψ. Naturally, a theory where processes can have effects on those outside their causal future is called *signalling*.

This is a non-trivial requirement of a process theory. In other words, there exist theories where processes can have an effect on other processes not in their causal future, even when we restrict to diagrams that respect the causal structure, as in (6.59). To see this, consider the following N-shaped causal structure:

and any process theory that admits string diagrams (e.g. possibly non-causal **quantum maps**). We can clearly violate the causal structure by introducing a cup at b and a cap at a':

While in the causal structure there is no edge from a to b', one can still send data from a to b since we have:

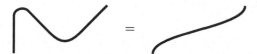

In fact, this is exactly what happens in quantum teleportation:

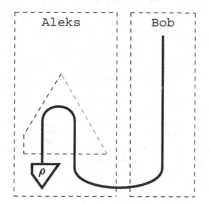

So, what's going on here? Is quantum theory signalling because it allows teleportation? Absolutely not! By just looking at a single 'branch' of the teleportation protocol, that is, a fixed error $\widehat{U}_i$, we are overlooking an important part: the communication of Aleks' value i to Bob, which is necessary to correct that error. If we account for this classical communication, the causal structure is again respected:

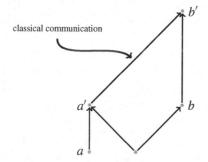

We will see in Section 6.4.4 that if Aleks does not send Bob his value i, the state Bob gets at the end will just be noise, with no trace of ρ whatsoever.

6.3.2 Causality Implies Non-signalling

To show this, we will rely on the simple V-shaped causal structure:

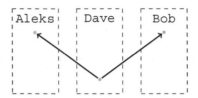

That is, Aleks and Bob may have some shared history, perhaps when Dave gave them two halves of an entangled state. However, now they have moved far away from each other, so far in fact that they can no longer communicate with each other without sending messages faster than the speed of light, which, of course, they can't do.

We can decorate the causal structure with a diagram of processes:

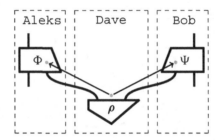

When we discussed teleportation before, we considered very specific processes, but now we will merely assume that all of the processes obey causality. In fact, the following argument goes through even if Φ, Ψ and ρ are not quantum processes, as long as they still satisfy the causality equation with respect to the discarding processes.

The causal structure allows Aleks and Bob to have local inputs and outputs, so that they can control their processes and receive data from them. Technically, it allows inputs/outputs for Dave as well, but we won't need them. But crucially, since there is no arrow in the causal structure from Aleks to Bob or vice versa, it should not be possible for either of them to send a signal to the other. This means that neither of them should be allowed to derive something about the other one's input using just their own inputs and outputs. Otherwise, this could be exploited to communicate some data, i.e. signal, in a way that violates the causal structure.

Our claim is that non-signalling follows from nothing but the causality postulate (6.55). We start by considering the case of Bob learning something about Aleks' input. From Bob's perspective, we should discard Aleks' output since it is not accessible to Bob:

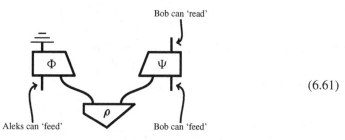

(6.61)

Let's now see if Bob can learn anything about Aleks' input from his own input-output pair. By causality we have:

and hence it follows that:

So from Bob's perspective, his input-output pair is disconnected from Aleks' input, which is discarded. Thus no signalling from Aleks to Bob can take place. By symmetry it also follows that Bob cannot signal to Aleks. Thus we have the following.

Theorem 6.83 If a process theory has a discarding process for each type and it satisfies causality, then it is non-signalling.

Exercise 6.84 Above we only established non-signalling for the V-shaped causal structure. What if we extend this to a diamond-shaped one, that is, we add a point in the joint future of Aleks and Bob? More generally, show that non-signalling holds for any causal structure.

6.3.3 Causality and Covariance

We will now show that distinct observers Aleks and Bob will always experience the same laws of physics. More precisely, we will show that states restricted to a single location in spacetime (known as *local states*) look the same to both. Consequently, whatever laws of physics made the system end up in that state must be the same too.

So, what does it mean to ask what a local state 'looks like' for a given observer? Recall from Section 6.3.1 that the notion of simultaneity depends on the observer. Suppose we chop up a diagram into a series of layers indicating what processes have happened by the time a clock ticks for a single observer:

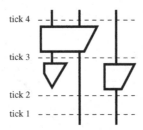

This way of chopping up a diagram is called a *foliation*, and since the notion of 'simultaneous' depends on the observer, so too do foliations. Thus a single diagram admits many different ones:

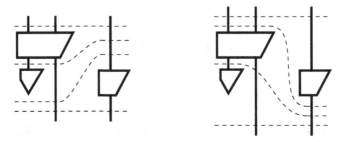

Another way of saying the laws of physics are the same for all observers is to say they are foliation-independent, or *covariant*.

If we input an initial state into a foliated diagram, we can compute the resulting state at each layer, so we can see it evolving in time according to a particular foliation. For, example the left foliation above yields:

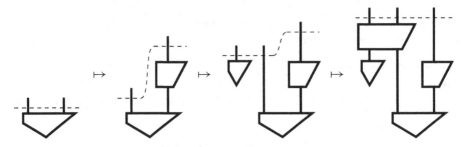

A local state is obtained by fixing a single spacetime point, and discarding all of the other systems in a single layer, e.g.

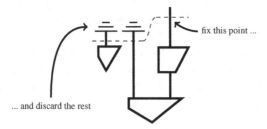

Of course, there are typically many different layers that include a single spacetime point, coming from different possible foliations:

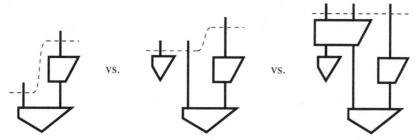

But the causality postulate guarantees that the local state does not depend on the choice of layer:

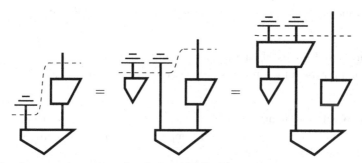

Hence it also does not depend on the choice of foliation.

6.4 Quantum Processes

In the introduction to this chapter we announced that causality will be a defining constraint on general quantum processes. These quantum processes can have more than one quantum map, which, as a whole, obey a version of the causality postulate that generalises the one for quantum maps. A single quantum map satisfying causality is then a special case, namely, that of a deterministic quantum process. However, deterministic quantum processes are scarce. For example, by Theorem 6.54 there is only one causal quantum effect for any type: discarding. The true potential of quantum processes only becomes apparent when also considering the non-deterministic ones. Without these, things like teleportation are simply not possible, since the cap effect that we relied on to perform teleportation is not causal, nor is any non-separable bipartite effect.

Proposition 6.85 All causal quantum effects are separable.

Proof Since by equation (6.16) discarding multiple systems is the same as discarding each system individually, from Theorem 6.54 it follows that:

$$\text{(diagram)}$$

which, in particular, is separable. □

Now, suppose we try to do quantum teleportation with the only causal bipartite quantum effect available:

$$
\tag{6.62}
$$

The best that Bob can obtain is the maximally mixed state, that is, nothing but noise. In particular, there is absolutely no trace left of the state ρ.

The fact that we don't have caps around is bad news, as indicated by the following clanger.

Proposition 6.86 Circuits of causal quantum maps are again causal quantum maps, so causal quantum maps form a process theory. However, that theory does <u>not</u> admit string diagrams.

Proof The proof that circuits of quantum maps are again quantum maps is included in the proof of Proposition 6.89 below. Now, suppose causal quantum maps did admit string diagrams. Then, by Proposition 6.85:

$$
\cap \;:=\; \overline{\overline{\top}}\;\overline{\overline{\top}}
$$

from which it follows that the identity is disconnected:

$$
| \;=\; \cap\!\cup \;=\; \top\,\widetilde{\top}\cup \;=\; \underset{\top}{\overline{\overline{\bot}}}
$$

and hence the process theory must be trivial. Since the process theory of causal quantum maps is not trivial, it cannot admit string diagrams. □

Fortunately, by allowing for non-determinism, we will be able to reintroduce all that was lost, and hence rescue string diagrams!

6.4.1 Non-deterministic Quantum Processes

An (uncontrolled) *quantum process* is a collection of quantum maps:

$$
\Phi_1 \quad,\quad \Phi_2 \quad,\ldots,\quad \Phi_n
$$

which we shall denote:

$$\left(\boxed{\Phi_i}\right)^i \tag{6.63}$$

that together satisfy the *causality postulate*:

$$\sum_i \boxed{\Phi_i} = \top \tag{6.64}$$

We refer to the elements in the set (6.63) as *branches*. If there is only one branch, then we call the quantum process *deterministic*; otherwise, we call it *non-deterministic*. When a system undergoes a quantum process, one of the branches actually happens. We call this branch (or simply its index i) the *outcome* of the process.

Remark 6.87 We say 'uncontrolled' quantum process because in Section 6.4.5, we will generalise this definition to allow a quantum process to depend on (i.e. 'be controlled by') the outcome of an earlier process.

Physics is all about making predictions, so if many alternatives are possible, one wants to know how probable each of these alternatives is. This is precisely what quantum processes consisting of numbers:

$$\left(\langle p_i \rangle\right)^i$$

do for us. We already know that doubling guarantees that the numbers are positive, and this new multibranch version of causality guarantees that these numbers moreover form a probability distribution:

$$\sum_i \langle p_i \rangle = \boxed{}$$

We can also associate a probability distribution to quantum processes consisting of states. Given such a quantum process:

$$\left(\boxed{\rho_i}\right)^i$$

the weight of each state provides its probability:

$$P(\rho_i) := \boxed{\rho_i} \tag{6.65}$$

As mentioned in Section 6.2.3, this is nothing but an instance of the Born rule. Again, causality means that these probabilities add up to one:

$$\sum_i \; \raisebox{-0.5em}{\rule{0pt}{2em}} \; \overset{\displaystyle \equiv}{\underset{\rho_i}{\nabla}} \; = \; \boxed{}$$

Unlike states, we cannot associate a fixed probability distribution to general quantum processes. This should come as no surprise, since in the case that a quantum process is applied to a system in a certain state, the probabilities may of course depend on that state. For example, although we haven't yet said what a quantum measurement is, it should be evident that the outcome of performing anything called a 'measurement' will depend on what is actually being measured. However, once we apply a quantum process to a state, we can again assign probabilities, and the Born rule tells us what they are:

$$P(\Phi_i \mid \rho) \; := \; \left. \begin{array}{c} \overset{\displaystyle \equiv}{\boxed{\Phi_i}} \\ \underset{\rho}{\nabla} \end{array} \right\} \begin{array}{l} \text{effect} \\ \\ \text{state} \end{array}$$

and once again causality guarantees that they add up to one:

$$\sum_i \; \overset{\displaystyle \equiv}{\underset{\rho}{\boxed{\Phi_i}}} \; = \; \overset{\displaystyle \equiv}{\underset{\rho}{\nabla}} \; = \; \boxed{}$$

Thus, the causality postulate (6.64) for a generic process guarantees that:

probabilities can be consistently assigned to branches.

In the case of deterministic processes, causality reduces to the form in which we already encountered it in Definition 6.52:

$$\overset{\displaystyle \equiv}{\boxed{\Phi}} \; = \; \overset{\displaystyle \equiv}{\rule{0pt}{1.5em}\mid}$$

So quantum processes generalise <u>causal</u> quantum maps, and in particular, a deterministic quantum process consists of a single, causal quantum map.

Remark 6.88 In the form of (6.64), the causality postulate doesn't seem to admit the interpretation we had provided it with in the case of quantum states and quantum maps, namely, that discarding the outputs of a process is the same as discarding its inputs without the process ever happening. This is because knowledge of which branch Ψ_i happened should actually be treated as a *classical output* of the process. Once we have a diagrammatic account on classical data, it will become clear that summing together all of the branches is indeed the result of 'discarding' this classical output, in which case (6.64) says the process might as well have never happened.

Now, consider what happens if we compose two non-deterministic processes sequentially. If the first process has, say, four branches (indexed by i), and we feed its output to another process that has three branches (indexed by j), then any combination of the i-braches and the j-branches could happen. So, the resulting process:

$$\left(\boxed{\Psi_j} \right)^j \left(\boxed{\Phi_i} \right)^i := \left(\boxed{\Psi_j} \quad \boxed{\Phi_i} \right)^{ij}$$

has $4 \cdot 3 = 12$ branches (indexed by ij). We can use a similar rule to compose processes in parallel, and even make arbitrary circuits:

$$\left(\boxed{\Psi_j} \right)^j \left(\boxed{\Psi'_l} \right)^l \left(\boxed{\Phi_i} \right)^i \left(\boxed{\Phi'_k} \right)^k := \left(\boxed{\Psi_j} \; \boxed{\Psi'_l} \quad \boxed{\Phi_i} \; \boxed{\Phi'_k} \right)^{ijkl} \tag{6.66}$$

That is, we can pull these braces to the outside of the picture, much like we could do with sums, which, as we will see in Chapter 8, is not entirely a coincidence. The result is again a quantum process because of the following.

Proposition 6.89 Causality is preserved when forming circuits of quantum processes.

Proof We show this for the composition structure in (6.66), which encompasses both parallel and sequential composition, and hence arbitrary circuits. By causality of the quantum processes $(\Psi_j)^j$ and $(\Psi'_l)^l$ we have:

$$\sum_{i,j,k,l} \boxed{\Psi_j} \; \boxed{\Psi'_l} \; \boxed{\Phi_i} \; \boxed{\Phi'_k} \;\; = \;\; \sum_j \boxed{\Psi_j} \; \sum_l \boxed{\Psi'_l} \; \sum_i \boxed{\Phi_i} \; \sum_k \boxed{\Phi'_k} \;\; = \;\; \sum_i \boxed{\Phi_i} \; \sum_k \boxed{\Phi'_k}$$

and by causality of the quantum processes $(\Phi_i)^i$ and $(\Phi'_k)^k$ we have:

$$\sum_i \boxed{\Phi_i} \; \sum_k \boxed{\Phi'_k} \;\; = \;\; \overline{\top} \;\; \overline{\top}$$

that is, causality of the four quantum processes in the LHS of (6.66) implies causality of the RHS of (6.66). □

Remark 6.90 Quantum processes 'almost' form a process theory. When we defined process theories back in Chapter 3, we said they give an interpretation for <u>diagrams</u> of processes, not diagrams with these funny parentheses ( )^i in them. We'll fix this in Chapter 8 when we replace parentheses with classical wires and get a bonafide process theory. But in the meantime, there's no harm in treating this 'almost' process theory as if it really were one.

Example 6.91 Thanks to composition of quantum processes, the three cases we gave earlier of assigning probabilities to branches all boil down to the first case of quantum processes consisting of numbers:

$$
\left(\widehat{\rho_i} \right)^i = \left(\widehat{\rho_i} \right)^i \qquad \left(\widehat{\Phi_i} \right)^i = \left(\widehat{\Phi_i} \right)^i
$$

6.4.2 Non-deterministic Realisation of All Quantum Maps

Since non-deterministic quantum processes are going to rescue our beloved string diagrams, then at the very least they should provide us with some effects other than discarding. The following is an example of a quantum process that does just that.

Proposition 6.92 For any ONB:

$$
\left\{ \widehat{i} \right\}_i \tag{6.67}
$$

the corresponding doubled effects form a quantum process:

$$
\left(\widehat{i} \right)^i \tag{6.68}
$$

Proof Vertically reflecting (6.46) without the $\frac{1}{D}$-factor yields:

$$
\sum_i \widehat{i} = \overline{\overline{\top}}
$$

so (6.68) is indeed causal, and hence a quantum process. □

This is in fact one of the most important examples of a quantum process, which is called an *ONB measurement*. We will study this and related processes extensively in the next chapter. For now, it will provide us with a sufficient dose of non-determinism to generate arbitrary quantum maps.

Since the Bell basis is an ONB (cf. Section 5.3.6), by Proposition 6.92 there is a corresponding quantum process:

$$\left(\begin{array}{c} B_i \end{array} \right)^i$$

which up to a number includes the doubled cap as a branch:

$$\begin{array}{c} B_0 \end{array} := \tfrac{1}{2} \; \cap$$

This holds not just for the cap for qubits, but also for arbitrary dimension, which, in turn, will allow us to recover all quantum maps.

Lemma 6.93 Bell effects can be realised non-deterministically. Explicitly, there exists a quantum process:

$$\left(\begin{array}{c} \widehat{\phi_i} \end{array} \right)^i \qquad \text{such that} \qquad \begin{array}{c} \widehat{\phi_1} \end{array} := \tfrac{1}{D} \; \cap \qquad (6.69)$$

Proof Since by Proposition 5.79 any normalised state can be regarded as a member of an ONB, there exists an ONB containing the normalised cup:

$$\left\{ \begin{array}{c} \phi_1 \end{array} := \tfrac{1}{\sqrt{D}} \; \cup \; , \; \begin{array}{c} \phi_2 \end{array} \; , \; \ldots \; , \; \begin{array}{c} \phi_{D^2} \end{array} \right\} \qquad (6.70)$$

Then, by Proposition 6.92 it follows that (6.69) is a quantum process. □

Theorem 6.94 Every quantum map Φ can be realised non-deterministically, up to a number. Explicitly, there exists a quantum process:

$$\left(\begin{array}{c} \Psi_i \end{array} \right)^i \qquad \text{such that} \qquad \begin{array}{c} \Psi_1 \end{array} := r \begin{array}{c} \Phi \end{array} \qquad (6.71)$$

for some $r > 0$.

Proof Choose k such that the following state is causal:

When we compose this quantum process with the quantum process (6.69) that we constructed in Lemma 6.93:

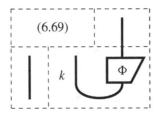

we obtain the quantum process:

$$\left(\boxed{\Psi_i}\right)^i := \left(k\;\boxed{\widehat{\phi_i}}\;\boxed{\Phi}\right)^i$$

where:

$$\boxed{\Psi_1} = \frac{k}{D}\;\boxed{\Phi} = \frac{k}{D}\boxed{\Phi}$$

which completes the proof, for $r := \frac{k}{D}$.   □

Exercise 6.95 A quantum map can be realised by either encoding it as a state or an effect:

$$\text{effect}\;\;\boxed{\Phi} = \boxed{\Phi} = \boxed{\Phi}\;\;\text{state}$$

Construct an alternative proof of Theorem 6.94 by encoding Φ as an effect that occurs as a branch of a non-deterministic process.

6.4.3 Purification of Quantum Processes

By Proposition 6.46 we know that each quantum map can be purified, and hence, all deterministic quantum processes can be purified. In fact, this was pretty much built in to the notion of a quantum map, which is defined to be the result of composing pure quantum maps with discarding. Purification has the appealing interpretation that every quantum map arises from a pure one by discarding an output, and, by causality, this pure quantum map can be assumed to be an isometry (Stinespring dilation).

At first glance, it seems to trivially follow that we can also purify non-deterministic quantum processes. Indeed, each branch in any quantum process:

$$\left(\begin{array}{c} {}^{\widehat{B}} \\ \boxed{\Phi_i} \\ {}_{\widehat{A}} \end{array} \right)^i$$

can be purified, that is, there exist:

$$\left(\begin{array}{c} {}^{\widehat{B}} \quad {}^{\widehat{C}_i} \\ \boxed{\widehat{g}_i} \\ {}_{\widehat{A}} \end{array} \right)^i \tag{6.72}$$

such that:

$$\begin{array}{c} {}^{\widehat{B}} \\ \boxed{\Phi_i} \\ {}_{\widehat{A}} \end{array} = \begin{array}{c} {}^{\widehat{B}} \quad {}^{\widehat{C}_i} \\ \boxed{\widehat{g}_i} \\ {}_{\widehat{A}} \end{array}$$

However, this is where we hit a bit of a snag. For each i the type $\widehat{C}_i$ might be different, in which case the processes $\widehat{g}_i$ may have different output types, meaning (6.72) is no longer a well-defined quantum process.

Fortunately, this problem can be fixed by finding a suitable *joint purification* of the branches of a quantum process:

Lemma 6.96 For any quantum process:

$$\left(\begin{array}{c} {}^{\widehat{B}} \\ \boxed{\Phi_i} \\ {}_{\widehat{A}} \end{array} \right)^i$$

there exists a single type $\widehat{C}$ and a quantum process:

$$\left(\begin{array}{c} {}^{\widehat{B}} \quad {}^{\widehat{C}} \\ \boxed{\widehat{f}_i} \\ {}_{\widehat{A}} \end{array} \right)^i$$

such that:

$$\begin{array}{c} {}^{\widehat{B}} \\ \boxed{\Phi_i} \\ {}_{\widehat{A}} \end{array} = \begin{array}{c} {}^{\widehat{B}} \quad {}^{\widehat{C}} \\ \boxed{\widehat{f}_i} \\ {}_{\widehat{A}} \end{array} \tag{6.73}$$

Proof We know that for each quantum map Φ_i there exists a purification:

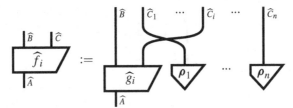

For every type C_j with $j \neq i$ pick a causal state ρ_j. To each pure quantum map $\widehat{g}_i$ we then associate another pure quantum map $\widehat{f}_i$ as follows:

So in particular we have:

$$\widehat{C} := \widehat{C}_1 \otimes \cdots \otimes \widehat{C}_n$$

where n is the number of outcomes of the quantum process $(\Phi_i)^i$. By causality of all the states ρ_i we then have:

and hence (6.73) indeed holds.   □

The choice of purification, and the choice of purifying system $\widehat{C}$ in particular, is not unique. While the proof of Lemma 6.96 gives a simple, diagrammatic way to construct $\widehat{C}$, it is not optimal in the sense that the dimension of C in general won't be the smallest possible one. In particular, when the maps Ψ_i are not already pure, the dimension of C from Lemma 6.73 grows exponentially with the number of maps.

On the other hand, we can stick an upper bound on how big C needs to be, which depends only on the dimensions of A and B. This upper-bound result is sometimes known as *Choi's theorem*.

Exercise 6.97 Show that one can always purify a quantum process:

using the auxiliary system $\widehat{C}$ to be $\widehat{A} \otimes \widehat{B}$, by showing that the linear map of the form:

$$\sqrt{\;\;f\;\;}^{A\;B\;B}_{\quad A} \;=\; \sum_i \frac{1}{\sqrt{D}} \; \bigcup \; \boxed{i}^{A\;B} \; r_i \; \boxed{i}^{B}_{\;A}$$

yields a purification of any quantum map Φ_i:

for a suitably chosen Hilbert–Schmidt basis. (Hint: rely on Corollary 6.69.)

Remark 6.98 While Exercise 6.97 gives an upper bound on the dimension needed for C, it can in general be smaller. For example, even applying Lemma 6.96 to a set of pure quantum maps will yield a smaller (in this case, trivial) system C. On the other hand, sometimes the full system $\widehat{A} \otimes \widehat{B}$ is necessary. For example, in:

$$\text{[diagram]}$$

the auxiliary system $\widehat{C}$ clearly at least needs to be $\widehat{A} \otimes \widehat{B}$ for any joint purification procedure.

So far so good. We saved the idea of purification for the case of non-deterministic quantum processes. But what about Stinespring dilation? In particular, one may wonder if each branch could be chosen to be an isometry (possibly multiplied by some probability). The answer is no. One counter-example is any quantum process of effects (e.g. the ONB of effects from Proposition 6.92), since clearly no isometry can exist from a non-trivial system to a trivial one.

However, what we can do is associate a <u>single</u> isometry to any quantum process, which in fact, as we shall see in Section 7.3.4, does provide a very satisfactory generalisation of Stinespring dilation to non-deterministic quantum processes, called *Naimark dilation*.

6.4.4 Teleportation Needs Classical Communication

With what we've learned in this chapter, let's now try to fill in the quantum teleportation diagram:

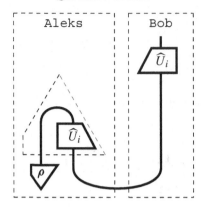

with quantum processes.

Starting with ρ and half of a Bell state, Aleks needs to perform some quantum process that produces a Bell effect, possibly with some error (represented by the pure quantum map $\widehat{U}_i$). Then, some time later Bob should perform another quantum process to correct the error.

We established in (6.62) that we definitely need non-deterministic quantum processes in order to realise teleportation, so Aleks should perform a non-deterministic process consisting of a set of effects:

$$\left(\frac{1}{D} \quad \boxed{\widehat{U}_i} \right)^{i} \tag{6.74}$$

such that causality is satisfied:

$$\sum_i \frac{1}{D} \quad \boxed{\widehat{U}_i} \quad = \quad \overline{\overline{\top}} \; \overline{\overline{\top}} \tag{6.75}$$

At this point Aleks communicates the value i produced by his quantum process to Bob and Bob performs his correction. Even though we know how to express non-deterministic quantum processes, we haven't yet seen an instance of a quantum process that <u>depends</u> on the outcome of another process. Before we go to the effort of defining such a thing, we should ask ourselves: is Bob's correction really necessary?

In other words, what happens if Aleks decides not to communicate the value i to Bob? Consider a version of teleportation without a correction:

$$\left(\boxed{\widehat{U}_i} \atop \rho \right)^{i} = \left(\boxed{\widehat{U}_i} \atop \rho \right)^{i}$$

The overall process is a non-deterministic quantum state (we've omitted the $\frac{1}{D}$ for clarity). However, Bob doesn't have access to the value i. To account for this lack of information, what Bob perceives is actually just a mixture of each of the possible branches. Then, by relying on causality of the quantum process we obtain:

$$\sum_i \left(\begin{array}{c} \widehat{U_i} \\ \rho \end{array} \right) = \left(\sum_i \begin{array}{c} \widehat{U_i} \\ \rho \end{array} \right) \overset{(6.75)}{=\!=} \begin{array}{c} \overline{\overline{}} \ \overline{\overline{}} \\ \rho \end{array} = \begin{array}{c} \overline{\overline{}} \end{array}$$

(6.75)

So Bob receives the maximally mixed state, with no sign of ρ anywhere. Hence we can conclude that without any classical communication, nothing can be teleported whatsoever.

6.4.5 Controlled Processes

Right, so instead of hiding the value i, Aleks must send it to Bob so that Bob can perform a correction depending on the outcome i:

$$\boxed{\widehat{U_i}}$$

Now, while it might be tempting to think of the set:

$$\left\{ \boxed{\widehat{U_i}} \right\}_i$$

(6.76)

as a non-deterministic quantum process, this is not the case! In fact, since <u>each</u> of the elements is itself a quantum process:

$$\forall i \; : \; \boxed{\widehat{U_i}} = \begin{array}{c} \overline{\overline{}} \end{array}$$

it follows the whole N-element set cannot be a quantum process:

$$\sum_i \boxed{\widehat{U_i}} = \sum_i \begin{array}{c} \overline{\overline{}} \end{array} = N \begin{array}{c} \overline{\overline{}} \end{array} \neq \begin{array}{c} \overline{\overline{}} \end{array}$$

All together there is only one index i in play. The key difference between the set (6.76) and a quantum processes is that a quantum process produces a value of i, whereas (6.76) depends on it. In fact, there is no reason why we should restrict to deterministic processes here: depending on i we may wish to perform one (non-deterministic) quantum process out of a set of quantum processes. We will refer to such a set of quantum processes depending on some classical index as *controlled* by i.

So, the overall quantum teleportation process should look something like this:

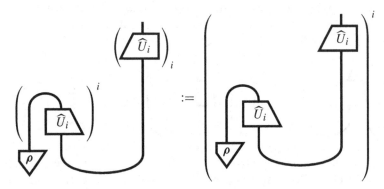

where the brackets still indicate that there are several branches, indexed by i, but these branches are related to the index i in two different ways. The process with an upper index i produces the value of i, whereas the process with a lower index i depends on, or consumes, the value of i.

One can also imagine scenarios where i is consumed several times. For example, in:

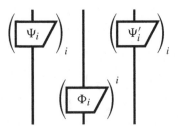

the quantum process $(\Phi_i)^i$ produces i, while the two controlled quantum processes $(\Psi_i)_i$ and $(\Psi'_i)_i$ consume it. Although we can consume i any number of times, we cannot force several non-deterministic processes to have the same outcome, so i must be produced by exactly one process.

Remark 6.99 The use of upper and lower indices to model classical inputs/outputs mirrors that of the diagram formulas from Section 3.1.3. The fact that one classical outcome can be used multiple times comes from the fact that classical data can be *copied* (i.e. 'cloned'). We will see how this unique feature of classical data manifests in Chapter 8.

We are now ready to give a full definition of a quantum process, which includes the possibility of being controlled by a classical input. By combining the conditions:

$$\forall i : \quad \overbrace{\Phi_i} = \overline{\overline{\top}} \quad \text{and} \quad \sum_j \overbrace{\Phi_j} = \overline{\overline{\top}}$$

into one, we get the full definition of a quantum process.

Definition 6.100 A *quantum process* is a set of quantum maps:

$$\left(\begin{array}{c} \Phi_{ij} \end{array} \right)^{j}_{i}$$

which satisfy:

$$\forall i : \quad \sum_{j} \begin{array}{c} \Phi_{ij} \end{array} \; = \; \overline{\top} \tag{6.77}$$

For $1 \leq i \leq m, 1 \leq j \leq n$, if $m = 1$ above, this definition yields our prior notion of an uncontrolled quantum process. If $n = 1$, this yields a controlled quantum process consisting entirely of causal quantum maps (i.e. without non-determinism), like Bob's unitary corrections:

$$\left(\begin{array}{c} \widehat{U}_i \end{array} \right)_{i}$$

Of course, if both m and n are 1, this is just a single causal quantum map.

You may wonder why we only defined individual quantum processes rather than the theory of **quantum processes**. At this point the indices i for a quantum process are treated as something external to the diagram, rather than something that lives on a wire. This doesn't really fit in with our definition of a process theory, which is purely diagrammatic. We will fix this in Chapter 8 by replacing these indices with *classical wires*.

6.4.6 *Quantum Teleportation in Detail*

And finally, the moment you've been waiting for: we now have all of the ingredients to fully describe quantum teleportation. In the case of qubits, quantum teleportation is realised as follows:

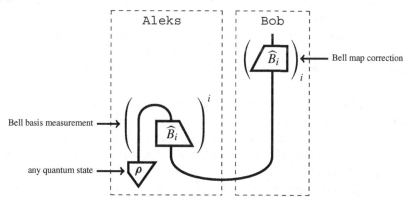

for B_i the ith Bell map. In words:

1. Aleks is in possession of an arbitrary quantum state ρ, which he wishes to send to Bob, and Aleks and Bob together share a Bell state.
2. Aleks performs a *Bell basis measurement*, that is, the following non-deterministic quantum process:

   on ρ and his half of the Bell state.
3. Aleks sends the outcome i to Bob.
4. Bob performs a *Bell map correction*, that is, the following controlled quantum process:

$$\left(\widehat{B_i}\right)_i$$

As we have seen many times by now, the result of performing this protocol is that, regardless of Aleks' measurement outcome, Bob will receive ρ:

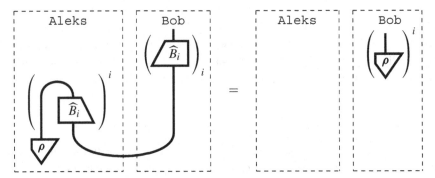

Back in Section 4.4.4 we also emphasised that teleportation does not enable one to magically beam something through space, but rather enables one to to use one kind of data (classical) to send another kind of data (quantum). Since there are four elements in the Bell basis, Aleks has to send a number from 0 to 3 (i.e. two classical bits) to Bob. In exchange, Bob receives Aleks' qubit, which could be any one of infinitely many points in the Bloch ball.

One remarkable thing about quantum teleportation is that, while it is essential that we rely on non-deterministic quantum processes, we produce a deterministic process overall. That is, no matter what outcome Aleks' measurement produces, the overall process will always give Bob the state ρ. The reason is of course that each of the branches, due to the correction made by Bob, gives rise to the same process. This shows that with clever tricks one can 'undo' the non-determinism of quantum processes.

While the protocol described just now is the simplest non-trivial case of a quantum teleportation protocol, there is nothing particularly special about the Bell matrices or qubits, in terms of the role they play in teleportation. To do teleportation, all we need is a set of unitaries:

$$\left\{\;\boxed{U_i}\;\right\}_i$$

such that:

$$\left\{\;\frac{1}{\sqrt{D}}\;\boxed{U_i}\;\right\}_i \tag{6.78}$$

forms an ONB. Then, there is a process consisting of ONB effects and corresponding corrections such that we have:

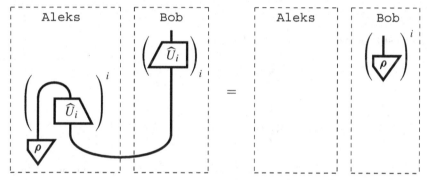

Relying on Theorem 5.32, the conditions for the set (6.78) forming an ONB can equivalently be stated as the following two equations:

$$\frac{1}{D}\;\boxed{\begin{array}{c}U_j\\U_i\end{array}} \;=\; \delta_i^j \tag{6.79}$$

$$\sum_i \frac{1}{D}\;\boxed{\begin{array}{c}U_i\\U_i\end{array}} \;=\; \Big\|\;\Big\| \tag{6.80}$$

Thus we can conclude the following.

Corollary 6.101 A set of unitaries:

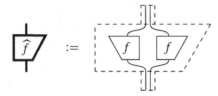

yields a teleportation protocol whenever (6.79) and (6.80) are satisfied.

6.5 Summary: What to Remember

1. Pure quantum maps arise from 'doubling' linear maps:

Composing pure quantum states and effects:

yields the *Born rule* for quantum theory. This process theory admits string diagrams, where the cups and caps are given by:

2. Let D and D' be arbitrary diagrams in **linear maps**, and $\widehat{D}$ and $\widehat{D}'$ be their doubled versions in **pure quantum maps**, then:

We refer to the number $e^{i\alpha}$ as a *global phase*.

3. *Discarding* is the linear map:

which fails to be a pure quantum map. The process theory of **quantum maps** arises from composing pure quantum maps with discarding:

Any quantum map Φ can be written in the following form:

$$\includegraphics{}$$

and we call $\widehat{f}$ a *purification* of Φ. For example, the *maximally mixed state* can be purified by means of the *Bell state*:

4. The *causality* postulate:

$$\text{(6.81)}$$

implements the following obvious (and necessary) fact:

> *If the output of a process is discarded, it may as well have never happened.*

This implies that we can discard everything that is not 'connected' to what we care about, which is an obvious prerequisite for basic scientific practice. Causality also imposes compatibility with the theory of relativity, notably this fact:

> *Nothing can attain speeds that are faster than light.*

5. Causality imposes the following restrictions:

- all <u>pure</u> quantum maps are <u>isometries</u>, and hence
- all pure quantum maps from a system to itself are <u>unitary</u>.

The following fact is known as *Stinespring dilation*:

- Any causal quantum map Φ can be purified as an isometry $\widehat{U}$:

$$\text{[diagram: } \Phi \text{]} = \text{[diagram: } \widehat{U} \text{ with discard]}$$

6. Since discarding decomposes for any ONB as follows:

$$\text{[diagram]} = \text{[diagram]} = \sum_i \text{[diagram]} = \sum_i \text{[diagram]}$$

every quantum map Φ admits a *Kraus decomposition*:

$$\text{[diagram: } \Phi \text{]} = \sum_i \text{[diagram: } \widehat{f_i} \text{]}$$

If such representation is moreover of the form:

$$\text{[diagram: } \Phi \text{]} = \sum_i p^i \text{[diagram: } \widehat{f_i} \text{]}$$

where $\{p^i\}_i$ is a probability distribution and all $\widehat{f_i}$ are causal, then we call it a *mixture*. Every causal quantum <u>state</u> can be regarded as a mixture of pure causal quantum states. However, this is <u>not</u> the case for every causal quantum <u>map</u>.

⚠ The notation for mixing involving sums is only a <u>temporary</u> one. In Chapter 8, we will picturalise these.

---

7. If the *reduced process* of any quantum process is pure:

$$\text{[diagram: } \Phi \text{ with discard]} = \text{[diagram: } \widehat{f} \text{]}$$

then the state Φ separates as follows:

$$\text{[diagram: } \Phi \text{]} = \text{[diagram: } \rho \text{][diagram: } \widehat{f} \text{]}$$

for some (causal) quantum state ρ.

8. Quantum states cannot be broadcast; i.e. there exists no quantum map Δ such that for all quantum states ρ we have:

$$
\vcenter{\hbox{\includegraphics{delta1}}} = \vcenter{\hbox{\includegraphics{delta2}}} = \vcenter{\hbox{\includegraphics{rho}}}
$$

The key ingredient for proving this is **7**.

9. **Quantum processes** are lists of quantum maps:

$$
\left(\vcenter{\hbox{\includegraphics{phii}}} \right)^{i}
$$

satisfying (generalised) *causality*:

$$
\sum_i \vcenter{\hbox{\includegraphics{phii}}} = \vcenter{\hbox{\includegraphics{ground}}}
$$

The quantum maps Φ_i are the *branches*, and if there is more than one branch, then the quantum process is *non-deterministic*. When a quantum process is applied to a state, then the probability of each branch is computed using the Born rule:

$$
P(\Phi_i \mid \rho) := \left. \vcenter{\hbox{\includegraphics{phii_effect}}} \right\} \text{effect} \atop \left. \vcenter{\hbox{\includegraphics{rho}}} \right\} \text{state}
$$

> ⚠ The notation involving braces and sums is only a <u>temporary</u> one. In Chapter 8, we will picturalise these too.

10. All quantum maps arise as branches of quantum processes. That is, for every quantum map Ψ, there exists a quantum process:

$$
\left(\vcenter{\hbox{\includegraphics{psii}}} \right)^{i} \qquad \text{such that} \qquad \vcenter{\hbox{\includegraphics{psi1}}} := r \vcenter{\hbox{\includegraphics{phi}}}
$$

for some $r \geq 0$.

6.6 Advanced Material*

By the time we started this chapter, we had already made a huge, (mostly) unmotivated assumption: that quantum processes should be built up from **linear maps**. In this section, we'll look at how one might go about getting around this assumption. In particular, we take a closer look at the *doubling construction* (which throughout this section includes adjoining discarding), and in particular, how it can be understood not just as a construction on **linear maps**, but something that can be applied to any process theory admitting string diagrams. By doing so, we can hope to understand or even reproduce all of the predictions of quantum theory in a way that is completely independent of Hilbert spaces and linear maps.

We end this chapter with something completely different.

6.6.1 Doubling General Process Theories*

There is nothing particularly special about **linear maps** that enabled us to construct a doubled process theory. Given any process theory **p** that admits string diagrams, we obtain a new process theory $\mathcal{D}(\mathbf{p})$, again admitting string diagrams, which is made up of all processes of the form:

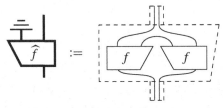

where f can be any process in **p**. Many of the theorems we proved in this chapter and later also don't depend on linear maps or have natural generalisations. For example, as we already noted in Remark 6.19, the proof that doubling eliminates 'generalised global phases' was purely diagrammatic, so it immediately generalises.

Theorem 6.102 Let f and g be processes in any process theory that admits string diagrams, then we have:

$$\widehat{f} = \widehat{g}$$

if and only if there exist λ and μ such that:

$$\lambda \boxed{f} = \mu \boxed{g} \qquad \text{and} \qquad \lambda \bar{\lambda} = \mu \bar{\mu}$$

Furthermore, any process theory of the form $\mathcal{D}(\mathbf{p})$ admits a discarding map, so things such as causality, purification, and Stinespring dilation all still make perfect sense.

Thus many of the concepts we develop apply not just to quantum theory, but can be used to study a whole family of 'quantum-like' theories. The upshot of doing so is that, first, by drawing contrasts with the other theories, this provides insights into the true identity of quantum theory. Second, as discussed in Section 1.2.4, the future theories of physics might not be based on **linear maps** but on some new kind of beast.

Exercise* 6.103 Characterise $\mathcal{D}($**relations**$)$. What are its states? What are its general processes? What does causality mean?

There are in fact many equivalent ways to build such a doubled process theory. Notably the 'twisted' version we use in this book:

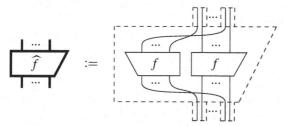

is easy to work with diagrammatically because it uses the same parallel composition as the non-doubled theory and keeps the 'two halves' of each system close together. This comes in particularly handy in Chapter 8 when we start modelling classical-quantum interaction.

The original version of this construction (which is usually called the *CPM-construction*) avoids this twisting:

but at the price of needing to define a new parallel composition in the doubled category:

Now, the 'two halves' of each system might be far apart, but you could imagine folding the paper along the dotted line to see which systems should be combined together to make thick wires. A discarding process then introduces a connection between the left half and the right half:

This version of doubling is in fact easier to define using category theory (where it originates), but its diagrams start to become messy fairly quickly.

6.6.2 Axiomatizing Doubling*

Can we tell if a process theory is the result of doubling? You might say: 'That's easy, just check if the wires are thick!' Okay, so maybe we should rephrase the question. What we are looking for here is an axiomatic characterisation of process theories of the form $\mathcal{D}(\mathbf{p})$ for some $\mathbf{p}$.

The crucial aspect of that theory is that it has discarding processes that can be used to purify any process. Note that, since we don't know (in advance) that a process theory $\mathbf{P}$ is of the form $\mathcal{D}(\mathbf{p})$, we needed to give a characterisation of 'pure' processes that doesn't refer to doubling. Here is the result.

Definition 6.104 A *discarding structure* for a process theory admitting string diagrams $\mathbf{P}$ consists of:

- a subtheory $\mathbf{p} \subseteq \mathbf{P}$ of 'pure processes' that includes all of the types from $\mathbf{P}$, and
- for each type A, a discarding effect:

which are subject to the following axioms:

1. All processes f, g in $\mathbf{p}$ satisfy:

2. Every process Φ in $\mathbf{P}$ can be *purified*; that is, there always exists a pure process f in $\mathbf{p}$ such that Φ arises by discarding part of its output:

3. Discarding two systems means discarding each of them; discarding nothing means doing nothing:

and the transpose of discarding is the same as its adjoint:

Each process theory with a discarding structure is equivalent to one that arises from doubling some process theory.

Theorem 6.105 If **P** has a discarding structure, then:

$$\mathcal{D}(\mathbf{p}) \cong \mathbf{P}$$

Proof We prove this by showing a bijection between the processes in $\mathcal{D}(\mathbf{p})$ and the processes in **P**. Since **p** has the same types as **P**, there is an obvious correspondence between the types $\widehat{A}$ in $\mathcal{D}(\mathbf{p})$ and A in **P**. Similarly, since **p** is a subtheory of **P**, there is an obvious correspondence between pure processes $\widehat{f}$ in $\mathcal{D}(\mathbf{p})$ and their counterparts f in **P**. But what about the impure processes? For these, we first purify, then interpret (the normal) discarding process as the discarding structure from **P**:

$$\tag{6.82}$$

For this map to be well defined, it should not depend on the particular choice of purification; i.e. it should be the case that:

and for it to be injective, the reverse implication should hold:

Both of these can be proven using Axiom 1 in Definition 6.104 and a healthy dose of process–state duality. Axiom 2 guarantees that the mapping (6.82) is surjective. From these facts and Axiom 3, it is then fairly straightforward to show that this mapping also preserves string diagrams. We leave it to the reader to fill in the details. $\qquad\square$

One would expect the converse to hold as well; namely, process theories of the form $\mathcal{D}(\mathbf{p})$ should carry a discarding structure, where the pure processes are precisely those of the form $\widehat{f}$ for f in **p**. It turns out this is almost true, provided that the pure processes in $\mathcal{D}(\mathbf{p})$ satisfy one extra property.

Exercise 6.106 Show that if for pure processes in $\mathcal{D}(\mathbf{p})$ we have that:

then $\mathcal{D}(\mathbf{p})$ carries a discarding structure. This condition can be read as: 'the doubled theory has no global phases'.

6.6.3 And Now for Something Completely Different*

We now know that string diagrams and the doubling construction are useful tools for expressing quantum processes. However, surprisingly, they have been employed in a totally different area: computational linguistics!

A typical problem in computational linguistics is to take a pair of words and compute how similar they are in meaning. For example, 'dog' and 'hound' should be very similar, 'dog' and 'cat' less similar (but at least they are still both animals), and 'dog' and 'taxes' should not be similar at all. The most common way to solve this problem is to compute vectors for the words 'dog', 'hound', and 'taxes', often by scanning a big body of text and 'learning' the vectors automatically using some techniques from artificial intelligence. Then, as we already saw back in Section 4.3.3, we can measure the similarity (i.e. 'commonality' or 'overlap') just by taking the inner product.

This method is now standard, and you will find it under the hood of basically any software that deals with human language somehow (web search, translation, targeted advertising, etc.). But where things really start to get interesting is in trying to extend this from words to sentences.

Consider this possible outcome of co-authoring a textbook:

Each word is a state, and different kinds of words (nouns, transitive verbs, intransitive verbs, etc.) have different types (hence the different numbers of wires). Then, to compute the meaning of this sentence, all we need to do is wire these words together. For this, we need a set of rules for how to combine the words together, but that's just called *grammar*:

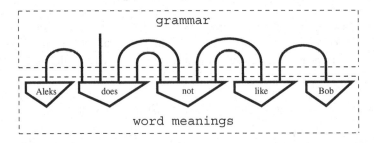

We treat some words as 'black boxes', whereas others may be represented as diagrams themselves, indicating how they interact with other words:

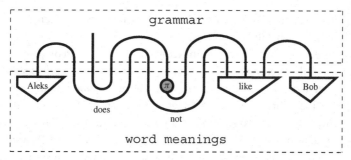

at which point we can start to do the types of diagrammatic calculations we've seen all throughout this book.

So what does doubling everything do for us? Well, just as in the quantum case, it makes room for 'impurity'. Does this have a place in this model of language? Of course it has! Many words are ambiguous; for example, Bob may refer to a folk singer, a reggae singer, a cartoon builder, a cartoon sponge, or one of the authors of this book. We can use mixing to represent this ambiguity.

6.7 Historical Notes and References

Max Planck (1900) is considered as the originator of quantum theory. He made his discovery when studying how to get the most light out of light bulbs with the minimal amount of electricity, a job he was doing for the electricity companies. For this purpose, he studied the radiation of light by a perfect absorber of light and realised that the energy carried by light could only be emitted in certain packages, called *quanta*. These packages of energy are given by *Planck's law*, $E = h\nu$, where h is *Planck's constant*, ν is the frequency of the light, and E is the corresponding energy package.

Then, Heisenberg (1925) formulated a predecessor of quantum theory using indexed sets of 'transition amplitudes'; Born and Jordan (1925) realised that Heisenberg was actually talking about (infinite) complex matrices; Schrödinger (1926) instead was speaking in waves; Dirac (1926) realised that these were actually one and the same thing; and finally von Neumann (1932) put everything together in the language of Hilbert space. The Born rule was introduced by Born (1926). Many more were involved in all of this at an earlier stage, and many obtained Nobel Prizes for their respective contributions, including Einstein, Bohr, Compton, de Broglie, Fermi, and Pauli. The Bloch sphere was introduced by Felix Bloch (1946) as a representation of the states of two-dimensional quantum systems, and spin-$\frac{1}{2}$ systems in particular.

Von Neumann (1927b) introduced density matrices, which were used to represent mixed states. Quantum maps, to which one usually refers as completely positive maps, were first proposed as the general dynamics of quantum systems in Sudarshan et al. (1961). As mathematical entities these maps had been around quite a bit longer as maps between C*-algebras (see e.g. Paulsen, 2002). It is in this context that Stinespring dilation first appeared (Stinespring, 1955). Kraus decomposition can be found in Kraus (1983) and the no-broadcasting theorem first appeared in Barnum et al. (1996). Another 'generalised 'no-broadcasting theorem' is Barnum et al. (2007), which, rather than process theories, concerns generalised probabilistic theories.

As we shall discuss in Section 7.3.2, von Neumann's formulation of quantum theory was quite different from the presentation of quantum theory as a process theory. The doubling construction of Section 6.1 that turns linear maps into pure quantum maps was introduced in Coecke (2007), including the proof of Theorem 6.17. The generalisation to all quantum maps from Section 6.2 was introduced by Selinger (2007) as the *CPM-construction*. The idea that this can be done by adding the discarding process was put forward by Coecke (2008), and the 'ground' notation for discarding was introduced in Coecke and Perdrix (2010). The axiomatisation of the CPM-construction of Section 6.6.2 also first appeared in Coecke (2008).

Quantum processes without classical input are also called *quantum instruments*, a notion originated by Davies (1976) and further developed by Ozawa (1984).

The causality postulate that governs the passage from quantum maps to quantum processes of Section 6.4 was only recently identified as a core principle of quantum theory, by Chiribella et al. (2010, 2011), where it was one of a series of axioms from which quantum theory was reconstructed. In fact, it was defined in terms of probabilities adding up to one, from which uniqueness of causal effects was derived. There is, however, a discrepancy between the terminology in those papers and in ours, in that those papers use the term 'deterministic' to mean 'causal' in the sense of equation (6.64), so this may also refer to what we call non-deterministic quantum processes.

The characterisation of all teleportation protocols as in Section 6.4.6 appeared first in Werner (2001). The first proof of the non-signalling theorem for quantum theory can be found in Ghirardi et al. (1980). The derivation of non-signalling from the causality principle as in Section 6.3.2 is taken from Coecke (2014b). A similar result is also in Fritz (2014) and Henson et al. (2014). The derivation of covariance from causality as in Section 6.3.3 is taken from Coecke and Lal (2013), in which earlier covariance results of Markopoulou (2000) and Blute et al. (2003) were generalised to causal process theories.

Abramsky and Coecke (2004) started the axiomatic approach that takes composition of systems as its starting point. There were earlier approaches that tried to take composition of systems as a starting point of a new axiomatic approach (see e.g. Coecke, 2000), but with no real success. Two recent reconstructions of quantum theory that take diagrams as their backbone are Chiribella et al. (2011) and Hardy (2012). We already provided references for the earlier axiomatic approaches to quantum theory at the end of Chapter 1. The alternative definition of purity hinted at in Remark 6.81 is taken from Chiribella (2014). In fact, Chiribella only defined purity of states in this way. It turns out that extending this definition to general processes only makes sense for quantum systems, and not classical ones. Since classical systems <u>can be broadcast</u>, not even the bare wire is 'pure' using this alternative definition, even though 'doing nothing' is about as 'pure' as it gets!

The use of quantum maps for describing meaning in natural language was initiated in Coecke et al. (2010c); a direct comparison with quantum teleportation can be found in Clark et al. (2014); and the interpretation of ambiguity as mixing is taken from Piedeleu et al. (2015).

7

Quantum Measurement

A new scientific truth does not triumph by convincing its opponents and making them see the light, but rather because its opponents eventually die, and a new generation grows up that is familiar with it.

– *Max Planck, 1936*

The only way quantum theory allows us to interact with quantum systems is via quantum processes. Thus, the only way we can extract information about the state of a quantum system is to apply some non-deterministic process and observe which of the branches happened. For that reason, many people refer to the act of applying certain quantum processes as *measurement* and sometimes even as *observation*. A consequence of this unfortunate terminology is that one of the most touted 'strange' features of quantum theory is often misleadingly described as follows: 'The mere act of observing a quantum state changes it.'

'Misleading', since the concept of measurement described in the previous paragraph is far from the passive concept of observation familiar from our macroscopic world, but is rather a non-trivial process that will almost always drastically affect the quantum state. So, the mysterious aspect of quantum theory is not that 'observation' alters the quantum state but, rather, that it is impossible, <u>even in principle</u>, to 'observe' a quantum system in the classical sense.

Given this fundamental restriction on how we can extract information from a quantum system, what can we learn about that system by means of a quantum process? The answer is, in fact, very little! First, performing a particular measurement could, for the vast majority of quantum states, yield any outcome i with a non-zero probability. In that case, obtaining an outcome i doesn't tell us much about what the state of the system was. Second, we say 'was', because the measurement will moreover irreversibly change the state according to the outcome. So rather than revealing the state of a system, a quantum measurement typically erases that state from history!

Nonetheless, these quantum measurements are crucial to quantum theory since they constitute the only interface between us, in our classical world, and the quantum world. The great insight of the quantum computing community is that the non-deterministic changes induced by these quantum measurements are not a nuisance, but rather an extremely useful resource. Indeed, in the previous chapter we saw how any quantum map can be realised by means of a non-deterministic quantum process. In fact, the quantum process we used to demonstrate this was the simplest kind of quantum measurement.

In the quantum teleportation protocol, it is Aleks' (non-deterministic) realisation of a cap via measurement that allows Bob to receive Aleks' state after performing a correction. We'll see in Section 7.2 that there are other very useful protocols that use a similar trick. For example, we can use measurements to glue short bits of entanglement into long bits of entanglement and give the crucial ingredient for *measurement-based quantum computation*, which we'll develop fully in Chapter 12.

When exploiting quantum measurement in these ways, the quantum algorithm and protocol designers have to be really clever in order to cancel out the resulting non-determinism (e.g. with unitary corrections in the case of quantum teleportation). This balancing act is pretty much the crux of the art of quantum algorithm and protocol design.

So what is a measurement exactly? From our point of view, there is nothing really to distinguish a measurement from any other non-deterministic quantum process, so in this chapter we will discuss three important families of processes that have been called measurements. These families are:

1. *ONB measurements*
2. *von Neumann measurements*
3. *POVM measurements*.

Each of these is more general than the previous one, and comes in two flavours, demolition and non-demolition measurements:

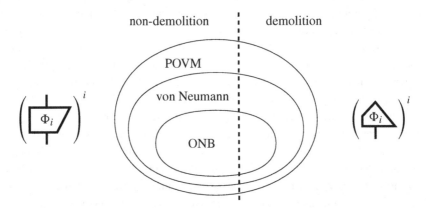

ONB measurements are the most specific, 'purest' kind of measurement, whereas POVM measurements, like impure quantum states, are much more general. In between these two extremal cases are von Neumann measurements, which over the past 80 years or so have been the most dominant notion of measurement. As the name suggests, they were the measurements that featured in von Neumann's original formulation of quantum theory.

One of the main reasons for this chapter is that we will be providing many concrete examples of quantum measurements, starting with the Stern–Gerlach device in Section 7.1.1. We give so many in fact that we won't bother putting them in special 'Example' sections.

7.1 ONB Measurements

The physicist working in the laboratory may be a bit confused by all this discussion about measurement. This is because, even though the result of performing a measurement on a state hardly resembles what one would expect from simple 'observation', in the laboratory many of these measurements boil down to observing something, for example, the position of a needle or a spot on a photo plate.

So there is a big mismatch between, on the one hand, how one acts on a system, i.e.:

- *one intends to observe it,*

and, on the other hand, what actually happens to that system, i.e.:

- *it undergoes a non-deterministic radical change.*

In order to see where this tension comes from, it helps to know what kinds of devices can actually perform quantum measurements.

7.1.1 A Dodo's Introduction to Measurement Devices

One of the earliest examples of a quantum measurement was performed by a *Stern–Gerlach apparatus*. We will explain here how this works in some detail.

A standard exercise in a course on classical electromagnetism is to study what happens to a spinning magnet as it passes through a magnetic field. If the field is the same every-where, i.e. *homogeneous*, all the forces cancel out, and the spinning magnet flies straight through. However, if it is stronger in some places than others, i.e. *inhomogeneous*, then it will get deflected according to the direction it is spinning. Classical physics predicts that it would be deflected in one direction (say up) if it is spinning clockwise (relative to the magnetic field), and in the other direction (say down) if it is spinning anticlockwise. The details of this calculation are not important. The important fact is that the state of our tiny magnet can encode (at least) one bit of information: say 0 for clockwise and 1 for anticlockwise. The way we can measure that bit is to send it through an inhomogeneous magnetic field and see which way it deflects.

We can now gradually tilt the rotation axis of the little spinning magnet, until the initial clockwise rotation has become an anticlockwise rotation:

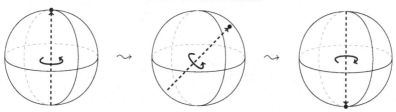

In this case, the deflection in an inhomogeneous magnetic field will also gradually change from up-deflection to down-deflection, and in particular, at some point passing through a state of no-deflection.

In the early twentieth century, it was discovered that electrons behave the same way as spinning magnets when they travel through a magnetic field. We can exploit this fact to design a device for measuring this 'deflect-up' versus 'deflect-down' property, called the Stern–Gerlach apparatus:

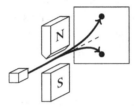

This device sends atoms through an inhomogeneous magnetic field (which is indicated by the fact that one of the magnets, here the N magnet, is pointed), then measures the direction they are deflected on a screen. Suppose, for simplicity, we choose hydrogen atoms, which have just one electron. If we send an atom whose electron is in the 'deflect-up' state, the atom is deflected upwards, and if it is 'deflect-down', the atom is deflected downwards. By analogy to the classical case, this property of electrons is called *spin*, though it doesn't have anything to do with something actually spinning.

By analogy to the classical case, we expect that, as we gradually tilt the spin-axis, we see a gradual change from up-deflection to down-deflection. So if we send a stream of atoms whose electrons have random spin-axes through such a device, we expect to get a continuous line of different possible deflections:

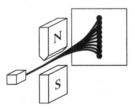

But the truly shocking thing that Stern and Gerlach discovered is that rather than getting a continuous line of outcomes, we always get one of precisely two outcomes, spin-up or spin-down:

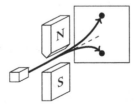

Thus, the spin-axis of an electron, which we can plot as an <u>arbitrary point</u> on a sphere, gets projected to one of just <u>two points</u>:

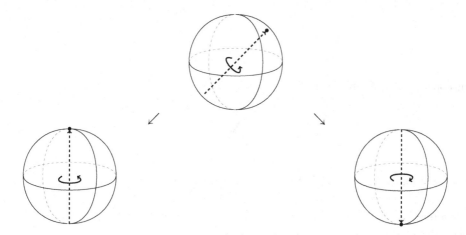

Does this look familiar? That's right, this is just like the dramatic travels of Dave the Dodo:

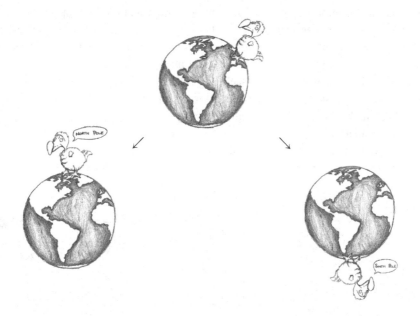

The fact that there are only two possible outcomes when measuring the spin of an electron with a Stern–Gerlach device is analogous to what Max Planck discovered (cf. Section 6.7) for energy, namely that it comes in certain packages called quanta. Here, there are only two such packages, namely *spin-up* and *spin-down*. This example is a member of the the simplest family of quantum processes referred to as measurements, which are called demolition ONB measurements.

7.1.2 Demolition ONB Measurements

Proposition 6.92 from the previous chapter guarantees that for any ONB:

$$\left\{ \ \vbox{\hbox{$\bigtriangledown$}} \right\}_i \tag{7.1}$$

the corresponding doubled effects:

$$\left(\ \vbox{\hbox{$\bigtriangleup$}} \ \right)^i \tag{7.2}$$

constitute a quantum process. We give such processes a special name.

Definition 7.1 A *demolition ONB measurement* is a quantum process of the form (7.2). The indices i are called the *measurement outcomes*.

    This is called a <u>demolition</u> measurement because after the process happens, the system no longer exists. A concrete physical example of a such a process is the 'observation' of a photon, by means of an old-fashioned photographic plate; an active-pixel sensor (APS), which can be found in most digital cameras; or a photomultiplier, which is often used in laboratories to detect single photons (see Fig. 7.1). Each of these absorbs the photon in order to produce an easily detectable witness for the photon: a spot on the photo plate, an activated pixel in the APS, or a click of the detector.

    The measurement outcome i corresponds to what we observe using our measurement device. For example, in the case of the APS, outcomes correspond to pixels that could be activated when they are struck by a photon. The number of pixels determines the number of

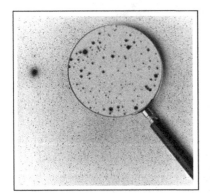

Figure 7.1 Three devices to measure photons (left to right): a photoplate; an active-pixel sensor (APS), as found in most digital cameras; and a photomultiplier.

basis states, and hence the dimension of the quantum system $\widehat{A}$. The process of the photon being detected at the i-th pixel corresponds to the effect:

As we will see later, all other measurements can be derived from ONB measurements by coarse-graining them (Section 7.3.1) or by applying them to part of a larger system (Section 7.3.4).

In Proposition 5.38, we saw that unitary processes are precisely those that send ONBs to ONBs. It furthermore implied that any ONB can be obtained by means of a unitary applied to some fixed ONB. This suggests that the set of ONB measurements on a system is tightly intertwined with the unitaries that one can apply to that system.

Consider a measurement performed by a Stern–Gerlach apparatus from the previous section. We can model this as measuring the ONB:

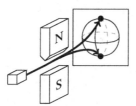

The outcomes can then be pictured as one of two antipodal points on the Bloch sphere, just like the corresponding ONB states (cf. Section 6.1.2):

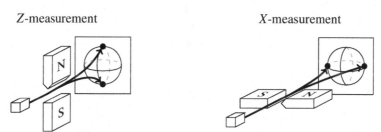

In this particular measurement, 0 means 'deflects up' and 1 means 'deflects down' along the Z-axis. We can now vary the ONB measurement simply by rotating the Stern–Gerlach apparatus relative to the particle source:

Z-measurement   X-measurement

On the Bloch sphere this corresponds to a rotation, and in our mathematical model, to a unitary by the following fact.

Proposition 7.2 Unitaries $\widehat{U}$ from $\widehat{C^2}$ to $\widehat{C^2}$ correspond exactly to rotations of the Bloch sphere:

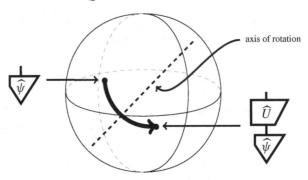

Proof (sketch) One can directly prove this fact by doing some horrible trigonometric mumbo-jumbo extending what we did in Section 6.1.2 or by some sophisticated representation theory for groups. However, intuitively this follows from two facts. The first is that unitaries send pure states to pure states. The second is that unitaries preserve the Born rule:

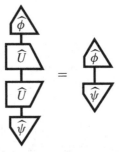

The Born rule measures the distances between points on the surface of the sphere, so unitaries must preserve those distances. The only things that preserve distances on a sphere are rotations and reflections, and of these two, the only things coming from linear maps are rotations.

Conversely, any rotation on a sphere is uniquely fixed by where any three distinct points on the surface go. We can choose for example:

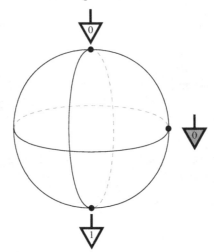

A rotation sends antipodes to antipodes, so the doubled Z-basis states will get sent to another doubled ONB $\{\widehat{\phi}_j\}_j$. The X-basis state will then get sent to an arbitrary location halfway between these two. Obviously we can build a unitary U such that $\widehat{U}$ does this (see Corollary 5.39). Less obviously, by choosing the phase $e^{i\alpha}$ in:

$$
\boxed{U} \quad :: \quad \left\{
\begin{array}{ccc}
0 & \mapsto & \phi_0 \\[10pt]
1 & \mapsto & e^{i\alpha}\, \phi_1
\end{array}
\right.
$$

we can send the doubled X-basis state to any location halfway between the two antipodes without affecting the doubled Z-basis states. □

Remark* 7.3 A much more general statement of Proposition 7.2 is called *Wigner's theorem*. It states that any function ξ on pure quantum states (with no a priori requirement of being a linear map) that preserves the Born rule, that is:

$$
\begin{array}{c}
\xi(\phi) \\
\xi(\psi)
\end{array}
=
\begin{array}{c}
\widehat{\phi} \\
\widehat{\psi}
\end{array}
$$

corresponds to either a unitary or a so-called *anti-unitary*, which is the composite of a unitary and conjugation. Now, on the Bloch sphere, the Born rule translates into the distance between points, and the maps on points that preserve this distance exactly correspond to either a rotation or a rotation composed with a reflection. It is then easy to see that the first corresponds to unitaries, while the latter corresponds to anti-unitaries.

From this intuitive geometric representation on the Bloch sphere it is clear that all ONB measurements on qubits can be obtained via unitaries by rotating two antipodal points to any other two antipodal points. In fact, this holds in general.

Proposition 7.4 Given an ONB measurement on a system:

$$
\left(\begin{array}{c} i \\ \triangle \\ i \end{array} \right)^{i}
\tag{7.3}
$$

all other ONB measurements on that system are obtained as follows:

$$
\left(\begin{array}{c} i \\ \triangle \\ \widehat{U} \end{array} \right)^{i}
\tag{7.4}
$$

where $\widehat{U}$ is a unitary quantum map.

Proof Given any such ONB measurement:

the linear map:

$$\boxed{U} := \sum_j \;\; \text{(diagram)}$$

is unitary by Corollary 5.39. We moreover have:

$$\text{(diagram)} \;=\; \sum_j \text{(diagram)} \;=\; \text{(diagram)}$$

so (7.4) follows by doubling. Conversely, by Proposition 5.38, (1) ⇔ (2), any quantum process of the form (7.4) is an ONB measurement. ☐

There is a clear relationship between rotations we could apply to the Stern–Gerlach device and rotations we could apply to the state of the incoming particles. Namely, we will observe the same probabilities regardless of whether we rotate our measurement by $\widehat{U}$ or we rotate our state by the adjoint of $\widehat{U}$.

Proposition 7.5 The probabilities for measurement (7.3) on the state:

are the same as those for measurement (7.4) on the state:

$$\text{(diagram: } \rho \text{)}$$

Proof We have:

$$P\left(\text{(diagram)} \;\middle|\; \text{(diagram)} \right) = \text{(diagram)} = P\left(\text{(diagram)} \;\middle|\; \text{(diagram)} \right)$$

so the probabilities indeed coincide. ☐

In fact, these are two equivalent ways to think about how a quantum system changes over time. Thinking of this change as something happening to a measurement device is called the *Heisenberg picture*, whereas considering it as something happening to the quantum state is called the *Schrödinger picture*. Of course, the only difference is where you 'draw the line':

Heisenberg   vs.   Schrödinger

7.1.3 Non-demolition ONB Measurements

Rather than just considering doubled ONB-effects, we can instead consider effect–state pairs:

$$\left(\begin{array}{c} \bigtriangledown_i \\ \bigtriangleup_i \end{array} \right)^i \tag{7.5}$$

Again we obtain a quantum process:

$$\sum_i \; \overline{\underset{i}{\bigtriangledown}} \; = \; \sum_i \; \bigtriangleup_i \; = \; \overline{\top}$$

After such a process is performed, we still have a quantum system left behind. Thus, this is an example of a *non-demolition process*.

Definition 7.6 A *non-demolition ONB measurement* is a quantum process of the form (7.5). The resulting state is called the *outcome state*.

Given a non-demolition ONB measurement, we can recover its related demolition measurement by simply discarding the resulting system:

$$\left(\overline{\underset{i}{\bigtriangledown}} \right)^i = \left(\overline{\underset{i}{\bigtriangledown}} \right)^i = \left(\bigtriangleup_i \right)^i \tag{7.6}$$

Conversely, any non-demolition ONB measurement can be regarded as a demolition ONB measurement followed by a controlled preparation:

$$
\frac{\left(\;\bigtriangledown_i \; \right)_i}{\left(\;\bigtriangleup_i \; \right)^i} = \left(\begin{array}{c} \bigtriangledown_i \\ \bigtriangleup_i \end{array} \right)^i
$$

Consequently, each branch of the non-demolition ONB measurement causes the state of the system to change into the corresponding outcome state of the measurement (ignoring numbers):

$$
\begin{array}{c} \bigtriangledown_i \\ \bigtriangleup_i \end{array} \quad :: \quad \widehat{\psi} \quad \mapsto \quad \bigtriangledown_i \tag{7.7}
$$

So while non-demolition measurements do not destroy the system itself, they do irreversibly destroy almost every state of the system. The only exceptions are the *eigenstates* of that measurement.

Definition 7.7 Given an ONB measurement:

$$
\left(\; \bigtriangleup_i \; \right)^i \qquad \text{or} \qquad \left(\begin{array}{c} \bigtriangledown_i \\ \bigtriangleup_i \end{array} \right)^i
$$

the states in the set:

$$
\left\{ \; \bigtriangledown_i \; \right\}_i
$$

are called the *eigenstates* for this ONB measurement. A state that is not an eigenstate for a given ONB measurement is called a *superposition state*.

We defined eigenstates of generic processes back in Definition 5.42 as those states that were left unchanged (up to a number) by the process. The spirit of Definition 7.7 is much the same, in that eigenstates are precisely those states that get sent to themselves with certainty by (the non-demolition form of) an ONB measurement.

7.1.4 Superposition and Interference

Let's first have a look at which outcomes one can obtain for an ONB measurement when the system is in a certain state. The probability of obtaining an outcome i is, as always, given by the Born rule:

$$P\left(i \mid \rho\right) \;:=\; \begin{array}{c} \triangle \\[-2pt] i \\[-2pt] \rho \\[-2pt] \triangledown \end{array}$$

This probability is 0 if and only if ρ and the i-th ONB state are orthogonal:

$$\begin{array}{c} \triangle \\[-2pt] i \\[-2pt] \rho \\[-2pt] \triangledown \end{array} \;=\; 0$$

In the case of qubits, there is only one such state, depicted on the Bloch sphere as the antipodal point. So for the vast majority of states, all of the outcomes i are possible! Hence, each state may lead to most of the outcomes, and each outcome may occur for most of the states. So, measurement seems to look more like playing the lottery than actual 'observation'.

In the case of an APS (Fig. 7.1) this means that for most states a photon may actually be detected by each of the pixels. But then, before it is destroyed, one may ask:

Where was that photon?

According to the probabilities that the Born rule gives, it's typically a bit everywhere at the same time, and then, when the measurement process is launched, a lottery decides where it ends up being detected. But hold on a second here, how do digital cameras (or our eyes, for that matter!) produce pictures that are not just pure randomness?

Thankfully, even though the photon is a bit everywhere, it tends to be 'mostly' where we expect it to be. For example, suppose we take this setup:

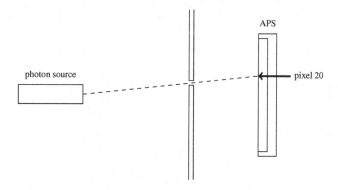

The state of the photon by the time it hits the APS can be written in this form:

$$\widehat{\psi}\!\!\!\bigtriangledown = \text{double}\left(\sum_j r_j e^{i\alpha_j}\; \widehat{j}\!\!\!\bigtriangledown\right)$$

Here, the basis state j corresponds to the j-th pixel on the APS, and we've written the complex numbers in polar form (see Section 5.3.1). For $\widehat{\psi}$, the real number r_j is large when $j = 20$, but falls off quickly as j moves away from 20. Of course, the phase $e^{i\alpha_j}$ also depends on j, but when we compute the probabilities for each pixel of the APS firing the phase disappears:

$$P\left(j\;\middle|\;\widehat{\psi}\right) := \;\;\underset{\widehat{\psi}}{\overset{i}{\bigtriangleup}\!\!\!\bigtriangledown}\;\; = \;\;\overline{(r_j e^{i\alpha_j})}(r_j e^{i\alpha_j}) \;=\; (r_j)^2$$

If we plot these probabilities, we will therefore see something like this:

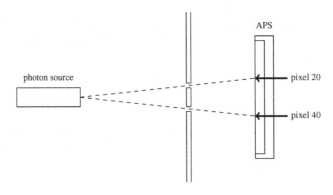

$$(7.8)$$

where 'white' means probability 1 at pixel j and 'black' means 0. Similarly, if we open a slit in front of pixel 40, we produce another state $\widehat{\phi}$. The probabilities for $\widehat{\psi}$ will be concentrated around 40:

$$(7.9)$$

So, why are (7.8) and (7.9) really bad 'observations' of the states $\widehat{\psi}$ and $\widehat{\phi}$? We saw above that $P(j \mid \widehat{\psi})$ only depends on r_j and similarly for $P(j \mid \widehat{\phi})$. Notably, there is <u>no trace</u> of the phases α_j. But so what? Maybe these phases, like the global phases we killed back in Section 6.1.2, don't effect anything anyway. We might think this is so, until we open both slits:

effectively forming the state:

$$\text{double}\left(\;\boxed{\psi}\; + \;\boxed{\phi}\;\right) \tag{7.10}$$

When we look at the probabilities for the APS, we get a bit of a surprise:

$$\tag{7.11}$$

Rather than two nice, even bumps, we get a bunch of alternating bands of light and dark. This is because the phases from $\widehat{\psi}$ and $\widehat{\phi}$, which were invisible in the first two measurements, start to *interfere* with each other in the third measurement. This produces a result that we would never have expected from (7.8) and (7.9) alone.

Remark 7.8 This kind of setup is called the *double slit* experiment, which is used to demonstrate how light behaves both like a particle and like a wave. Even though we get interference fringes like one would expect when studying waves, in each run of the experiment, the photon is only detected at a single pixel, like a particle (see Fig. 7.2).

Now, suppose instead we are able to detect which slit the photon goes through each time. In this case (by ignoring the outcome of the 'which slit' measurement), we get a mixture instead of a superposition:

$$\frac{1}{2}\left(\;\boxed{\widehat{\psi}}\; + \;\boxed{\widehat{\phi}}\;\right) \tag{7.12}$$

Of course, we have known since Section 6.1.5.2 that doubling does not preserve sums, and indeed this extra measurement has killed all of the interesting interference from (7.11), yielding probabilities:

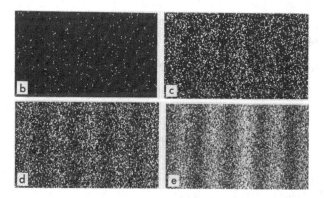

Figure 7.2 A more traditional double-slit experiment, involving a photographic plate. Note how interference fringes appear even when photons are sent one by one through a double slit. This exposes wave-like behaviour even for single photons.

$$20 \qquad\qquad 40 \qquad\qquad 60$$

So, ONB measurements will typically throw away lots of data about the quantum state and can even kill interesting quantum behaviours. As such, ONB measurements are really bad at 'observing' quantum states. In spite of this they are the next best thing to observation.

7.1.5 The Next Best Thing to Observation

The bottom line concerning quantum measurement is that it is impossible, even in principle, to 'observe' a quantum system in the classical sense:

Theorem 7.9 'Observing' is not a quantum process.

Let's provide Theorem 7.9 with some formal content. Ideally, an 'observation' would be a quantum process that tells us the exact state of a system. More explicitly, an observation:

$$\left(\ \vphantom{\phi}\right)^{\widehat{\phi}}$$

should be a process with the property that, for every pure state $\widehat{\phi}$, there is a corresponding outcome obtained with certainty if and only if the system is in the state $\widehat{\phi}$:

$$\quad = \quad \begin{cases} 1 & \text{if } \widehat{\psi} = \widehat{\phi} \\ 0 & \text{otherwise} \end{cases} \qquad (7.13)$$

Note that we restrict to pure states. We wouldn't expect this to hold for impure states because some (possibly important) part of the state has already been discarded. However, even after making this restriction, we can still prove the following.

Lemma 7.10 No quantum process satisfies (7.13).

Proof Let:

$$\left\{ \ , \ \right\} \qquad \text{and} \qquad \left\{ \ , \ \right\}$$

be pure states formed from two different (and disjoint) ONBs, e.g. the Z- and X-bases from Section 5.3.4. By reflecting (6.37) we have:

$$\ + \ = \ = \ + \ $$

Now, suppose we have an effect e_0 that identifies the white 0-state. Applying e_0 to each of the states in the LHS and the RHS we obtain:

which is a contradiction. Thus no such effect exists. $\qquad\qquad\square$

We might also wish to define the process of 'observing' as something that leaves the system unaffected after identifying its state. If such a process existed, then we could obtain (7.13) just by discarding its output. Thus, proving the non-existence of this 'demolition' observation process proves the non-existence of <u>any</u> observation process.

If there is no such thing as observing in quantum theory, what then is the the next best thing to observing? There is actually more than one possible answer to this question, depending on what we hope to accomplish by means of a quantum measurement (see e.g. Section 7.4.2 for an alternative point of view to the one outlined in this section). However, one choice for the next best thing to an 'observation' is a process that behaves like (7.13) for as many pairs of states as possible. This turns out to put a pretty steep restriction on a pair of states.

Exercise* 7.11 Show that two states are perfectly distinguishable by a quantum process if and only if they are orthogonal. That is, suppose there exists a quantum process containing an effect **e** as one of its branches, where:

for causal states ρ and ρ'. Show then that ρ and ρ' must be orthogonal.

As a consequence, ONB measurements are genuinely the next best thing to observing in the sense that they allow for the maximum number of states to be perfectly distinguished, i.e. 'observed', simultaneously.

7.2 Measurement Dynamics and Quantum Protocols

While ONB measurements may not be very good at observing the state of a system, which is what they were intended for, they are extremely good at something they were not intended for! We saw in Section 6.4.6 that, in the teleportation protocol, Aleks' non-deterministic effect does (almost) all of the work:

But what does 'work' precisely mean here? Work here means effectively acting on systems and thereby invoking radical, often non-local, changes. The use of 'radical' is by no means an overstatement, given that for a physical system to dynamically evolve from one state to another takes time, while these measurement-induced changes happen instantaneously. We'll see that small bits of entanglement can instantaneously become long bits of entanglement and that arbitrary linear maps can instantaneously be applied, all by means of quantum measurements.

7.2.1 Measurement-Induced Dynamics I: Backaction

We already saw an example in Section 6.4.2 of ONB measurements changing the state of a system in useful ways, where we used (what we now know is) an ONB measurement to realise any quantum map non-deterministically:

$$\text{ONB measurement} \longrightarrow \qquad \qquad (7.14)$$

Similarly, in the case of qubits, we can realise quantum teleportation by means of a *demolition Bell measurement*:

$$\qquad \qquad (7.15)$$

Since each of the B_i are unitary, we can appropriately correct the error:

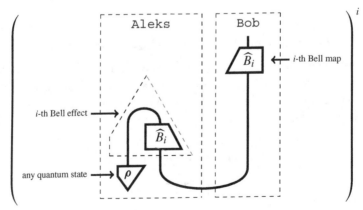

Let's break this into smaller steps and see how the state changes over time. When performing the measurement, the overall state at times t_0 and t_1 in the scenario:

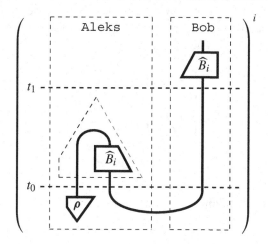

is altered radically. At time t_0 we have:

while at time t_1 we have:

That is, we have an <u>instant transition</u>:

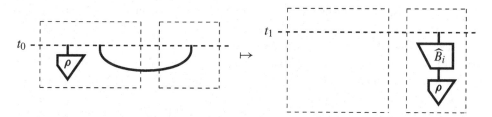

where the state ρ pops over from Aleks to Bob, and an error is created.

However, we already learned in Section 6.3.2 that due to some subtle balancing act between quantum theory and the theory of relativity, instant transitions like this cannot be exploited to achieve faster-than-light signalling. Nonetheless, they enable one to teleport and are very useful for other things too, as we will see shortly.

The above scenario is not the simplest one that exhibits the instant transition phenomenon. A simplification that only involves two systems consists of an ONB measurement performed on one of a pair of systems in a Bell state:

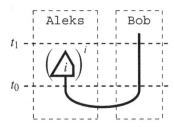

in which case we have the following instant transition:

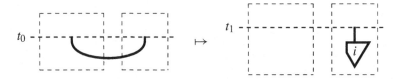

Here, Aleks' measurement outcome determines Bob's resulting state. So the simple performance of a measurement changes the state of the system so drastically that whatever effect Aleks observes ends up as a state at Bob's end.

Definition 7.12 The instant transition induced by a demolition measurement on part of a composite system is called its *backaction*.

We call this 'backaction' because the measurement's influence seems to travel backwards in time, according to the 'logical reading' of diagrams that we discussed back in Section 4.4.3. Importantly, this measurement-induced backaction is not just 'a bit' of dynamics, but some really mind-blowing whole lot of dynamics, which has no counterpart anywhere else in physics!

In spite of this, we can compare this situation with something much less surprising that occurs in classical probability theory. Probability theory can be seen as a way of modelling our state of knowledge about a system. As soon as some part of the system (e.g. a random variable) becomes known, this could induce a global change of our knowledge about the system (e.g. when that variable is correlated with others). This change to our state of knowledge is referred to as *Bayesian updating*.

Consider a situation similar to one we first encountered in Example 4.91 where Aleks and Bob each have a sealed envelope containing the same message, which is taken randomly from a set of possible messages $\{1, \ldots, n\}$. Then, we can treat Aleks opening his envelope as a non-deterministic process:

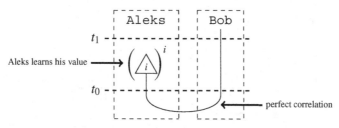

Now, since the content of both envelopes is the same, the content of Bob's envelope is known instantaneously once Aleks performs his 'measurement' process:

However, the crucial difference between quantum theory and classical probability theory is that in the case of the latter what is being altered is nothing more than our knowledge about the content of the envelope, not the content itself. In contrast, according to the standard conception, in quantum theory it is the (pure) state of the system itself that changes, not just our knowledge about it. Many have attempted to provide a similar explanation for the quantum case in terms of knowledge. However, quantum non-locality forbids any such model to be *local*, as we shall explain in Section 11.1. In other words, something must be happening instantaneously between Aleks and Bob, even though they are far apart.

7.2.2 Example: Gate Teleportation

Teleportation exploits the backaction of an ONB measurement to pass a state from one place to another, leaving it unchanged, i.e. applying the identity process to it. Would it also be possible to apply some other quantum map Φ to that state? In other words, can we encode an entire 'computation' in a quantum state, then 'perform' the computation just by measuring? Astoundingly, the answer turns out to be yes, although it does require some cleverness.

The key to solving this problem is process–state duality. The quantum map Φ can be encoded as a bipartite state. We can then modify the quantum teleportation protocol by using this bipartite state instead of the Bell state:

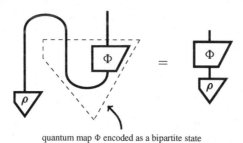

quantum map Φ encoded as a bipartite state

Accounting for the error, this becomes:

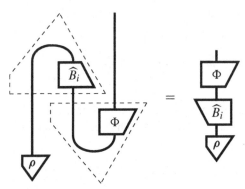

So, we have nearly applied the quantum map Φ that was encoded in the quantum state. But there's a bit of a problem: the error $\widehat{B}_i$ is stuck behind Φ, so it cannot be corrected directly. However, if there exists another unitary $\widehat{\beta}_i$ such that:

then we are back in business:

This might seem like quite a restriction on the kinds of maps we can realise this way, but it turns out that this trick suffices to produce a universal quantum computational model, called *measurement-based quantum computation*, where the 'dynamics' is entirely measurement induced. In Section 12.3 we will describe precisely how this model of computation works.

7.2.3 Measurement-Induced Dynamics II: Collapse

As compared with their demolition counterparts, non-demolition ONB measurements exhibit two kinds of measurement-induced dynamics:

- the backaction discussed in Section 7.2.1, and
- the *collapse* we saw in (7.7):

which could also be called 'forward action'.

For example, the *non-demolition Bell measurement*:

which has the same backaction as the corresponding demolition version, causes two unentangled systems to become entangled (where again we ignore numbers):

$$(7.16)$$

Just as there exists a counterpart to backaction in classical probability theory, the same is true for collapse. In classical probability theory, collapse:

simply means that the content of the envelope goes from being unknown (i.e. in some probability distribution p) to known (i.e. in a point distribution at i). As with backaction, nothing changes except our knowledge when a collapse occurs, whereas collapse is much more destructive in the quantum case. For example, consider what happens if after opening a sealed envelope, we decide to seal the envelope up again and forget the value we saw.

Forgetting the outcome of a non-deterministic process just corresponds to mixing the branches together:

$$\left(\begin{array}{c} \end{array} \right)^i \mapsto \sum_i \quad = \quad \Big| \qquad\qquad (7.17)$$

Since our knowledge hasn't changed, nothing changes. However, if we do the same for a quantum measurement, we get something quite different:

$$\left(\begin{array}{c} \end{array} \right)^i \mapsto \sum_i \quad \neq \quad \Big| \qquad\qquad (7.18)$$

The resulting quantum process will send any quantum state to a mixture of basis states. In particular, a pure state $\widehat{\psi}$ indeed becomes a mixture of the basis states:

weighted by the corresponding probabilities. The process:

is called *decoherence*, and we will study it extensively in Section 8.3.2, once we give it a simple (sum-free) diagrammatic presentation.

Exercise 7.13 Show that decoherence is indeed not the identity by writing both as sums of ONB states.

Despite its destructive nature for quantum states, collapse can also be very useful. One use is *state preparation*, where one obtains a state $\widehat{\psi}$ simply by performing an ONB measurement that has state $\widehat{\psi}$ as a possible outcome state. Performing the measurement a sufficient number of times on a state for which the outcome state $\widehat{\psi}$ is possible will yield $\widehat{\psi}$ at some point. This is exactly how a polarising filter produces polarised light, for example.

We will now see a protocol that exploits both the backaction and the collapse of a non-demolition Bell-basis measurement.

7.2.4 Example: Entanglement Swapping

In Section 7.2.1, we saw how teleportation depends on the backaction of a (demolition) Bell-basis measurement. That is, when the measurement is applied to systems **1** and **2**,

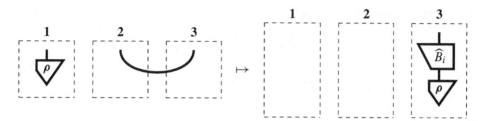

ρ shifts from system **1** to system **3**, despite the fact that **3** is never directly acted on (except later to make a correction). We also saw in (7.16) how the forward action of a non-demolition Bell measurement allows us to entangle two previously unentangled systems.

Let us now try something that combines these two ideas. Suppose we start with a pair of Bell states:

and we apply a non-demolition Bell-basis measurement to systems **1b** and **2a**. Then we obtain the following state:

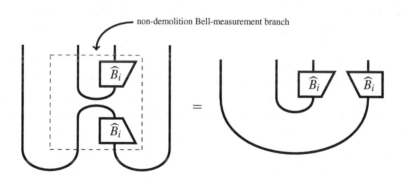

After applying the necessary corrections, we end up with:

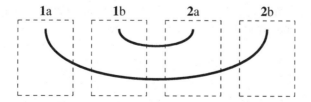

While originally system **1a** was entangled with **1b** and **2a** with **2b**, the final state has **1a** entangled with **2b** and **1b** with **2a**. In other words, the entanglements have been 'swapped'. We call this procedure *entanglement swapping*. The amazing bit about this is that **1a** becomes entangled with **2b**, although these systems were never acted on together, or in other words, quantum theory allows for:

> *entangling without touching*

A practical use of this procedure for quantum technologies is to generate entanglement over a large distance, given that one possesses some entangled states over shorter distances:

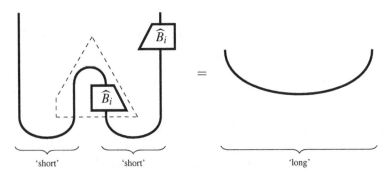

A device that performs entanglement swapping for this particular purpose is sometimes called a *quantum repeater* and is a crucial component to the feasibility of producing high-quality entangled states over long distances. It is called a repeater by analogy to classical signal repeaters, which make it possible to send messages long distances by occasionally capturing the signal and 'repeating' an amplified version of the signal down the wire.

Exercise 7.14 Given n short pieces of entanglement, design a protocol that produces a 'long' piece of entanglement, while minimising the amount of needed corrections.

Remark 7.15 While the teleportation protocol captures in essence the defining equation of string diagrams:

which involves two caps/cups in the LHS, entanglement swapping is the next one in line. It captures an equation involving three caps/cups in the LHS:

$$\bigcup\!\!\bigcap\!\!\bigcup \;=\; \bigcup$$

7.3 More General Species of Measurement

In this section, we will look at more general kinds of measurements that arise once one considers ONB measurements within the broader context of a process theory. For example, what happens if we measure just one sub-system of a quantum state or compose an ONB measurement with other processes?

7.3.1 Von Neumann Measurements

Consider the quantum process obtained by doing a non-demolition ONB measurement on one system, while doing nothing on another system:

$$\left(\begin{array}{c} \bigtriangledown \\ i \\ \bigtriangledown \\ i \\ \bigtriangleup \end{array} \right)^{i} \bigg| \; = \; \left(\begin{array}{c} \bigtriangledown \\ i \\ \bigtriangledown \\ i \\ \bigtriangleup \end{array} \bigg| \right)^{i}$$

This process cannot itself be an ONB measurement, simply because its branches are not o-separable:

$$\left(\begin{array}{c} \bigtriangledown \\ i \\ \bigtriangleup \end{array} \bigg| \right)^{i} \; \neq \; \left(\begin{array}{c} \widehat{\phi_i} \\ \widehat{\phi_i} \end{array} \right)^{i}$$

However, it does share an important characteristic with an ONB measurement, which we can see if we perform the process twice:

$$\left(\begin{array}{c} \bigtriangledown \\ j \\ \bigtriangleup \\ j \\ \bigtriangledown \\ i \\ \bigtriangleup \end{array} \right)^{j}_{\;\;i} \; = \; \left(\begin{array}{c} \bigtriangledown \\ j \\ \bigtriangleup \\ \bigtriangledown \\ i \\ \bigtriangleup \end{array} \right)^{ij} \; = \; \left(\delta^j_i \begin{array}{c} \bigtriangledown \\ i \\ \bigtriangleup \end{array} \right)^{ij} \; = \; \left(\begin{array}{c} \bigtriangledown \\ i \\ \bigtriangleup \end{array} \right)^{ii}$$

The reason why we could take $i = j$ is that for $i \neq j$ we get the impossible process, which, of course, will never happen. Hence, in the two consecutive processes we get the same outcome, and the overall process is the same as as what we would have gotten if we did the process only once.

It may be helpful to see what happens if these processes are applied to a state. Suppose we perform this process on a state ρ and get outcome i. Then the new state will be:

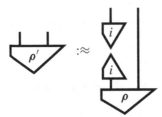

Then, if we immediately perform the process again, all of the branches $j \neq i$ occur with probability 0:

$$P\left(j \neq i \mid \rho'\right) \overset{(6.65)}{=} \quad \approx \quad = 0$$

so we are guaranteed to get outcome i again. Moreover, it doesn't matter if we perform this process 1 time, 2 times, or 100 times. After one measurement, the damage is done, so the state no longer changes after successive measurements.

This idea that a process can cause a quantum state to 'collapse' in such a way that repeating the process will always yield the same thing was considered by von Neumann to be the characterising feature of quantum measurements, yielding the following.

Von Neumann's collapse postulate: After a non-demolition measurement is performed and an outcome i is obtained, performing the same measurement again yields the same outcome with certainty, and does not further affect the state of the system.

This can be put in the language of quantum processes as follows.

Definition 7.16 A (non-demolition) *von Neumann measurement* is a quantum process:

$$\left(\boxed{\widehat{P}_i} \right)^i$$

such that:

$$\begin{array}{c} \boxed{\widehat{P}_j} \\ \boxed{\widehat{P}_i} \end{array} = \delta_i^j \, \boxed{\widehat{P}_i} \qquad (7.19)$$

Von Neumann measurements are also sometimes called *projective measurements*, for the following reason.

Proposition 7.17 For any von Neumann measurement, the branches $\widehat{P}_i$ are projectors.

Proof We already know from (7.19) that the $\widehat{P}_i$ are idempotent, so it only remains to show they are self-adjoint. The collection of quantum maps $\widehat{P}_i$ together forms a quantum process, so it is causal. Unfolding the causality condition yields:

$$\sum_i \; \boxed{P_i \quad P_i} \;\; = \;\; \bigcap$$

so we obtain:

$$\sum_i \; \begin{array}{c} \boxed{P_i} \\ \boxed{P_i} \end{array} \;\; = \;\; \Big| \qquad (7.20)$$

By Theorem 6.17 we can replace the doubled equations (7.19) by non-doubled equations, up to a global phase $e^{i\alpha_{jk}}$ for all values of j and k:

$$\begin{array}{c} \boxed{P_k} \\ \boxed{P_j} \end{array} = e^{i\alpha_{jk}} \, \delta_j^k \, \boxed{P_j} \qquad (7.21)$$

Hence:

$$\boxed{P_j} \;\; \overset{(7.20)}{=} \;\; \sum_k \; \begin{array}{c} \boxed{P_k} \\ \boxed{P_k} \\ \boxed{P_j} \end{array} \;\; \overset{(7.21)}{=} \;\; \sum_k e^{i\alpha_{jk}} \delta_j^k \; \begin{array}{c} \boxed{P_k} \\ \boxed{P_j} \end{array} \;\; = \;\; e^{i\alpha_{jj}} \; \begin{array}{c} \boxed{P_j} \\ \boxed{P_j} \end{array}$$

Doubling then eliminates the global phase:

$$\widehat{P}_i = \begin{array}{c} \widehat{P}_i \\ \widehat{P}_i \end{array}$$

So, by Proposition 4.70 each $\widehat{P}_i$ is a projector.   □

Projectors are called *orthogonal* precisely when they satisfy (7.19), so we could have just as well defined a von Neumann measurement as a quantum process consisting of mutually orthogonal projectors. Moreover, since we are dealing with projectors, causality as in (7.20) simplifies to the underlying projectors (cf. Proposition 6.21) forming a resolution of the identity:

$$\sum_i \boxed{P_i} \;=\; \Big|$$

Alternatively, one can obtain a von Neumann measurement from an ONB measurement, by combining multiple measurement outcomes into one, or *coarse-graining*. For example, suppose we have a three-dimensional quantum system; then perhaps rather than performing a measurement to tell which of these states:

the system is in, we just devise a measurement to check whether the system is in state 1. We can do this by fixing projectors:

$$\boxed{P_0} := \begin{array}{c} \sqrt{1} \\ \times \\ \underline{1} \end{array} \qquad\qquad \boxed{P_1} := \begin{array}{c} \sqrt{2} \\ \times \\ \underline{2} \end{array} + \begin{array}{c} \sqrt{3} \\ \times \\ \underline{3} \end{array}$$

then measuring:

$$\left(\boxed{\widehat{P}_i} \right)^i \tag{7.22}$$

More generally, fix any *partition* of the set of outcomes:

$$I := \{1, \dots, D\}$$

that is, a collection of subsets of I:

$$\{I_1, \dots, I_n\}$$

which satisfy:

$$I_1 \cup \cdots \cup I_n = I \qquad \text{and} \qquad \forall i \neq j : I_i \cap I_j = \emptyset$$

Then we obtain a von Neumann measurement as follows:

$$\left(\boxed{\widehat{P_i}} \right)^i \qquad \text{with} \qquad \boxed{P_i} := \sum_{j \in I_i} \bigvee_{\substack{j \\ \triangle_j}} \tag{7.23}$$

Clearly each of the $\widehat{P}_i$ are projectors, and mutual orthogonality is a consequence of the fact that each of the sets I_i is disjoint.

Exercise 7.18 Show that for any partition of an ONB, (7.23) is a quantum process and hence a von Neumann measurement. Conversely, use the spectral theorem from Section 5.3.3.1 to show that any von Neumann measurement can be expressed as (7.23) for some partition of an ONB.

A coarse-grained measurement teaches us less than an ONB measurement, but it also does less damage. For example, any state of the form:

$$\bigvee_{\psi} = \text{double} \left(\lambda_2 \bigvee_2 + \lambda_3 \bigvee_3 \right)$$

is kept intact by the measurement (7.22) because $\widehat{P}_0$ will never occur and:

$$\frac{\boxed{\widehat{P}_1}}{\bigvee_{\psi}} = \bigvee_{\psi}$$

This is because we were careful to coarse-grain at the level of undoubled processes P_i. If we instead performed a quantum process consisting of:

$$\bigvee_1 \hspace{-0.1cm} \bigwedge_1 \qquad \text{and} \qquad \bigvee_2 \hspace{-0.1cm} \bigwedge_2 + \bigvee_3 \hspace{-0.1cm} \bigwedge_3$$

then $\widehat{\psi}$ would no longer even get sent to a pure state (unless λ_2 or λ_3 is 0), much less itself. On the other hand, if we consider *demolition von Neumann measurement*, that is quantum processes of the form:

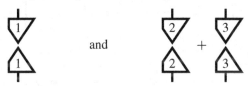

such that (7.19) holds, the distinction disappears. Consequently, unlike demolition ONB-measurements, the branches of a demolition von Neumann measurement need not be pure.

Proposition 7.19 A quantum process is a demolition von Neumann measurement if and only if it is of the form:

$$\left(\begin{array}{c} \pi_i \end{array} \right)^i \qquad \text{with} \qquad \begin{array}{c} \pi_i \end{array} := \sum_{j\in I_i} \begin{array}{c} j \end{array}$$

Proof Let the projections P_i be defined as in (7.23). Then we have:

$$\widehat{P_i} \quad = \quad P_i \; P_i \quad = \quad \sum_{j,k\in I_i} \begin{array}{c} j \; k \\ j \; k \end{array} \quad \overset{(*)}{=} \quad \sum_{j\in I_i} \begin{array}{c} j \; j \end{array} \quad = \quad \pi_i$$

where in (∗) we rely on orthonormality of basis states.   □

The final thing to note is that for both demolition and non-demolition von Neumann measurements, this coarse-graining is not unique. That is, a single von Neumann measurement can arise from coarse-graining more than one ONB.

Exercise 7.20 For an ONB:

$$\mathcal{B} := \left\{ \begin{array}{c} 1 \end{array}, \begin{array}{c} 2 \end{array}, \begin{array}{c} 3 \end{array}, \ldots, \begin{array}{c} D \end{array} \right\}$$

set:

$$\begin{array}{c} + \end{array} = \frac{1}{\sqrt{2}} \left(\begin{array}{c} 1 \end{array} + \begin{array}{c} 2 \end{array} \right) \qquad \text{and} \qquad \begin{array}{c} - \end{array} = \frac{1}{\sqrt{2}} \left(\begin{array}{c} 1 \end{array} - \begin{array}{c} 2 \end{array} \right)$$

in order to obtain another ONB:

$$\mathcal{B}' := \left\{ \begin{array}{c} + \end{array}, \begin{array}{c} - \end{array}, \begin{array}{c} 3 \end{array}, \ldots, \begin{array}{c} D \end{array} \right\}$$

Now show that we have:

$$\begin{array}{c} + \end{array} + \begin{array}{c} - \end{array} = \begin{array}{c} 1 \end{array} + \begin{array}{c} 2 \end{array}$$

and hence that the measurement:

$$\left(\begin{array}{c} 1 \end{array} + \begin{array}{c} 2 \end{array}, \begin{array}{c} 3 \end{array}, \ldots, \begin{array}{c} D \end{array} \right)$$

arises by coarse-graining ONB measurements for $\mathcal{B}$ or $\mathcal{B}'$. Can you characterise all of the ONBs that yield this measurement via coarse-graining?

7.3.2 Von Neumann's Quantum Formalism

Von Neumann measurements make up the core of the quantum formalism as it is still found in most textbooks, which is quite different from the one that we have presented. For one thing, one typically distinguishes between *pure state quantum theory* and *mixed state quantum theory*. Pure states are taken to be primitive, while mixed states are considered to be an (optional) derived concept. Pure state quantum theory is given as three postulates.

Postulate 1: systems. A *quantum system* is represented by a Hilbert space. The *state* of a quantum system then corresponds to an equivalence class of normalised vectors that are equal up to a global phase (cf. Section 6.1.2). *Composite systems* are represented by the tensor product of the Hilbert spaces representing the sub-systems (cf. Section 5.4.2).

Postulate 2: evolution. *Deterministic, reversible quantum processes* are represented by unitaries acting on the Hilbert space (cf. Corollary 6.58).

Postulate 3: measurements. Quantum *measurements* are represented by self-adjoint linear maps acting on the Hilbert space.

But wait a minute here. The third postulate looks totally different from anything we have called a measurement, much less a von Neumann measurement, which presumably lies at the heart of the von Neumann formalism. However, with a bit of help from the spectral theorem (Section 5.3.3.1), we can see that this isn't so different. Starting with a von Neumann measurement, we can wrap up a set of projectors into a single map:

$$
\boxed{f} = \sum_i r_i \boxed{P_i} \tag{7.24}
$$

where all r_i are distinct real numbers. The resulting map is self-adjoint, and we can always recover the projectors P_i via the spectral theorem, which guarantees the existence of an ONB such that:

$$
\boxed{f} = \sum_i r_i \; \bigvee_{\substack{i}}^{\substack{i}}
$$

In general, some numbers r_i might be repeated. This induces a partition $\{I_1, \ldots, I_n\}$ where each <u>distinct</u> real number r_i corresponds to a particular set I_i. We can then use (7.23) to recover the decomposition (7.24).

Example 7.21 The Pauli maps:

$$
\boxed{\sigma_X} \leftrightarrow \begin{pmatrix} 0 & 1 \\ 1 & 0 \end{pmatrix} \qquad \boxed{\sigma_Y} \leftrightarrow \begin{pmatrix} 0 & -i \\ i & 0 \end{pmatrix} \qquad \boxed{\sigma_Z} \leftrightarrow \begin{pmatrix} 1 & 0 \\ 0 & -1 \end{pmatrix}
$$

whose matrices we first encountered in Remark 5.99, are self-adjoint linear maps that represent measurements for the X-basis, Y-basis (see Exercise 6.7), and Z-basis, respectively:

So, the resulting self-adjoint linear map should be considered not as a process itself, but rather as one way to present the actual processes, i.e. the projectors.

Postulate 3: measurements (continued). When a measurement takes place the state of the system changes (i.e. 'collapses') according to the action of one of the projectors, and the probability of each of the projectors for doing so is given by the Born rule:

(7.25)

where (∗) comes from the fact that P_i is a projector.

Remark 7.22 In the case of impure states, the Born rule becomes:

where $\tilde{\rho} := g \circ g^\dagger$ is the density operator associated with ρ (see Remark 6.43). So in the more traditional notation, the Born rule probabilities for pure and mixed states, respectively, become:

$$\langle \psi | P_i | \psi \rangle \qquad \text{and} \qquad \text{tr}\left(P_i \tilde{\rho} \right)$$

What is most notable about this formulation of pure quantum theory is that unitaries and von Neumann measurements are singled out as very special processes. One reason for singling out unitaries is Stinespring's dilation theorem in the form of Corollary 6.63, which states that any deterministic quantum process can be realised by means of a state, a unitary, and discarding. Moreover, in the following section we will see that all quantum processes can be realised if we additionally consider ONB measurements (see Section 7.3.4).

On the other hand, to obtain a quantum process theory, we necessarily have to consider more general processes than unitaries and von Neumann measurements because these operations are simply not closed under composition. For example, composing a state and a unitary in parallel yields a proper isometry. Similarly, when composing a non-demolition von Neumann measurement with a unitary, either in sequence or in parallel, one gets a quantum process that is no longer a von Neumann measurement, for example:

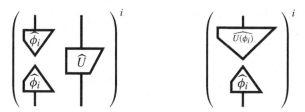

Also when composing two non-demolition von Neumann measurements sequentially the result is usually not a von Neumann measurement. For example, composing two distinct ONB measurements yields:

$$
\left(\begin{array}{c} \phi_j \\ \phi_j \\ \psi_i \\ \psi_i \end{array} \right)^{ij} = \left(\langle \lambda_{ij} \rangle \begin{array}{c} \phi_j \\ \psi_i \end{array} \right)^{ij}
$$

Exercise 7.23 Prove that when composing two non-demolition von Neumann measurements:

$$
\left(\widehat{P_i} \right)^{i} \qquad \text{and} \qquad \left(\widehat{Q_j} \right)^{j}
$$

sequentially, one again gets a von Neumann measurement if and only if the measurements *commute*. That is, for all i and j we have:

$$\frac{\widehat{Q_j}}{\widehat{P_i}} = \frac{\widehat{P_i}}{\widehat{Q_j}}$$

So we can conclude the following.

Theorem 7.24 Von Neumann's formulation of quantum theory is not closed under forming diagrams. In particular, it is not closed under parallel and sequential composition of processes.

Remark 7.25 One advantage of presenting von Neumann measurements as self-adjoint linear maps is that the r_is can be taken to be actual physical quantities associated with a projector, e.g. a position or a momentum. If we replace P_i in the RHS of (7.25) above with a self-adjoint linear map:

$$\boxed{f} := \sum_i r_i \boxed{P_i}$$

instead of a probability, we get a weighted-average of these numbers r_i:

$$E_f\left(\widehat{\psi}\right) := \frac{\widehat{\psi}}{\boxed{f}} = \sum_i r_i \frac{\widehat{\psi}}{\boxed{P_i}} = \sum_i r_i P\left(i \,\middle|\, \widehat{\psi}\right)$$

which is called the *expectation value*.

Remark* 7.26 Von Neumann considered projectors on a Hilbert space to be the quantum analogue to the more familiar notion of 'proposition' from classical logic. This insight led him to co-found the field of *quantum logic*, which we discuss in Section* 7.6.2.

7.3.3 POVM Measurements

The most general kind of demolition measurement is simply <u>any</u> quantum process into the trivial system, i.e. a collection of effects that is only constrained by being jointly causal.

Definition 7.27 Any quantum process of the form:

$$\left(\widehat{\varphi_i}\right)^i$$

is called a *demolition POVM measurement*.

The abbreviation 'POVM' stands for 'positive operator-valued measure'. We briefly explain this terminology, as it is a bit of a departure from the terminology used in this book.

- Why 'positive operator'? While for projective measurements we have:

now we have:

$$(7.26)$$

So, this just refers to the fact that the elements of this quantum process are represented by positive operators, a.k.a. positive linear maps.

- Why 'measure'? In probability theory, a finite 'probability measure' is an assignment of positive numbers $P(i)$ to each element $i \in \{1, \ldots, D\}$ of a finite set, such that these numbers add up to 1. A 'positive operator-valued measure' is a generalisation, in that it assigns to each i a positive map, such that these maps add up to the identity. In other words, this just means causality, since we have:

$$(7.27)$$

Remark 7.28 As with von Neumann measurements, we can write the Born rule in terms of a density matrix and the trace:

$$P(i \mid \rho) = \quad (7.28)$$

where:

$$
\boxed{E_i} \;:=\; \text{(7.29)}
$$

In non-graphical notation (7.28) becomes:

$$
P(i \mid \rho) \;=\; \mathrm{Tr}(E_i\,\tilde{\rho}\,)
$$

which is the Born rule for POVMs found in the standard literature.

Though it's not a standard concept, we could also consider 'non-demolition' POVM measurements. That is, we look at a family of quantum processes such that we can obtain any demolition POVM measurement by discarding the output. Such a family can be obtained via purification (cf. Section 6.4.3):

$$
\left(\widehat{\varphi_i}\right)^{i} = \left(\overline{\widehat{f_i}}\right)^{i}
$$

Thus, it suffices to consider just quantum processes with pure branches to recover all demolition POVM measurements. So, 'non-demolition POVM measurement' is just a synonym for pure quantum process:

$$
\left(\widehat{f_i}\right)^{i}
$$

Remark 7.29 Though it won't be necessary for what follows, to really be called a non-demolition measurement, the maps $\widehat{f}_i$ should have the same output system-type as input. In other words, for a POVM measurement on $\widehat{A}$, we should be able to chose the $\widehat{f}_i$ to have output type $\widehat{A}$:

$$
\left(\widehat{\varphi_i}\right)^{i} = \left(\overline{\widehat{f_i}}\right)^{i}
$$

By Exercise 6.97, this is always possible.

Of course, many non-demolition POVM measurements produce the same demolition POVM measurement when discarding the output. This comes from the fact that there are many ways to decompose φ_i as in (7.26). So the ultimate fate of the quantum state depends on the $\widehat{f}_i$ themselves, not just φ_i. From this, we can deduce that many different non-demolition POVM measurements could reproduce the same probabilities for outcomes, but may act differently on the quantum system itself.

7.3.4 Naimark and Ozawa Dilation

Now, recall that by Stinespring dilation (cf. Theorem 6.61) every causal quantum map Φ (i.e. every deterministic quantum process) arises from some isometry $\widehat{U}$ by discarding one of its outputs:

$$\mathbf{\Phi} \quad = \quad \widehat{U} \qquad (7.30)$$

A similar dilation result holds for POVM measurements. We mentioned that any quantum process can be associated with a single isometry. We first show this for pure quantum processes, where it is called *Naimark dilation*.

Lemma 7.30 For any pure quantum process with D branches:

$$\left(\boxed{\widehat{f}_i} \right)^i$$

and any ONB with D basis states:

$$\left\{ \boxed{i} \right\}_i$$

the following is an isometry:

$$\boxed{U} \;:=\; \sum_i \boxed{f_i}\; \boxed{i}$$

and hence so is the quantum map obtained by doubling.

Proof We have:

$$\sum_j \boxed{f_j}\,\boxed{j} \;\bigg/\; \sum_i \boxed{f_i}\,\boxed{i} \quad = \quad \sum_i \boxed{f_i} \quad = \quad \bigg|$$

where the last equality holds by causality as in (7.27). $\qquad\qquad\square$

Theorem 7.31 (Naimark dilation) Every non-demolition POVM measurement arises as an isometry $\widehat{U}$ with a ONB measurement at one of its outputs:

$$\left(\boxed{\widehat{f_i}}\right)^i = \begin{array}{c}\left(\triangle_i\right)^i \\ \boxed{\widehat{U}} \end{array} \tag{7.31}$$

Consequently, every demolition POVM measurement arises as an isometry $\widehat{U}$ with one of its outputs discarded and an ONB measurement at the other output:

$$\left(\triangle_{\varphi_i}\right)^i = \begin{array}{c}\overline{\equiv}\left(\triangle_i\right)^i \\ \boxed{\widehat{U}} \end{array} \tag{7.32}$$

Proof For the isometry constructed in Lemma 7.30 we have:

$$\begin{array}{c}\triangle_j \\ \boxed{U}\end{array} = \sum_i \boxed{f_i}\ \begin{array}{c}\triangle_j \\ \triangledown_i\end{array} = \boxed{f_j}$$

which, when doubled, yields (7.31). □

Remark 7.32 We can obtain (7.30) from (7.31) by summing over the branches of the ONB measurement:

$$\sum_i \boxed{\widehat{f_i}} = \sum_i \begin{array}{c}\triangle_i \\ \boxed{\widehat{U}}\end{array}$$

which corresponds to measuring then forgetting the outcome:

$$\boxed{\Phi} = \begin{array}{c}\overline{\equiv} \\ \boxed{\widehat{U}}\end{array}$$

So, we saw that 'demolition POVMs' are just arbitrary quantum processes into the trivial system, and 'non-demolition POVMs' are just pure quantum processes from a system to itself. Thus, when we introduced POVMs, all we really did was give a fancier name to stuff we already knew about. But we did establish something important, namely, that we can reproduce all probabilities generated by arbitrary quantum processes in terms of nothing more than an isometry and an ONB measurement. That is the one important thing to remember from this section.

If we combine this with Stinespring dilation, we get a general form for all quantum processes:

Theorem 7.33 (Ozawa dilation) Every quantum process arises from some isometry $\widehat{U}$ with one of its outputs discarded and an ONB measurement at another output:

$$
\left(\begin{array}{c} \Phi_i \end{array} \right)^i = \frac{\left(\begin{array}{c} i \end{array} \right)^i}{\widehat{U}} \tag{7.33}
$$

Exercise 7.34 Prove Theorem 7.33.

So one reason to consider general POVM measurements is that they arise naturally when considering an ONB measurement on part of a larger system, as in Theorem 7.31. The following are two more reasons.

- POVM measurements arise due to imperfections in the measurement procedure due to noise or limited access to the physical system. From this perspective, it is natural to think of (proper) POVM measurements as *mixed measurements*.
- POVM measurements can perform tasks that no von Neumann measurement can. We will see an example of this in Section 7.4.2, where special measurements called *informationally complete* POVM measurements are used to perform a task called *quantum state tomography*.

7.4 Tomography

A single measurement typically destroys a quantum state without giving us much information in return. However, if we have lots of identical copies of a quantum state, we can do better. By performing a series of carefully chosen measurements, one can actually infer the state of a system from the probability distributions over all of the measurement outcomes. This procedure is known as *tomography*. In this section, we will look at how tomography works, and what sorts of measurements are required to achieve it.

7.4.1 State Tomography

Obviously, 'observing', as we defined it in Section 7.1.5, is the ultimate form of tomography. So for process theories where an observation process is available the notion of tomography is more or less redundant. On the other hand, given that 'single-shot observing' is not available in quantum theory, tomography is a very relevant concept.

Since a single measurement isn't good enough, the next thing we might try are many applications of the same ONB measurement. However, that doesn't work either. Since a doubled ONB is not an ONB (see Section 6.1.5.3), the probabilities:

$$P(i \mid \rho) \;=\; \underset{\rho}{\overset{i}{\triangledown}}$$

will never suffice to uniquely fix the state ρ.

However, there do exist larger sets of states that serve as (non-orthonormal) bases for quantum systems. We even explicitly constructed one in Theorem 6.24:

$$\left\{ \text{double}\left(\overset{\,}{\underset{j}{\triangledown}} + \overset{\,}{\underset{k}{\triangledown}} \right) \right\}_{j \leq k} \;\cup\; \left\{ \text{double}\left(\overset{\,}{\underset{j}{\triangledown}} + i\,\overset{\,}{\underset{k}{\triangledown}} \right) \right\}_{j > k}$$

In other words, there exists a set of quantum states

$$\left\{ \overset{\,}{\underset{\phi_i}{\triangledown}} \right\}_i$$

such that:

$$\left(\forall i : \quad \underset{\widehat{\phi_i}}{\overset{\Phi}{\square}} = \underset{\widehat{\phi_i}}{\overset{\Phi'}{\square}} \right) \implies \overset{\Phi}{\square} = \overset{\Phi'}{\square}$$

It follows from this that the associated set of quantum effects

$$\mathcal{E} := \left\{ \overset{\widehat{\phi_i}}{\triangle} \right\}_i$$

suffices to distinguish states:

$$\left(\forall i : \quad \underset{\rho}{\overset{\widehat{\phi_i}}{\bigtriangleup}} = \underset{\rho'}{\overset{\widehat{\phi_i}}{\bigtriangleup}} \right) \implies \overset{\,}{\underset{\rho}{\triangledown}} = \overset{\,}{\underset{\rho'}{\triangledown}}$$

Fixing the probabilities:

$$\underset{\rho}{\overset{\widehat{\phi_i}}{\bigtriangleup\!\!\bigtriangledown}} \tag{7.34}$$

for every $\widehat{\phi_i}$ in $\mathcal{E}$ thus uniquely fixes the state ρ. Of course, we cannot realise effects deterministically, but for any quantum map we can find a non-deterministic process that realises it. So in particular, we can find a collection of measurements that together include all of the effects in $\mathcal{E}$.

Example 7.35 For qubits, the following four effects will uniquely fix any state:

so we can perform state tomography on qubits via measurements in the X-, Y-, and Z-bases (defined in Exercise 6.7):

In fact, no smaller set of ONB measurements than the one given in Example 7.35 does the job. To prove this, we can again exploit the geometry of the Bloch sphere.

Proposition 7.36 Qubit state tomography by means of ONB measurements requires measurements in at least three different ONBs.

Proof We will give a geometric proof, recalling from Section 6.1.2 that the inner product (7.34) determines the distance between two states on the Bloch sphere. It suffices to show that for any two ONB measurements, there exist at least two states ρ and ρ' that cannot be distinguished. Assume without loss of generality that the first ONB is:

$$\left\{ \overset{\perp}{\underset{0}{\bigtriangledown}}, \overset{\perp}{\underset{1}{\bigtriangledown}} \right\}$$

since otherwise we could just express everything to follow in terms of some other basis, rather than the Z-basis. Fix a second ONB $\{\phi_0, \phi_1\}$. Then $\widehat{\phi}_0$ is an arbitrary point on the Bloch sphere, and $\widehat{\phi}_1$ is its antipode:

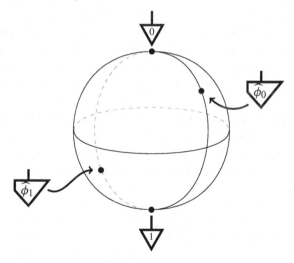

To find two states that cannot be distinguished, it suffices to find two states that are equally far from all four of the basis states. To do this, draw a line perpendicular to the plane made by the four points, and mark the two places this line crosses the Bloch sphere:

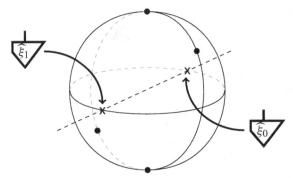

Then, measuring ξ_i in either ONB yields either outcome with probability $1/2$. Thus, with just two basis measurements, we cannot possibly distinguish ξ_0 from ξ_1. □

7.4.2 Informationally Complete Measurements

For a qubit one needs three distinct ONB measurements to realise state tomography. Somewhat surprisingly, one can do better when not restricting to ONB measurements and allowing for general POVM measurements. In fact, there is a kind of POVM measurement called *symmetric informationally complete* (SIC), which can do state tomography all by itself. It may be somewhat counterintuitive that a 'mixed' process would be better at anything than a pure one, but here you have a compelling example of that fact.

To understand how this can be the case, assume that we consider the three ONB measurements required for qubit tomography. Now, define a (single) new quantum process as follows. Roll a die, and if the outcome is 1 or 2, then perform the first ONB measurement; if the outcome is 3 or 4, then perform the second ONB measurement; and if the outcome is 5 or 6, then perform the third ONB measurement. Clearly, if we do this new quantum process many times, each of the three ONB measurements will have been performed a sufficient number of times for the sake of doing qubit tomography. A SIC-POVM measurement improves on this quantum process by doing the same job in a more direct, and geometrically elegant, way.

A SIC-POVM measurement consists of pure (but unnormalised) quantum effects. In order to be *informationally complete*, that is, sufficing for tomography, we need to require that this quantum process has (at least) D^2 branches, since this is the number of quantum states we need to form a basis for the doubled system (cf. Theorem 6.24). For such a set consisting of D^2 effects:

$$\left\{ \hat{\phi}_i \right\}_{i=1}^{i=D^2}$$

it is impossible for these to all be orthogonal. In the case of qubits, this is like trying to find four points on the Bloch sphere that are all antipodal! Thus, we do the 'next best thing' and require that all effects overlap by the same amount:

$$
\begin{array}{c}
\widehat{\phi_j} \\
\hline
\widehat{\phi_i}
\end{array}
=
\begin{cases}
1 & \text{if } i = j \\
\lambda & \text{if } i \neq j
\end{cases}
\tag{7.35}
$$

for some fixed number λ. Geometrically, this means that we choose them to be evenly spaced out across the state space. For the Bloch sphere, a set of four evenly spaced points is always a tetrahedron:

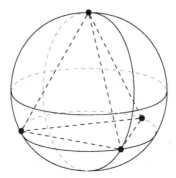

This condition is what the *symmetric* part of 'symmetric informationally complete' stands for. For this set of effects to also be causal, we will need to scale down each effect, thus obtaining the quantum process:

$$
\left(\tfrac{1}{D} \, \widehat{\phi_i} \right)^i
\tag{7.36}
$$

where causality becomes:

$$
\sum_{i=1}^{i=D^2} \tfrac{1}{D} \, \widehat{\phi_i} = \;\bar{\top}
\tag{7.37}
$$

Definition 7.37 A *SIC-POVM measurement* is a causal quantum process of the form (7.36) that satisfies (7.35) for some fixed number λ.

Finally, and most importantly, one can prove that SIC-POVM measurements are informationally complete. That is, the probabilities produced by a SIC-POVM measurement uniquely characterise a quantum state.

Exercise* 7.38 Show that for any SIC-POVM measurement the probabilities:

$$\left(\frac{1}{D} \; \begin{array}{c} \widehat{\phi_i} \\ \rho \end{array} \right)^i$$

totally characterise the state ρ. Hint: since a SIC-POVM measurement consists of D^2 effects, an equivalent characterisation of being a basis is the following condition, known as 'linear independence':

$$\sum_i c_i \; \begin{array}{c} \widehat{\phi_i} \end{array} \; = 0 \qquad \Longrightarrow \qquad \forall i : c_i = 0$$

To show all of the c_i must be zero, start by showing they must all be the same.

If you find a solution to the next exercise, let us know!

Exercise* 7.39 Is there a (nice) diagrammatic characterisation of SIC-POVM measurements?

7.4.3 *Local Tomography = Process Tomography*

Quantum theory allows enough ONB measurements to identify quantum states. A related question is: can we identify quantum states on multiple systems using only *local measurements*? That is, can we identify a state:

$$\begin{array}{cc} \widehat{A} & \widehat{B} \\ & \rho \end{array}$$

just by measuring each system individually? Following the previous section, this amounts to asking whether there exist sets of effects:

$$\left\{ \begin{array}{c} \varphi_i \\ \widehat{A} \end{array} \right\}_i \qquad \text{and} \qquad \left\{ \begin{array}{c} \varphi_j \\ \widehat{B} \end{array} \right\}_j$$

such that:

$$\left(\forall i,j : \; \begin{array}{c} \varphi_i \quad \varphi_j \\ \rho \end{array} \; = \; \begin{array}{c} \varphi_i \quad \varphi_j \\ \rho' \end{array} \right) \quad \Longrightarrow \quad \begin{array}{c} \\ \rho \end{array} \; = \; \begin{array}{c} \\ \rho' \end{array}$$

One usually refers to this property as *local tomography*.

The answer for quantum theory is a clear 'yes'. Such local measurements do exist, and it is very easy to see why. Since we can find bases:

$$\left\{ \begin{array}{c} \hat{A} \\ \varphi_i \end{array} \right\}_i \qquad \text{and} \qquad \left\{ \begin{array}{c} \hat{B} \\ \varphi'_j \end{array} \right\}_j$$

we can define the desired local measurements in terms of the product basis:

$$\left\{ \begin{array}{cc} \hat{A} & \hat{B} \\ \varphi_i & \varphi'_j \end{array} \right\}_{ij}$$

We showed in Section 5.2 that products of ONBs give ONBs, but in fact for **linear maps** this works for arbitrary bases as well. Hence we have the following theorem.

Theorem 7.40 Quantum theory obeys local tomography.

Once one becomes accustomed to the concept of product bases, local tomography seems like such a natural assumption that one wonders how it can possibly fail. However, we do not need to stray too far from quantum theory for this to no longer hold. For instance, a variation of quantum theory built on matrices of <u>real</u> instead of complex numbers does not have this property. This is shown in Section* 7.6.3.

A related, but (seemingly) quite different tomographic notion is *process tomography*. Process tomography is a procedure by which we try to identify a black-box process:

$$\begin{array}{c} \hat{B} \\ \Phi \\ \hat{A} \end{array}$$

by feeding in states and performing measurements. It amounts to finding states and effects:

$$\left\{ \begin{array}{c} \hat{A} \\ \rho_i \end{array} \right\}_i \qquad \text{and} \qquad \left\{ \begin{array}{c} \varphi_j \\ \hat{B} \end{array} \right\}_j$$

such that:

$$\left(\forall i,j : \begin{array}{c} \varphi_j \\ \Phi \\ \rho_i \end{array} = \begin{array}{c} \varphi_j \\ \Phi' \\ \rho_i \end{array} \right) \implies \begin{array}{c} \Phi \end{array} = \begin{array}{c} \Phi' \end{array}$$

For process theories described by string diagrams, these two notions of tomography are actually equivalent.

Theorem 7.41 A process theory that admits string diagrams has local tomography if and only if it has process tomography.

Proof This simply follows from process–state duality. Assume a process theory admits local tomography. Then for any processes f and g we have:

$$
\left(\forall i,j : \quad \vcenter{\hbox{[diagram ψ_i ϕ_j f]}} = \vcenter{\hbox{[diagram ψ_i ϕ_j g]}} \right) \implies \vcenter{\hbox{[diagram f]}} = \vcenter{\hbox{[diagram g]}}
$$

from which we get:

$$
\left(\forall i,j : \quad \vcenter{\hbox{[diagram ϕ_j f ψ_i]}} = \vcenter{\hbox{[diagram ϕ_j g ψ_i]}} \right) \implies \vcenter{\hbox{[diagram f]}} = \vcenter{\hbox{[diagram g]}}
$$

Thus the processes f and g are distinguishable via the following set of states and effects:

$$
\left\{ \vcenter{\hbox{[diagram ψ_i]}} \right\}_i \qquad \text{and} \qquad \left\{ \vcenter{\hbox{[diagram ϕ_j]}} \right\}_j
$$

The converse is proved similarly.   □

7.5 Summary: What to Remember

1. 'Observing' is not a quantum process.

2. Several species of quantum processes are called *quantum measurements*.

- A *demolition ONB measurement* is a quantum process of the form:

$$
\left(\vcenter{\hbox{[diagram]}} \right)^i
$$

  where:

$$
\left\{ \vcenter{\hbox{[diagram i]}} \right\}_i
$$

is any ONB, and a corresponding *non-demolition ONB measurement* is a quantum process of the form:

Any doubled ONB state is called an *eigenstate*, and any other state a *superposition* for that measurement.

- A *von Neumann measurement* is a quantum process of the form:

$$\left(\boxed{\widehat{P_i}\diagdown} \right)^i$$

where the branches are mutually orthogonal projectors. Equivalently, von Neumann measurements coarse-grain ONB measurements:

$$\boxed{P_i\diagdown} := \sum_{\alpha \in I_i} \begin{matrix} \boxed{\alpha\diagdown} \\ \boxed{\alpha\diagup} \end{matrix}$$

- A *POVM measurement* is any quantum process of the form:

$$\left(\boxed{\varphi_i\diagdown} \right)^i$$

The probability distribution produced by a POVM measurement can also be achieved by means of an isometry and an ONB measurement:

$$\frac{\left(\boxed{\triangle_i} \right)^i}{\boxed{\widehat{U}\diagdown}}$$

which one refers to as *Naimark dilation*. More generally, every quantum process arises from some isometry $\widehat{U}$ with one of its outputs discarded and an ONB measurement at another output:

$$\left(\boxed{\Phi_i\diagdown} \right)^i = \frac{\stackrel{=}{} \left(\boxed{\triangle_i} \right)^i}{\boxed{\widehat{U}\diagdown}}$$

3. Quantum measurements induce two kinds of 'dynamics':

1. *Backaction*, which causes radical instantaneous changes in the state of systems other than those measured. For example, in the scenario:

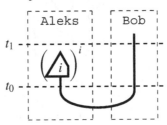

we have the following dynamics between times t_0 and t_1:

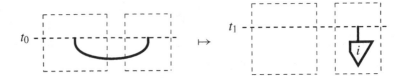

2. *Collapse*, which causes the state being measured to instaneously become an eigenstate of the measurement:

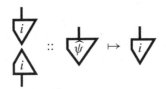

4. These dynamics are exploited in quantum protocols, including *quantum teleportation*:

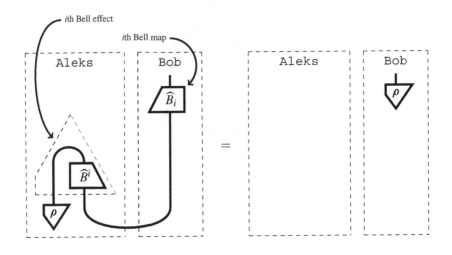

entanglement swapping:

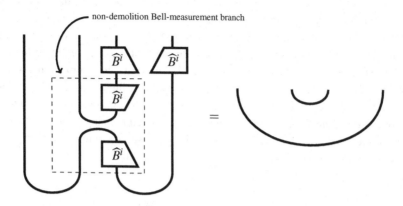

which allows one to create long bits of entanglement from short ones, and *gate teleportation*:

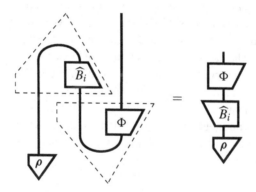

which allows one to apply an arbitrary quantum map Φ, encoded in a bipartite state, to ρ. This last trick forms the basis of *measurement-based quantum computation*.

5. If we have many copies of a system, all in the same quantum state, then we can infer that state by means of *tomography*. Certain POVM measurements, called SIC-POVM measurements, allow one to do that by only relying on one particular measurement. For process theories that admit string diagrams, local tomography and process topography coincide:

7.6 Advanced Material*

We'll now have a look at some foundational aspects that have surrounded quantum measurement. We start with a more philosophical discussion of the *measurement problem*. As the name already indicates, there are indeed some sticky issues to discuss. That's also where logic entered the picture for the first time, under the name *quantum logic*, which takes as its starting point the idea that projectors (as used in von Neumann measurements) should be treated as the analogue to propositions in ordinary logic. We contrast this 'logic' with our 'logic of interaction' (cf. Section 1.2.3). We end with an example – taken from work in quantum foundations – showing how something as practical as tomography identifies a fundamental difference between a theory based on complex numbers and one based on real numbers.

*7.6.1 Do Quantum Measurements Even Exist?**

Quantum measurements are clearly indispensable, as they provide our only access to the quantum world. But what are they as compared with their classical counterparts? In particular, is the name 'measurement' truly justified? In Section 7.3.4 we saw that all quantum processes can be seen as some kind of measurement, which seems to render the name 'measurement' redundant. However, in this section we embark on a more philosophical path, contemplating what a quantum measurement is truly about.

Why is there no observation process in quantum theory? We have already seen a proof that such a thing doesn't exist in Section 7.1.5, but this assumes quite a lot of mathematical structure from quantum theory. What we would like is a conceptual justification for why one should not expect there to be an 'observation' process. An answer that traces back to the early days of quantum theory, mostly associated with Niels Bohr and Werner Heisenberg, is:

Any attempt to observe is bound to disturb.

Many would argue that in prequantum theories observing comes for free, and thus the absence of an observation process in quantum theory is a clear departure from our everyday intuition. But is it actually true that in prequantum theories observing comes for free? Consider just bare-bones Newtonian mechanics, before we add any fancy stuff like electromagnetism (and, in particular, light). Since there is no light, we don't have things like cameras (or eyes!) for making observations. If this stripped-down version of Newtonian mechanics admits an observation process, then we should be able to find an object in a dark room without disturbing it. Clearly, if that object is very light, say a balloon, then it is virtually impossible to locate it without moving it a bit. More specifically, in order for us to locate an object, it must exert a force on us, and according to the action–reaction principle, it will experience a force too, and hence move a bit.

Of course, if we now bring electromagnetism, i.e. light and eyes, back into the game, then we can effectively observe the balloon without disturbing it in any noticeable way.

In contrast, for quantum systems there is unfortunately no analogue to the role light and eyes play for mechanical objects. It is unfortunate indeed, but not that surprising anymore. One could say that we suffer from *quantum blindness* and that we can only probe that quantum world by means of some invasive interaction, just like what we have to do in search of an object in a dark room.

Many scientists do find it really hard to give up on the idea that a measurement genuinely represents an observation, and out of pure desperation even invoke the conscious human act of deciding to make an observation within quantum theory. This gave rise to things like the famous *Schrödinger's cat* and the perhaps less famous *Wigner's friend* paradoxes.

Given that there is no observation process in quantum theory, the difficulty of accessing the quantum state then raises the questions about what a quantum state actually represents. For example: does a quantum state represent actual properties of a system, or is it more like a probability distribution, i.e. something that merely reflects the state of our knowledge about the system? In many ways, the second interpretation would be more palatable, because like quantum states, probability distributions can never be observed perfectly, and can even 'collapse' when new information is gained, via *Bayesian updating* (Section 7.2). However, several no-go theorems substantially obstruct this interpretation of the quantum state. Most notably, the recent *Pusey–Barrett–Rudolph theorem* states that, under a few assumptions about quantum systems (which many consider reasonable), this 'state of knowledge' interpretation is totally incorrect.

A closely related problem to that of interpreting the quantum state is the following.

Definition 7.42 The *measurement problem* comprises two questions concerning a (typically von Neumann) measurement:

m1 What causes the measurement process and, in particular, the collapse of the quantum state, to take place?

m2 What 'decides' the outcome i of this process?

Attempts to address these questions are usually referred to as *interpretations*. Some of these deny that there is a change of the state, hence rejecting von Neumann's collapse postulate, in a desire to rescue the idea of there being an observation process. This usually comes at a very high cost; for example, in the case of *Everett's many-worlds* interpretation one has to accept the existence of a humongous number of completely independent parallel universes. We refer to the vast body of existing literature for these interpretations, and provide some pointers in Section 7.7. Here we will restrict ourselves to recasting the measurement problem when viewing quantum theory as a process theory, and how some natural strands for resolving it present themselves.

Early solutions to **m1** included the idea that being coupled to a *measurement device* causes the collapse or, more generally, that any *macroscopic system* causes the collapse. However, this requires a clear-cut definition of what a measurement device is and where the micro-world ends and the macro-world starts. Both of these ideas proved very difficult, if not impossible, to fully develop. Nothing in our experience suggests the existence of a 'wall' that separates micro and macro levels, and it is even more ridiculous to think that a particular human-made machine would play a leading role in fundamental physics which pre-existed humanity.

Alternatively, one could say that a measurement process takes place when the system is in an <u>environment</u> that causes this particular process to happen. Nothing more, nothing less. Taking this stance, **m1** amounts to providing an explicit description of such an environment that causes a measurement process to happen. This point of view fits particularly well with the description of quantum systems as a wire in a diagram, where a system is characterised by its behaviour in context. This is also closely related to the late-Wittgensteinian concept of *meaning in context*, which states that the meaning of something only comes about when one also considers the context in which this thing is considered. Translated to quantum theory, this means that we should conceive of quantum systems not as isolated entities, but rather as entities in interaction with a context, i.e. the rest of a diagram, which in particular includes the measurement.

A related question concerns the conceptual status of an ONB in a measurement process. If we want to learn something about a quantum system we can only do so in terms of things that we are able to perceive, e.g. locations in spacetime. But maybe spacetime is not the natural habitat of a quantum system, and maybe the measurement process is then all about 'forcing' the system into the spacetime theatre. One could refer to such a process as *classicization*. The idea that spacetime may not be the theatre in which all of reality takes place, but rather a form of human experience, traces back to philosopher Immanuel Kant and was further refined by mathematician Henri Poincaré, who went as far as attributing the role of geometry in physics in part to human intuition.

Turning our attention now to **m2**, we address what causes a particular measurement outcome i to occur. The first attempted solutions, under the impetus of Einstein, who famously stated that 'God does not play dice', suggested that there is more to a quantum system than just the quantum state. This 'more' was referred to as *hidden variables*. The main idea is that additional variables associated to the quantum system could determine the measurement outcomes. Again, a string of no-go theorems excludes the existence of many kinds of hidden variables. Most notably, the *Bell–Kochen–Specker theorem* rules out *non-contextual* hidden variables, whose values are in some sense 'real' and independent of measurement choices. Some hidden variable theories survive these no-go theorems, notably *Bohm's hidden variable model*. However, these hidden variables are very different from what Einstein had in mind. In particular, they are necessarily non-local, as we will demonstrate in Section 11.1.

On the other hand, if one associates additional variables to the context of a state rather than the state itself, which includes the interaction between the quantum system and its environment, then there is no obstruction to attributing the outcomes to the pair consisting of the quantum state and these additional variables. Somewhat surprisingly, this option has never really entered the mainstream. While all of this sounds very reasonable to us, again driven by a desire to rescue the process of observation, many scientists refuse to accept the idea that there may be environmental interference to what one learns about a system.

A major obstacle to swallowing this idea is that science has traditionally been built on the presupposition that any system subject to scientific investigation must be sufficiently isolated from its environment when being probed. Maybe the key lesson from quantum theory is that this stance cannot be retained and that one has to opt for something more along the lines of *relationalism*, in that the fabric of reality is about relationships between things, rather than their individual attributes. Diagrams of course provide the natural language for such a relational universe.

7.6.2 Projectors and Quantum Logic*

In Remark 7.26 we mentioned that projectors can be thought of as propositions about quantum systems, resulting in the field of *quantum logic*. The starting point of quantum logic is not to follow Schrödinger's vision that the beating heart of quantum theory is the manner in which quantum systems compose, but rather to follow von Neumann's vision that understanding quantum measurements is the key.

In classical logic, we typically consider systems to be sets of states X. Then, a *proposition* is just a subset of states $P \subseteq X$, which should be thought of as the set of all states for which P holds. For example, if our system is a potato, P might be 'is boiled', which we represent formally as the set of all possible states of a potato where the potato is boiled.

As subsets, propositions come with a natural ordering, namely subset inclusion. So, for instance, if Q stands for 'is cooked', any boiled potato is also cooked, so we have:

$$P \subseteq Q$$

In other words, we can deduce Q from P. This ordering, which represents 'deduction', is the cornerstone of any logical system. We can also represent conjunction ('P and Q'), disjunction ('P or Q'), and negation ('not P') straightforwardly as operations on subsets, respectively:

$$P \cap Q \qquad\qquad P \cup Q \qquad\qquad P^\perp := X \backslash P \tag{7.38}$$

These operations give the set of all propositions the mathematical structure of a *boolean lattice*. Boolean lattices are sets with operations $\cup, \cap, (\ )^\perp$ satisfying various equations. Most notably, they are *distibutive*:

$$P \cap (Q \cup R) = (P \cap Q) \cup (P \cap R)$$

Projectors in some sense play the role of propositions in quantum theory, in that they are the 'verifiable propositions'. Namely, for any $\widehat{P}$, we can fix a von Neumann measurement that checks whether $\widehat{P}$ holds. As with propositions, we can give a 'deduction' ordering to projectors:

$$\left(\boxed{\widehat{P}} \ \leq \ \boxed{\widehat{Q}} \right) \ := \ \left(\begin{array}{c} \boxed{\widehat{Q}} \\ \boxed{\widehat{P}} \end{array} = \boxed{\widehat{P}} \right) \tag{7.39}$$

So, how is this like the ordering on propositions from before? We can associate to each projector $\widehat{P}$ a set of states for which $\widehat{P}$ holds (with certainty):

$$S_{\widehat{P}} := \left\{ \psi \;\middle|\; \frac{\widehat{P}}{\psi} = \psi \right\}$$

Then it's easy to check that (7.39) is equivalent to:

$$S_{\widehat{P}} \subseteq S_{\widehat{Q}}$$

So, for a given state ψ, 'satisfying' $\widehat{P}$ implies 'satisfying' $\widehat{Q}$. However, these $S_{\widehat{P}}$ aren't just arbitrary sets, but rather *subspaces*.

Exercise 7.43 Show that, for any projector $\widehat{P}$ on a quantum system $\widehat{H}$, there exists a subspace $H_P \subseteq H$ such that $S_{\widehat{P}}$ consists of all the states $\widehat{\psi}$ for $\psi \in H_P$. Conversely, show that every subspace of H corresponds to a (unique) projector in this way.

We can get some mileage out of this. For example:

$$S_{\widehat{P}} \cap S_{\widehat{Q}}$$

also comes from a subspace, so we can let '$\widehat{P}$ and $\widehat{Q}$' just be its associated projector, which we denote by:

$$\widehat{P} \wedge \widehat{Q}$$

However:

$$S_{\widehat{P}} \cup S_{\widehat{Q}}$$

is <u>not</u> a subspace, so we need to get a bit creative. Letting:

$$\boxed{P^\perp} := \left| \; - \; \boxed{P} \right.$$

we get negation. Then by de Morgan's law, 'not (not P and not Q)' should be the same as 'P or Q', so let:

$$\widehat{P} \vee \widehat{Q} := (\widehat{P}^\perp \wedge \widehat{Q}^\perp)^\perp$$

Unfortunately, $\wedge$ and $\vee$ are *not distributive*.

Exercise 7.44 Show that:

Therefore, instead of distributivity, quantum logicians usually assume something a lot weaker, called *orthomodularity*:

$$\widehat{P} \leq \widehat{Q} \quad \Longrightarrow \quad \widehat{Q} = \widehat{P} \vee \left(\widehat{P}^{\perp} \wedge \widehat{Q} \right)$$

The goal of quantum logic is to reason about quantum systems to the greatest extent possible just using this very weak logical structure.

This perspective clears away a lot of the 'noise' associated with Hilbert spaces and can provide new insights, and in this way chimes well with the goals of this book. However, as we stressed in Section 1.2.4, quantum logic aims to characterise quantum theory by the <u>failure</u> of something, which explains the failure of this research program. Interestingly, the issue on which quantum logic failed mostly was the description of composite systems, which is precisely where we started off in this book, and that's perhaps the most important lesson to be taken from quantum logic.

7.6.3 Failure of Local Tomography*

More recently, much of the interesting quantum foundations research has embraced the fact that many characteristic features of quantum theory are tightly intertwined with the interaction of multiple systems. One very remarkable such result is the fact that one can affirm the key role complex numbers play by means of tomography on two systems.

Suppose we define a new process theory called $\mathbb{R}$-**quantum maps**, consisting of only those quantum maps that involve real numbers. More precisely, rather than doubling **matrices**($\mathbb{C}$) one doubles **matrices**($\mathbb{R}$). One would expect this theory to be quite similar to that of **quantum maps**, but the fact that we leave out the imaginary part of complex numbers has some dramatic consequences.

Theorem 7.45 The theory of $\mathbb{R}$-**quantum maps** does <u>not</u> admit process tomography and, hence, does <u>not</u> admit local tomography.

Proof As with **quantum maps**, any state ρ in the theory of $\mathbb{R}$-**quantum maps** is a $\otimes$-positive linear map; so in particular, it is self-conjugate. Thus any two-dimensional state in $\mathbb{R}$-**quantum maps** (a.k.a. 'rebit' state) can be written as:

for $a, b, c \in \mathbb{R}$. The Bell maps have real matrices, so $\widehat{B}_3$ is a process in $\mathbb{R}$-**quantum maps**. Applying it to ρ yields:

When we add the resulting state to ρ itself, we get:

$$\text{(diagram)} + \text{(diagram }\widehat{B_3}\text{)} = (a+c)\left(\text{(diagram }_0\,_0\text{)} + \text{(diagram }_1\,_1\text{)}\right) = \text{(diagram)}$$

Thus, the processes:

$$\Big| + \boxed{\widehat{B_3}} \qquad \text{and} \qquad \text{(diagram)} \qquad\qquad (7.40)$$

agree on all states in ℝ-**quantum maps**, yet they are clearly not equal.   □

We cannot tell the difference between the processes in (7.40) just by applying these processes to states. So what? It could have been the case that we will just never see the difference in these two processes, in which case they would be, for all intents and purposes, the same. But this is not the case! We can see this immediately if we turn these processes into bipartite states. The equivalence of process tomography and local tomography comes from process–state duality. Applying process–state duality to the processes in (7.40) tells us that we cannot distinguish the bipartite states:

$$\text{(diagram }\mu\text{)} := \text{(diagram }\widehat{B_0}\text{)} + \text{(diagram }\widehat{B_3}\text{)} \qquad \text{and} \qquad \text{(diagram)}$$

by means of local effects. However, we can still distinguish them by means of a <u>global</u> effect, namely any of the other Bell-basis effects:

$$\text{(diagram }\widehat{B_2}\text{ over }\mu\text{)} = 0 \neq 1 = \text{(diagram }\widehat{B_2}\text{)}$$

So we can conclude that ℝ-**quantum maps** is truly a very different theory from **quantum maps**. In particular, tomography of processes and composite systems would be an entirely different ballgame.

Exercise 7.46 Give a diagram in which the two processes in (7.40) are distinguishable (i.e. yield different probabilities via the Born rule) in the theory of ℝ-**quantum maps**.

7.7 Historical Notes and References

Quantum measurements as described here, and von Neumann measurements in particular, are evidently due to von Neumann (1932), in particular including the idea of the collapse, and that this collapse is induced by projectors.

Naimark's theorem (which is also sometimes spelled Neumark's theorem) for POVMs first appeared in Neumark (1943). Ozawa (1984) stated a more general version of this theorem for what he calls 'quantum instruments', which include quantum processes as we have defined them. Entanglement swapping was first proposed in Zukowski et al. (1993), and teleportation-based universal quantum computing was proposed in Gottesman and Chuang (1999). Wigner's theorem, which was referred to in Remark* 7.3, is taken from Wigner (1931).

The local tomography axiom in the form that we presented it here is taken from Chiribella et al. (2010). A similar formulation appeared earlier in Barrett (2007), which in turn traced back to a reconstruction of quantum theory in Hardy (2001). SIC-POVMs trace back to Lemmens and Seidel (1973). Exercise 7.39 was suggested to us by Chris Fuchs, who strongly believes that SIC-POVMs have a fundamental role in quantum theory (Fuchs, 2002). This stance is one aspect of an interpretation of quantum theory called *quantum Bayesianism*, or *QBism* (Fuchs et al., 2014).

The measurement problem grew out of the Bohr–Einstein debate, which mainly followed onwards from the EPR papers (Einstein et al., 1935; Einstein, 1936) and Niels Bohr's reply (Bohr, 1935). Heisenberg's and Bohr's positions were already stated much earlier in their respective books Heisenberg (1930) and Bohr (1931). Schrödinger's cat paradox appeared first in Schrödinger (1935), and Wigner's friend in Wigner (1995a). Many textbooks have been dedicated to the measurement problem, a selection being Jammer (1974), Redhead (1987), and Bub (1999). For original texts on interpretations that are currently still popular we refer to Everett (1957) and Bohm (1952a,b), respectively, for the many-worlds interpretation and an example contextual hidden variable theory. Constraint theorems on the interpretation of the quantum state are Jauch and Piron (1963), Kochen and Specker (1967), and Pusey et al. (2012). In fact, Kochen and Specker (1967) is a fairly straightforward corollary to the theorem of Gleason (1957), as explained in Belinfante (1973).

Planck's quote at the beginning of this chapter dates back to 1936, from his address on the twenty-fifth anniversary of the Kaiser Wilhelm Society for the Advancement of Science (see Macrakis, 1993), which after being implicated in Nazi scientific operations, was dissolved and had its functions taken over by the Max Planck Society. Evidently, it was motivated by the aforementioned interpretational difficulties that caused many physicists not to accept the theory.

Poincaré's modified Kantian views mentioned in Section 7.6.1 are taken from Poincaré (1902). Wittgenstein's 'meaning in context' appeared first in Wittgenstein (1953). Relational views on space and time trace back to Leibniz (see e.g. Rickles, 2007).

Already in his book on quantum theory, von Neumann (1932) attributed a fundamental significance to projectors, in that they represented the propositions of the quantum world. This then became the basis of so-called quantum logic (Birkhoff and von Neumann, 1936). For von Neumann, quantum logic was a path towards a better formalism for quantum theory. This in part drove him to the introduction of *von Neumann algebras* and in particular to the study of *Type II factors* therein. All of this is detailed in Redei (1996). Interestingly, while quantum logicians adopted the orthomodular law, von Neumann insisted on the stronger *modular law*:

$$\widehat{P} \leq \widehat{Q} \quad \Longrightarrow \quad \widehat{P} \vee \left(\widehat{R} \wedge \widehat{Q}\right) = \left(\widehat{P} \vee \widehat{R}\right) \wedge \widehat{Q}$$

which holds only for finite-dimensional quantum theory. His reason was that this law holds for projective geometries and that the lattice of closed subspaces of any Hilbert space naturally embeds in a modular one, which then, via the fundamental theorem of projective geometry, yields a vector space representation (Piron, 1976; Stubbe and van Steirteghem, 2007).

For many quantum logicians, quantum logic wasn't really about logic, but rather about probability theory and algebra. An operational variant was initiated by Mackey (1963) and was further conceptually and philosophically underpinned in Piron (1976) and Moore (1999). Constantin Piron (1964) also proved what could be considered as the first reconstruction of quantum theory from operational principles.

As already hinted at in the preface to this book, one of the authors of this book ended up realising the importance of processes through quantum logic. In particular, rather than thinking of static propositions, passing to how propositions evolve allows one to derive the linearity of processes (Faure et al., 1995). The same argument provides the most compelling interpretation for orthomodularity too, corresponding to the fact that projectors are actual processes (Coecke et al., 2001; Coecke and Smets, 2004), via an argument that generalises *weakest precondition semantics* from computer science (Dijkstra, 1968; Hoare and He, 1987). This line of research is further developed in Baltag and Smets (2005). Our love for quantum logic is not entirely gone, as witnessed by a recent attempt to reconcile diagrams and quantum logic (Coecke et al., 2013b).

8

Picturing Classical-Quantum Processes

... she sprinkled her with the juice of Hecate's herb, and immediately at
the touch of this dark poison, Arachne's hair fell out. With it went her
nose and ears, her head shrank to the smallest size, and her whole body
became tiny. Her slender fingers stuck to her sides as legs, the rest is
belly, from which she still spins a thread, and, as a spider, weaves her
ancient web.

– Ovid, The Metamorphoses, 8 AD

Most quantum protocols rely on the interplay between quantum systems and classical
data. For instance, measurements extract classical data from a quantum system, whereas
controlled operations use classical data to affect a quantum system. Moreover, given that
truth is in the eye of the beholder, we want to understand quantum theory relative to our
perception of reality, which is classical, and hence want to understand how the two relate.
Somewhat surprisingly, it turns out to be much easier to represent the classical world
relative to quantum processes, than the other way around.

One way to get a handle on this interaction is to express as much of it as possible in a
purely diagrammatic form. Previously, we have drawn diagrams of quantum processes, then
used some sort of 'external' means of describing the classical data flow, i.e. with indices
and brackets, which can't really be plugged together like pieces of a diagram. Even worse,
in most standard textbooks, classical data is not even part of the actual formalism, but is
described in words.

Rather than describing this interplay of quantum systems and classical data using lots of
'blah blah blah', or a cross-breed between diagrams and symbols, can we instead just give
a diagram of all of the devices involved and how they are wired together? For example,
suppose we have a device 'Bell' that prepares Bell states, another device 'Bell-M'
that performs Bell measurements, and a third device 'Bell-C' that does Bell corrections,
depicted very realistically as follows:

Bell   Bell-M   Bell-C

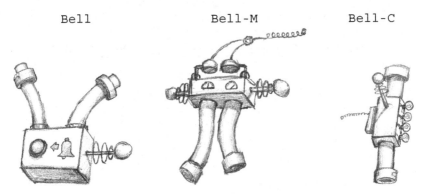

Now, suppose we want to describe to a technician how to wire these devices together to do teleportation:

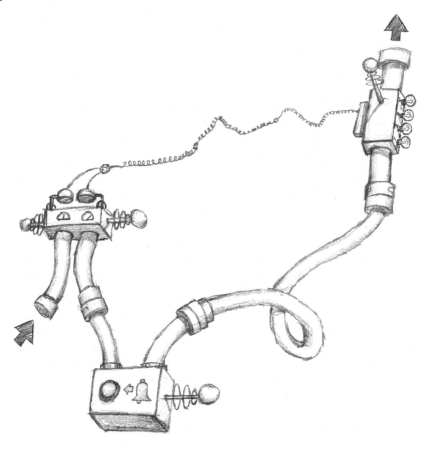

We could describe this using a *specification language*, that is, a diagrammatic language where the boxes correspond to devices and the wires correspond to literally 'wiring up' the devices:

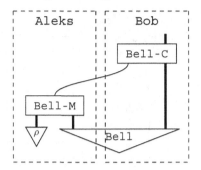

where we now distinguish quantum wires and classical wires.

Every well-formed specification of a protocol, that is, a diagram in the specification language, should have a corresponding mathematical model, which the theorist can use to predict what a protocol will actually <u>do</u>. We can picture this mapping between specification and model via *interpretation brackets* $[\![\ ]\!]$. Then, for teleportation we have:

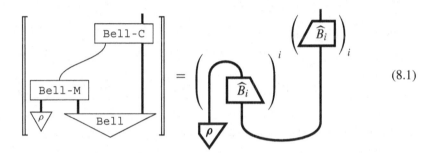

$$(8.1)$$

So, the LHS describes how the technician sets up the equipment, whereas the RHS is what the theorist predicts the equipment will do. One could say the real predictive power of quantum theory amounts to giving a definition of $[\![-]\!]$.

As we have noted before, there is a slight disconnect here, since quantum systems are wires, but classical systems are something else entirely:

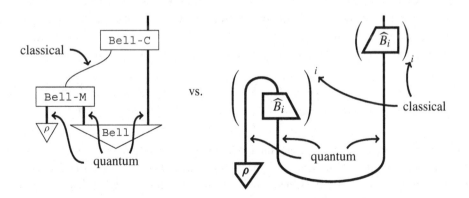

What we would really like is to treat classical and quantum systems both as wires, at which point the recipe for interpreting a specification diagram amounts to simply interpreting each of its components:

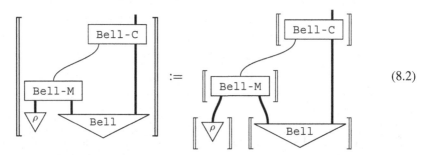

$$(8.2)$$

That is, modelling the overall process amounts to modelling these four smaller processes:

and then composing them together.

We already know how to do this for purely quantum boxes and wires:

$$\left[\!\!\left[\; \rho \;\right]\!\!\right] = \rho \qquad \left[\!\!\left[\; \text{Bell} \;\right]\!\!\right] = \cup$$

so to complete the picture, we need to figure out how to interpret boxes that have some classical wires as well:

$$\left[\!\!\left[\; \text{Bell-M} \;\right]\!\!\right] = \; ? \qquad \left[\!\!\left[\; \text{Bell-C} \;\right]\!\!\right] = \; ?$$

We have already benefitted from the fact that doubling makes space for some extra processes, namely impure quantum maps. Now we will see how it makes space for some extra, classical systems as well. By extending **quantum maps** with classical wires, we obtain a new process theory of **classical-quantum maps**, or **cq-maps** for short. After imposing causality, we obtain the full theory of **quantum processes**, which is really what this whole book is about. The blanks above can then be filled in as follows:

$$\left[\!\!\left[\; \text{Bell-M} \;\right]\!\!\right] = \widehat{U} \qquad \left[\!\!\left[\; \text{Bell-C} \;\right]\!\!\right] = \widehat{U}$$

Just as in the specification language, the thick wires represent quantum systems, whereas the thin wires represent classical systems. A state of a quantum system is the same as before, but as we will see shortly, a state of a classical system is a probability distribution.

Thus, cq-maps simultaneously generalise quantum maps and the natural notion of mappings between classical probability distributions (i.e. *stochastic maps*):

$$\text{quantum} \qquad \text{classical-quantum} \qquad \text{classical}$$

As classical data can be copied and deleted, it is natural to allow classical wires to split and merge in various ways. These points where multiple classical wires meet are called *spiders*:

Spiders can be seen as a generalisation of caps and cups, whose behaviour is dictated by a single 'fusion' rule:

While a wire is something that connects two ends together, a spider is a generalisation that connects many ends together. Consequently, the spider-fusion rule, which subsumes the yanking equations of caps and cups, is the embodiment of the concept that 'only connectivity matters'.

While we introduce spiders in order to reason about classical data, they will also be used to construct new quantum maps and classical-quantum hybrids. In fact, these spiders are so powerful that we will be able to give the entire story of quantum theory in terms of them. Now-familiar concepts such as *measurement, classical control*, and *mixedness*, as well as new notions such as *classical copying*, (non-pure state) *entanglement*, and *decoherence* will all be expressed in terms of these new kinds of processes. Indeed, from this chapter onwards, this book will be full of spiders!

8.1 Classical Systems as Wires

In this section we show how classical systems, just like quantum systems, can be represented in terms of linear maps. Thus, rather than needing to treat classical inputs and outputs of a quantum process 'externally' with indices, we can simply express them as wires that can be plugged together. This allows us to give a simple presentation for the theory of **quantum processes** and recover the elegance of the original (sum-free) version of the causality postulate.

8.1.1 Double versus Single Wires

The theory of **quantum maps** can be seen as restricting the theory of **linear maps** such that we only allow maps of a very particular form:

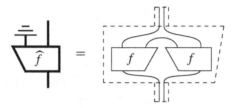

We now wish to construct a theory that includes maps where quantum and classical systems interact with each other. Quantum systems are of a fundamentally different type from classical systems, so there should be two distinct kinds of wires. Since quantum systems are already represented by thick doubled wires, to introduce a new kind of system, we simply put some single wires back in:

$$
\left(\text{quantum} := \; \| \; \right) \quad \neq \quad \left(\text{classical} := \; | \; \right)
$$

And as we will soon see, this turns out to be a perfect fit for representing classical systems. So:

$$
\frac{\text{classical}}{\text{quantum}} = \frac{\text{thin/single wires}}{\text{thick/double wires}}
$$

We encode classical data on a single wire by representing classical values as basis states of a fixed ONB. So for an ONB on a classical system of dimension D, this means that the corresponding classical data has D possible values. For example, a bit corresponds to an ONB on a two-dimensional system, i.e.:

$$
\text{`bit'} := \left\{ \overset{|}{\underset{0}{\bigtriangledown}}, \overset{|}{\underset{1}{\bigtriangledown}} \right\}
$$

More precisely, we interpret the states of this ONB as:

- $\overset{|}{\underset{i}{\bigtriangledown}}$:= 'providing classical value i'

and for the corresponding effects we then have:

- $\overset{i}{\underset{|}{\bigtriangleup}}$:= 'testing for classical value i'

Orthonormality:

$$\frac{j}{i} = \delta_i^j \tag{8.3}$$

then makes perfect sense in light of this interpretation. We obtain probability 0 (i.e. 'impossible') when we are testing for the value $j \neq i$ on the value i, and we obtain probability 1 (i.e. 'certain') when we are testing for the value i on the value i.

Remark 8.1 We'll use self-conjugate ONBs for classical data, since conjugation has no classical counterpart; e.g. 'conjugating a bit' is meaningless.

When one applies a quantum process such as this one:

$$\left(\boxed{\Phi_i}\right)^i$$

two things happen. Of course, one of the branches Φ_i gets applied to the quantum system, but also a classical value i pops out, telling us which branch happened. To capture this with a classical wire, we can represent a quantum process like this:

$$\left(\boxed{\Phi_i}\right)^i \quad \rightsquigarrow \quad \sum_i \boxed{\Phi_i} \tag{8.4}$$

that is, we make the creation of classical output explicit and represent the whole process as one big linear map.

Similarly, if the outcomes of quantum processes are controlled by a classical input, we can rely on the i-effects to represent the fact that it tests for an input value:

$$\left(\boxed{\Psi_i}\right)_i \quad \rightsquigarrow \quad \sum_i \boxed{\Psi_i} \tag{8.5}$$

It is easy to see that these give us faithful representations of the respective quantum processes. We can still access the individual branches of this process by composing it with the appropriate classical effect, i.e. subjecting it to the appropriate test:

$$\sum_i \boxed{\Phi_i}^{\;\; \text{test for value } j} \;=\; \sum_i \boxed{\Phi_i}\, \delta_i^j \;=\; \boxed{\Phi_j}$$

and for a controlled process, we can access each component by providing the corresponding input:

$$
\sum_j \Psi_j \Bigg[\text{input value } i \Bigg] \;=\; \sum_j \Psi_j \, \delta_i^j \;=\; \Psi_i
$$

— input value i

So we can indeed reconstruct (8.4) via:

$$
\left(\boxed{\Phi_j} \right)^j \;=\; \left(\sum_i \boxed{\Phi_i} \right)^j
$$

and (8.5) as:

$$
\left(\boxed{\Psi_i} \right)_i \;=\; \left(\sum_j \boxed{\Psi_j} \right)_i
$$

What's more, we can now connect the i-state to the i-effect by a wire to indicate the actual classical communication:

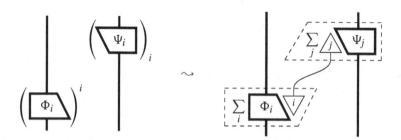

Note how, in the diagram on the right, the indices i and j only occur locally. In other words, the two linear maps in the dashed boxes are completely independent processes. However, by the presence of a classical wire connecting the two, i and j are forced to become the same, thanks to orthonormality:

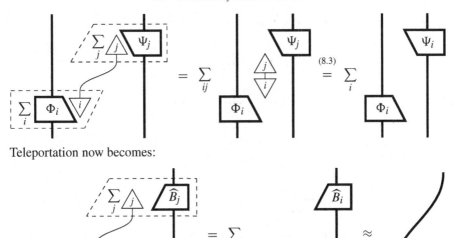

Teleportation now becomes:

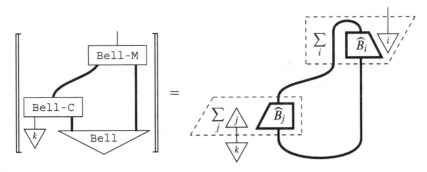

In particular, the entire picture is now a linear map of a certain kind, which we will define shortly. Clearly, such maps properly generalise quantum maps, since now there are also classical systems involved. So we have now been able to eliminate syntactic garbage that lived 'outside' **linear maps** and return to working just with diagrams of processes, which has some big advantages.

8.1.2 Example: Dense Coding

Now that classical systems have their own wires, we can consider protocols with classical inputs and classical outputs. While teleportation relied on classical communication (with the help of a Bell state) to send a quantum state, we now ask the converse question: can we use a quantum state to send a classical one? Also, do we gain anything by doing so?

Of course, the magic of teleportation wouldn't have been possible without the Bell state, so we will assume that here too. This yields the following specification diagram and corresponding interpretation:

Then, by orthonormality of the Bell basis:

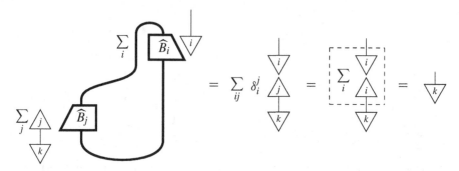

we obtain:

So what does this protocol achieve? When viewed as a communication protocol, communicating classical data by means of quantum systems seems a bit heavy handed. However, the quantum systems we are dealing with (i.e. the bold wires in the pictures) are two dimensional. On the other hand, there are four Bell states, so the measurement generates one of four values as classical output (i.e. the thin wires in the pictures). So, while Aleks only sends one qubit, he succeeds in communicating two classical bits. This result straightforwardly generalises to D-dimensional bold wires, which enable one to communicate D^2-valued classical data.

Remark 8.2 It is sometimes said that teleportation and dense coding are dual. This is because there is an exchange of roles played by the classical and the quantum channels (while the entangled state stays quantum of course). However, it is worth noting that the two aspects of the Bell matrices – the fact that (i) they form unitaries and (ii) their associated states form an ONB – play different roles in the two protocols. Whereas teleportation most obviously depends on (i), the key to dense coding is (ii). What is true is that if one possesses an entangled state in advance, then:

- a 'D-dimensional quantum channel' and
- a 'D^2-dimensional classical channel'

become equivalent, in that teleportation allows us to convert the classical channel into a quantum channel, while dense coding allows us to convert the quantum channel into a classical channel.

8.1.3 Measurement and Encoding

We now convert the quantum process for an ONB measurement into this new representation using classical wires:

$$\left(\Big/\!\!\bigtriangleup\!\!\Big\backslash\right)^{i} \quad \rightsquigarrow \quad \sum_{i} \; \overset{\underset{i}{\bigtriangledown}}{\bigtriangleup}$$

This linear map is so important that we give it its own notation:

$$\phi \; := \; \sum_{i} \; \overset{\underset{i}{\bigtriangledown}}{\underset{i}{\bigtriangleup}}$$

So what does this map do? Let's apply it to an arbitrary quantum state:

$$\underset{\rho}{\overset{\phi}{\bigtriangledown}} \; = \; \sum_{i} \; \overset{\underset{i}{\bigtriangledown}}{\underset{i}{\bigtriangleup}} \underset{\rho}{} \; = \; \sum_{i} P(i \mid \rho) \; \overset{\underset{i}{\bigtriangledown}}{}$$

Unsurprisingly, these numbers are exactly the probabilities for the ONB measurement according to the Born rule. So the linear map representing an ONB-measurement sends a quantum state to a probability distribution:

$$\underset{\rho}{\bigtriangledown} \; \mapsto \; \sum_{i} P(i \mid \rho) \; \overset{\underset{i}{\bigtriangledown}}{}$$

over all of the possible measurement outcomes. Seen as a matrix, this state is:

$$\underset{\rho}{\overset{\phi}{\bigtriangledown}} \quad \leftrightarrow \quad \begin{pmatrix} P(1 \mid \rho) \\ P(2 \mid \rho) \\ \vdots \\ P(D \mid \rho) \end{pmatrix}$$

Now that an ONB measurement is represented by a linear map, which from now on we'll call *measure*, we can also consider its adjoint:

$$\psi \quad := \quad \sum_i \quad \bigvee\hspace{-1.1em}\bigtriangleup$$

Applying this linear map to an arbitrary probability distribution, we obtain:

$$\phi_p \quad = \quad \sum_i \quad \bigvee\hspace{-1.1em}\bigtriangleup_p \quad = \quad \sum_i p^i \quad \bigvee$$

that is, we obtain the encoding of a probability distribution as a quantum state given in Proposition 6.74. For that reason, we'll call this linear map *encode*.

Measuring undoes this operation. Hence, these two processes interconvert the two representations of a classical probability distribution:

$$\sum_i p^i \; \bigvee_i \quad \overset{\longrightarrow}{\underset{\longleftarrow}{}} \quad \sum_i p^i \; \bigvee_i$$

In particular, they interconvert classical values and their corresponding quantum states:

$$\bigvee_i \quad \overset{\longrightarrow}{\underset{\longleftarrow}{}} \quad \bigvee_i \qquad\qquad (8.6)$$

8.1.4 Classical-Quantum Maps

Though they may look special, it turns out that these measure/encode processes are in fact all we need to add to get a full theory of classical-quantum maps. Just as we defined **quantum maps** as the process theory where we took **pure quantum maps** and added one new ingredient (discard), we can define classical-quantum maps by adding two new ingredients to **quantum maps**.

Definition 8.3 A classical-quantum map (cq-map) is a linear map obtained by composing quantum maps, encoding, and measurement:

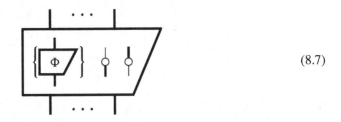

$$(8.7)$$

We call the associated process theory **cq-maps**.

Before we do anything else, we first show the following theorem.

Theorem 8.4 The theory of **cq-maps** admits string diagrams.

Proof Since measure is the adjoint of encode, the theory of **cq-maps** inherits adjoints from **linear maps**. We saw back in Chapter 6 how to build cups and caps for the quantum systems, so we only need to show them for the classical systems. These arise as follows:

In the next section we will see a very easy way to prove such equations, so for now we leave these (still pretty easy) proofs as an exercise. $\square$

Also, just like what we established in Proposition 6.46 for quantum maps, we can always put cq-maps into a 'normal form'.

Proposition 8.5 All cq-maps are of the form:

$$(8.8)$$

up to reordering some input/output wires.

Proof The proof goes much like that of Proposition 6.46 giving a standard form for quantum maps. First, note that any time the classical wires from a measure and an encode connect, this results in a quantum map:

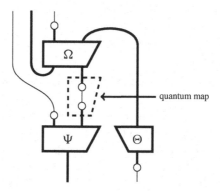

Hence any cq-map can be written as a quantum map with some measures on its outputs and some encodes on its inputs:

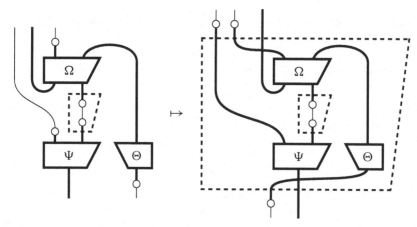

Then, by reordering some inputs/outputs, we can bring all of the measure/encode processes together:

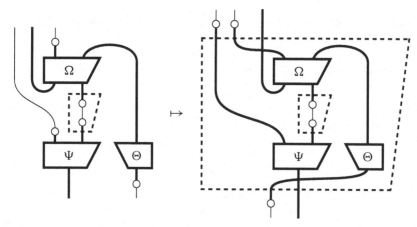

which gives a process of the form:

$$(8.9)$$

Just as with discarding, multiple measure processes can be combined into one process on a larger system:

$$
\begin{array}{c} \text{\scriptsize A} \quad \text{\scriptsize B} \\ \text{[diagram]} \end{array}
= \sum_i \text{[diagram]}_i \sum_j \text{[diagram]}_j = \sum_{ij} \text{[diagram]}_i \text{[diagram]}_j = \begin{array}{c} \text{\scriptsize A}\otimes B \\ \text{[diagram]} \\ \text{\scriptsize } \hat{A}\otimes\hat{B} \end{array}
\tag{8.10}
$$

where the RHS is the measure process for the ONB:

$$
\left\{ \; \text{[diagram]}_i \; \text{[diagram]}_j \; \right\}_{ij}
$$

Encode processes can be combined similarly. Thus, combining the measure/encode processes in (8.9) yields a process of the form (8.8). $\qquad\square$

Then, just by purifying the quantum map Φ above, we have:

Corollary 8.6 Every cq-map is of the form:

Remark 8.7 In this chapter, we'll assume that every classical system carries its own, 'preferred' ONB, which we'll always write using white triangles. Hence, the ONBs on the systems A and B in equation (8.10) will be different if $A \neq B$. In Chapter 9, where we study interaction between different ONBs for the same system, we will remove this assumption, but we will then need to be a bit careful with the classical wires. We'll explain this in Section 9.2.

So what does Definition 8.3 have to do with our efforts from the previous sections? Before, we showed how to express quantum processes as single linear maps:

$$
\left(\; \boxed{\Phi_{ij}} \; \right)_i^{\,j} \quad \rightsquigarrow \quad \sum_{ij} \; \text{[diagram]}_i^{\,j} \; \boxed{\Phi_{ij}}
\tag{8.11}
$$

We can now show that such linear maps are indeed cq-maps. First roll all of the Φ_{ij} together with doubled ONB states/effects:

$$
\boxed{\Phi} \quad := \quad \sum_{ij} \; \text{[diagram]}_i^{\,j} \; \boxed{\Phi_{ij}}
\tag{8.12}
$$

Since this is a sum of quantum maps, by Theorem 6.67, Φ is itself a quantum map. So, the linear map (8.11) is indeed a cq-map:

Conversely, any cq-map is of the form (8.11):

where:

So for a generic cq-map, the quantum maps Φ_{ij} are precisely the branches of that map, which have been selected via classical inputs/outputs:

However, Definition 6.100 of quantum processes included one important extra ingredient, which so far cq-maps are lacking. Namely, there is nothing to guarantee that these branches, taken together, satisfy the causality postulate (6.77). To treat this in an elegant, sum-free fashion, we will require one more diagrammatic ingredient.

8.1.5 Deleting and Causality

Back before non-determinism came into the picture, we had a beautifully simple equation that captured causality:

$$\Phi = \quad$$

If we discard everything coming out of a process, we might as well just not do the process and discard everything coming in. To extend this to cq-maps, we just need to say what it means to 'discard' a classical system. For this, we will introduce a new process called *deleting*:

It is clear from the name what such a process should do to any classical value *j*:

$$\frac{}{j} = \boxed{} \tag{8.13}$$

but then, of course, since all of these classical values together form an ONB this totally fixes deleting:

$$\bigvee := \sum_i \bigwedge_i \tag{8.14}$$

One might think that we need to put this into the theory of **cq-maps** by hand, but in fact, it's already there:

$$\frac{}{} \overset{(5.37)}{=} \sum_i \bigwedge_i \overset{(8.6)}{=} \sum_i \bigwedge_i \overset{(8.14)}{=} \quad \tag{8.15}$$

which of course makes perfect sense. If we encode some classical data as a quantum system, then discard the system, this is the same as just deleting the data. The following definition then pretty much writes itself.

Definition 8.8 A cq-map is *causal* if:

$$\Phi = \quad \tag{8.16}$$

And this is indeed equivalent to how we defined causality before.

Proposition 8.9 For a cq-map:

$$
\text{(diagram)} \tag{8.17}
$$

Definition 6.100 and Definition 8.8 for causality coincide.

Proof The cq-map (8.17) encodes a quantum process:

$$
\left(\boxed{\Phi_{ij}} \right)^{j}_{i} \qquad \text{where} \qquad \boxed{\Phi_{ij}} := \boxed{\Phi}
$$

Assuming causality as in Definition 8.8, we obtain the following for all i:

$$
\sum_{j} \boxed{\Phi_{ij}} \;=\; \sum_{j} \boxed{\Phi} \;\overset{(8.14)}{=}\; \boxed{\Phi} \;\overset{(8.16)}{=}\; \;=\;
$$

which is precisely the statement of causality from Definition 6.100. The proof of the converse proceeds similarly. □

Causal quantum maps now arise as a special case of this definition. Just like with discarding for quantum systems, deleting is the unique causal effect for a classical system. That is, for any causal effect on a classical system, equation (8.16) reduces to:

$$
\boxed{\rho} \;=\;
$$

Measure and encode are also causal. We already 'accidentally' showed encoding was causal in equation (8.15). The proof for measuring is very similar:

$$
\overset{(8.14)}{=}\; \sum_{i} \overset{i}{\triangle} \;\overset{(8.6)}{=}\; \sum_{i} \overset{i}{\triangle} \;\overset{(5.37)}{=}\;
$$

An important additional benefit of this new presentation of causality is that many proofs about causality can now be done in exactly the same way as the purely quantum case. For example, proving that the sequential composition of two causal cq-maps is again causal is simply:

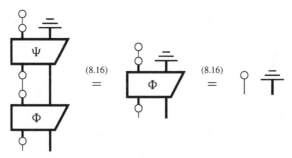

More generally, it is therefore also easily seen that any circuit diagram of causal cq-maps is again a causal cq-map. So, at last we are ready to define the most important process theory in the book.

Definition 8.10 Quantum processes is the subtheory of **cq-maps** consisting of all causal cq-maps.

Since deleting is just the classical counterpart to discarding:

$$\overline{\overline{\top}} \quad \leftrightarrow \quad \varphi$$

Definition 8.8 is just a minor update to our original slogan for causality:

> *If we discard/delete all of the quantum/classical outputs of a quantum process, it may as well have never happened.*

Thus we have succeeded (as promised) in extending the interpretation of causality for quantum maps of Section 6.2.4, to quantum processes that may also involve classical inputs and outputs. So classical and quantum inputs, in the light of causality, are now on equal footing. Consequently, statements about causality for quantum maps transfer straightforwardly to quantum processes. For example, the results about non-signalling established in Section 6.3.2 now also apply to general quantum processes.

8.2 Classical Maps from Spiders

We now have a proper process theory of **quantum processes** capturing both classical and quantum systems as wires. This allows us to reason about these processes using diagrams:

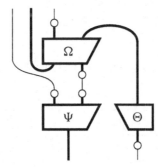

Unfortunately, these diagrams contain some very specific linear maps, namely measure and encode, and in order to establish all equations between diagrams we will unavoidably have to use their explicit forms:

$$\quad := \sum_i \quad\quad\quad\quad\quad := \sum_i$$

which involve ONB-states and sums. This is a bit of a pain.

In this section and the next we will do better than this. We will establish measure and encode as purely diagrammatic entities that allow us to derive equations between diagrams of **quantum processes** without ever needing the explicit matrix forms of measure and encode again.

The key to doing so, it turns out, is to better understand what classical processes are. While we have built up a pretty good repertoire of quantum maps, we have still said fairly little about their classical counterparts. We have seen one classical map already (deleting), and we will see a few more of these before establishing that they all emerge as special cases in the next diagrammatic revolution in this book: the rule of the spiders!

8.2.1 Classical Maps

By just restricting Definition 8.3, we have the following definition.

Definition 8.11 A *classical map* is a cq-map with only classical inputs and classical outputs, that is, a linear map f of the form:

$$\boxed{f} \quad := \quad \boxed{\Phi} \tag{8.18}$$

and a *classical process* is a classical map that is causal:

$$\boxed{f} \quad = \quad$$

Writing out a classical map as a sum over its branches as in (8.11), we obtain:

$$\boxed{\Phi} \quad = \quad \sum_{ij} \quad \langle\!\langle \Phi_{ij} \rangle\!\rangle$$

Of course, quantum maps with no inputs or outputs are just positive numbers, so a classical map is really nothing more than a matrix of positive numbers:

$$\sum_{ij} p_i^j \;\; \vee\!\!\wedge \quad \leftrightarrow \quad \begin{pmatrix} p_1^1 & p_2^1 & \cdots & p_m^1 \\ p_1^2 & p_2^2 & \cdots & p_m^2 \\ \vdots & \vdots & \ddots & \vdots \\ p_1^n & p_2^n & \cdots & p_m^n \end{pmatrix}$$

In particular, classical states are just vectors of positive real numbers:

$$\sum_j p^j \;\; \vee \quad \leftrightarrow \quad \begin{pmatrix} p^1 \\ p^2 \\ \vdots \\ p^n \end{pmatrix}$$

For classical processes, causality reduces to:

$$\forall i \; : \; \sum_j p_i^j = 1$$

That is, classical processes have matrices with positive entries where each column sums to 1. For classical states causality becomes:

$$\sum_j p^j = 1$$

so causal classical states are exactly probability distributions!

The matrices of classical processes are often referred to as *stochastic matrices*, and the linear maps themselves as *stochastic maps*. 'Stochastic' is essentially a synonym for 'random', and stochastic maps correspond to processes acting on probability distributions where there can be some element of randomness involved. Much like quantum processes are the most general maps that send causal quantum states to other causal quantum states, stochastic maps are the most general maps that send probability distributions to probability distributions.

Example 8.12 Imagine a process that takes in a classical bit, and with probability 1/3 flips the bit (thus leaving the bit fixed with 2/3 probability). We could describe this with the following stochastic map:

$$\boxed{f} \quad \leftrightarrow \quad \begin{pmatrix} \frac{2}{3} & \frac{1}{3} \\ \frac{1}{3} & \frac{2}{3} \end{pmatrix} \qquad (8.19)$$

If we input bit zero:

$$\frac{f}{0} = \frac{2}{3}\,\frac{}{0} + \frac{1}{3}\,\frac{}{1}$$

we get a bit 0 out with probability $\frac{2}{3}$ and a bit 1 with $\frac{1}{3}$. If we input bit 1, we get the opposite:

$$\frac{f}{1} = \frac{1}{3}\,\frac{}{0} + \frac{2}{3}\,\frac{}{1}$$

A deterministic classical process sends each ONB state to one (and only one) ONB state, and hence acts like a function on classical values. Thus, they take the following form.

Definition 8.13 A classical process f is called *deterministic* if there exists an underlying function:

$$f : \{1, \ldots, m\} \to \{1, \ldots, n\}$$

such that:

$$\frac{f}{i} = \frac{}{f(i)} \tag{8.20}$$

We will refer to deterministic classical processes as *function maps* for short. To see that any linear map described by (8.20) is automatically a classical process, it suffices to examine its matrix. Entries are all either 0 or 1, so of course they are all positive. Furthermore, there is precisely one 1 in each column, so causality follows immediately.

Restricting Definition 8.13 to states, we conclude that *deterministic classical states* are just ONB states, a.k.a. point distributions (cf. Section 5.1.4). Deterministic states are the classical analogue to pure quantum states, since they don't arise from (non-trivial) probabilistic mixing.

Exercise 8.14 Show that when:

$$\frac{}{p} = \sum_i \frac{}{q_i}$$

p is a deterministic classical state if and only if for all i we have:

$$\frac{}{p} \approx \frac{}{q_i}$$

Exercise 8.15 Recall from Remark 8.7 that all of the notions in this section depend on a particular choice of ONB for each classical type. Indeed, if we express the matrix of a classical map or state in a different ONB, typically the entries will no longer be positive. Find an ONB in which the matrix of the stochastic map (8.19) has non-positive entries.

We'll now have a look at some very special classical processes.

8.2.2 Copying and Deleting

In addition to deleting, which we have seen already, we can also copy classical data. What may come as a surprise is that some discarding-related features of quantum processes fail to have a classical deleting-related counterpart, most notably purification. In other words, the ability to purify is a characteristic feature of quantum processes. On the other hand, copying has no counterpart for quantum systems (cf. Section 4.4.2), so this process witnesses classicality. Hence, rather than considering no-cloning to be a shortcoming of quantum systems, we will consider *copiability* (or *cloneability*) to be the characterising feature for classical ones.

8.2.2.1 Deleting

As we saw above, deleting is the classical counterpart to discarding and is therefore the unique causal effect for a classical system.

Similarly, the adjoint of deleting:

$$\frac{1}{D} \; \overset{\displaystyle\text{\Large $\circ$}}{\vert} \;\; = \;\; \frac{1}{D} \sum_i \; \overset{\displaystyle i}{\overset{\displaystyle\vee}{\vert}}$$

is the classical counterpart to the maximally mixed state:

$$\frac{1}{D} \; \overset{\displaystyle\perp}{\underline{\underline{}}} \;\; = \;\; \frac{1}{D} \sum_i \; \overset{\displaystyle\vee}{\vert}$$

This classical state has a standard name.

Definition 8.16 The classical state:

$$\frac{1}{D} \; \overset{\displaystyle\circ}{\vert}$$

is called the *uniform probability distribution*.

Just as we had a notion of reduced quantum states (see Proposition 6.33):

given a state on two classical systems, if we delete one of the systems:

$$ \text{(8.14)} $$

we obtain a familiar operation from probability theory, called *marginalisation*. The classical state (a.k.a. probability distribution) x with one of the systems deleted is called the *marginal distribution*.

While reduced states have natural classical counterparts, which have many practical applications, here's a remarkable shortcoming of the classical world:

Purification of quantum states has no classical counterpart.

In other words, it is not the case that any classical state can be realised as the marginal of a deterministic classical state. The only deterministic classical states on two systems are of the form:

Then, clearly deleting one system will again result in a deterministic state. Hence, the only states that can be 'purified' to a deterministic state were deterministic in the first place!

Comparing this situation with the quantum case, we can represent probability distributions as quantum states (cf. Proposition 6.74):

$$ \sum_i p^i $$

This is indeed a quantum state and hence can be purified:

$$ \sum_i p^i \quad = $$

Explicitly, for:

$$ \psi := \sum_i \sqrt{p^i} $$

then we indeed have:

$$ \sum_i \sqrt{p^i} \quad \sum_j \sqrt{p^j} \quad = \quad \sum_i p^i $$

So the magic of purification comes from the fact that quantum states have two ways of combining processes via summation, mixing, and superposition:

$$\sum_i p^i \; \bigtriangledown_{\psi_i} \qquad \text{vs.} \qquad \text{double}\left(\sum_i \sqrt{p^i} \; \bigtriangledown_{\psi_i}\right)$$

and it is the latter that gives us enough extra flexibility to purify any quantum state.

8.2.2.2 Copying

Definition 8.17 Copying is the following classical map:

This map also behaves as advertised on basis states:

(8.21)

In fact, a state is copied by $\curlyvee$ if and only if it is a basis state.

Theorem 8.18 Copying uniquely fixes an ONB. More specifically, for any non-zero state ψ:

Proof First note that copying is an isometry:

(8.22)

Assuming (∗), this implies that:

$$\psi \; \overset{(8.22)}{=} \; \bigcirc \; = \; \psi\psi$$

The only numbers such that $p = p^2$ are 0 and 1, so ψ must be normalised. By Theorem 4.85 we know that normalised states that are jointly cloneable by an isometry must be orthogonal. Since by (8.21), the ONB states are also copied, it follows that ψ must either be an ONB state or be orthogonal to every ONB state. But the only state that is orthogonal to every state in an ONB is 0, so ψ must be equal to exactly one of the ONB states. □

So, not only does fixing an ONB of classical states yield a copying map, the copying map uniquely fixes an ONB. Moreover, copiability gives a diagrammatic characterisation for deterministic classical processes.

Proposition 8.19 A linear map f is a function map (i.e. a deterministic classical process) if and only if it satisfies the following two equations:

$$= \qquad\qquad = \qquad (8.23)$$

Proof First, assume f is a function map. By Definition 5.2 linear maps are equal if they agree on an ONB, so we can prove the two equations above by composing with ONB states:

$$\overset{(8.20)}{=} \quad \overset{(8.21)}{=} \quad \overset{(8.20)}{=} \quad \overset{(8.21)}{=}$$

The second equation is shown similarly. Conversely, we have:

$$\overset{(8.23)}{=} \quad \overset{(8.21)}{=}$$

So $f \circ i$ is copied by the copy map, and it is furthermore non-zero by the second equation in (8.23). Thus, by Theorem 8.18, the state $f \circ i$ must be an ONB state. Thus, we can define the underlying function f as follows:

The input and output systems of f (and hence the copy operations on the LHS and RHS above) could be different. In particular, copying ONB states is just the special case where the input of f is trivial.

One would expect that if we copy classical data, then the two identical copies can be swapped freely. Also, if we copy one of the two resulting copies again, both ways of making three copies are equal. This is indeed the case.

Proposition 8.20 We have:

(8.24)

(8.25)

Proof Unfolding the definitions, we have:

and:

$\square$

8.2.2.3 Copying and Deleting

One would also expect that if we copy classical data, then delete one copy, this is the same as doing nothing. Again, this is true.

Proposition 8.21 We have:

$$
\bigcup = \Big| = \bigcup \tag{8.26}
$$

Proof   Unfolding the definitions, we have:

$$
\bigcup = \frac{\sum_j}{\sum_i} = \sum_i = \Big|
$$

$$\square$$

Note that this immediately implies that copying is causal, since deleting the outputs is indeed the same as deleting the input:

$$
\bigcup = \phi
$$

Remark* 8.22 Earlier in Remark 3.17 we said that copying is an example of a *comultiplication*. Equations (8.24) and (8.25) tell us that copying is *coassociative* and *cocommutative*, which are the coalgebraic counterparts to associativity and commutativity in algebra, respectively. Equation (8.26) tells us that deleting is the corresponding *counit*, which is the coalgebraic counterpart to the usual notion of a unit in algebra. The equations in Proposition 8.19 then tell us that function maps are *comonoid homomorphisms*.

In (6.48) we saw that cloning fails for general probability distributions, and indeed:

$$
\boxed{\sum_i p^i} = \sum_i p^i \neq \boxed{\sum_i p^i}\ \boxed{\sum_j p^j}
$$

However, probability distributions **can** be broadcast. Indeed, we can interpret *broadcasting* as the existence of a map we can apply to any classical state p such that Aleks and Bob can each recover p just by deleting the other system:

So, we have two equivalent ways to express probability distributions, as well as the associated copying and deleting operations:

8.2.2.4 Matching

Matching is the adjoint of copying:

This classical map takes in two ONB states. If those states are the same, it sends that state out; otherwise it goes to zero:

$$
\text{} = \delta_i^j \;\; \text{}
\tag{8.27}
$$

Remark 8.23 Matching, unlike its adjoint, is not causal. In particular, it does not send probability distributions to probability distributions. It has a clear meaning and will be a useful operation nonetheless.

We can now simply take the adjoint of equations (8.24), (8.25), and (8.26) to obtain corresponding equations for matching.

Proposition 8.24 We have:

$$
\text{} = \text{}
\tag{8.28}
$$

$$\tag{8.29}$$

$$\tag{8.30}$$

where the adjoint to deleting is:

$$\; := \; \sum_i \tag{8.31}$$

Remark* 8.25 The equations (8.28), (8.29), and (8.30) are the less exotic, *algebraic* versions of the coalgebraic equations explained in Remark 8.22, namely *associativity*, *commutativity*, and *unitality*.

On arbitrary states:

$$\boxed{\psi} \; := \; \sum_i \psi^i \qquad\qquad \boxed{\phi} \; := \; \sum_j \phi^j$$

matching multiplies matrix entries pointwise:

$$= \quad \sum_i \psi^i \; \sum_j \phi^j \quad = \; \sum_{ij} \psi^i \phi^j \delta_i^j \quad = \; \sum_i \psi^i \phi^i$$

Written as an operation on matrices:

$$\leftrightarrow \quad \begin{pmatrix} \psi^1 \\ \vdots \\ \psi^D \end{pmatrix} \star \begin{pmatrix} \phi^1 \\ \vdots \\ \phi^D \end{pmatrix} := \begin{pmatrix} \psi^1 \phi^1 \\ \vdots \\ \psi^D \phi^D \end{pmatrix} \tag{8.32}$$

This $\star$-operation is sometimes called the *Hadamard product* or the *Schur product*, and it extends to arbitrary matrices.

Exercise 8.26 Show that for any two linear maps f and g of the same type, the diagram:

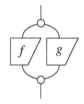

yields the Hadamard product of matrices:

$$\begin{pmatrix} f_1^1 & \cdots & f_D^1 \\ \vdots & \ddots & \vdots \\ f_1^D & \cdots & f_D^D \end{pmatrix} \star \begin{pmatrix} g_1^1 & \cdots & g_D^1 \\ \vdots & \ddots & \vdots \\ g_1^D & \cdots & g_D^D \end{pmatrix} = \begin{pmatrix} f_1^1 g_1^1 & \cdots & f_D^1 g_D^1 \\ \vdots & \ddots & \vdots \\ f_1^D g_1^D & \cdots & f_D^D g_D^D \end{pmatrix}$$

Example 8.27 Another use for the copying map is turning causal classical maps into causal bipartite classical states. In probability theory, the former are called *conditional probability distributions*, whereas the latter are called *joint probability distributions*. Suppose we write down some probabilities in terms of a classical map and a particular input state:

In the language of probability theory, the probabilities $P(i)$ are known as *priors*. Along with the conditional probabilities, they are used to compute the joint probabilities $P(ij)$, i.e. those of 'i and j both happening':

Since this gives a probability distribution, we can form a new classical state on two systems as follows:

which gives us an expression of the joint distribution. The resulting state is causal because f, p, and copying are.

We can also go the other direction and turn a joint distribution into a conditional distribution. For this, we need the *inverse state*:

$$\widehat{p^{-1}} \;:=\; \sum_i \frac{1}{p^i}\, \widehat{i}$$

or equivalently, the unique classical state satisfying:

 $$=\;\; \bullet \qquad\qquad (8.33)$$

(If you are worried about dividing by zero, see Convention 8.29 below.) Then, thinking of:

as a 'cup', we can build the associated 'cap' using p^{-1}:

$$=\;\; \bigg| \qquad\qquad (8.34)$$

This equation follows directly from (8.33) and the 'spider-fusion' rule, which we will soon unveil. But in the meantime one can use the definitions of copying/merging to prove (8.34) concretely. Combining the cap and the cup gives us a way to turn $P(j \mid i)$ into $P(i \mid j)$, which is called *Bayesian inversion*:

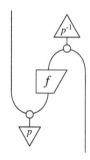

Exercise* 8.28 For the classical map:

$$\boxed{f} \;\leftrightarrow\; \begin{pmatrix} 1/3 & 1/2 \\ 2/3 & 1/2 \end{pmatrix}$$

compute the associated joint distributions for the following priors:

$$\begin{pmatrix} 1 \\ 0 \end{pmatrix} \qquad \begin{pmatrix} 0 \\ 1 \end{pmatrix} \qquad \begin{pmatrix} 1/2 \\ 1/2 \end{pmatrix}$$

Then, compute f's Bayesian inverse for these states.

Convention 8.29 In forming the inverse state above, we have implicitly assumed that our probability distribution has *full support*, that is:

$$\bigvee_{p} := \sum_i p^i \bigvee_i \qquad \text{where} \qquad \forall i: \; p^i \neq 0 \qquad (8.35)$$

When we are free to choose the dimension of our classical system, this is no loss of generality: we can always pass from any probability distribution to one with full support in a lower dimension by getting rid of classical values that occur with probability zero.

8.2.3 Spiders

We encountered the following classical maps, which all admit a natural interpretation as a classical data operation:

- *copying* and *deleting*:

- *matching* and the (unnormalised) *uniform state*:

There is no need to stop here. We could also add, e.g.:

- states/effects representing *perfect correlation* of two classical systems:

We can derive various equations (and their adjoints), which have natural interpretations similar to (8.24), (8.25), and (8.26), e.g.:

- Copying followed by matching equals doing nothing:

- 'Copying' the uniform state yields the perfectly correlated state:

Some of these equations even look familiar:

- Classical data admits yanking:

$$\qquad\qquad\qquad\qquad\qquad\qquad\qquad (8.36)$$

Exercise 8.30 Prove the above equations between classical maps.

Imagine now a slowly growing crescendo in the background:

It would be natural to aim for all the equations that hold between these classical maps. How many would there be? Maybe hundreds? Maybe an infinite number? Now imagine a blast of a gong:

There is only one equation! Indeed, all the equations that we have have seen thus far are in fact instances of one and the same equation. To see this, one first needs to realise that all the classical maps that we have seen thus far are special cases of one family of classical maps, which we call spiders.

Definition 8.31 *Spiders* are linear maps of the form:

$$
\begin{array}{c}
\overbrace{}^{n} \\
\text{(spider diagram)}
\end{array}
:= \sum_i
\begin{array}{c}
\overbrace{}^{n} \\
\text{(copy diagram)} \\
\underbrace{}_{m}
\end{array}
\tag{8.37}
$$

Intuitively, spiders force all of the inputs and outputs to be the same basis element. As such, sometimes it is helpful to think of them as a 'big Kronecker delta', which is exactly what we get if we compute the matrix of a spider:

$$
\delta^{j_1 \cdots j_n}_{i_1 \cdots i_m} =
\begin{cases}
1 & \text{if } i_1 = \cdots = i_m = j_1 = \cdots = j_n \\
0 & \text{otherwise}
\end{cases}
$$

The usual Kronecker delta:

$$
\delta^{j}_{i} =
\begin{cases}
1 & \text{if } i = j \\
0 & \text{otherwise}
\end{cases}
$$

arises as a special case, as the matrix of the identity map.

Exercise 8.32 Prove the *generalised copy rule* for spiders:

$$
\begin{array}{c}
\text{(spider with triangles diagram)}
\end{array}
= \delta^{j_1 \cdots j_n}_{i_1 \cdots i_m}
\begin{array}{c}
\text{(separated spiders diagram)}
\end{array}
\tag{8.38}
$$

which, for example, generalises equation (8.21) as well as (8.27).

From the definition, we can conclude a couple of things about spiders. First, a spider with only two legs is just a wire:

$$
\begin{array}{ccc}
\text{(diagram)} = \text{(wire)} & \text{(cup diagram)} = \text{(cup)} & \text{(cap diagram)} = \text{(cap)}
\end{array}
\tag{8.39}
$$

and second, spiders exhibit a lot of symmetry.

Proposition 8.33 All spiders are invariant under 'leg-swapping':

$$\tag{8.40}$$

and conjugation (i.e. horizontal reflection):

$$\tag{8.41}$$

Proof Both equations follow directly from (8.37).   □

And now the grand finale ...

All of the equations for classical maps we have encountered so far can now be subsumed by one simple rule:

> *If two spiders touch, they fuse together.*

More precisely, we have the following.

Theorem 8.34 Spiders compose as follows:

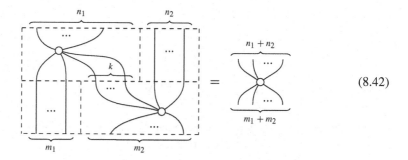

$$(8.42)$$

for $k \geq 1$.

Proof Unfolding the two spiders, we obtain:

$$\sum_i \quad = \sum_{ij} \delta_i^j \quad = \sum_i$$

$\square$

We can now prove any of the equations from the previous sections by just squashing the LHS and the RHS down to a single spider. For instance, this equation involving copying:

$$\overset{(8.42)}{=} \quad \overset{(8.42)}{=}$$

More generally, by (8.39) cups and caps are also spiders, so we can squash diagrams of spiders into one big spider.

Corollary 8.35 Any <u>connected</u> string diagram consisting only of spiders is equal to a <u>single</u> spider:

$$= \qquad (8.43)$$

Hence, such a diagram is uniquely determined by its number of inputs and outputs, and we can establish equality simply by counting them.

We refer to this rule as *spider fusion*.

Corollary 8.36 Spiders satisfy 'leg flipping', that is, if we bend one of a spider's legs up or down, we get again a spider:

$$= \qquad (8.44)$$

and hence the transpose of a spider is again a spider.

Of course, combining this with the fact that spiders are self-conjugate, it follows that taking the adjoint of a spider also yields a spider:

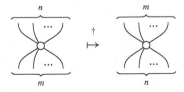

Exercise 8.37 Prove using just the properties of spiders that the spider with no legs equals the 'circle' (i.e. the dimension):

$$\circ = \bigcirc$$

Exercise 8.38 As already pointed out in Remark 8.23, some spiders are not causal. For correlating to be causal it suffices to introduce a normalisation factor:

$$\frac{1}{D}$$

However, for comparing and matching no number would do the job, given that for $i \neq j$ we have:

$$
\text{(spider diagram)} = 0 \qquad\qquad \text{(spider diagram)} = 0
$$

More generally, which spiders are causal, which can be made causal by normalising, and which cannot?

In Section 5.2.2 we saw how one can encode bit strings as basis states. We can also associate spiders with them.

Exercise 8.39 The spiders:

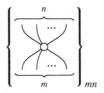

for the two-dimensional basis:

$$
\left\{ \bigtriangledown_0 , \bigtriangledown_1 \right\}
$$

are associated with *bits*. Show that the following family of classical maps:

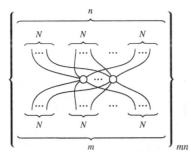

is also a family of spiders. Furthermore, show that it is associated with the ONB of *N-bitstrings*:

$$
\left\{ \bigtriangledown_0 \cdots \bigtriangledown_0 \bigtriangledown_0 , \bigtriangledown_0 \cdots \bigtriangledown_0 \bigtriangledown_1 , \bigtriangledown_0 \cdots \bigtriangledown_1 \bigtriangledown_0 , \ldots , \bigtriangledown_1 \cdots \bigtriangledown_1 \bigtriangledown_1 \right\}
$$

Remark 8.40 The equations:

$$
\smile\!\!\!\!\circ\!\!\!\!\smile = \smile \qquad\qquad \overset{\circ}{\frown} = \frown \tag{8.45}
$$

which are required, e.g. for 'leg-flipping', come from the fact that we have chosen a self-conjugate ONB (cf. Section 5.2.3). In a few (rare) cases, it is useful to define spiders for non-self-conjugate bases, in which case we should either drop these equations or fix them somehow (cf. Section* 8.6.3).

8.2.4 If It behaves like a Spider It Is One

Given the importance of spiders, we need to figure out how to recognise them. That is, how can we distinguish a real spider from, say, a dodo in a spider costume desperately trying to fight extinction?

In Theorem 8.18 we showed that the 'copy spider' defines an ONB in terms of the states that it 'copies':

$$\left\{ \left. \vcenter{\hbox{$\bigtriangledown_i$}} \;\middle|\; \vcenter{\hbox{$\bigtriangledown_i$}} = \vcenter{\hbox{$\bigtriangledown_i$}}\;\vcenter{\hbox{$\bigtriangledown_i$}} \right\} \tag{8.46}$$

So consequently, a family of spiders as defined in Definition 8.31 always fixes an ONB. Surprisingly, the converse is also true, and this is a non-trivial result: any collection of linear maps that composes like a family of spiders is in fact a family of spiders:

Theorem 8.41 Any collection of linear maps:

$$\left\{ \vcenter{\hbox{$\overbrace{}^{n}\; f^n_m \; \underbrace{}_{m}$}} \right\}_{mn}$$

which satisfy:

$$
\boxed{f_m^n} = \boxed{f_n^m} \qquad \boxed{f_m^n} = \boxed{f_m^n} \qquad \boxed{f_1^1} = \Big|
$$

and compose as follows:

$$
\boxed{f_{m'+k}^{n'}} \atop \boxed{f_m^{n+k}} \;=\; \boxed{f_{m+m'}^{n+n'}}
$$

is a family of spiders. That is, there exists an ONB:

$$
\left\{ \; \boxed{i} \; \right\}_i
$$

such that:

$$
\overbrace{\boxed{f_m^n}}^{n} \;=\; \sum_i \; \overbrace{\boxed{i}\,\boxed{i}\,\cdots\,\boxed{i}}^{n} \atop \underbrace{\boxed{i}\,\boxed{i}\,\cdots\,\boxed{i}}_{m}
$$

Since we are mostly interested in spiders for self-conjugate ONBs (cf. Remark 8.1) we will now specialise this result to this case. Back in Proposition 5.60 we saw that an ONB is self-conjugate if and only if:

$$
\smile \;=\; \sum_i \boxed{i}\,\boxed{i}
$$

We now know that the RHS is just a spider with two outputs, so we have the following.

Corollary 8.42 The collection of linear maps from Theorem 8.41 represents a <u>self-conjugate</u> ONB if and only if it additionally satisfies:

$$
\boxed{f_0^2} \;=\; \smile
$$

The proof of Theorem 8.41 uses some techniques from *representation theory* and goes beyond the scope of this book. We do give some indication of how it goes in Section* 8.6.1. But the punchline is that the equations that we identified as holding for spiders actually 'axiomatise' these spiders; that is, spiders and nothing but spiders can satisfy all of these. Therefore we call these the *spider equations*. An important consequence is the fact that now we can define ONBs, which at first seem totally undiagrammatic, using the purely diagrammatic concept of spiders.

Corollary 8.43 An ONB can be defined totally in terms of diagram equations, namely, the spider equations.

Furthermore, since spiders make sense in any process theory, copiable states/effects do as well. Surprisingly, orthonormality is (nearly) automatic.

Exercise 8.44 Assuming that the numbers in a process theory satisfy:

$$\lambda^2 = \lambda \quad \Longrightarrow \quad \lambda \in \{0, 1\}$$

(which is true e.g. for real numbers, complex numbers, and booleans), show that the copiable states (8.46) for any family of spiders are always orthonormal:

$$\begin{array}{c} \triangle^{j} \\ \vdots \\ \triangledown_{i} \end{array} = \delta_i^j$$

but they don't always form a basis.

Exercise* 8.45 Give a family of spiders in **relations** whose copiable states do <u>not</u> form an ONB.

8.2.5 All Linear Maps as Spiders + Isometries

The spectral theorem (Theorem 5.71) lets one decompose self-adjoint linear maps in terms of ONBs. Now that we know that ONBs are really all about spiders, we can see that the spectral theorem means any self-adjoint linear map has a spider hiding inside.

Theorem 8.46 Any self-adjoint linear map f admits a *spectral decomposition*:

$$(8.47)$$

If f is moreover positive, then r is a classical state, so any positive f decomposes as:

$$(8.48)$$

Proof By the spectral theorem f decomposes as follows:

$$\vcenter{\hbox{\includegraphics{f}}} \;=\; \sum_i r_i \; \vcenter{\hbox{\includegraphics{spider}}}$$

for some ONB and real numbers r_i. Then, let U be the unitary that sends the ONB associated with the spider to the ONB above:

$$\vcenter{\hbox{\includegraphics{U}}} \;::\; \vcenter{\hbox{\includegraphics{i}}} \;\mapsto\; \vcenter{\hbox{\includegraphics{i}}}$$

and let:

$$\vcenter{\hbox{\includegraphics{r}}} \;:=\; \sum_i r_i \; \vcenter{\hbox{\includegraphics{i}}}$$

Then:

$$\vcenter{\hbox{\includegraphics{d1}}} \;=\; \vcenter{\hbox{\includegraphics{d2}}} \;=\; \sum_i r_i \, \vcenter{\hbox{\includegraphics{d3}}} \;=\; \sum_i r_i \, \vcenter{\hbox{\includegraphics{d4}}} \;=\; \vcenter{\hbox{\includegraphics{f}}}$$

Also by Theorem 5.71, if f is positive, then the numbers r_i are positive. Hence r becomes a classical state. $\qquad\square$

Decomposition (8.48) tells us what's 'inside' any positive linear map:

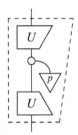

namely, nothing but spiders, unitaries, and classical states.

A slight variation of this decomposition allows us to express a positive linear map in terms of a classical state with full support (cf. Convention 8.29). This can be realised by making the system in the middle smaller and replacing the unitaries with isometries:

$$
\boxed{f} \quad = \quad \begin{array}{c} \boxed{U} \longleftarrow \text{isometry} \\ \circ \\ \bigtriangledown_{p} \longleftarrow \text{classical state} \\ \text{(with full support)} \\ \boxed{U} \end{array} \tag{8.49}
$$

Exercise 8.47 Prove that any positive linear map decomposes as (8.49).

Furthermore, by relaxing the requirement that the isometry on the bottom and the top be the same, we can obtain a decomposition which applies to <u>all</u> linear maps:

Theorem 8.48 Any linear map f admits a *singular value decomposition*:

$$
\boxed{f} \quad = \quad \begin{array}{c} \boxed{V} \longleftarrow \text{isometry} \\ \circ \\ \bigtriangledown_{p} \longleftarrow \text{classical state} \\ \boxed{U} \longleftarrow \text{adjoint of isometry} \end{array}
$$

for some isometries U and V and a classical state p with full support.

Proof Since $f^{\dagger} \circ f$ is positive, we can rely on the spectral theorem to decompose it as in (8.49):

$$
\begin{array}{c} \boxed{f} \\ \boxed{f} \end{array} \quad = \quad \begin{array}{c} \boxed{U} \\ \circ \\ \bigtriangledown_{q} \\ \boxed{U} \end{array} \tag{8.50}
$$

Now note that, since U is an isometry, the following is a projector (i.e. positive and idempotent, as in Definition 4.69):

$$
\boxed{P} \quad := \quad \begin{array}{c} \boxed{U} \\ \boxed{U} \end{array}
$$

Then, by the form of (8.50), it immediately follows that:

Hence, by Exercise 5.77 (which also follows directly from the spectral theorem), we have:

(8.51)

Now, for:

$$\bigtriangledown_{q} \; := \; \sum_{i} q^{i} \; \bigtriangledown_{i}$$

with all $q^{i} \neq 0$, we define two additional states:

$$\bigtriangledown_{q^{\frac{1}{2}}} \; := \; \sum_{i} \sqrt{q^{i}} \; \bigtriangledown_{i} \qquad\qquad \bigtriangledown_{q^{-\frac{1}{2}}} \; := \; \sum_{i} \frac{1}{\sqrt{q^{i}}} \; \bigtriangledown_{i}$$

which by (8.32) satisfy:

(8.52)

(8.53)

We can now show that:

$$(8.54)$$

is an isometry:

Letting $p := q^{\frac{1}{2}}$ in:

completes the proof. $\square$

By bending the wire, we also discover what's 'inside' any bipartite state.

Corollary 8.49 Any bipartite state ψ decomposes as follows:

Exercise 8.50 Show that when the two output systems of a bipartite state ψ are the same, it can be decomposed as:

Remark 8.51 The 'sideways' version of the singular value decomposition for a bipartite state is often called the *Schmidt decomposition*.

8.2.6 Spider Diagrams and Completeness

In our definition of spiders we already singled out cups and caps as special cases of spiders, which we referred to as correlating and comparing. But of course, what defines cups and caps is the relationship between them, namely that if we compose them we get an identity. This is in fact a direct instance of the spider fusion rule (8.34):

This is kind of funny, to think about 'yanking wires' as a special case of 'fusing spiders', but that's exactly what we established here, that:

> *Reasoning with string diagrams is an instance of reasoning with spiders!*

Whereas a wire connects two ends together, a spider connects many ends together. Spider fusion is then all about connecting many things together by means of multiple spiders.

And just like we could treat string diagrams either as 'circuit diagrams + caps/cups' or as a new kind of diagram (cf. Theorem 4.19), we can do the same with diagrams containing spiders.

Definition 8.52 A *spider diagram* consists of boxes and wires that are allowed to connect any number of inputs and outputs together.

Diagrammatically, we represent these 'multiwires' as spiders, for example:

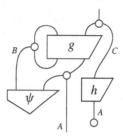

In the associated diagram formula (cf. Definitions 3.8 and 4.21), we can represent them just by repeating wire names as many times as we like:

$$\longleftrightarrow \quad \psi^{B_1 \check{A}_1} g^{B_1 \hat{C}_1}_{B_1 \check{A}_1} h^{\hat{C}_1}_{A_2} \qquad\qquad (8.55)$$

For 'multiwires', it's no longer clear which wire names correspond to inputs/outputs, so we mark inputs with a check $(\check{-})$ and outputs with a hat $(\hat{-})$. This for example allows one to distinguish:

$$f \quad \longleftrightarrow \quad f^{A_2}_{A_1} \qquad \text{vs.} \qquad f \quad \longleftrightarrow \quad f^{\hat{A}_2}_{\check{A}_1}$$

Theorem 8.53 The following two notions are equivalent:

 (i) spider diagrams and
(ii) circuit diagrams to which we adjoin spiders for each type

in the sense that (ii) can be unambiguously expressed as (i) and vice versa.

Proof We can translate a circuit diagram with spiders to a spider diagram just by fusing all of the connected spiders into single spiders:

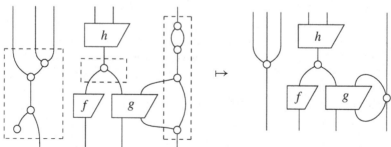

In the opposite direction, if we replace cup- and cap-shaped wires in a spider diagram with the corresponding spiders, we obtain a circuit diagram with spiders:

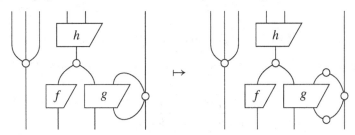

□

Recall from Section 5.4.1 that string diagrams are complete for **linear maps**. That is, an equation between string diagrams holds for all Hilbert spaces and linear maps if and only if the string diagrams are the same.

It is always possible that by enriching the diagram language, we could break completeness. Since spider-diagram language is richer, we can write down more equations (i.e. equations involving spiders), but can we still prove them all? Thankfully, the answer is yes.

Theorem 8.54 Spider diagrams are complete for **linear maps**. That is, for any two spider diagrams D and E, the following are equivalent:

- $D = E$
- For all interpretations of D, E into **linear maps**, $[\![D]\!] = [\![E]\!]$.

We can thus make a statement analogous to the one about string diagrams:

> *An equation between spider diagrams holds for all Hilbert spaces and linear maps if and only if the spider diagrams are the same.*

8.3 Quantum Maps from Spiders

Readers of the previous two sections might notice a suspicious similarity between the notation for spiders:

and for measuring and encoding:

This is of course no accident. In this section, we'll see how these are also a certain species of spider. We'll then take advantage of that fact to do some pretty cool stuff. In particular, this will let purely diagrammatic rules such as spider fusion do most of the hard work from now on.

8.3.1 Measuring and Encoding as Spiders

The whole point of cq-maps is to allow classical and quantum systems to interact. Using our paradigm:

$$\frac{\text{classical}}{\text{quantum}} = \frac{\text{thin/single wires}}{\text{thick/double wires}}$$

we can express boxes whose inputs and outputs consist of both classical and quantum wires. We saw in Section 8.1.3 that quantum states can be turned into classical data, and vice versa, using measure and encode. Can we give these two processes a form that is as elegant as the spider form for classical processes? Yes we can!

We have already seen maps that have a pair of wires in and a single wire out and vice versa: the copying and matching spiders. However, rather than treating these as operations on classical data, we can treat a pair of wires as a single quantum wire, hence forming a bridge from classical to quantum:

$$\text{(8.56)}$$

Unfolding the definition of spiders, we see this indeed gives us measuring:

and taking the adjoint we obtain encoding.

Hence, we can understand what these processes do in terms of spiders. For example, on ONB states, encoding unfolds as copying, so we obtain:

Similarly, measuring can be understood in terms of matching, so to see what measuring does to an arbitrary quantum state ρ:

we can use the fact that measurement unfolds as matching:

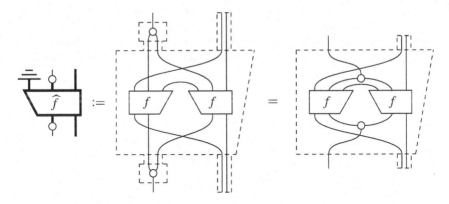

That is, all <u>non-diagonal</u> entries of the matrix of ρ in the product basis are gone, and the diagonal ones (which by Corollary 5.41 are all positive) are retained unaltered. As we already saw in Section 8.1.3, these diagonal entries are the probabilities for the ONB measurement according to the Born rule:

$$\rho^{ii} := \quad = P(i \mid \rho)$$

Corollary 8.6 stated that we can always express a cq-map in terms of a <u>pure</u> quantum map via measure, encode, and discarding. We can now give this generic form in terms of spiders:

In the special case of classical maps we have:

But in fact, this simplifies further.

Exercise 8.55 Show that all classical maps are of the form:

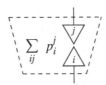

Since this simplification relies on particular properties of linear maps, you will need to use the following form for classical maps:

$$\sum_{ij} p_i^j \; \bigtriangledown_j^{}\!\!\bigtriangleup_i$$

Clearly Exercise 8.55 does not extend to more general cq-maps, since this would imply in particular that all quantum maps are pure. As a special case of Exercise 8.55, we now know that we can always write a classical state p as an ONB-measurement of <u>some</u> pure quantum state:

$$\raise2pt\hbox{$\bigtriangledown_p \;=\; \bigtriangledown_\psi$} \quad \left.\begin{matrix}\\ \end{matrix}\right\}\ \text{measurement} \quad \left.\begin{matrix}\\ \end{matrix}\right\}\ \text{pure quantum state}$$

Moreover, now that we know that encoding arises from the copying spider, we can show that every quantum state arises from encoding a classical state, thanks to the diagrammatic form of the spectral theorem.

Proposition 8.56 Any quantum state ρ encodes a classical state as follows:

$$\bigtriangledown_\rho \;=\; \begin{matrix}\widehat{U} \\ \text{encoding} \\ \bigtriangledown_p\end{matrix} \quad \begin{matrix}\}\ \text{unitary} \\ \}\ \text{encoding} \\ \}\ \text{classical state}\end{matrix}$$

Proof   Unfolding the equation above yields:

$$\bigtriangledown_f\,\bigtriangledown_f \;=\; \begin{matrix}U \quad U \\ \bigtriangledown_p\end{matrix} \tag{8.57}$$

for some f. This is just the bent-over version of decomposition (8.48), which exists thanks to the spectral theorem. $\square$

Remark 8.57 The presence of the unitary U is necessary since we opted for self-conjugate ONBs throughout this book. Otherwise, as we show in Section* 8.6.3, the decomposition can be simplified:

$$\text{encoding} \quad \text{classical state}$$

We can also exploit the fact that measuring and encoding both are spiders 'in disguise' to produce some more equations.

Proposition 8.58 We have:

1. Encoding followed by measuring is equal to doing nothing:

$$\qquad\qquad\qquad\qquad\qquad\qquad\qquad\qquad\qquad (8.58)$$

2. Encoding's transpose is measuring:

$$\qquad\qquad\qquad\qquad\qquad\qquad\qquad\qquad\qquad (8.59)$$

3. Measuring followed by deleting yields discarding:

$$\qquad\qquad\qquad\qquad\qquad\qquad\qquad\qquad\qquad (8.60)$$

4. Encoding followed by discarding yields deleting:

$$\qquad\qquad\qquad\qquad\qquad\qquad\qquad\qquad\qquad (8.61)$$

Proof All of these equations follow from unfolding the doubled parts and applying spider fusion:

$$
\dot{\varphi} := [\hat{A}] = \cap = [\hat{A}] =: \overline{\mp}
$$

$$
\overline{\mp} := [\hat{\theta}] = \hat{Q} = \dot{\varphi}
$$

□

The fact that discarding decomposes as in the first equation of (8.60) allows us to produce a new version of Proposition 6.79, which said that a quantum map separates when the reduced map is pure. Besides discarding quantum outputs, this also applies to deleting classical outputs.

Proposition 8.59 If the *reduced map* of a cq-map is pure:

$$
\boxed{\Phi} = \boxed{\hat{f}} \tag{8.62}
$$

then the process Φ separates as follows:

$$
\boxed{\Phi} = \boxed{p}\,\boxed{\hat{f}}
$$

for some (causal) classical state p (a.k.a. probability distribution).

Proof By equation (8.60), equation (8.62) is equivalent to:

$$
\boxed{\Phi} = \boxed{\hat{f}}
$$

so by Proposition 6.79 we have:

$$
\boxed{\Phi} \overset{(6.52)}{=} \boxed{\rho}\,\boxed{\hat{f}} = \boxed{p}\,\boxed{\hat{f}}
$$

where we set:

$$
\boxed{p} := \boxed{\rho}
$$

and which is causal by causality of measure and ρ.

□

Also, several things that were derived earlier using sums can now be derived diagrammatically, for example the following.

Exercise 8.60 Prove diagrammatically that classical maps are self-conjugate:

$$
\bar{f} \;=\; f
$$

Use this result to show that for function maps we have:

$$
\widehat{f} \;=\; f
$$

8.3.2 Decoherence

This section is dedicated to a very important (and infamous!) quantum process. We say 'infamous' because it has a bad habit of messing up the nice pure quantum states physicists try to prepare in their labs, which makes building things like quantum computers very difficult indeed.

Equation (8.58) indicates that if we go from classical data to classical data via a quantum system, the classical system remains unchanged. However, measurement is only a one-sided inverse of encoding. That is, if we compose the maps in the opposite order, this most certainly does not leave the quantum system unchanged, due to the invasive nature of measurement, so:

$$
\neq \quad\Big| \qquad\qquad \text{while} \qquad\qquad = \quad\Big|
$$

Definition 8.61 *Decoherence* relative to an ONB is the quantum process:

$$
:=
$$

In fact, we already encountered this process in the proof of Proposition 8.5, which gave a normal form for cq-maps. There, we relied crucially on the fact that this process was a quantum map. Now that we know measuring and encoding are spiders, this follows just from spider fusion:

$$
\text{(diagram)} \qquad (8.63)
$$

We still need to show that decoherence is not equal to the identity, but we can actually say something much stronger. Not only is it not equal to the identity, it is not even pure.

Proposition 8.62 Decoherence is not a pure quantum map.

Proof Suppose decoherence is a pure quantum map. Then, for some linear map f we would have:

$$
\text{(diagram)}
$$

But then, using spider fusion, we have:

$$
\text{(diagram)}
$$

i.e. the identity is disconnected, so decoherence cannot be pure.   □

When looking at the diagrammatic form of decoherence, the fact that it is not an identity should not come as a surprise. While the input and output are doubled, in the middle the quantum system seems to be 'squeezed' through a single (classical) wire. Hence, something that lives in two wires is forced into one wire (which will in general come with some data loss) before being injected back into two wires. In physical terms, this means that a quantum state is forced to become classical, and then quantum again.

On the other hand, once decoherence is applied to a quantum system, the damage is done. That is, a second application will leave everything unchanged.

Lemma 8.63 Decoherence is a projector.

Proof Decoherence is clearly self-adjoint, and also idempotent:

$$\text{(8.58)}$$

□

In particular, we can now identify a subset of *decoherent* states ρ, i.e. those that are unaffected by (further) decoherence:

What are these states? Since decoherence is the composite of measure and encode, we already know from the previous section what it does to the matrix of a quantum state:

$$= \sum_{ij} \rho^{ij} \quad = \quad \sum_i \rho^{ii} \quad = \quad \sum_i \rho^{ii}$$

Hence it preserves exactly those states whose non-diagonal entries are all zero. This is stated equivalently.

Theorem 8.64 Decoherence preserves a quantum state ρ if and only if it encodes a probability distribution, i.e. is of the form:

$$\boxed{p} = \sum_i p^i \quad (8.64)$$

Exercise 8.65 When defining decoherence as a completely positive map (cf. Remark 6.50), what does it do to density matrices?

For qubits, the Bloch ball (see Section 6.2.7) provides a nice geometrical picture of what decoherence does to states, and which states are left unchanged by (further) decoherence. Decoherence projects every state onto the axis passing through the two ONB states:

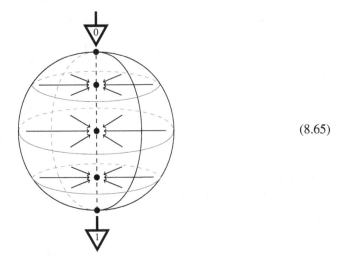

(8.65)

Recall the correspondence:

which in sum-free terms can be rewritten as:

This correspondence says that a quantum output to which decoherence is applied behaves just like a classical output. In fact, this correspondence extends to arbitrary quantum processes. By plugging measuring and encoding maps at its classical inputs and outputs we can turn any cq-map:

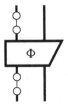

into a quantum map:

where the presence of a classical input/output of the cq-map we started from is now witnessed by the presence of decoherence. We also easily recover this cq-map by plugging encoding and measuring maps:

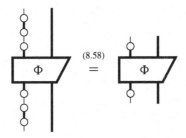

The moral of the story is:

Decoherence forces quantum systems to behave as classical systems.

For a quantum map, this can be directly seen by drawing a picture. When we compose decoherence with the inputs and outputs of any quantum process, at its core it becomes classical:

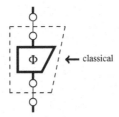

Since decoherence forces a quantum system to behave like a classical one it shouldn't come as a surprise that decoherence is no friend to anyone wishing to exploit quantum features to, for example, build a quantum computer. In practice, what often happens to quantum systems in the real world is that they undergo some *partial decoherence*:

$$(1-p) \left|\; + p \;\right.$$

where the longer we try to store a system, the greater the value of p becomes. The time it takes for p to become 1 is called the *decoherence time*, which typically is very small. One of the biggest challenges in building a quantum computer is to get this time to be as long as possible while still being able to interact with the quantum system in interesting ways.

Exercise* 8.66 Characterise partial decoherence without using sums.

8.3.3 Classical, Quantum, and Bastard Spiders

In this section, we will see how measure and encode arise as members of a new species of spider. In fact, all of the calculations in the rest of this book consist of letting different spider species meet and interact with each other. In this section, we perform a bit of spider taxonomy by distinguishing three types of spiders: classical spiders, quantum spiders, and (most interestingly) bastard spiders.

We already know what classical spiders are.

Definition 8.67 A *classical spider* is a spider with only thin legs:

And we also know how they compose:

Now, since these classical spiders are linear maps, we can turn them into pure quantum maps by doubling them.

Definition 8.68 A *quantum spider* is a quantum map of the form:

Since equations between linear maps carry over into the doubled world (cf. Corollary 6.16) we also know how they compose.

Corollary 8.69 Quantum spiders compose as follows:

$$(8.66)$$

Example 8.70 (The GHZ state) Important examples of quantum spiders are quantum spider states. We have encountered one already: the two-system quantum spider state, which is just the *Bell state*:

The three-system quantum spider state:

is called the *Greenberger–Horne–Zeilinger (GHZ) state* and has many important applications. For example, in Section 11.1 we will provide a proof of quantum non-locality based on it. Written in terms of an ONB, the qubit GHZ state is:

Since it is related to the (doubled) copying spider as follows:

by Theorem 8.18 it characterises a (doubled) ONB.

Simply by relying on copying it follows that if any of the three systems is in one of the basis states of this ONB, then so are the other two:

$$\bigcup_{i} = \underset{i}{\vee}\underset{i}{\vee} \qquad (8.67)$$

The other n-system quantum spider states:

are usually referred to as *generalised GHZ states*.

The third type of spiders contains both classical and quantum wires. We have already encountered two extremely important examples:

More generally, we can use measure and encode to connect any classical spider to any quantum spider:

$$(8.68)$$

in order to obtain a whole family of classical-quantum hybrids. When unfolding the doubled spider we see that we can fuse all the dots together into a single dot:

Therefore, the most general spiders that we can obtain using (8.68) are the following ones, which involve only a single (non-bold) dot.

Definition 8.71 A *bastard spider* is a cq-map of the form:

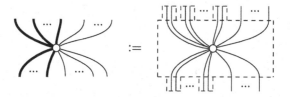

That is, they are the spiders obtained by interpreting some pairs of legs of a classical spider together as doubled systems (a.k.a. *folding*), while leaving others single.

Owing to the fact that it only consists of a single (non-bold) dot, a bastard spider with no quantum legs is the same as a classical spider, but a bastard spider with no classical legs is <u>not</u> a quantum spider:

For example, decoherence is a bastard spider with one quantum input and one quantum output:

$$\tag{8.69}$$

while the corresponding quantum spider is the identity:

Discarding is also a bastard spider:

$$\tag{8.70}$$

whereas the quantum spider with one quantum input is a pure (non-causal) quantum effect:

Of course, bastard spiders also fuse together:

However, the result may not always be a bastard spider, e.g.:

To understand bastard spider fusion, including fusion with classical and quantum spiders, one should think in terms of two species of spiders rather than three, namely:

- 'single-dot' spiders := classical spiders + bastard spiders
- 'double-dot' spiders := quantum spiders

The following is the resulting theorem.

Theorem 8.72 Any composition of spiders involving at least one single-dot spider yields a single-dot spider, for example:

This can be put somewhat differently.

Corollary 8.73 Any connected diagram of classical, quantum, or bastard spiders must be equal to one of the following:

1. a <u>quantum</u> spider if it contains only double dots,
2. a <u>classical</u> spider if it has only classical inputs/outputs, or
3. a <u>bastard</u> spider otherwise.

The paradigmatic example of a single-dot spider that is not a classical spider is decoherence, which forces quantum systems to behave like classical systems. Theorem 8.72 generalises this fact to the world of spiders, where being single-dot can be interpreted as 'being infected by classicality'.

Example 8.74 In equation (8.63), we showed that decoherence is a quantum map. We can now see this proof as an instance of bastard spider fusion:

$$\text{(8.70)}$$

Example 8.75 When we measure one system of three systems in a GHZ state, we obtain a bastard spider:

If we measure all three systems, we end up with a classical spider:

This classical spider describes the classical outcomes of the measurement. Writing it in an ONB in the case of a qubit:

we see that in all three measurements we will obtain the same outcome, although this outcome may be either 0 or 1. This again follows from the generalised copy rule:

which also extends to quantum and bastard spiders:

$$= \delta_{i_1\ldots i_m}^{j_1\ldots j_n} \qquad \qquad \text{(8.71)}$$

$$= \delta_{i_1\ldots i_m}^{j_1\ldots j_n} \qquad \qquad \text{(8.72)}$$

Also as with classical spiders, quantum/bastard spiders can be combined to form product quantum/bastard spiders on compound systems:

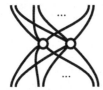 

In the case of bastard spiders, we only allow classical legs to pair with classical legs and quantum legs with quantum legs. In particular, the measurement/encoding maps for compound systems are just the obvious things:

8.3.4 Mixing with Spiders

Spiders enable us to define mixing, for which we previously relied on sums, entirely diagrammatically. First, let's take a moment to recall what 'mixing' means for quantum processes. A mixture (cf. Definition 6.70) gives us a way to represent a situation where we have one of several possible quantum processes, but we aren't sure which one. In other words, there is some classical randomness associated with how we obtained this process, which can be expressed as some probability distribution fed into a controlled quantum process. In fact, we can easily show that every mixture arises this way:

$$\sum_i p^i \boxed{\Phi_i} = \boxed{\sum_j \Phi_j \quad \sum_i p^i} = \boxed{\Phi}$$

The quantum process with a classical input represents the things being mixed, and the probability distribution p represents the mixing itself. We can obtain the components of the mixture by means of ONB-states:

$$\boxed{\Phi} = \boxed{\Phi_i}$$

Let's now revisit some of the results from Chapter 6 on mixing. First, the fact that every mixture of causal quantum maps is again a causal quantum map (cf. Theorem 6.71) follows from the fact that composing causal cq-maps:

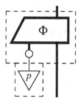

yields again a causal cq-map. In this case there are no classical inputs/outputs, so it is indeed a causal quantum map. The fact that every mixture can be interpreted in terms of discarding part of a system now follows from bastard spider fusion:

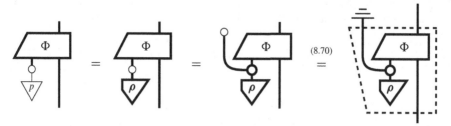

The fact that every causal quantum state can be regarded as a mixture of pure causal quantum states (Theorem 6.72) is simply the fact that any quantum state encodes a classical state (Proposition 8.56):

$$\underset{\rho}{\bigtriangledown} = \underset{p}{\overset{\widehat{U}}{\bigtriangledown}}$$

and the fact that these states can always be chosen to be orthonormal is already built in:

$$\overset{j}{\underset{i}{\overset{\widehat{U}}{\underset{\widehat{U}}{}}}} = \overset{j}{\underset{i}{\bigcirc}} = \overset{j}{\underset{i}{}} = \delta_i^j$$

We'll now use this diagrammatic form of mixing to prove something we didn't prove yet, namely, that if the result of mixing quantum processes is pure, then all the processes that have been mixed must be pure, and in fact equal to the result of mixing. For this, we will assume that the probability distribution p used in the mixture has full support (cf. Convention 8.29), which makes sense, since a component with probability 0 contributes nothing to the mixture anyway.

Proposition 8.76 If a mixture is pure:

$$\Phi \begin{array}{c} \\ p \end{array} = \widehat{f}$$

then:

$$\Phi = \quad \widehat{f} \tag{8.73}$$

Proof First, we represent any mixture as a reduced cq-map:

$$\Phi \begin{array}{c} \\ p \end{array} = \begin{array}{c} \\ \Phi \\ p \end{array}$$

By Proposition 8.59 the cq-map must separate as follows:

$$\begin{array}{c} \Phi \\ p \end{array} = p' \quad \widehat{f}$$

We assume p has full support, so apply p^{-1} to both sides:

$$\begin{array}{c} p^{-1} \\ \Phi \\ p \end{array} = \begin{array}{c} p^{-1} \\ p' \end{array} \widehat{f}$$

Using spider fusion the LHS becomes:

$$\begin{array}{c} p^{-1} \\ \Phi \\ p \end{array} = \begin{array}{c} \Phi \\ p \quad p^{-1} \end{array} \overset{(8.33)}{=} \begin{array}{c} \Phi \end{array}$$

so bending the classical wire down gives:

$$
\text{[diagram]} \qquad (8.74)
$$

Finally, the fact that the classical effect in (8.74) must be deleting follows from causality of Φ (which also implies causality of $\widehat{f}$). This can be seen by plugging any causal state ρ into the quantum input and discarding the output:

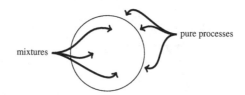

$\square$

While this proof is entirely diagrammatic, it establishes a geometric property of the space of quantum states, namely that each pure state is 'extremal'; i.e. no pure state can be decomposed as a non-trivial mixture (a.k.a. convex combination). For example, in the Bloch ball we have:

pure processes

mixtures

8.3.5 Entanglement for Impure States

Mixing with spiders will in turn allow us to provide a full characterisation of quantum entanglement, which applies not only to pure states but to all quantum states. We will furthermore give a diagrammatic proof that this reduces to $\otimes$-separability for pure states.

The tricky part about entanglement for impure states is that, by means of mixing, we can obtain $\otimes$-non-separable processes even in cases where there is nothing like quantum entanglement going on. Suppose for example we have a mixture like this:

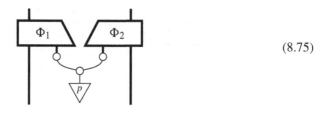

$$
\qquad\qquad (8.75)
$$

Now, we have two processes in a mixture, but rather than being controlled by two independent classical states, these processes are controlled by the <u>same</u> classical state, as indicated by the presence of a copying map. So, even though we don't know which process actually happens, we at least know that whenever:

happens on the left,

must happen on the right. That is, they are *classically correlated*. While the diagram as a whole is connected, the part that connects the two components is entirely classical. Such a connection is of a completely different nature from, for example, the quantum cups that we have been exploiting for deriving all kinds of quantum features.

Thus, to properly define entanglement, we should say not only that a quantum state doesn't separate, but also that it is not merely connected by classical correlations.

Definition 8.77 A bipartite quantum state ρ is *entangled* if it <u>cannot</u> be written in the following form for some quantum maps Φ_1 and Φ_2:

$$
\raisebox{-0.5em}{ρ} \quad = \quad \Phi_1 \quad \Phi_2 \tag{8.76}
$$

If a state is not entangled, then we call it *disentangled*.

Even though we define disentangled states using the classical cup (i.e. 'perfect correlations'), we could equally well use any classical correlations, as we did in (8.75) above.

Proposition 8.78 A bipartite quantum state ρ is entangled if it <u>cannot</u> be written in the following form for some quantum maps Φ_1 and Φ_2 and probability distribution p:

$$
\raisebox{-0.5em}{ρ} \quad = \quad \Phi_1 \quad \Phi_2 \quad p \tag{8.77}
$$

Proof If a state is disentangled as in Definition 8.77, then:

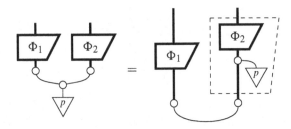

so it matches the form (8.77). Conversely, we have:

so it is indeed in the form (8.76). □

Example 8.79 Using process–state duality, decoherence gives rise to a classically correlated state of the form (8.76):

Back in Chapter 6, when we defined entanglement for pure states as ⊗-non-separability, we didn't know anything about classical wires. However, this still arises as a special case.

Proposition 8.80 If a <u>pure</u> bipartite quantum state is disentangled in the sense of Definition 8.77, then it is ⊗-separable.

Proof First, just as in the proof of Proposition 8.76, we represent a process that is assumed to be pure as a reduced process using spider fusion. In this case, we can represent any pure disentangled state as the following reduced state:

Since the bipartite state obtained from deleting a classical system is pure, by Proposition 8.59 the whole state must separate as follows:

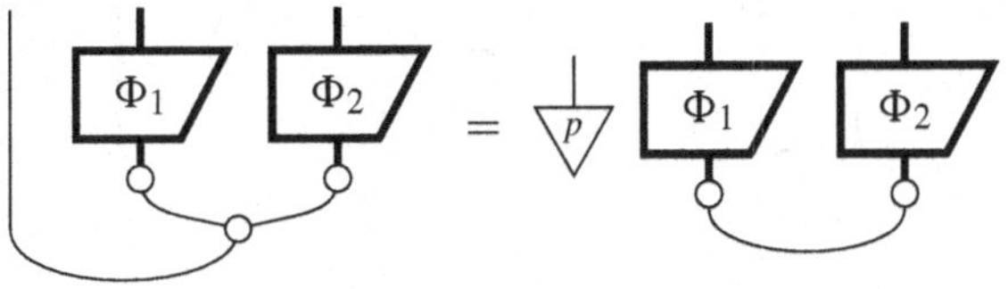

(8.78)

for some causal classical state p. So there exists at least one basis state i such that:

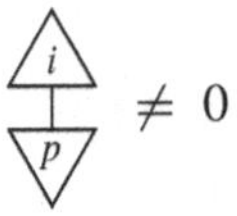

Hence we have:

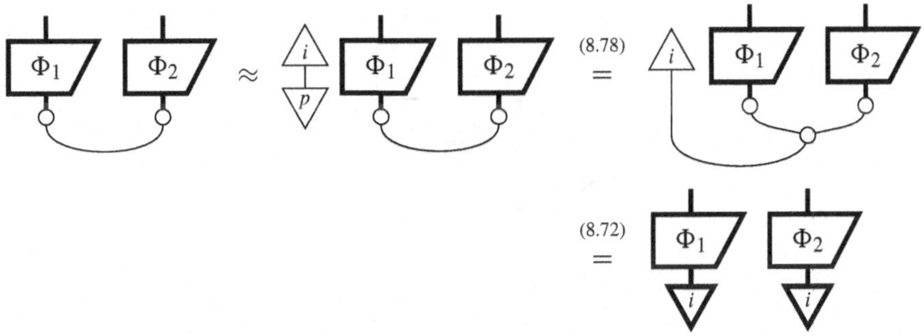

$\square$

This yields the following fact as corollary.

Corollary 8.81 The quantum cup is entangled.

Having diagrammatic forms at hand for many quantum features, we can now easily investigate how these relate. For example, what happens if we apply decoherence to one of two systems in a Bell state? It disentangles:

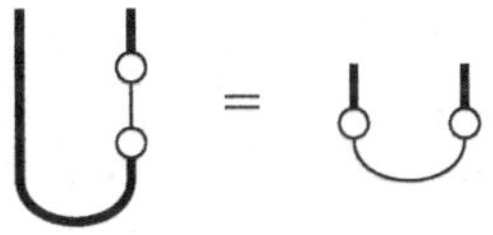

This is an instance of a more general fact.

Theorem 8.82 Decoherence destroys entanglement; that is, if we apply decoherence to one of two systems in an entangled state, then it disentangles.

Proof We have:

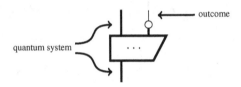

□

8.4 Measurements and Protocols with Spiders

We will now use spiders to give diagrammatic presentations of all of the families of quantum measurements and quantum protocols we have encountered so far, without all of the branches and sums.

Non-demolition measurements are cq-maps of this shape:

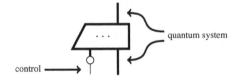

while controlled unitaries have a dual shape:

By wiring together maps like these, we will give graphical presentations of teleportation, dense coding, and entanglement swapping.

We top off this section by revisiting Naimark dilation. By translating it to the language of cq-maps and spiders, this now becomes a tautology.

8.4.1 ONB Measurements

For the measurement:

we used the same basis for the measurement effects as we did for the classical outcomes:

Of course, there is no reason that we should assume these bases are the same, so arbitrary (demolition) ONB-measurements are quantum processes of this form:

But by Proposition 7.4, we know that we can obtain measurements in any ONB by applying a unitary to some fixed ONB, so we have the following corollary.

Corollary 8.83 Every demolition ONB measurement is of the form:

$$(8.79)$$

where $\widehat{U}$ is a unitary quantum process.

The simple measuring process:

is of course a special case, where $\widehat{U}$ is the identity. The measuring process also has a non-demolition counterpart, which leaves the quantum system intact but sends every state to an eigenstate of the measurement, depending on the measurement outcome. We can picture this as measuring, followed by encoding:

However, we should also get the classical value out at the end, so before feeding it in to encoding, we make a copy:

$$\text{(8.80)}$$

If we expand this as a sum, we see it captures the expression of a non-demolition ONB measurement as a cq-map:

$$\text{outcome state} \qquad \text{classical outcome}$$

$$\text{measurement effect}$$

In particular, it is a bastard spider, and we can show causality by bastard spider fusion:

$$\overset{\text{(8.70)}}{=} \quad \overset{\text{(8.70)}}{=} \quad$$

If we discard the quantum output, we obtain the demolition measurement we had before:

$$\overset{\text{(8.70)}}{=} \quad =$$

On the other hand, if we discard the classical output, we get decoherence:

$$= \quad = \qquad\qquad \text{(8.81)}$$

As in the demolition case, we express a general non-demolition ONB measurement in terms of bastard spiders and a unitary.

Proposition 8.84 Every non-demolition ONB measurement is of the form:

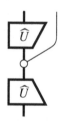

where $\widehat{U}$ is a unitary quantum process.

Exercise 8.85 First show that Exercise 8.32 extends to quantum spiders and bastard spiders. Then, use the following instance of this result:

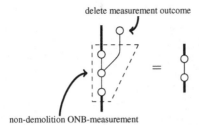

to prove Proposition 8.84.

Remark 8.86 Equation (8.81) gives us another way to understand decoherence:

We made a big point in the previous chapter that quantum measurement is just some sort of quantum process. It represents an interaction between a quantum system and, well, us. A happy result of this interaction is that we get some information about the quantum state: a measurement outcome. Decoherence is what happens to a quantum state when some interaction with its environment causes the state the collapse, as if it had been measured. Unfortunately, it's a particularly bad kind of measurement, because it happens spontaneously and we don't even get to know the outcome!

8.4.2 Controlled Unitaries

In order to diagrammatically present the protocols over the next few sections, we need to represent a controlled isometry as a single cq-map:

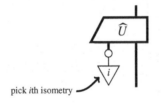

Since we can recover the individual isometries as follows:

a controlled isometry is a cq-map such that, for all i:

$$\tag{8.82}$$

A controlled unitary then additionally satisfies:

$$\tag{8.83}$$

With the help of spiders, we can roll these indexed sets of equations into single equations.

Proposition 8.87 A cq-map:

is a controlled isometry if and only if it satisfies:

$$\tag{8.84}$$

and is moreover a controlled unitary if and only if it additionally satisfies:

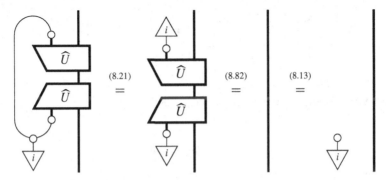

$$(8.85)$$

Proof We will show the equivalence of (8.82) and (8.84). First, assume (8.82) holds for all i. Then:

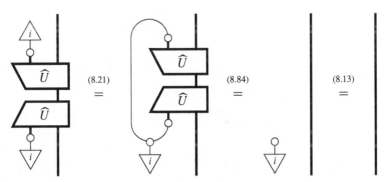

Since the LHS and RHS agree on all ONB states, (8.84) follows. Conversely, assuming (8.84), for all i we have:

The equivalence of (8.83) and (8.85) follows similarly. □

Exercise 8.88 Complete the proof of Proposition 8.87. That is, extend it to the case of controlled unitaries.

We can also ask if a controlled isometry as a whole is itself an isometry. It turns out this is the case.

Proposition 8.89 A quantum process:

$$\widehat{U}$$

(8.86)

satisfying equation (8.85) is an isometry up to a number.

Proof We have:

$$\widehat{U} \quad = \quad \widehat{U} \quad \overset{(8.85)}{=} \quad \approx$$

(8.87)

□

Exercise 8.90 Is (8.86) unitary if we also have (8.84)?

8.4.3 Teleportation

General teleportation can now be presented as follows:

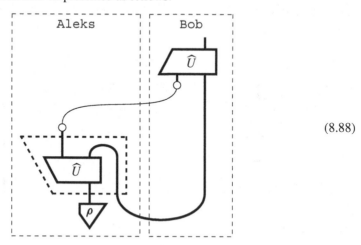

(8.88)

The key is that both:

and its adjoint occur in the diagram, which will enable us to invoke Proposition 8.89 to cancel them out. However, in each of those two occurrences the cq-map is playing a very different role. On the one hand, it is a controlled unitary obeying equations (8.84) and (8.85). On the other hand, it is used to construct an ONB measurement:

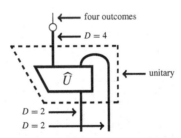

We now know, by Corollary 8.83, that we obtain an ONB measurement precisely when we have a unitary followed by the measure process:

$$\text{(8.89)}$$

Unitarity of the marked map means that the following two equations hold:

$$\text{(8.90)}$$

In particular, if each of the input systems is D-dimensional, the output system must be D^2-dimensional for the above map to be unitary. So, if each input system is a qubit, the output must be four-dimensional:

Remark 8.91 Note that unitarity has taken the place of equations (6.79) and (6.80), which we used in Section 6.4.6 to give a generalised teleportation protocol. In either case, these guarantee an ONB of measurement effects. This is not necessary for teleportation to 'work'. For that, we only need (8.89) to be causal, and hence only the first of equations (8.90) needs to hold. This corresponds to having more measurement outcomes than strictly necessary; e.g. Aleks could perform a POVM measurement by flipping a coin and deciding to perform one of two ONB measurements.

We can now show that (8.88) correctly implements teleportation:

$$(8.88) = \qquad \qquad \approx \qquad \qquad$$

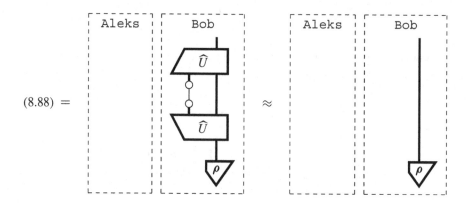

where the last step uses Proposition 8.89.

In case you didn't notice yet, there is something slightly weird about the manner in which we have described teleportation until now. Aleks' measurement outcome is used by Bob to do the appropriate correction, but then this measurement outcome seems to have been deleted from everyone's notebooks and memories. In reality, we expect to get some classical data out at the end, corresponding to the measurement outcomes Aleks got. We can fix this by making a copy of the classical data before it is 'consumed' by the controlled unitary:

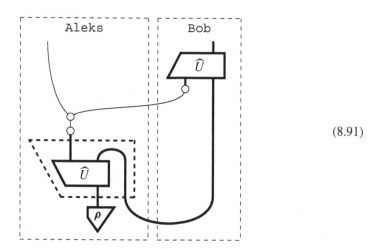

$$(8.91)$$

Interestingly, now we need the full power of equation (8.85) to prove correctness, rather than its reduced version as in (8.87):

$(8.91) =$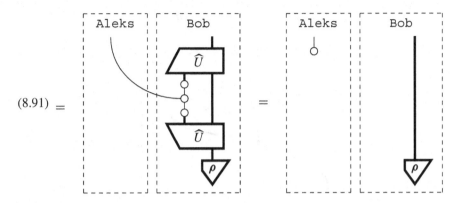

We can now also give a clearer (sum-free) picture of what happens if we delete the classical data immediately after the measurement and don't do any correction at all, as we did in Section 6.4.4. That would now look this:

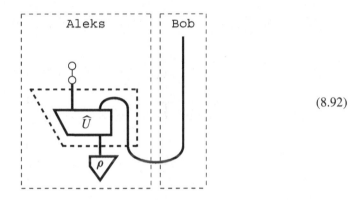

(8.92)

From causality for cq-maps it follows that we have:

$(8.92) =$

8.4.4 Dense coding

Recall from Section 8.1.2 that dense coding is the protocol where Aleks uses quantum systems to send classical data. He does this by using his classical data to perform a controlled unitary on half a Bell state, then sending his half to Bob, who can then recover Aleks' data by measuring both quantum systems together. As a diagram, dense coding looks like this:

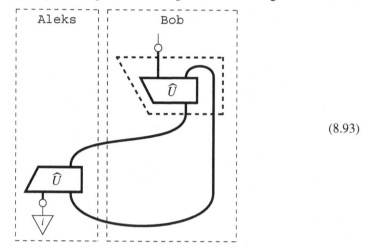

$$(8.93)$$

which simplifies to:

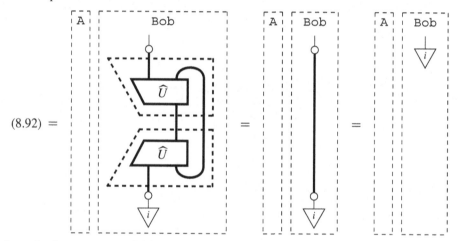

where the first step uses (8.90), i.e. unitarity of the marked boxes. So, in order to prove correctness we rely on an equation different from the one for teleportation, something that we already pointed out in Remark 8.2.

8.4.5 Entanglement Swapping

Recall from Section 7.2.4 that entanglement swapping is a protocol that swaps the entanglement among four quantum systems, by means of a non-demolition measurement on two of those systems. To realise the protocol, we start with Aleks and Bob each sharing a Bell state with a third party (in this case Dave the dodo). Dave then performs a non-demolition variant of the measurement we were using before:

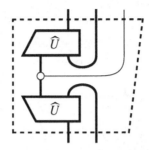

This is indeed a non-demolition ONB measurement by Proposition 8.84. The outcome of this measurement then needs to be copied to two controlled unitaries, which perform corrections. So, the whole protocol looks like this:

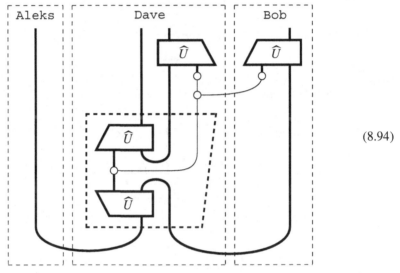

(8.94)

We can simplify by using bastard spider fusion and the controlled-isometry equations to eliminate all of the $\widehat{U}$ maps:

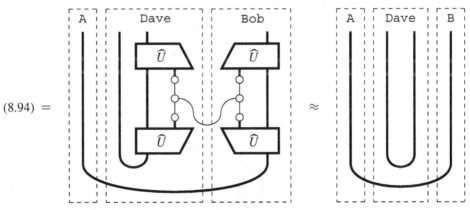

where we use Proposition 8.89 twice.

8.4.6 Von Neumann Measurements

In Section 7.3.1, we showed that a von Neumann measurement can be defined as a quantum process obeying the collapse postulate; that is, for all i,j:

$$\frac{\boxed{\widehat{P_j}}}{\boxed{\widehat{P_i}}} \;=\; \delta_i^j \; \boxed{\widehat{P_i}} \tag{8.95}$$

We can now write a von Neumann measurement as a single cq-map:

whose associated projectors can be recovered as follows:

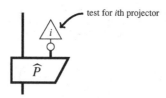

test for *i*th projector

Hence (8.95) now becomes, for all i and j:

$$= \; \delta_i^j \tag{8.96}$$

Proposition 8.92 A quantum process:

is a von Neumann measurement if and only if it satisfies:

$$(8.97)$$

Proof First, assume (8.97), then using:

$$(8.98)$$

we obtain:

Conversely, by assuming (8.96), we can show similarly that the LHS and RHS of (8.97) agree on all classical ONB effects. Hence they are equal (cf. the proof of Proposition 8.87). □

Equation (8.97) has a direct operational reading. Von Neumann measurements have the property that if we measure once, then measure again, we should always get the same result the second time. Equation (8.97) captures this as follows: if we measure twice, we will get precisely the same output as if we measure once, then copy the measurement outcome.

Example 8.93 The non-demolition ONB measurement given by (8.80) is a von Neumann measurement, letting:

Then, using bastard spider fusion we have:

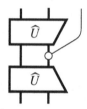

so, again using bastard spider fusion:

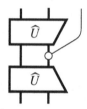

Exercise 8.94 Show that the more general form of a non-demolition ONB measurement, as presented in Proposition 8.84, is a von Neumann measurement and that, more generally, for any unitary $\widehat{U}$:

is a von Neumann measurement.

Just as before, we obtain demolition von Neumann measurements by discarding the quantum output of the associated non-demolition measurement:

8.4.7 POVMs and Naimark Dilation

In the previous chapter, we defined a demolition POVM measurement as any quantum process consisting of effects. Thus, a 'demolition POVM measurement' is really just a generic process from a quantum to a classical system:

As we saw in Section 7.3.3, 'non-demolition POVM measurements' are just quantum processes where each of the branches is pure. As a cq-map, this becomes:

$$ \widehat{U} \tag{8.99} $$

We showed in Section 7.3.3 that these are already generic enough to recover all demolition POVM measurements by discarding the quantum system.

It is easy to see how causality for the cq-map (8.99) translates into causality for the underlying pure quantum map $\widehat{U}$:

$$ \widehat{U} \;=\; \widehat{U} \;=\; \tag{8.100} $$

Since $\widehat{U}$ is pure and causal, it is an isometry by Theorem 6.56.

Exercise 8.95 Extend (8.100) to the case of arbitrary cq-maps. That is, show that for any quantum process:

$$ \Phi \tag{8.101} $$

we can always choose the quantum map Φ to be causal. Conversely, show that for any causal quantum map Φ, the associated cq-map (8.101) is causal.

Let's now have another look at Naimark's dilation theorem (cf. Theorem 7.31), which states that any non-demolition POVM measurement can be expressed in terms of an isometry and an ONB measurement. The non-demolition POVM measurement (8.99) is a causal cq-map, hence $\widehat{U}$ is an isometry. Thanks to our representation of quantum processes as cq-maps:

$$ \text{POVM-measurement} \longrightarrow \widehat{U} \;=\; \widehat{U} \quad \substack{\longleftarrow \text{ONB-measurement} \\ \\ \longleftarrow \text{isometry}} $$

there is nothing left to prove!

Combining this result with Stinespring dilation for causal quantum maps, we obtain a simple alternative presentation of quantum theory.

Theorem 8.96 Quantum processes are linear maps of the form:

where $\widehat{U}$ is an isometry.

8.5 Summary: What to Remember

1. The theory of **quantum processes** (a.k.a. quantum theory) is the processes theory of *classical-quantum maps*:

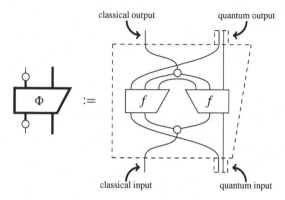

which are moreover *causal*:

$$
\Phi = \quad
$$

where:

$$
\; := \sum_i \quad\quad\quad \; := \sum_i \quad\quad\quad \; := \sum_i
$$

Equivalently, by Stinespring dilation, quantum processes are linear maps of the form:

where $\widehat{U}$ is an isometry.

2. The theory of all **classical-quantum maps** admits string diagrams. It includes **quantum maps** and **classical maps** as sub-theories, where the latter consists of processes of the form:

$$\widehat{f} := \begin{array}{c} f \quad f \end{array}$$

3. Particularly well-behaving classical maps are *classical spiders*:

$$\cdots \; := \; \sum_i \; \cdots$$

They compose as follows:

$$= $$

4. An example of a classical spider is *copying*:

$$= \; \sum_i $$

In particular, copying determines an ONB:

$$\psi \in \left\{ i \right\}_i \qquad \text{if and only if} \qquad \overset{(*)}{=} \; \psi \; \psi$$

Another example is *deleting*:

$$= \; \sum_i$$

which, as we saw in **1** above, plays a key role in stating causality. Namely, it is the classical counterpart to discarding:

$$\leftrightarrow$$

5. By doubling classical spiders or pairing certain legs:

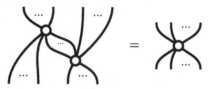

we obtain *quantum spiders* and *bastard spiders*, respectively. Quantum spiders compose just like classical spiders:

whereas any composition of spiders involving at least one single-dot spider yields again a single-dot spider:

6. An example of a quantum spider is the *GHZ state*:

Examples of bastard spiders are *measure* and *encode*:

which we also already encountered in **1** above. In particular, all cq-maps can be obtained by composing pure quantum maps, measuring, and encoding:

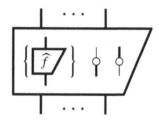

Another bastard spider is *decoherence*:

which models a quantum state degrading to a classical state.

7. Any linear map f decomposes as:

for an appropriately chosen spiders, and bipartite states decompose as:

All quantum states ρ encode classical states, via:

$$(8.102)$$

8. Mixing a set of processes of the same type by means of a probability distribution p can be represented as a cq map:

By (8.102) every causal quantum state can be regarded as a mixture of pure causal quantum states. If a mixture is pure:

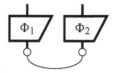

then:

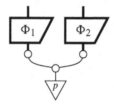

9. A bipartite quantum state is *entangled* if it <u>cannot</u> be written in the form:

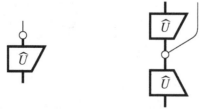

or, equivalently, if it <u>cannot</u> be written as a mixture as follows:

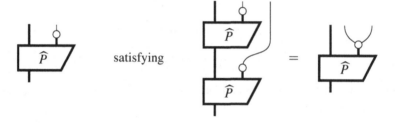

10. Demolition ONB measurements and non-demolition ONB measurements are of the following forms, respectively, for some unitary $\widehat{U}$:

Non-demolition von Neumann measurements are quantum processes:

satisfying

Demolition von Neumann measures are the same, but with the quantum output discarded. Demolition and non-demolition POVM measurements are quantum processes of the following forms:

In particular, $\widehat{U}$ must be an isometry by causality, so Naimark dilation now boils down to two different readings of the same diagram:

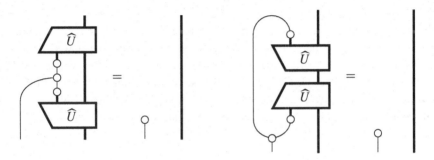

11. Controlled unitaries can be defined as follows:

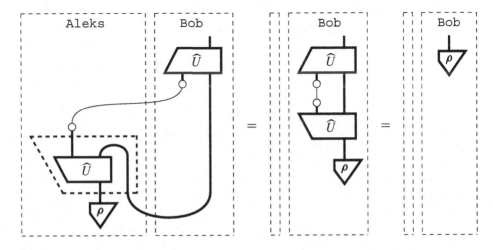

This enables us to give totally diagrammatic presentations of quantum protocols such as *quantum teleportation*:

and *dense coding*:

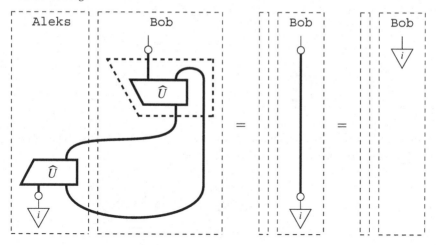

8.6 Advanced Material*

In this section, we elaborate a bit on what spiders are, from a more (co)algebraic perspective. In fact, if we drop the commutativity requirement, we discover how they become a graphical tool to study C*-algebras, which are a popular structure to study for algebraists and mathematical physicists. Then we briefly discuss how spiders for non-self-conjugate ONBs look. They turn out to have some hairs on their legs! And if that wasn't scary enough, we'll see in the last section that we've got spiders coming out of our mouths too!

8.6.1 Spiders Are Frobenius Algebras*

We introduced spiders as a species of creature with any number of legs that fuse when they 'shake legs':

However, this is quite different from the usual way one defines these processes, and in fact, not the way we originally encountered them. Usually they are defined in terms of something that may be more familiar to mathematicians.

Definition 8.97 An *associative algebra* on a vector space V consists of a pair of linear maps:

$$\, : V \otimes V \to V \qquad\qquad \, : \mathbb{C} \to V$$

such that Λ is associative and has unit $\circ$:

$$\text{[diagram: } \Lambda = \Lambda \qquad \Lambda = | = \Lambda \text{]}$$

This is the same as having a multiplication operation on elements of V that is linear in both arguments, associative, and unital. However, the benefit of writing it this way is it's very easy just to turn everything upside-down! It's a common convention in category theory to call something a 'co-Thing', if it is a Thing with all of the maps turned upside-down.

Definition 8.98 A *coassociative coalgebra* on a vector space V consists of a pair of linear maps:

$$\text{Y} : V \to V \otimes V \qquad \text{P} : V \to \mathbb{C}$$

such that Y is *coassociative* and has *counit* P :

$$\text{[diagram: } \text{Y} = \text{Y} \qquad \text{Y} = | = \text{Y} \text{]}$$

The most obvious way to get such a thing is to let V be a Hilbert space rather than just a vector space, and take the adjoint of an associative algebra:

$$\text{Y} := \left(\Lambda\right)^{\dagger} \qquad \text{P} := \left(\circ\right)^{\dagger}$$

While *algebraic* structures may be quite familiar, *coalgebraic* structures might be less so. Perhaps one of the reasons for this is they tend not to be very interesting in process theories where $\otimes$ behaves like a Cartesian product. For example, if we replace **linear maps** with **functions** in Definition 8.98, the only coassociative coalgebras are the 'universal' copying functions:

$$\text{Y} : X \to X \times X :: x \mapsto (x,x)$$

However, when $\otimes$ is non-Cartesian, as with **linear maps**, we have lots of interesting coalgebras, and, more importantly, we can define interesting structures that have both an algebraic and a coalgebraic part. Our key example is the following.

Definition 8.99 A *Frobenius algebra* consists of an associative algebra (Λ, $\circ$) and a coassociative coalgebra (Y, P) that additionally satisfy the *Frobenius equations*:

$$\text{[diagram: Frobenius equations]}$$

We can define 'spider-like' maps using a Frobenius algebra:

$$(8.103)$$

However, for these spiders to fuse as in Theorem 8.34 and for the species to be closed under taking adjoints we need a particularly 'special' kind of Frobenius algebra.

Definition 8.100 A dagger special commutative Frobenius algebra (†-SCFA) is a Frobenius algebra that additionally satisfies:

Evidently, the †-part of †-SCFA takes care of horizontal reflection, while the SC-part guarantees the following.

Proposition 8.101 Spiders defined as in (8.103) for a †-SCFA compose as:

Proof (sketch) It suffices to show that any connected diagram consisting of (, , ,) can be transformed into a canonical form, just using the Frobenius algebra equations:

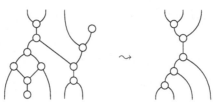

In particular, two connected spiders can be transformed into a canonical form, which will be one big spider. This can be shown by induction over the number of dots in the diagram. □

One thing that is notable is that none of the definitions in this section makes use of any vector space structure, so they actually make sense in any dagger symmetric monoidal category.

The algebraic presentation of spiders enables us to use standard theorems from algebra to prove that they always define an orthonormal basis.

Theorem 8.102 For any †-SCFA, the set of states $\{\phi_i\}_i$ such that:

$$\raisebox{-1em}{[diagram]} \;=\; \raisebox{-1em}{[diagram]}$$

always forms an ONB. Thus, every family of spiders uniquely determines (and is uniquely determined by) an ONB.

Proof (sketch) The proof goes in two stages. First, one can show that the algebra part of a †-SCFA is always *semi-simple*. Semi-simple algebras (which we won't define here) are well understood, especially in finite dimensions, thanks to *Wedderburn's theorem*. This theorem implies in particular that any associative algebra that is semi-simple and *commutative* is actually isomorphic to a direct sum of copies of the trivial algebra on $\mathbb{C}$:

$$\raisebox{-1em}{[diagram]} \;\cong\; \raisebox{-1em}{[diagram]}^{\mathbb{C}} \oplus \raisebox{-1em}{[diagram]}^{\mathbb{C}} \oplus \cdots \oplus \raisebox{-1em}{[diagram]}^{\mathbb{C}} \qquad (8.104)$$

This 'trivial algebra' might look strange, since we don't usually draw wires for $\mathbb{C}$. It's actually just:

$$\raisebox{-1em}{[diagram]} \;:=\; \raisebox{-1em}{[dashed box diagram]}$$

which is obviously associative and has as its unit also the empty diagram. The algebra (8.104) always has a basis of copyable states, which look like this:

$$\langle 0 | \oplus \cdots \oplus \langle 0 | \oplus \langle 1 | \oplus \langle 0 | \oplus \cdots \oplus \langle 0 |$$

Then, the fact that $(\,\raisebox{-0.3em}{[diagram]}\,)^\dagger = \raisebox{-0.3em}{[diagram]}$ suffices to show that any basis of copyable states must be orthogonal, and furthermore:

$$\raisebox{-1em}{[diagram]} \;=\; \raisebox{-1em}{[diagram]}$$

implies that it must consist of normalised states. $\square$

8.6.2 Non-commutative Spiders*

In Theorem 8.102 we saw that a †-SCFA fixes a unique ONB. So what happens if we drop some of the letters in '†-SCFA'? If we drop the S of 'special', then, rather than ONBs, we obtain orthogonal bases. If we drop the †, but retain the S, then we obtain arbitrary bases. That's quite interesting. However, if we start to mess around with the C of 'commutative', then even more interesting things happen. We won't just drop C but instead we'll replace it with something weaker.

Definition 8.103 A dagger special *symmetric* Frobenius algebra (†-SSFA) is a Frobenius algebra that additionally satisfies:

This definition looks almost the same as before, but note that the third equation not longer has an output. Despite this seemingly minor change, †-SSFAs are actually much more general than their commutative cousins.

Theorem 8.104 Every finite-dimensional C*-algebra is isomorphic to a †-SSFA, and vice versa.

Thus, the (less familiar) notion of †-SSFA is actually just a diagrammatic version of the (more familiar) notion of C*-algebra. Crucially, this allows non-commutative, *quantum algebras*, in addition to the classical, commutative ones. The most important of these is the Frobenius algebra we associate with a double wire.

Exercise 8.105 Show that the following linear maps define a †-SSFA:

which we will call a *pants algebra*.

Applying map–state duality (and ignoring the $\frac{1}{\sqrt{D}}$), we can see that the pants algebra just takes a pair of linear maps and composes them:

These pants algebras are often called M_n in the literature, referring to the fact that they essentially perform matrix composition. Among all of the †-SSFAs, they play a very special role.

Recall that classical maps are cq-maps of the form:

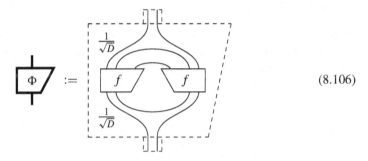

$$(8.105)$$

where the dots on the top and bottom are spiders (a.k.a. †-SCFAs). If we now generalise this to †-SSFAs, then we could, for instance, take these to be pants algebra. In that case, something very nice happens.

Proposition 8.106 A linear map Φ is a quantum map if and only if there exists some linear map f such that:

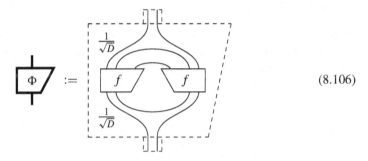

$$(8.106)$$

Proof The RHS of (8.106) is already in the form of a quantum map. Conversely, any quantum map takes the form:

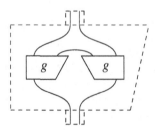

for some g. This can be put in the form of (8.106) by taking:

The numbers in (8.106) then cancel with the resulting circle. $\qquad\square$

It may not be immediately obvious what we win here, given the equivalent form (8.106) of a quantum map is actually more complicated than the one we started with. However, now the condition of being a classical map and being a quantum map is always (8.105), where the only thing that varies is the algebra. In other words, we can treat classical and quantum types on the same footing. What's more, if we have two †-SSFAs on systems A and B:

gives us a †-SSFA on the system $A \otimes B$. So, we can additionally treat arbitrary $\otimes$-compositions of classical and quantum systems in the same way as our basic types by means of (8.105). All the information about which parts of the system are classical and which parts are quantum is then encoded in the resulting algebra.

Hence, it makes sense to think of not just a Hilbert space A as a type, but the pair $(A, \circ)$ consisting of A and a †-SSFA on A. Using this as a guide, we define a new process theory from an old one.

Definition 8.107 The process theory **CP*[linear maps]** has as types pairs $(A, \circ)$ consisting of a Hilbert space A and a †-SSFA on A, and processes from $(A, \text{\AA})$ to $(B, \text{\AA})$ are linear maps Φ from A to B of the form:

for some linear map f.

In fact, cq-maps are precisely those maps on **CP*[linear maps]** whose types are $\otimes$-compositions of classical and quantum systems, thus:

$$\text{cq-maps} \;\subset\; \textbf{CP*[linear maps]}$$

The full process theory of **CP*[linear maps]** contains 'fully classical' and 'fully quantum' systems, as well as some extra stuff in between:

fully quantum   fully classical

This 'extra stuff' includes $\otimes$-compositions of classical and quantum systems, as well as more general semi-classical (or sometimes called 'super-selected') systems, which arise as direct sums of quantum systems. This general direct sum form for systems in **CP*[linear maps]** follows from Wedderburn's theorem, which was mentioned briefly in the proof of Theorem 8.102.

Note that the only thing we used about the process theory of **linear maps** to define **CP*[linear maps]** is that it admits string diagrams, so we can play this 'CP*' game with other process theories (or monoidal categories) and see what comes out. The results can sometimes be surprising! For example, **CP*[relations]** yields a process theory whose types are *groupoids* (i.e. certain mathematical objects that generalise groups) and whose maps are relations preserving some of the groupoid structure.

8.6.3 Hairy Spiders*

In Sections 8.2.4 and 8.3.1 we made reference to spiders for non-self-conjugate ONBs. So how would those look? In Remark 8.40, we said that one solution was simply to drop the equations:

 (8.107)

or 'fix' them to account for conjugation. In fact, we already done this for certain 'two-legged' spiders in Section* 4.6.2:

We can throw in dots:

and add some extra legs:

In Section 4.6.2, we used arrows to distinguish a system A from its dual system A^*. This has the handy side effect that the conjugate of a basis state has a different type from the state itself:

and, similarly, the adjoint has a different type from the transpose. Hence the arrows on the legs of a 'hairy' spider tell use which basis states/effects should be conjugated:

We have, by definition, fixed the cup/cap equations (8.107):

and whenever a spider has one 'in-arrow' leg and one 'out-arrow' leg it is a wire:

The rules of the game are now that spiders can only 'shake legs' when the orientations of the legs match. Note in particular that now a new kind of cups/caps arises:

which still satisfy yanking equations:

Similarly, there are several variations on our good old copying spider:

which can fuse with other hairy spiders, e.g.:

In Remark 8.57, we stated that the unitary in the decomposition:

$$\rho = \widehat{U} \begin{cases} \text{unitary} \\ \text{encoding} \\ \text{classical state} \end{cases}$$

of any quantum state ρ was necessitated by the restriction to self-conjugate ONBs. Now, setting:

we can drop $\widehat{U}$ and obtain a perfect symmetry between classical-as-quantum and quantum-as-classical decompositions:

8.6.4 Spiders as Words*

In Section* 6.6.3, we say that sentences can be represented using diagrams. Spiders play a role in this story as well, since they can be used to represent relative pronouns:

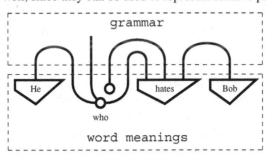

Here, the spider combines 'he' with 'hates Bob', in order to return whatever obeys these two properties, that is, being a (probably) human male, as well as hating Bob.

So, not only is <u>this</u> book is full of spiders, but in fact <u>every</u> book is!

8.7 Historical Notes and References

The diagrammatic representation of classical data was initiated in Coecke and Pavlovic (2007), in which spider-less but fully diagrammatic definitions of controlled unitaries, von Neumann measurements, and teleportation were given. The passage to spiders took place in Coecke and Paquette (2008). Much earlier, in Davies and Lewis (1970), the classical data produced by quantum measurements were also represented in terms of an ONB. However, no distinction was made between the spaces in which the classical data was represented and those in which the quantum systems were described. The two versus one wire paradigm and an early form of the resulting classical-quantum maps were introduced in Coecke et al. (2010a). Representing ONB-measurements as bastard spiders was first done in Coecke et al. (2012).

'Bastard' was the original name of the band Motörhead, but Lemmy changed it after being told that a band by the name of Bastard would never get a slot on Top of the Pops. This won't matter for our spiders given that Top of the Pops doesn't exist anymore. RIP Lemmy and Lil' Philthy.

These days, there are several other diagrammatic presentations of quantum theory on the market, e.g. Chiribella et al. (2010) and Hardy (2012). However, in these presentations, classical data is always treated non-diagrammatically. On the other hand, the fundamental importance of purification as a characteristic of quantum theory (cf. Section 8.2.2) was put forward in Chiribella et al. (2010, 2011). A study of the connection between distinguishability and copiability within process theories was done by Chiribella (2014).

Dense coding was first proposed in Bennett and Wiesner (1992), and its diagrammatic treatment appeared first in Coecke and Pavlovic (2007). The paper of Coecke and Paquette (2008) gives a diagrammatic proof of Naimark dilation, but that proof is unnecessarily complicated, as compared with the one that we presented here. The diagrammatic representation of Bayesian inversion of Example* 8.27 is taken from Coecke and Spekkens (2012).

The notion of a Frobenius algebra is due to Brauer and Nesbitt (1937) but was first presented in its modern, categorical form by Carboni and Walters (1987). The fact that (special) commutative Frobenius algebras have canonical forms (i.e. spiders) that 'fuse' together comes from a 'folk theorem' relating Frobenius algebras to geometrical objects called *cobordisms*. A standard reference is Kock (2004). An explicit proof of the 'spider' form for special commutative Frobenius algebras was given using distributive laws by Lack (2004), where what we call spiders are represented as cospans of functions between finite sets. The fact that spiders characterise bases is from Coecke et al. (2013c). Though the authors there rely on the spectral theory of C*-algebras to show that copiable states form a basis, our presentation relies on the (much older) classification theorem of Wedderburn (1906). Completeness for spider diagrams was given by Kissinger (2014b).

The connection between †-Frobenius algebras and C*-algebras (cf. Remark 8.104) is from Vicary (2011). The CP*-construction, a categorical construction that gave rise to the process theory **CP*[linear maps]** of Section 8.6.2, was given in Coecke et al. (2013a). An axiomatization of this construction, similar to the axiomatization of **quantum maps** of Section* 6.6.2, was done by Cunningham and Heunen (2015); the relationship between this process theory and quantum logic (cf. Section* 7.6.2) is discussed in Coecke et al. (2013b). Characterisations of classical and quantum systems in terms of information-theoretic constraints were investigated using the CP*-construction in Heunen and Kissinger (2016), generalising the C*-algebraic results of Clifton et al. (2003).

There is also a body of work that aims to classify spiders in **relations** rather than in **linear maps**. This is actually not so weird, given that Carboni and Walters (1987) introduced Frobenius algebras in order to axiomatise **relations**. This effort started with Coecke and Edwards (2011) where it was observed that there were some unexpected spiders, which don't arise from ONBs. After that Pavlovic (2009) classified all spiders in **relations**, and Heunen et al. (2012b) extended this to the case of non-commutative Frobenius algebras.

Within the context of natural language meaning, relative pronouns in terms of spiders appeared in Sadrzadeh et al. (2013, 2014). More recent work in this area that emphasises even more the structural connection with quantum theory can be found in Piedeleu et al. (2015), Balkir et al. (2016), and Bankova et al. (2016).

The last sentences of the introduction to this chapter and Section* 8.6.4 are in reference to David Wong's comedy horror novel *This Book Is Full of Spiders*.

9

Picturing Phases and Complementarity

When spider webs unite, they can tie up a lion.

– Ethiopian proverb

In the previous chapter spiders entered the picture. Their initial role seemed essentially just to shuttle classical data around or to provide transit to and from 'planet quantum', leaving all of the interesting, fully quantum stuff to happen inside some generic quantum process:

Since we can't apply any of our funky rules like spider fusion, this 'black box' is basically a diagrammatic dead end. This chapter is about 'opening up' those boxes. We already half-opened the boxes in Section 8.2.5 when we showed that all linear maps, and hence all quantum maps, consist of spiders and 'black box' isometries. We will now finish opening those boxes.

What do we find inside? An arachnophobe's worst nightmare: more spiders, of course! Unlike here on earth, where the arthropods' role has been reduced primarily to food or fertiliser, in this book, they become the dominant species and, in fact, the only one!

Indeed, by the end of this chapter, we will be able to build arbitrary maps using just spiders. But before we get there, we need to further diversify the spider population. These (final) additions to the graphical language are motivated by two key notions in quantum theory: *phases* and *complementarity* (a.k.a. mutual unbiasedness).

Phases are 'decorations' that can be carried by a spider:

These decorations have two important features. First, when spiders fuse, their decorations combine together:

Second, these decorations are the stuff that doesn't survive the passage from the quantum to the classical realm. Indeed, when a decorated <u>quantum</u> spider makes any attempt to make contact with the classical realm, its decoration <u>vanishes</u>:

As foreshadowed in the previous chapter, we will now also consider spiders of different colours:

which represent different ONBs. These spiders of different colours no longer fuse, but they still should interact in a simple way. In fact, how they interact is kind of the opposite to fusing. Whereas spiders of the same family like each other, *complementary spiders* do not, and the resulting spider wars will cause some serious loss of limbs:

(9.1)

That is, spiders of the same family fuse together, while, when complementary spiders 'shake legs', those legs fall off (but always in pairs). If we take the essential part of the above equation:

(9.2)

and write it in terms of bastard spiders:

there is a clear operational reading for complementary basis:

(encode in ○) THEN (measure in ●) = (no data transfer)

While complementarity seems to be saying something about what we 'can't do', equation (9.2) proves itself to be quite powerful. In particular, we will sketch out how to exploit complementarity for *quantum cryptography* in Section 9.2.6. Even more useful than equation (9.2) are these equations:

which are satisfied by particularly nice pairs of complementary spiders called *strongly complementary spiders*. While they are only a very recent player in the field of quantum research, these new equations aren't ad hoc at all and have already been around for quite a while in a variety of disciplines of pure mathematics, where they are the defining equations of a *bialgebra*.

These new equations provide significantly more proving power, so much more in fact that these, along with the (decorated) spider-fusion laws, form the core of a set of equations called the *ZX-calculus*, which are *complete* for proving equations between a large class of quantum maps called **Clifford maps**. The ZX-calculus will become our graphical Swiss Army knife in the following chapters as we study applications in quantum computing, quantum foundations, and theories of (quantum) resources.

9.1 Decorated Spiders

So, let's start decorating.

9.1.1 Unbiasedness and Phase States

From now on, we will refer to a family of spiders (or equivalently, an ONB via Theorem 8.41) simply by a dot of the appropriate colour, e.g. $\bigcirc$.

Definition 9.1 A normalised pure state is *unbiased* for $\bigcirc$ if we have:

$$\text{LHS} = \frac{1}{D} \; \bigg| \qquad \qquad (9.3)$$

or equivalently:

$$\text{LHS} = \frac{1}{D} \; \bigg| \qquad \qquad (9.4)$$

So what does this mean? In the LHS of (9.3) we see a quantum state $\widehat{\psi}$ being measured. In the RHS of (9.3) we see the uniform probability distribution. So, what is required here is that measurement of $\widehat{\psi}$ yields a uniform probability distribution over all outcomes, or in other words: the quantum state $\widehat{\psi}$ has no bias towards any of the measurement outcomes and, hence, is 'unbiased'.

We can also restate Definition 9.1 in terms of the Born rule:

$$
\overset{i}{\underset{\widehat{\psi}}{\triangle}} \overset{(8.6)}{=} \overset{i}{\underset{\widehat{\psi}}{\triangle}} \overset{(9.3)}{=} \frac{1}{D} \overset{i}{\underset{}{\triangle}} = \frac{1}{D}
$$

That is, for each outcome the Born rule gives the same probability. The converse also holds.

Exercise 9.2 Show that a normalised pure state is *unbiased* for an ONB-measurement if for all i we have:

$$
\overset{i}{\underset{\psi}{\triangle}} = \frac{1}{D}
$$

Example 9.3 We could ask what the analogue would be in probability theory. There, an 'unbiased probability distribution' would be one that has the same probability for each 'outcome' i. Of course, there is only one, namely, the uniform probability distribution itself. Hence, the notion of an unbiased probability distribution doesn't give us anything new.

Evidently, unbiasedness for quantum states does give us something new, otherwise we wouldn't have defined it. In order to establish that for an ONB-measurement there exist many unbiased states, let's look at what the matrix form of an unbiased state is. Using the correspondence given in (8.32), the LHS of (9.4) is the Hadamard product of ψ with its conjugate. Hence, written in terms of matrices, equation (9.4) becomes:

$$
\begin{pmatrix} \overline{\psi^0}\,\psi^0 \\ \vdots \\ \overline{\psi^{D-1}}\,\psi^{D-1} \end{pmatrix} = \begin{pmatrix} \frac{1}{D} \\ \vdots \\ \frac{1}{D} \end{pmatrix}
$$

that is, for all i we have:

$$
\overline{\psi^i}\,\psi^i = \frac{1}{D}
$$

We saw in Section 6.1.2 that numbers satisfying:

$$
\overline{\psi^i}\,\psi^i = 1
$$

can always be written as $e^{i\alpha}$ for some angle $\alpha \in [0, 2\pi)$. Thus there exist complex phases $\alpha_1, \ldots, \alpha_D$ such that:

$$
\begin{pmatrix} \psi^0 \\ \vdots \\ \psi^{D-1} \end{pmatrix} = \begin{pmatrix} \frac{1}{\sqrt{D}}\,e^{i\alpha_0} \\ \vdots \\ \frac{1}{\sqrt{D}}\,e^{i\alpha_{D-1}} \end{pmatrix}
\tag{9.5}
$$

So there are indeed many unbiased quantum states!

The reason for the $\frac{1}{\sqrt{D}}$ in the matrix entries of the RHS of equation (9.5) is the presence of $\frac{1}{D}$ in equation (9.3), which itself comes from the normalisation of the quantum state $\widehat{\psi}$:

$$
\overset{\overline{\underline{}}}{\underset{\psi}{\triangledown}} \quad = \quad \overset{\circ}{\underset{\psi}{\triangledown}} \quad \overset{(9.3)}{=} \quad \tfrac{1}{D}\,\overset{\circ}{\underset{\circ}{}} \quad = \quad \boxed{}
$$

We can get rid of these numbers if, rather than normalising $\widehat{\psi}$, we set:

$$
\overset{\overline{\underline{}}}{\underset{\psi}{\triangledown}} \quad = \quad D
$$

These unnormalised unbiased states $\widehat{\psi}$ will play such a crucial role in the diagrammatic language that we have a special name for them.

Definition 9.4 A *phase state* for $\bigcirc$ is a pure state $\widehat{\psi}$ that satisfies:

$$
\overset{|}{\underset{\psi}{\triangledown}} \quad = \quad \overset{|}{\underset{\circ}{}} \tag{9.6}
$$

For phase states, equation (9.5) gets replaced by:

$$
\begin{pmatrix} \psi^0 \\ \vdots \\ \psi^{D-1} \end{pmatrix} = \begin{pmatrix} e^{i\alpha_0} \\ \vdots \\ e^{i\alpha_{D-1}} \end{pmatrix} \tag{9.7}
$$

Since doubling eliminates global phases (cf. Proposition 6.6), we can assume without loss of generality that $\alpha_0 = 0$ (otherwise just multiply the whole state by $e^{-i\alpha_0}$). Consequently, a phase state $\widehat{\psi}$ is uniquely fixed by a list of the remaining $D - 1$ complex phases:

$$
\vec{\alpha} := (\alpha_1, \ldots, \alpha_{D-1})
$$

Because of this fact, from now on we denote phase states as follows:

$$
\overset{|}{\underset{\vec{\alpha}}{\ominus}} \tag{9.8}
$$

Convention 9.5 Since our notation for phases is invariant under horizontal reflection, we indicate conjugation by introducing a minus sign:

$$
\overset{|}{\underset{\vec{\alpha}}{\ominus}} \quad \overset{\overline{(-)}}{\longmapsto} \quad \overset{|}{\underset{-\vec{\alpha}}{\ominus}}
$$

We denote the transpose as:

and hence the adjoint becomes:

We will see in Section 9.1.4 why we have chosen this notation.

The defining equality (9.6) for phase states now becomes:

$$\text{(9.9)}$$

or equivalently:

$$\text{(9.10)}$$

By (9.9), the *phase* data $\vec{\alpha}$ is totally obliterated as soon as it comes into contact with the classical world via measurement. Thus:

> phase := the data destroyed by the quantum-classical passage

While the ONB-states represent purely classical data, phase states represent the opposite notion: they are extremely non-classical, or 'maximally quantum'. This goes hand in hand with the fact that unbiased states, and hence phase states, have no meaningful classical counterpart (cf. Example 9.3). It should then also come as no surprise that they will play a crucial role in many quantum features that also have no classical counterpart. We'll elaborate more on this essential non-classicality of phases in the following section.

Before we do so, let's have a look at where phase states live on the Bloch sphere. In the case where the dimension $D = 2$, a phase state depends only on a single complex phase α. In that case, we have:

$$\leftrightarrow \begin{pmatrix} 1 \\ e^{i\alpha} \end{pmatrix} \qquad \text{(9.11)}$$

so the form of a phase state simplifies to:

$$= \text{double}\left(\frac{}{0} + e^{i\alpha} \frac{}{1} \right) \qquad \text{(9.12)}$$

Recall from Section 6.1.2 that any two-dimensional pure state can be written in Bloch sphere coordinates as:

$$\text{double}\left(\cos\frac{\theta}{2} \frac{}{0} + \sin\frac{\theta}{2} e^{i\alpha} \frac{}{1} \right)$$

Then, states of the form of (9.12) are precisely those where $\theta = \pi/2$. In other words, they live on the equator of the Bloch sphere:

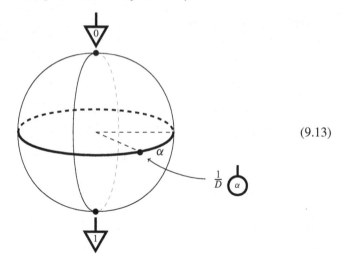

(9.13)

Remark 9.6 In most other texts on quantum theory, phases are introduced in this geometric manner. However, we defined phase states entirely in the language of spiders, without using any other ingredients of **linear maps**:

So they are intrinsic to the diagrammatic language. One consequence of this is that they are also meaningful in many other process theories, as we shall see in Section 11.2.2.

Example 9.7 When comparing the representation of phase states on the Bloch sphere with the picture of Exercise 6.7, we see that the X-basis states are in fact phases for $\bigcirc$:

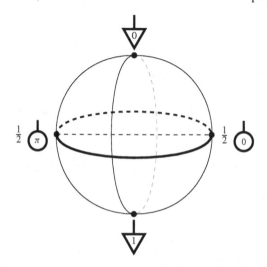

Hence we have:

$$\vcenter{\hbox{▽}}_0 = \tfrac{1}{\sqrt{2}}\,\vcenter{\hbox{◯}}_0 \qquad\qquad \vcenter{\hbox{▽}}_1 = \tfrac{1}{\sqrt{2}}\,\vcenter{\hbox{◯}}_\pi$$

This can be seen as the first indication of the gradual spider takeover happening through the course of this chapter.

Exercise* 9.8 The D-dimensional generalisation of this ONB of phases is called the *Fourier basis*:

$$\left\{ \tfrac{1}{\sqrt{D}}\,\vcenter{\hbox{◯}}_{\bar\kappa_j} \right\}_j, \qquad \text{where} \qquad \vcenter{\hbox{◯}}_{\bar\kappa_j} := \sum_k e^{\frac{2\pi i}{D}jk}\,\vcenter{\hbox{▽}}_k$$

Prove that the Fourier basis is in fact an ONB. During your calculations, the formula for a finite geometric series could come in handy:

$$\sum_{k=0}^{D-1} r^k = \frac{r^D - 1}{r - 1}$$

9.1.2 Phase Spiders

Since phases constitute 'maximally quantum' data, we will primarily use them to 'decorate' quantum spiders. To decorate a quantum spider, we simply plug in a phase state at one of its legs.

Definition 9.9 A *phase spider* is a pure quantum map of the form:

$$\vcenter{\hbox{spider}}_{\bar\alpha} := \vcenter{\hbox{spider}}_{\bar\alpha} \tag{9.14}$$

Note that by Proposition 8.33 it doesn't matter which leg we pick. Furthermore, it doesn't matter if we plug a phase state into an input or the transpose of a phase state into an output:

$$\vcenter{\hbox{spider}}_{\bar\alpha} = \vcenter{\hbox{spider}}_{\bar\alpha} = \vcenter{\hbox{spider}}_{\bar\alpha}$$

As an immediate consequence we have the following.

Proposition 9.10 The transpose of a phase spider is a phase spider with the same phase. In particular, if a phase spider has the same number of inputs as outputs, it is self-transposed.

Using the matrix form of a phase state:

$$\vcenter{\hbox{spider}}_{\bar\alpha} = \text{double}\left(\sum_j e^{i\alpha_j}\,\vcenter{\hbox{spider}}_j \right)$$

and the generalised copy rule (8.38), we obtain the matrix form of a spider:

$$
\text{(spider diagram)} \;=\; \text{double} \left(\sum_j e^{i\alpha_j} \;\text{(diagram)}\right)
$$

As before, we can remove a global phase and assume $\alpha_0 = 0$. So, in two dimensions, this simplifies to:

$$
\text{(spider diagram)} \;=\; \text{double} \left(\text{(diagram)} + e^{i\alpha}\,\text{(diagram)} \right)
$$

Whenever a decorated quantum spider attempts to make contact with the classical realm, it loses its decoration.

Theorem 9.11 If any leg of a phase spider is measured, its phase vanishes:

$$
\text{(diagram)} \;=\; \text{(diagram)}
\tag{9.15}
$$

Proof Using bastard spider fusion we have:

$$
\text{(diagram)} \overset{(9.14)}{=} \text{(diagram)} \;=\; \text{(diagram)} \overset{(9.9)}{=} \text{(diagram)} \;=\; \text{(diagram)}
$$

$\square$

In Section 8.3.2 we showed that decoherence witnesses classicality, in that decoherence-invariant inputs/outputs behave the same as classical inputs/outputs. Hence one would expect decoherence to be pretty destructive as well.

Corollary 9.12 Decoherence erases phases:

$$
\text{(diagram)} \;=\; \text{(diagram)}
\tag{9.16}
$$

A phase state itself is a special case of a phase spider, in which case equation (9.16) specialises to:

$$\text{(diagram)} = \text{(diagram)} = \text{(diagram)}$$

In Section 8.3.2, we showed that we can picture decoherence on a two-dimensional system as projecting on to the centre line of the Bloch ball. This gives us a geometric picture of how decoherence destroys phases. The phase states on the equator all get projected to the middle, i.e. the maximally mixed state:

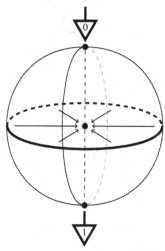

Exercise 9.13 Show that, more generally, whenever a phase spider fuses with any bastard spider, the phase vanishes:

$$\text{(diagram)} = \text{(diagram)} \qquad (9.17)$$

9.1.3 Phase Spider Fusion

So, we now know what happens when a phase spider fuses with a bastard spider, but what about when two phase spiders fuse together? Using quantum spider fusion we have:

$$\text{(diagram)} = \text{(diagram)} = \text{(diagram)}$$

The RHS is again a phase spider.

Lemma 9.14 Let $\vec{\alpha}$ and $\vec{\beta}$ be phases. Then:

(9.18)

is a phase state, and hence.

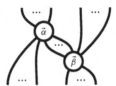

is a phase spider.

Proof Using bastard spider fusion we have:

$$
\vcenter{\hbox{(diagram)}} \;=\; \vcenter{\hbox{(diagram)}} \;\overset{(9.9)}{=}\; \vcenter{\hbox{(diagram)}} \;=\; \vcenter{\hbox{(diagram)}}
$$

So equation (9.9) is indeed satisfied for the state (9.18).   □

By introducing some new notation for this combined phase, we have the following.

Theorem 9.15 Phase spiders fuse as follows:

$$
\vcenter{\hbox{(diagram)}} \;=\; \vcenter{\hbox{(diagram with $\vec{\alpha}+\vec{\beta}$)}}
$$

(9.19)

where we used the shorthand:

$$
\boxed{\vec{\alpha}+\vec{\beta}} \;:=\; \vcenter{\hbox{(diagram with $\vec{\alpha}$ and $\vec{\beta}$)}}
$$

Clearly the order of $\vec{\alpha}$ and $\vec{\beta}$ is irrelevant:

$$
\boxed{\vec{\alpha}+\vec{\beta}} \;=\; \vcenter{\hbox{(diagram $\vec{\alpha}\;\vec{\beta}$)}} \;=\; \vcenter{\hbox{(diagram $\vec{\beta}\;\vec{\alpha}$)}} \;=\; \boxed{\vec{\beta}+\vec{\alpha}}
$$

and we can extend this notation straightforwardly to n phases, obtaining a 'decorated' version of Corollary 8.35.

Corollary 9.16 Any diagram consisting only of phase spiders and that is moreover connected is itself a phase spider, whose phase is the sum of the phases of each of the component spiders:

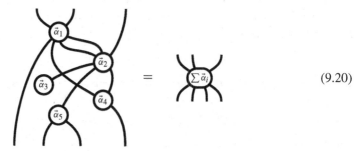

$$(9.20)$$

where we used the shorthand:

So why did we choose to write this 'phase mingling' as a sum? Let's have a look at what is happening with the underlying linear maps. First, note that when we multiply two complex phases, the angles add together:

$$e^{i\alpha} e^{i\beta} = e^{i(\alpha+\beta)}$$

Then, by expressing (9.18) as a Hadamard product (8.32), we have:

$$
\leftrightarrow
\begin{pmatrix}
1 \\
e^{i\alpha_1} e^{i\beta_1} \\
\vdots \\
e^{i\alpha_{D-1}} e^{i\beta_{D-1}}
\end{pmatrix}
=
\begin{pmatrix}
1 \\
e^{i(\alpha_1+\beta_1)} \\
\vdots \\
e^{i(\alpha_{D-1}+\beta_{D-1})}
\end{pmatrix}
$$

Hence the resulting phase is indeed the pointwise sum of the two phases we started with:

$$\vec{\alpha} + \vec{\beta} := (\alpha_1 + \beta_1, \ldots, \alpha_{D-1} + \beta_{D-1})$$

In two dimensions, this simplifies to:

$$
\leftrightarrow
\begin{pmatrix} 1 \\ e^{i\alpha} e^{i\beta} \end{pmatrix}
=
\begin{pmatrix} 1 \\ e^{i(\alpha+\beta)} \end{pmatrix}
\leftrightarrow
$$

We conclude this section with an analogue to Corollary 8.35.

Corollary 9.17 When phase spiders are composed, then, if the resulting diagram is connected, it depends only on:

- the number of inputs and outputs and
- the total sum of the phases.

9.1.4 The Phase Group

In the previous section, we saw how spiders can be used to define a 'sum' operation for phase states. In this section, we will see that the set of phase states in fact forms a *commutative group*. You may have already encountered groups in algebra, but in case you haven't, here's a quick review.

Definition 9.18 A *commutative group* is a set A with:

- a *group-sum* operation that returns $a + b \in A$ given any $a, b \in A$,
- a distinguished element $0 \in A$ called the *unit*, and
- an operation *inverse* that returns $-a \in A$ given any $a \in A$,

which satisfy the following equations, for all $a, b, c \in A$:

$$a + (b + c) = (a + b) + c \qquad a + b = b + a \qquad a + 0 = a \qquad -a + a = 0$$

These equations are typically referred to as *associativity*, *commutativity*, *unitality*, and the *inverse law*, respectively.

Taking phase states to be group elements:

we have already identified a candidate for the group-sum:

So it only remains to find the unit and the inverse.

Lemma 9.19 Let $\vec{\alpha}$ be a phase. Then:

are phase states.

Proof Using spider fusion we have:

So equation (9.9) is satisfied. We can show that the conjugate of a phase state is again a phase state by conjugating both sides of (9.9):

□

Theorem 9.20 For any family of spiders $\bigcirc$, the set of phase states:

$$\left\{ \,\begin{matrix} \vec{\alpha} \end{matrix}\, \right\}_{\vec{\alpha}}$$

form a commutative group where:

- the group-sum is:

$$\begin{matrix} \vec{\alpha}+\vec{\beta} \end{matrix} \; := \; \begin{matrix} \vec{\alpha} \quad \vec{\beta} \end{matrix}$$

- the unit is:

$$\begin{matrix} \vec{0} \end{matrix} \; := \; \begin{matrix} \end{matrix}$$

- the inverse is:

$$\begin{matrix} -\vec{\alpha} \end{matrix}$$

Proof In Lemmas 9.14 and 9.19 we already established the candidate group-sum, unit, and inverse. So it only remains to verify the group equations. Associativity, commutativity, and unit laws follow from spider fusion:

$$\begin{matrix} \end{matrix} = \begin{matrix} \end{matrix} \qquad \begin{matrix} \end{matrix} = \begin{matrix} \end{matrix} \qquad \begin{matrix} \end{matrix} = |$$

For example, in the case of associativity we have:

$$\underbrace{\overset{\overbrace{\vec{\beta}+\vec{\gamma}}}{\vec{\alpha} \quad \vec{\beta} \quad \vec{\gamma}}}_{\vec{\alpha}+(\vec{\beta}+\vec{\gamma})} \; = \; \underbrace{\overset{\overbrace{\vec{\alpha}+\vec{\beta}}}{\vec{\alpha} \quad \vec{\beta} \quad \vec{\gamma}}}_{(\vec{\alpha}+\vec{\beta})+\vec{\gamma}}$$

Then, the inverse law arises by doubling (9.10):

$$\underbrace{\begin{matrix} -\vec{\alpha} \quad \vec{\alpha} \end{matrix}}_{-\vec{\alpha}+\vec{\alpha}} \; = \; \underbrace{\begin{matrix} \end{matrix}}_{\vec{0}} \qquad\qquad (9.21)$$

which completes the proof.   □

So, using just spider rules and the definition of unbiasedness we showed that the phase states always form a group. In particular, we never made any reference to the explicit form of phases:

$$
\phi_{\vec{\alpha}} \quad := \quad \text{double}\left(\sum_j e^{i\alpha_j} \bigtriangledown_j \right)
$$

Exercise 9.21 Show using the following properties of complex numbers:

$$
e^{i0} = 1 \qquad\qquad \overline{e^{i\alpha}} = e^{-i\alpha}
$$

that $\vec{0}$ and $-\vec{\alpha}$ can be given explicitly as:

$$
\vec{0} := (0, \ldots, 0) \qquad \text{and} \qquad -\vec{\alpha} := (-\alpha_1, \ldots, -\alpha_{D-1})
$$

respectively.

In the case of $D = 2$, we can represent a phase by just a single angle. In that case, the phase group has elements represented by angles $\alpha \in [0, 2\pi)$, and the group-sum is addition of angles, i.e. addition modulo 2π:

The inverse is just sending an angle to its opposite angle:

For obvious reasons, this is sometimes called the *circle group*. In slightly more sophisticated language, this group is also called $U(1)$, owing to the fact that phases are the same thing as 1×1 unitary matrices:

$$
e^{i\alpha} \qquad \leftrightarrow \qquad \left(e^{i\alpha} \right)
$$

In higher dimensions, we just get more copies of $U(1)$. That is, the phase group is always of the form:

$$
\underbrace{U(1) \times \cdots \times U(1)}_{D-1}
$$

9.1.5 Phase Gates

We already encountered phase states as one particularly important example of phase spiders. We will now study another one that plays a central role in the remainder of this book.

Definition 9.22 A *phase gate* is a quantum process of the form:

$$\phi_{\vec{\alpha}} := \phi_{\vec{\alpha}}$$ (9.22)

As with all phase spiders, taking the adjoint (or equivalently the conjugate) of a phase gate introduces a minus sign to its phase:

$$\phi_{-\vec{\alpha}} = \phi_{-\vec{\alpha}} = \phi_{-\vec{\alpha}} = \phi_{-\vec{\alpha}}$$

As a consequence, we have the following proposition.

Proposition 9.23 Phase gates are unitary.

Proof Using spider fusion we have:

$$\phi_{-\vec{\alpha}}^{\vec{\alpha}} = \phi_{\vec{0}} = \phi = |$$

Composing in the other order similarly yields the identity. □

Hence, phase gates are prime examples of *quantum gates* (cf. Section 5.3.4 and Example 6.13).

Example 9.24 Picking up where we left off in Example 6.13, phase gates allow us to write quantum circuits that have no classical counterpart:

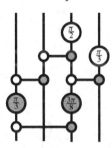

A set of quantum gates is *universal* when arbitrary unitaries can be obtained as a quantum circuit including only gates of that set, and we shall show in Section 12.1.3 that a set consisting of the quantum CNOT-gate and phase gates is universal. Consequently, phase gates play a central role in quantum computing.

Since phase gates are examples of phase spiders, the group structure from the previous sections carries immediately over into phase gates, e.g.:

Corollary 9.25 For any family of spiders ○, the set of phase gates:

$$\left\{ \phi_{\vec{\alpha}} \right\}_{\vec{\alpha}}$$

forms a commutative group where:

- the group-sum is:

- the unit is:

- the inverse is:

So what do these phase gates actually do? We can figure this out by looking at how they act on phase states:

$$\phi_{\vec{\beta}} \;::\; \phi_{\vec{\alpha}} \;\mapsto\; \phi_{\vec{\alpha}}^{\vec{\beta}} \;=\; \phi_{\vec{\alpha}+\vec{\beta}}$$

So a phase gate with phase $\vec{\beta}$ sends a phase state with phase $\vec{\alpha}$ to another phase state with phase $\vec{\alpha} + \vec{\beta}$. In two dimensions, this corresponds to a Bloch sphere rotation about the axis fixed by the two basis states:

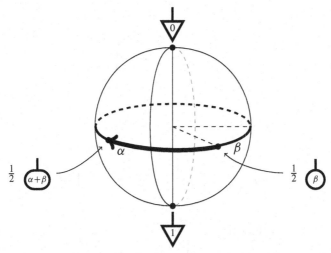

Since phase gates are unitary, and unitaries correspond to rotations on the Bloch sphere (cf. Proposition 7.2), it immediately follows that phase gates rotate all states around the Z-axis:

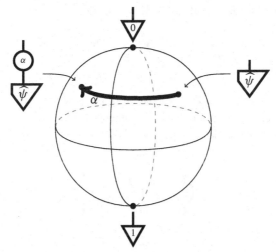

Unsurprisingly, the matrices of phase gates are closely related to the matrices of their associated phase states. For:

$$
\overset{|}{\underset{\vec{\alpha}}{\bigcirc}} := \sum_j e^{i\alpha_j} \overset{|}{\underset{j}{\bigtriangledown}}
$$

(where we take $\alpha_0 := 0$) we obtain:

$$
\overset{\circ}{\underset{\vec{\alpha}}{|}} = \overset{\circ}{|} \sum_j e^{i\alpha_j} \overset{}{\underset{j}{\bigtriangledown}} = \sum_j e^{i\alpha_j} \overset{\underset{j}{\bigtriangledown}}{\underset{j}{\bigtriangleup}}
$$

Hence, for a phase state:

$$\overset{\mid}{\underset{\bar{\alpha}}{\bigcirc}} \quad \leftrightarrow \quad \begin{pmatrix} 1 \\ e^{i\alpha_1} \\ \vdots \\ e^{i\alpha_{D-1}} \end{pmatrix}$$

the matrix of the associated phase gate is:

$$\overset{\mid}{\underset{\bar{\alpha}}{\bigcirc}}\overset{}{\underset{\mid}{}} \quad \leftrightarrow \quad \begin{pmatrix} 1 & 0 & \cdots & 0 \\ 0 & e^{i\alpha_1} & \cdots & 0 \\ \vdots & \vdots & \ddots & \vdots \\ 0 & 0 & \cdots & e^{i\alpha_{D-1}} \end{pmatrix}$$

For $D = 2$, the matrix of a phase map is of this form:

$$\begin{pmatrix} 1 & 0 \\ 0 & e^{i\alpha} \end{pmatrix}$$

Example 9.26 In Example 8.70 we already encountered the three-system GHZ state. When we apply a phase gate to each of the systems we obtain:

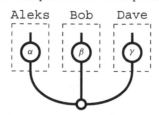

While this is a seemingly innocent application of phase spider fusion, when we interpret this equation physically the implications are somewhat shocking!

Assume that the three parties that perform the three phase gates:

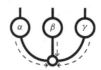

are so far apart that light does not have the time to travel between them.

While the choices of the angles α, β, and γ are made independently, at very distant locations, the resulting state depends only on the group-sum of the three phases. So if, for instance, these phases would have been permuted, the resulting state would be the same. This hints at the fact that these processes are interacting instantaneously over a long distance. The diagram literally shows that it is <u>as if</u> the three phases are travelling backwards in time to meet up with each other:

and in contrast to our discussion in Section 4.4.3, which involved (non-causal) caps, all the processes involved here are perfectly causal. Of course, all of this happens at the level of a quantum state, and naïve measurement kills the phases, and hence all the magic:

Below in Section 9.3.3 we will pick up this story again and show that an alternative choice of measurements won't kill the magic.

9.2 Multicoloured Spiders

Spiders of the same species happily mate and mingle their phases. However, it is not all love and peace in spider land. Now we introduce spiders of distinct species, and expose the carnage that happens when these species clash.

As we did in Section 5.3.5 we will represent two distinct ONBs, and now their associated spiders, using different colours:

This will enable us to consider measurement and encoding operations in two different bases and, in particular, to study how these operations interact.

9.2.1 Complementary Spiders

The concept of a phase state (a.k.a unbiasedness) has a very clear interpretation as proper quantumness and gives rise to diagrammatic creatures called phase spiders. A simple diagrammatic rule that concerns how spiders of different colours interact yields a related and perhaps even more important concept.

Definition 9.27 Spiders ○ and ◉ are *complementary* if:

$$(9.23)$$

or equivalently:

$$
\begin{array}{c} \end{array} \;=\; \frac{1}{D} \; \begin{array}{c} \end{array} \tag{9.24}
$$

So what does this mean? In the LHS of (9.23) we are encoding classical data in the white basis then measuring in the grey one. Then, (9.23) says this is equivalent to just deleting the classical input:

and outputting a uniform probability distribution:

$$
\frac{1}{D}
$$

Thus the classical data at the input vanishes and is replaced by randomness. In summary:

$$
\boxed{(\text{encode in } \bigcirc) \text{ THEN } (\text{measure in } \bullet) = (\text{no data flow})} \tag{9.25}
$$

Remark 9.28 Note that the uniform probability distribution arising from a $\bullet$-measurement is given as a $\bullet$-spider. We'll say more about this in Section 9.2.4.

Condition (9.25) says that the white basis is a very poor way of encoding classical data with respect to the grey measurement. For example, if we encode a classical value i as the i-th white ONB state of a quantum system and we measure it with the white measurement, we will get outcome i with certainty. On the other hand, if we measure it with the grey measurement, we are equally likely to obtain any outcome, so this gives us no information whatsoever about the encoded value. Another way of saying this is that if we have maximal information about a system with respect to one basis (i.e. it is in a pure ONB state), we have no information with respect to the other basis.

Example* 9.29 A well-known instance of this phenomenon in quantum mechanics is that if one perfectly knows the position of a quantum system, then one cannot know anything about its momentum. One usually treats position as a continuous variable, which necessitates moving to infinite-dimensional systems. However, the fundamental principle is the same as the discrete version of complementarity we present here.

While equation (9.23) required a separation into a particular state and effect, namely (white) deleting followed by the (grey) uniform probability distribution, it actually suffices to say that the LHS above simply separates:

$$
\underset{\text{wire}}{} \;=\; \underset{\text{no wire}}{}
$$

Thus, (9.25) can be taken as a literal statement of complementarity. More precisely, we can give several variations on this theme.

Proposition 9.30 The following are equivalent for ○ and ●:

(i) complementarity:

$$\text{[diagram]} \quad = \quad \frac{1}{D} \; \text{[diagram]}$$

(ii) complementary measurements on a Bell state give a uniform probability distribution for two classical systems:

$$\frac{1}{D} \; \text{[diagram]} \quad = \quad \frac{1}{D} \; \text{[diagram]} \quad \frac{1}{D} \; \text{[diagram]}$$

(iii) the existence of an effect p and a state q such that:

$$\text{[diagram]} \quad = \quad \text{[diagram] } q \text{ over } p$$

(iv) there exist states p and q such that:

$$\frac{1}{D} \; \text{[diagram]} \quad = \quad \text{[diagram] } p \quad q$$

Proof Equivalence of **(i)** and **(ii)** is trivial, and so is equivalence of **(iii)** and **(iv)**, and since **(i)** implies **(iii)**, it only remains to be shown that **(iii)** implies **(i)**. First, we can show that q is ●, up to some number λ:

$$\text{[diagram]} = \text{[diagram]} \overset{(8.70)}{=} \text{[diagram]} \overset{(8.70)}{=} \text{[diagram]} = \text{[diagram]} = \text{[diagram]} = \lambda \; \text{[diagram]} \, q$$

Then, by a similar argument:

$$\text{[diagram]} \quad = \quad \lambda' \; \text{[diagram] } p$$

Note that λ and λ' must both be non-zero, so we can deduce that:

$$\text{[diagram]} \quad = \quad \frac{1}{\lambda \lambda'} \; \text{[diagram]} \tag{9.26}$$

We can figure out what this number is by means of the recipe outlined at the end of Section 3.4.3. Pre- and post-composing with dots yields:

$$ D \;=\; \underset{\equiv}{\overset{\equiv}{\bot}} \;\overset{(8.60)}{=}\; \left| \right| \;=\; \tfrac{1}{\lambda\lambda'} \;\left| \right| \;=\; \tfrac{D^2}{\lambda\lambda'} $$

so $\lambda\lambda' = D$. Plugging this into (9.26) yields (i). □

We will now contrast the behaviour of complementary spiders with the usual spider-fusion rules. Rather than fusing, complementary spiders do the opposite: they break apart. We will see the complementarity equation taking on several guises in the next few chapters. The first is for bastard spiders.

Proposition 9.31 For complementary ○ and ● we have:

$$ \text{(9.27)} $$

Proof A bit of spider fusion:

and a bit of complementarity:

$$ \overset{(9.23)}{=}\; \tfrac{1}{D} $$

and another bit of spider fusion:

$$ \tfrac{1}{D} \qquad = \qquad \tfrac{1}{D} $$

□

The second guise of complementarity is for classical spiders. Here we think of ● as classical maps for ○, or vice versa (see Section 9.3.5 below).

Proposition 9.32 For complementary ○ and ● we have:

$$\text{(diagram)} = \frac{1}{D} \text{(diagram)} \tag{9.28}$$

The proof is nearly identical; just replace (9.23) with (9.24). The third guise of complementarity doesn't have any direct connection to measurements or classical data at all. Nonetheless, it will be remarkably useful for quantum protocols and algorithms. It is just the doubled form of (9.28).

Proposition 9.33 For complementary ○ and ● we have:

$$\text{(diagram)} = \frac{1}{D^2} \text{(diagram)} \tag{9.29}$$

One thing you might have noticed from this section is we are starting to get a lot of Ds popping up all over the place. Well, we have some bad news and some good news. The bad news is, this is basically unavoidable. If we renormalise spiders to get rid of these numbers, the spider-fusion rule becomes really horrible.

Exercise* 9.34 How would the spider-fusion rule look when we renormalise spiders such that we have the following?

$$\text{(diagram)} = \text{(diagram)}$$

However, the good news is that (as explained in Section 3.4.3) we can pretty much always ignore these numbers. So we can rewrite complementarity as:

$$\text{(diagram)} \approx \text{(diagram)} \tag{9.30}$$

and the derived 'spider detachment' rules become:

$$\text{(diagram)} \approx \text{(diagram)} \tag{9.31}$$

$$\text{(diagram)} \approx \text{(diagram)} \tag{9.32}$$

$$(9.33)$$

Isn't that nicer? And we can recover the ignored numbers too.

Proposition 9.35 For spiders ◯ and ●:

Proof Assuming:

and pre- and post-composing with dots just as we did in the proof of Proposition 9.30, we obtain:

so $\lambda = \frac{1}{D}$. ∎

Remark 9.36 The spiders we use in this chapter will continuously hop between being **linear maps** and **quantum maps/cq-maps**. This means there could be some ambiguity whether ≈ means 'up to a complex number' (i.e. the numbers in **linear maps**) or 'up to a positive number' (i.e. the numbers in **cq-maps**). We will always specify this explicitly when there could be some confusion.

Exercise 9.37 Show that equations (9.32) and (9.33) extend to phase spiders; that is, prove:

$$(9.34)$$

$$(9.35)$$

9.2.2 Complementarity and Unbiasedness

Both unbiasedness and complementarity have something to do with an obstruction of information flow across measurements, so it shouldn't come as a surprise that they are closely connected. In fact, complementarity can be formulated entirely in terms of unbiasedness. First, note that, for complementary $\bigcirc$ and $\bullet$, we have the following:

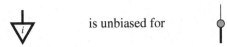

$$(9.36)$$

Hence, in the light of Exercise 9.2, the basis state:

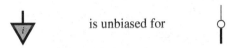

and by vertically reflecting, we can also show that:

This particular mutual relationship has a standard name.

Definition 9.38 Two ONBs:

are *mutually unbiased* if every state of one ONB is unbiased for the other ONB; that is, if for all i,j we have:

$$ = \tfrac{1}{D} \qquad\qquad (9.37)$$

Remark 9.39 The defining equation (9.37) for mutually unbiased ONBs can also be written in undoubled form, which may look more familiar to some:

$$ = \tfrac{1}{\sqrt{D}}$$

Theorem 9.40 Spiders $\bigcirc$ and $\bullet$ are complementary:

$$ = \tfrac{1}{D}$$

if and only if their associated ONBs are mutually unbiased:

$$\forall i,j \ : \quad \bigvee\!\!\!\bigtriangleup \ = \ \tfrac{1}{D} \tag{9.38}$$

Proof For ○ and ⦿ complementary, (9.36) already showed that the associated ONBs are mutually unbiased. Conversely, if the two ONBs are mutually unbiased, then we obtain:

$$\overset{(8.6)}{=} \quad \overset{(9.23)}{=} \quad \tfrac{1}{D} \quad \overset{(8.13)}{=} \quad \tfrac{1}{D}$$

So the matrix entries of the LHS and RHS of the complementarity equation (9.23) agree for these bases, and hence the equation holds. □

This relationship with mutual unbiasedness gives us another bunch of alternative characterisations of complementarity.

Corollary 9.41 The following are equivalent for ○ and ⦿:

(i) complementarity:

$$= \ \tfrac{1}{D}$$

(ii) for all i:

$$= \ \tfrac{1}{D} \tag{9.39}$$

(iii) for all i there exists a phase $\vec{\kappa}$ such that:

$$\bigvee_{i} \ = \ \tfrac{1}{D} \ \widehat{\vec{\kappa}} \tag{9.40}$$

(iv) for all j:

$$= \ \tfrac{1}{D} \tag{9.41}$$

(v) for all j there exists a phase $\vec{\kappa}$ such that:

$$\bigvee_{j} \ = \ \tfrac{1}{D} \ \widehat{\vec{\kappa}} \tag{9.42}$$

(vi) mutual unbiasedness:

$$\forall i,j \ : \ \text{(spider diagram)} \ = \ \tfrac{1}{D}$$

Example 9.42 One advantage of the characterisation of complementarity in terms of mutual unbiasedness is that it easy to check that a pair of ONBs induces complementary spiders. For example, the Z-basis and the X-basis, which, as we already saw a while back, can be expressed in terms of the Z-basis as:

$$\text{(grey 0)} := \tfrac{1}{\sqrt{2}}\left(\text{(white 0)} + \text{(white 1)} \right) \qquad \text{(grey 1)} := \tfrac{1}{\sqrt{2}}\left(\text{(white 0)} - \text{(white 1)} \right)$$

are indeed mutually unbiased:

$$\text{(spider diagram with } j, i) \ = \ \tfrac{1}{2}$$

Hence they induce complementary spiders, and hence we are entitled to represent them by means of complementary colours:

$$\text{(white spider)} := \ Z\text{-measurement} \qquad\qquad \text{(grey spider)} := \ X\text{-measurement}$$

Exercise 9.43 In order to establish equivalence in Theorem 9.40 we relied on explicitly plugging in white and grey basis states. Instead, establish equivalence between **(i)** and **(ii)** above by plugging in states from only one basis.

Example 9.44 A set of ONBs is called *pairwise mutually unbiased* if each pair of distinct ONBs in the set is mutually unbiased. Up to an overall unitary, there are three ONBs for qubits that are pairwise mutually unbiased. The Z-basis, the X-basis, and the Y-basis:

$$\text{(black 0)} := \tfrac{1}{\sqrt{2}}\left(\text{(white 0)} + i\,\text{(white 1)} \right) \qquad \text{(black 1)} := \tfrac{1}{\sqrt{2}}\left(\text{(white 0)} - i\,\text{(white 1)} \right)$$

For all i, j, we have:

$$\text{(grey/white spider)} \ = \ \text{(black/white spider)} \ = \ \text{(grey/black spider)} \ = \ \tfrac{1}{2}$$

and hence these ONBs are indeed pairwise mutually unbiased. (Note that, while the Y-basis does not consist of self-conjugate basis states, our definition of mutual unbiasedness straightforwardly extends to these.) On the Bloch sphere, they mark the three main axes:

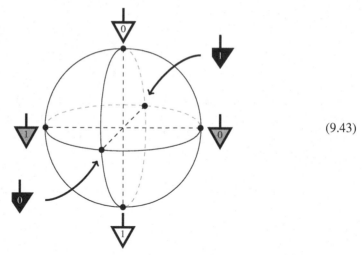

$$(9.43)$$

A set of pairwise mutually unbiased bases is called *maximal* if it cannot be extended. Determining the size of such maximal sets for all dimensions is an (extremely hard) open problem, comparable in difficulty to the problem of classifying SIC-POVMs mentioned in Section 7.4.2. In dimension 2, these maximal sets are always of size 3, as above. More generally, if $D = p^N$ for some prime number p, the answer is $p^N + 1$. However, at the time of this writing, the answer for other dimensions, e.g. $D = 6$, is completely unknown.

In (9.43) we depicted the basis states of some very important measurements on the Bloch sphere. In diagrammatic terms, the characteristic property of such a basis state is that they are copied by the relevant spiders:

$$
\text{\small (diagram)} \quad = \quad \text{\small (diagram)} \qquad \text{\small (diagram)} \quad = \quad \text{\small (diagram)} \qquad \text{\small (diagram)} \quad = \quad \text{\small (diagram)}
$$

Since these basis states satisfy several mutual unbiasedness relationships, by Corollary 9.41 we moreover know that each of these states is also a phase state for its complementary spiders, yielding more useful equations. Let's figure out what these phases precisely are for these six states.

We already saw in Example 9.7 that:

$$
\text{(diagram)} = \tfrac{1}{2}\,\text{(diagram)}_0 \qquad\qquad \text{(diagram)} = \tfrac{1}{2}\,\text{(diagram)}_\pi \tag{9.44}
$$

and comparing (9.43) with (9.13) we also have:

$$
\text{(diagram)} = \tfrac{1}{2}\,\text{(diagram)}_{\frac{\pi}{2}} \qquad\qquad \text{(diagram)} = \tfrac{1}{2}\,\text{(diagram)}_{-\frac{\pi}{2}} \tag{9.45}
$$

But we can of course also do something similar in terms of X-phase states, which according to equation (9.12) have the following matrix form:

$$\bigodot_{\alpha} \;=\; \text{double}\left(\underset{0}{\bigtriangledown} \;+\; e^{i\alpha}\, \underset{1}{\bigtriangledown} \right)$$

These grey phase states lie on the equator for the axis that goes through the doubled grey basis states:

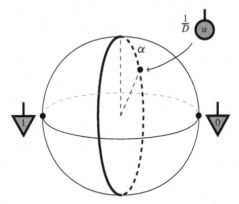

and comparing with (9.43) yields:

$$\underset{0}{\bigtriangledown} \;=\; \tfrac{1}{2}\; \underset{0}{\bigcirc} \qquad\qquad\qquad \underset{1}{\bigtriangledown} \;=\; \tfrac{1}{2}\; \underset{\pi}{\bigcirc} \tag{9.46}$$

$$\underset{0}{\blacktriangledown} \;=\; \tfrac{1}{2}\; \underset{-\frac{\pi}{2}}{\bigcirc} \qquad\qquad\qquad \underset{1}{\blacktriangledown} \;=\; \tfrac{1}{2}\; \underset{\frac{\pi}{2}}{\bigcirc} \tag{9.47}$$

We could do the same for ● too, but we won't bother, since later we will see that for our purposes, we will only need <u>pairs</u> of complementary spiders (see Theorem 9.66). To summarise all the above we present the phase states representing basis states (up to $\frac{1}{2}$) on the Bloch sphere:

$$\tag{9.48}$$

Example 9.45 We noted back in Section 5.3.5 that the adjoint of the X-copying map behaves as an XOR operation on the Z basis, more precisely:

$$\text{(diagram)} \quad = \quad \tfrac{1}{\sqrt{2}} \boxed{\text{XOR}}$$

This fact can now be verified simply by using phase spider fusion:

$$\text{(diagram)} \quad = \quad \tfrac{1}{2} \quad \text{(diagram)} \quad = \quad \tfrac{1}{2} \quad \text{(diagram with 0)} \quad = \quad \tfrac{1}{\sqrt{2}} \quad \text{(diagram with 0)}$$

$$\text{(diagram)} \quad = \quad \tfrac{1}{2} \quad \text{(diagram)} \quad = \quad \tfrac{1}{2} \quad \text{(diagram with }\pi\text{)} \quad = \quad \tfrac{1}{\sqrt{2}} \quad \text{(diagram with 1)}$$

$$\text{(diagram)} \quad = \quad \tfrac{1}{2} \quad \text{(diagram)} \quad = \quad \tfrac{1}{2} \quad \text{(diagram with }\pi\text{)} \quad = \quad \tfrac{1}{\sqrt{2}} \quad \text{(diagram with 1)}$$

$$\text{(diagram)} \quad = \quad \tfrac{1}{2} \quad \text{(diagram)} \quad = \quad \tfrac{1}{2} \quad \text{(diagram with 0)} \quad = \quad \tfrac{1}{\sqrt{2}} \quad \text{(diagram with 0)}$$

9.2.3 The CNOT-Gate from Complementarity

In previous chapters, we have been assuming the existence of certain 'black boxes' that have useful properties (e.g. measurements and controlled unitaries). Now that we have a variety of spiders at hand:

- classical, quantum, and bastard spiders:

- phase spiders:

- complementary spiders:

we can start constructing these things. For our first trick, we will show how a pair of complementary spiders always induces a unitary quantum map. In the case of the Z and X complementary pair of Example 9.42, this unitary is the CNOT-gate that we saw in Section 5.3.4:

$$
\text{(9.49)}
$$

The notation that we used to denote the CNOT-gate:

includes a white and a grey dot, which clearly points in the direction of two families of spiders. However, what does a horizontal line mean?

Lemma 9.46

$$
\text{(9.50)}
$$

Proof The proof relies on the fact that:

We prove the first equation of (9.50):

The remainder of the proof is left as an exercise. $\qquad\square$

Exercise 9.47 Complete the proof of Lemma 9.46.

So since it doesn't really matter how we make the passage from the white to the grey dot with a wire, we can just depict it as a horizontal line. The following now also straightforwardly follows.

Lemma 9.48 For spiders $\bigcirc$ and $\bullet$ the map:

is always self-adjoint.

And we indeed recover the CNOT-gate, at least up to a number.

Exercise 9.49 Show that if we choose ○ and ● to be spiders for the Z-basis and the X-basis as in Example 9.42, then:

$$\sqrt{2}\ \text{⟿}$$

is the CNOT-gate (9.49).

The fact that ○ and ● induce a unitary map gives us yet another collection of alternative characterisations for complementarity.

Proposition 9.50 The following are equivalent for ○ and ●:

(i) The spiders are complementary:

$$\text{⟿} = \tfrac{1}{D}\ \text{⟿}$$

(ii) The following linear map is unitary:

$$\sqrt{D}\ \text{⟿} \tag{9.51}$$

(iii) The following quantum map is unitary:

$$D\ \text{⟿} \tag{9.52}$$

Proof First, we show that **(i)** implies **(ii)**. Since (9.51) is self-adjoint, it suffices to show that this map composed with itself is the identity:

$$\sqrt{D}\ \sqrt{D}\ \text{⟿} \quad = \quad D\ \text{⟿} \quad \overset{(9.28)}{=} \quad \text{⟿} \quad = \quad ||$$

(ii) implies **(iii)** by doubling the unitarity equations. Conversely, if (9.52) is unitary, then (9.51) is unitary by Theorem 6.20. Thus, it suffices to show that **(ii)** implies **(i)**:

$$\text{⟿} \quad = \quad \tfrac{1}{D}\ \sqrt{D}\ \sqrt{D}\ \text{⟿} \quad = \quad \tfrac{1}{D}\ \text{⟿} \quad = \quad \tfrac{1}{D}\ \text{⟿}$$

where the first step is just spider fusion (and some number juggling), and the second step uses unitarity of (9.51).   □

Exercise 9.51 Give an alternative proof that **(i)** implies **(iii)** by first showing causality of **(iii)**.

Remark 9.52 We will see in Section 12.2 that complementarity is equivalent to unitarity of:

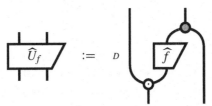

for <u>any</u> (doubled) function map f (cf. Definition 8.13). Then, Proposition 9.50 is just the special case where f is the identity. Whereas $\widehat{f}$ may not be causal (and hence not a quantum process), $\widehat{U}_f$ is always a quantum process. Such a unitary quantum process $\widehat{U}_f$ is called a *quantum oracle* and forms a crucial component to many quantum algorithms.

9.2.4 'Colours' of Classical Data

The careful reader may have noticed that we have thus far glazed over an important point about multicoloured spiders and classical wires. Let's have a look at the matrix form of the measure processes for two different spiders:

$$
\begin{array}{cc}
\phi = \sum_i \triangledown_i^i \triangle_i & \phi = \sum_i \blacktriangledown_i^i \blacktriangle_i
\end{array}
$$

It is clear that these measurements both produce classical states, but these classical states don't look exactly the same:

$$
\sum_i p^i \,\triangledown_i \qquad \text{vs.} \qquad \sum_i p^i \,\blacktriangledown_i
$$

That is, the classical states are encoded in different ONBs. This is a feature rather than a bug. The numbers that make up a probability distribution are totally useless information, unless we also know what these numbers are probabilities of. In the two classical states above, the ONBs tell us what the numbers p^i are about; namely, they are the probabilities for the outcomes of a particular measurement. More generally, a classical wire carries a specification of how that classical data was produced, i.e. by which measurement:

probabilities w.r.t. $\left\{\triangledown_i\right\}_i$ probabilities w.r.t. $\left\{\blacktriangledown_i\right\}_i$

(the same) quantum system

Similarly, for encoding we have:

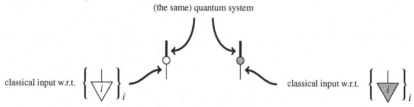

The bottom line is: classical wires carry additional type information about the basis in which the classical data is encoded. We could make this information explicit if, for instance, we labeled wires as follows:

but this won't really be necessary since the type will always be clear from the context. Most importantly, any thin wire connected to a bastard spider carries classical data of that 'colour'.

As with any kind of system-types, we don't allow wires of different types to be plugged together:

(9.53)

On the other hand, both of these measurement choices (crucially) operate on the same type of quantum system; so, for instance, these compositions are allowed:

(9.54)

The golden rule one should take from (9.53) is that bastard spiders of different colours should never be connected via classical wires. On the other hand, it might sometimes be the case that a <u>classical</u> spider of one colour is actually a valid classical map for another colour. We already encountered such an example in Proposition 5.88 where we saw that an XOR gate in the white basis is matching in the grey basis. In that case, a linear map such as:

could indeed be a valid cq-map. However:

violates the golden rule, since now bastard spiders of different colours are touching the same type of classical wire.

Remark 9.53 We could instead choose to use one fixed basis for classical outcomes and define ONB measurements via a single dot and a unitary, as we did in Section 8.4.1:

However, as we have already seen in the past three sections, representing classical data in different bases substantially simplifies our diagrams and calculations.

9.2.5 *Complementary Measurements*

Let's use complementarity to show some interesting quantum features. We return to the Stern–Gerlach apparatus that we studied in Sections 7.1.1 and 7.1.2. Recall that we could measure in different ONBs just by rotating the whole device:

Z-measurement   X-measurement

From example 9.42 we also know that the Z- and X-measurements can be described by means of complementary spiders:

with respect to the Z- and X-bases from Example 9.42. If, rather than letting the particles hit a screen, we allow them to pass on through, we obtain complementary non-demolition measurements (cf. Section 8.4.1):

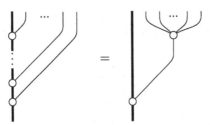

If we pretend we don't know anything about quantum measurements, it could be the case that the particles produced by ⌐⌐⌐ have a 'Z-property' that tells us which way they will deflect in a Z-measurement and an 'X-property' that tells us which way they will deflect in an X-measurement, and the operations above correspond to just 'observing' those properties. When we perform these measurements in isolation, this interpretation seems okay. For example, no matter how many times we 'observe the Z-property', we'll get the same result:

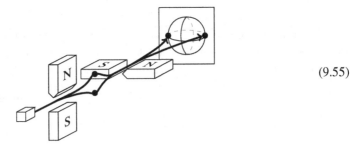

However, when we start combining Z-measurements and X-measurements, this measurement-as-observing idea breaks down. First, suppose we perform a Z-measurement before an X-measurement:

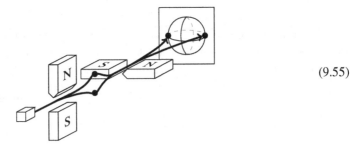

(9.55)

In diagrams, the experimental setup of (9.55) becomes:

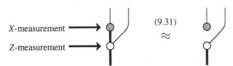

That is, the X-measurement always produces a uniform probability distribution, no matter what state we input into the apparatus. If we just observed these probabilities in a lab, without having complete control over the states we input, we might not think this is so strange. It could just be that our particle source ⌐⌐⌐ is producing particles that will yield an X-value of 0 half the time and 1 the other half of the time. However, we can definitely convince ourselves that there is something fishy going on if we add a third Stern–Gerlach device:

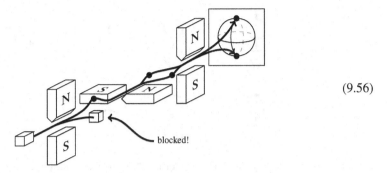

(9.56)

and block one of the exits of the first Z-measurement. Supposing this is the exit corresponding to outcome 1; the particle will only hit the screen at the end if the first measurement yields 0. We can restrict to just the scenarios where the first measurement yields outcome 0 as follows:

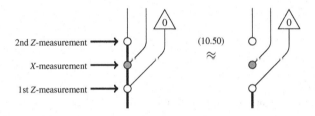

(10.50)

Even when we only allow particles through that yield 0 in the first Z-measurement, by the time we measure Z again, rather than getting 0 again, we get the uniform probability distribution! Of course, it is the presence of the X-measurement in the middle that causes this to happen. This setup is one of the most famous demonstrations of how measurements in quantum theory behave very differently from just 'observing' some property of a system. The fact that one measurement affects the outcomes of another one in this manner is referred to as *incompatibility* of measurements. Since complementary measurements will cause each other's outcomes to be complete noise, they are *maximally incompatible*.

In addition to the fact that it gives one of the most striking physical manifestations of the 'properly quantum' feature of complementarity, we have also included this example to show the unreasonable effectiveness of diagrammatic reasoning with spiders! The arguments above each merely required a single application of a spider detachment rule.

Thanks to this simplicity, we can easily consider other similar situations. From Example 9.44 we know that also Y-measurements should be maximally incompatible both with X-measurements and Y-measurements, so we should be able to derive similar results involving Y-measurements along the lines of what we derived above. One way to do so could be introducing a third colour, but that's not even necessary. It suffices to throw in some decorations.

Since both the X-basis states and the Y-basis states lie on the equator of the Bloch sphere (cf. picture (9.43)), we can use a white phase gate to turn one basis into the other:

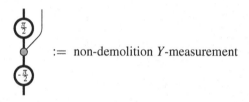

Recalling from Section 8.4.1 that general ONB-measurements consist of a unitary composed with a measure spider, we can transform the X-measurement into a Y-measurement as follows:

$$:= \text{demolition } Y\text{-measurement}$$

with a corresponding:

$$:= \text{non-demolition } Y\text{-measurement}$$

Using phase/bastard spider fusion we have:

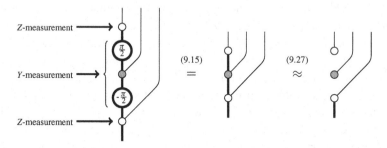

Exercise 9.54 Compute the result of first performing a non-demolition Y-measurement, then a Z-measurement, and then again a Y-measurement diagrammatically.

Similarly, one can also represent Y-measurement using X-phase gates:

$$:= \text{demolition } Y\text{-measurement}$$

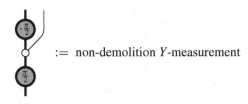

:= non-demolition Y-measurement

and the same results can of course also be derived in this alternative form.

9.2.6 Quantum Key Distribution

Something very closely related to our analysis of incompatibility in the previous section is *quantum key distribution* (QKD). In particular, we will see how a complementary pair of measurements can be used to establish a common secret (e.g. a cryptographic key) between Aleks and Bob in such a way that they will always be able to tell if someone is eavesdropping (hence justifying the use of the term 'secret'). The most well-known protocol for QKD is called BB84. The actual quantum part of the protocol is extremely simple. We'll use bits and qubits to describe the protocol, but it works just fine for $D > 2$. All that is required is complementarity.

Essentially, Aleks has a bunch of random bits that he wants to send to Bob. He doesn't particularly care if they all get there. He will already be happy if enough of the bits get there to make a cryptographic key, which he can then use to do the 'real' secret communication with Bob.

To do so, Aleks and Bob first fix a pair of complementary spiders ○ and ● and Aleks selects with equal probability to encode his classical data either using ○-encode or ●-encode. Afterwards, Bob chooses with equal probability to either ○-measure or ●-measure. If the two choices agree, Bob gets Aleks' bit:

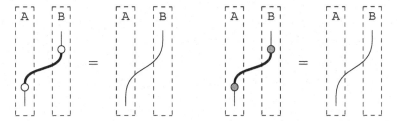

If the two choices are different, Bob gets noise, i.e. the uniform distribution:

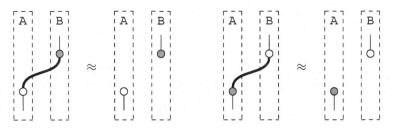

They repeat this for all of the bits Aleks wants to send Bob, knowing that there is a 50% chance Bob will get Aleks' bit, and a 50% chance he will get garbage. Afterwards, to know which is which, Aleks and Bob simply announce which colour of spider they used to encode/measure each bit. For the bits where the colour was the same, they know they have the same value, and the rest they throw away. Since Aleks' encoding and Bob's measurement choices were random, and have nothing to do with the key they are trying to establish, they can broadcast this data publicly, and don't have to worry about any sketchy dodos getting hold of the private key.

On the other hand, what if sketchy Dave is listening in on the channel Aleks is using to send his quantum systems to Bob?

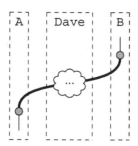

If it were possible to clone a quantum system, Dave could simply keep a copy of every state Aleks sends to Bob, then, once Aleks and Bob announce their bases, Dave can simply measure his copy to get the message. Of course, we've known since Chapter 4 that this isn't possible, so the best Dave can do is perform a measurement of some kind.

First, for simplicity we'll assume that Dave also chooses his measurements from ○/●. We'll look at the instances where Aleks and Bob intend to keep the bit Aleks is sending, i.e. when their choices of ○/● are the same. Since Aleks is choosing his encoding bases randomly, the best Dave can do is measure in the correct basis half of the time. When that happens, he does indeed get a copy of Aleks' bit:

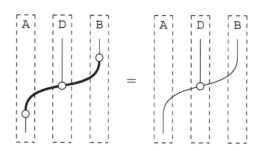

However, the other half of the time, Bob will get noise, just like when he made the wrong measurement choice:

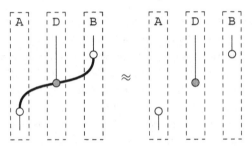

Rather than receiving Aleks' bit correctly half the time, Bob sometimes gets noise instead, so the probability that Bob gets the correct bit goes down. This suggests an easy strategy for detecting eavesdroppers. Every once in a while, Aleks and Bob randomly pick out a bunch of spare bits (i.e. bits Aleks sent but that they don't intend to use) and compare them. If significantly fewer than 50% of the bits are correct, someone must be eavesdropping, so they call it a day.

Exercise* 9.55 What is the probability of Bob getting Aleks' bit when Dave is eavesdropping? Note that, assuming a process is fed a classical state in the uniform probability distribution, the probability that it produces the same input as output can be computed via the Born rule as follows:

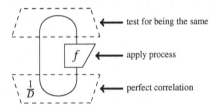

Remark 9.56 It can be shown that it is possible to detect Dave even if he is allowed to perform arbitrary quantum processes, not just $\bigcirc$/$\bullet$-measurements. However, the analysis is quite a bit trickier (cf. the references in Section 9.7).

One of the benefits of a graphical presentation is we can simply bend the wires around and read off a seemingly different protocol. Suppose instead of Aleks sending systems to Bob along a quantum channel, Aleks and Bob instead share a Bell state. Then, if they both make random choices of measurements, they establish perfect correlations when the measurements are the same, and no correlations when they are different:

This gives a (simplified version of) a different QKD protocol, called E91. Even though the actual steps in the protocol are different, the result is the same. Ultimately, Aleks and

Bob end up with a perfectly correlated string of random bits. Furthermore, since we are working with essentially the same diagrams, we can check for eavesdropping just as we did before. However, reading BB84 'sideways' has some additional benefits. First, Aleks and Bob do not need to be continuously in (quantum) contact to share the key. They just need to establish some shared Bell states at some point, then 'use them up' as needed. Second, Aleks and Bob can be sure they have real Bell states (i.e. they haven't been tampered with) by checking that their measurement outcomes actually exhibit quantum non-locality, which we will discuss at length in Section 11.1.

Remark* 9.57 In the full version of E91, Aleks and Bob check for non-locality by incorporating a third measurement and checking that the correlations between their measurement outcome violate what's known as a *Bell inequality*. Such a violation guarantees non-locality. A nice thing about this technique is they only have to share their outcomes for the 'garbage' bits, i.e. where their measurement choices disagree, so they don't need to sacrifice any usable bits to check security.

9.2.7 Teleportation with Complementary Measurements

As promised, we can replace the 'black box':

used in teleportation by something constructed entirely by means of spiders. Teleportation relies initially on Aleks making a joint measurement on his two systems:

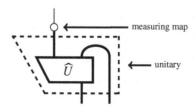

If the quantum systems have dimension D, the classical wire will take D^2 different values. Therefore, we could just as well represent it with two classical wires that each take D possible values:

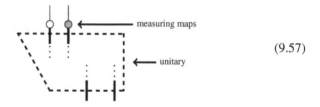

(9.57)

(The fact that we used two different colours anticipates what will follow.) Now, to get some teleportin' done, this measurement better not ⊗-separate, otherwise Bob will, at best, get some decoherent version of Aleks' state, since everything would then have to pass through a classical wire:

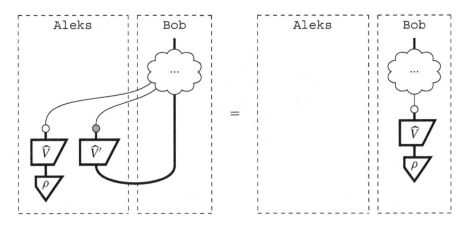

Given that we have a complementary pair of spiders around, we can try to use the induced unitary from Proposition 9.50:

to produce a non-separable measurement. Here's our first attempt:

$$= \qquad \overset{(9.31)}{\approx}$$

Oops, that didn't work! Applying the CNOT didn't change the measurement at all! In particular, it is still separated. However, since we are trying to do some science, let's experiment some more. What happens if we plug the induced unitary in the other way?

$$(9.58)$$

This flipped version is of course also unitary, so (9.58) is still an ONB-measurement, and furthermore it doesn't ⊗-separate in any obvious way. That isn't a proof that it doesn't separate, of course, but we will know that's the case if it gives us a working teleportation protocol. So, lets play 'fill in the boxes' for this teleportation diagram:

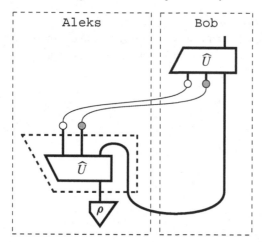

Since we already have a candidate for Aleks' measurement, we just need to find $\widehat{U}$, which is needed for constructing Bob's correction. This amounts to finding a solution to the following equation:

Bending up the wire on both sides yields:

so a valid solution is:

Hence, the correction is:

(9.59)

Exercise 9.58 Prove (9.59) is causal and a controlled unitary (up to a number).

Filling in the boxes yields the following candidate teleportation protocol:

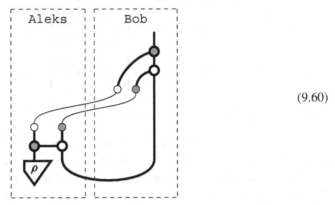

(9.60)

So it only remains to show that the thing actually works for our choice of $\widehat{U}$. Using the spider-fusion rules and complementarity, we can indeed show that (9.60) performs quantum teleportation. We will in particular make use of the following fact:

$$\approx \qquad \overset{(9.33)}{\approx} \qquad \approx$$ (9.61)

which uses bastard spider fusion and the <u>quantum</u> 'spider detachment' rule. Simply swapping colours we also have:

$$\approx$$ (9.62)

From this, we can conclude that this instance of teleportation indeed works:

$$= \quad = \quad \overset{(9.61)}{\approx} \quad = \quad \overset{(9.62)}{\approx}$$

Exercise 9.59 Show, in the case where ○ and ● represent the Z- and X-bases, that:

1. The Bell basis is:

(ignoring numbers) and hence (9.58) is a Bell-basis ONB measurement.

2. The Bell maps are:

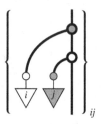

So (9.60) gives us a fully comprehensive diagrammatic presentation of the same quantum teleportation protocol we have been studying all along, but now with a general recipe for 'filling in' the boxes. All we need to do is find a complementary pair of spiders for any dimension, and the teleportation protocol (9.60) (along with the proof of correctness!) goes through unmodified.

The same ingredients allow us to do this for dense coding and for entanglement swapping. In the latter case, the required non-demolition measurement becomes:

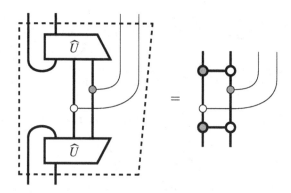

Exercise 9.60 Prove correctness for dense coding and entanglement swapping only using spider-fusion and spider-detachment rules.

As we have seen may times before, a single diagram can have more than one possible reading. This is also the case of diagram (9.58):

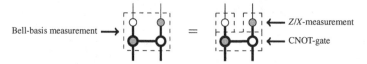

While for our purposes, i.e. unveiling quantum features, this distinction is not of any importance, for someone implementing quantum processes in a laboratory it may make a huge difference. For example, it may be really hard to make any non-separable measurement on two systems; it may be easier to perform a quantum CNOT-gate and perform single-system measurements, which turns out to be the case for most laboratory realisations of quantum systems.

Example 9.61 Quantum teleportation allows us to transfer the state from one system to another, using:

- an ancillary system
- a Bell state
- a quantum CNOT-gate
- two single-system measurements, and
- two single system corrections.

One may wonder if there is a manner of doing the same thing requiring fewer resources, and this is indeed the case. In fact, we only need:

- a ○-state
- a quantum CNOT-gate
- one single-system measurement, and
- one single system correction.

In particular, no ancillary system is needed. This is how this goes:

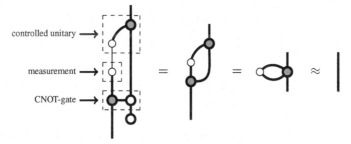

So instead of having two systems prepared in a Bell state, we now only need to have one system prepared in a 0-phase state. The price we pay is that, unlike teleportation, state transfer requires the source and the target system to be in the same place so a CNOT can be applied to both, since at the moment there's no such a thing as a 'non-local CNOT-gate'.

It is also insightful to compare this correctness proof with the one of teleportation; in particular, this proof is essentially the same as the two last steps of the teleportation proof.

9.3 Strong Complementarity

We've already made great progress in proving things by only using simple diagrammatic rules such as (phase) spider fusion and spider detachment. But how powerful are the rules we have so far? In other words, what sort of diagrams can we simplify just using these rules?

Suppose we have some reasonably complex diagram involving complementary spiders:

and we want to start simplifying it. We can of course apply spider rules to fuse any dots of the same colour:

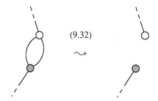

After that, we end up with a diagram that has cycles of spiders with alternating colours, which we can try to get rid of. These are of course always even, so we can ask how to deal with 2-cycles, 4-cycles, etc. For 2-cycles, we can just apply complementarity:

$$(9.32)$$

This allows us to reduce the complexity to the above example a bit:

$$(9.63)$$

However, now those pesky 4-cycles trip us up, since none of the rules we have so far will apply:

$$?$$

In rewriting terminology, having no rules that apply means you have arrived at a *normal form*, which is sometimes just a nice way of saying 'you're stuck'. What should we do? Call it a day? Alternatively, we can try to find the 'missing rule(s)' to carry on.

9.3.1 The Missing Rules

Let us focus for a moment on the Z and X spiders. In Example 9.45 we recalled that this pair of spiders satisfies a distinctive property; namely, the adjoint of the X-copying map behaves as an XOR operation on the Z basis:

$$\boxed{\text{XOR}} \approx$$

For one thing, this gives us another way to see why these two spiders are complementary, simply by thinking about classical bits. Namely, XOR-ing any bit with itself always gives the zero bit, so if we copy-then-XOR, that's the same as ignoring the input and outputting a zero:

$$\begin{array}{ccc} & & (9.46) \\ \approx & = & \approx \end{array}$$

In addition to this property, a (seemingly) unrelated consequence is that the grey dot defines a function map (see Definition 8.13) for the white basis, whose underlying function is of course XOR. In Proposition 8.19 from the last chapter, we gave a succinct characterisation of function maps in terms of copying and deleting. Applying this to XOR yields:

$$= \qquad\qquad =$$

and we already saw that the ○-spider with a single output is, up to a number, part of the ○-ONB, so:

$$\begin{array}{ccccc} & (9.46) & & & (9.46) \\ & \approx & = & & \approx \end{array}$$

Replacing the first two of these equations with X-spiders, we obtain:

$$\approx \tag{9.64}$$

It now becomes obvious that the role played by Z and X in these equations is interchangeable; that is, the above equations also hold with all of the colours reversed, just by taking the adjoint of each equation. The last two equations also imply one more relating the two single-legged spiders:

$$\approx \tag{9.66}$$

which just amounts to saying the number above is non-zero.

Exercise 9.62 Show that spiders satisfying (9.65) also satisfy (9.66).

While (9.64) and (9.65) seem like an 'accidental' property of Z and X, if we assume these equations for an arbitrary pair of spiders, we see that equations (9.64) and (9.65) imply complementarity.

Theorem 9.63 Equations (9.64) and (9.65) imply complementarity.

Proof By Proposition 9.35, it suffices to show the complementarity equation up to $\approx$. We have:

So, equations (9.64) and (9.65) yield very special pairs of complementary spider families, which suggests the following name.

Definition 9.64 A pair of spiders $\bigcirc$ and $\bullet$ are *strongly complementary* if they satisfy the following equations:

$$\tag{9.67}$$

$$\tag{9.68}$$

While they aren't too important, we have included the numbers in the above equations for the sake of completeness. In fact, as with complementarity, they are already fixed by the number-free versions.

Exercise 9.65 Prove the numbers in Definition 9.64 are uniquely fixed by (9.64) and (9.65), as we did for complementarity in Proposition 9.35. Note in particular that it is important here that $\approx$ means equivalence up to a positive number (cf. Remark 9.36).

So what does all of this have to do with our introductory chat about 4-cycles? Well, that's exactly what this is about, since the LHS of equation (9.67) is nothing but a 4-cycle with a twist:

Hence, we now know what to do with a 4-cycle, and pick up where we left off in (9.63):

$$\text{[diagram]} \quad = \quad \text{[diagram]} \quad \approx \quad \text{[diagram]} \quad = \quad \text{[diagram]} \quad \approx \quad \text{[diagram]}$$

Much better!

On the other hand, while we introduced complementarity by means of a very crisp canonical interpretation, there isn't really any compelling counterpart to this in the case of strong complementarity. However, despite the lack of a canonical interpretation, strong complementarity will have important consequences in the rest of this chapter (and book). We will look at some of these consequences in the next section. None of these holds for general (not strongly) complementary pairs of spider families.

9.3.2 Monogamy of Strong Complementarity

In Example 9.44, we remarked that for most dimensions, the maximum number of pairwise complementary spiders (a.k.a. mutually unbiased ONBs) is unknown. Since strong complementarity puts tighter constraints on which spiders can be related, this number should be smaller. In fact, we can show that strongly complementary spiders only come in pairs.

Theorem 9.66 At most two spiders can be pairwise strongly complementary. That is, for any non-trivial system if $\bigcirc/\!\!\text{◉}$ and $\bigcirc/\!\!\text{●}$ are both strongly complementary, then $\text{◉}/\!\!\text{●}$ cannot be strongly complementary.

Proof Suppose $\bigcirc/\!\!\text{◉}$ and $\bigcirc/\!\!\text{●}$ are both strongly complementary. Then:

$$\text{[diagram]} \approx \text{[diagram]} \qquad \text{and} \qquad \text{[diagram]} \approx \text{[diagram]}$$

So, by Theorem 8.18:

$$\text{[diagram]} \qquad \text{and} \qquad \text{[diagram]}$$

are both $\bigcirc$-ONB states, up to a number. Thus, they must be equal (up to a number) or orthogonal. Now, assume $\text{◉}/\!\!\text{●}$ are also strongly complementary, then by (9.66):

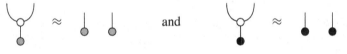

so:

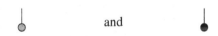

But then:

$$\cup \;=\; \bigvee \;\approx\; \bigvee \;\approx\; \bullet\; \bullet$$

which cannot be the case for any non-trivial system. Hence no non-trivial system can have three pairwise strongly complementary spiders. □

9.3.3 Faces of Strong Complementarity

Earlier we encountered a number of equivalent operational characterisations of complementarity: in terms of measuring after encoding in Definition 9.27, in terms of measuring two systems in a Bell state in Proposition 9.30, and in terms of CNOT-gates in Proposition 9.50. Since strong complementarity implies complementarity, all of these still hold for strongly complementary pairs, but obviously there will be many more consequences. In this section we go through the most important of these.

In Proposition 9.50 we learned that complementarity boils down to unitarity for 'generalised-CNOT' gates:

$$\boxminus \;\approx\; \| \; \|$$

or, equivalently, in terms of 'generalised-quantum-CNOT' gates:

$$\boxminus \;\approx\; \| \; \|$$

This equation can be seen as one between quantum circuits (cf. Examples 6.13 and 9.24). Strong complementarity can be put in a similar form.

Proposition 9.67 Strong complementarity is equivalent to:

$$\boxminus \;\approx\; \bowtie \tag{9.69}$$

$$\boxminus \;\approx\; \Big|\; \Big| \qquad\qquad \boxminus \;\approx\; \Big|\; \Big| \tag{9.70}$$

Doubling equation (9.69), we obtain:

$$\boxminus \;\approx\; \bowtie \tag{9.71}$$

Consequently, three CNOT-gates make a swap:

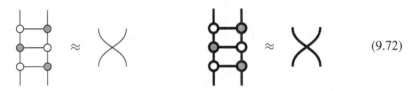

$$(9.72)$$

Exercise 9.68 Show that, when assuming complementarity, equations (9.69) and (9.71) are indeed equivalent to equations (9.72).

We initiated our search for strong complementarity in order to get rid of 4-cycles, so one might be tempted to think that strong complementarity is really only saying something specifically about 4-cycles. However, it turns out to imply equations between a much more general family of diagrams. As a simple example, we can combine the copying equations:

with spider fusion, to prove the *n*-ary version of these copying rules:

Proposition 9.69 Strong complementarity implies:

$$(9.73)$$

A more complicated example involves the following diagrams:

Definition 9.70 A *complete bipartite diagram* of spiders is a diagram with the property that every spider of one colour is connected to every spider of the other colour, and nothing else:

Even though this diagram seems to be highly connected, we can use strong complementarity to simplify it to two spiders connected by just a single wire.

Theorem 9.71 Strong complementarity of ○ and ● is equivalent to:

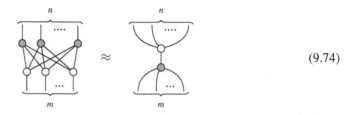

$$(9.74)$$

Proof We'll prove this using two inductions. First we'll show the special case where $n = 2$ and m is arbitrary:

$$(9.75)$$

by induction on m. For the base case, $m = 0$, this is simply:

For the step case, we will assume (9.75) for fixed m and prove it for $m + 1$:

Thus we have shown (9.75). From this, we can show (9.74) by induction on n. The base case is just an m-ary copy, like the ones proved in Proposition 9.69. For the step case, we assume (9.74) for fixed n, and show it for $n + 1$. The proof is almost the mirror image of the previous one, except we use (9.75), which is a 'beefed up' version of (9.64):

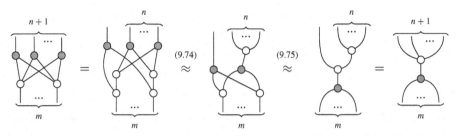

Conversely, the three strong complementarity rules from (9.64) and (9.65) all arise as special cases of (9.74):

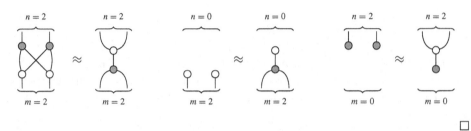

Moreover, restricting equation (9.74) to $m = 0$ yields one of the 'multi copy' laws from Proposition 9.69, when read from right to left:

$$\approx \qquad\qquad (9.76)$$

and similarly, restricting to $n = 0$ yields:

$$\approx \qquad\qquad (9.77)$$

Remark* 9.72 At first sight, it might seem like we are just pulling equation (9.74) out of a hat. Thankfully this is not the case, but rather comes from the fact that for any strongly complementary pair, these maps:

form an algebraic structure called a *bialgebra*, where 'bi' refers to the fact that both algebra and coalgebra are involved (cf. Remarks 3.17 and 8.22). The defining equation represents a somewhat funky commutation of the multiplication (i.e. algebra) and the comultiplication (i.e. coalgebra):

multiplication { ⬚ } comultiplication

comultiplication { ⬚ } multiplication

$$\approx$$

These structures are well understood, and it is possible to construct equations involving them using a technique called *path counting*. This is detailed in Section* 9.6.2.

Theorem 9.71 gives us a second equivalent presentation of strong complementarity. However, none of the presentations of strong complementarity given thus far resembles our initial definition of complementarity, which involved both classical and quantum systems:

$$\approx$$

Just by combining some doubled parts of (9.74) into cq-maps, strong complementarity provides us with more equations that help us reason about classical-quantum interaction. The following one will play a crucial role in our derivation of quantum non-locality in Section 11.1.

Corollary 9.73 Strong complementarity of ○ and ● implies:

$$\approx \qquad\qquad (9.78)$$

Proof By applying Theorem 9.71, we have:

$$= \qquad \overset{(9.74)}{\approx} \qquad =$$

where the first equality is simply undoubling, and the third one is redoubling. □

The ●-spider in the LHS of (9.78) is a quantum process, whereas the ●-spider in the RHS is seen here as an operation on classical data. We'll see in Section 9.3.5 what precisely this process does. In the meantime, we can interpret equation (9.78) as follows: if we first apply the given quantum ●-spider and ○-measure all of its outputs, this is the same as performing a ○-measurement and then the classical ●-spider. This of course lacks some of the conceptual crispness of (ordinary) complementarity but is nonetheless very useful.

Example 9.74 In Example 9.26 we saw how phases interact in a very surprising way within a GHZ state. However, attempting to perform a measurement of the same colour as the phases simply destroys the phases. On the other hand, if we perform a measurement of a colour that is strongly complementary to that of the phases, things work out quite differently:

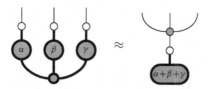

What we have now is that the three measurements are replaced by a single measurement, followed by some classical map. And since the phase is unbiased with respect to ⬤ and not ○, it does <u>not</u> vanish. Moreover, as we will show in Section 11.1, particularly clever choices of measurement phases α, β, and γ in:

will even leave hard evidence of this surprising 'backwards movement' of phase data and predict the existence of interactions between distant quantum systems that cannot be explained by mere probabilistic correlations, i.e. quantum non-locality.

Exercise 9.75 Show that strong complementarity implies:

Can you give an interpretation for this equation? Conversely, show that these equations imply strong complementarity (not so easy!):

9.3.4 The Classical Subgroup

In Section 9.2.2 we saw that for complementary pairs of spiders, the basis states of one colour live in the phase group (cf. Section 9.1.4) of the other colour. In this section we provide another characterisation of strong complementarity in terms of the special role basis states of one colour play in the phase group of the other colour. While the characterisation that we give in this section seems completely different from the one given in Definition 9.64, we will have to do remarkably little work to show it is equivalent.

We saw in Example 9.7 that X-basis states could be represented as phase states, with respect to the Z-basis:

and conversely, Z-basis states can be represented as X-phase states:

We also saw that we can compute their group-sums simply by means of spider fusion (cf. Theorem 9.20):

$$(9.79)$$

$$(9.80)$$

Notably, all these group sums result in one of the two phase states we started with. That is, the phase states that (up to a number) make up the Z-basis are *closed* under group-summation in the X-phase group. Moreover, they also contain the X-phase group unit and inverses of all of the Z-basis elements (since by (9.79), they are all self-inverse). Hence, they form a *subgroup* of the phase group. In fact, it is a standard result in group theory that a finite subset of a group that is closed under the group-sum is automatically a subgroup, so we actually don't even need to verify the unit and the inverses. Evidently, the same is also true if we exchange the roles of Z and X spiders.

Example 9.76 For the Z- and X-bases, the phase group is the circle group $U(1)$ and the classical subgroup is the two-element cyclic group $\mathbb{Z}_2$, which can be represented as no rotation and a half-rotation:

$$(9.81)$$

The appearance of this subgroup is a direct consequence of strong complementarity. In fact, it is equivalent to strong complementarity.

Theorem 9.77 Spiders ○ and ● are strongly complementary if and only if the basis states of ○ form a subgroup of the phase group of ●, and vice versa. Thinking of basis states as classical outcomes (cf. Section 8.1.1), we call this subgroup of a phase group the *classical subgroup*.

Proof Assuming ○ and ● are strongly complementary, we show that the ●-sum of two ○-basis states:

$$(9.82)$$

is again a ○-basis state. Recall from Theorem 8.18 that:

$$\psi \in \left\{ \vee_i \right\}_i \qquad \text{if and only if} \qquad = \overset{(*)}{\psi\,\psi}$$

We can show that (9.83) is a ○-basis state as follows:

$$\approx \quad = \quad = \quad$$

So the ○-basis is a finite subset of the phase group closed under the group-sum, and hence by standard group theory, it forms a subgroup.

The proof of the converse mimics the way we motivated the strong complementarity equations using the behaviour of XOR in Section 9.3.1. We repeat the argument here for the general case. Since ○-basis states form a subgroup of the ●-phase group, the ●-sum sends every pair of ○-basis states to another ○-basis state (up to a number):

$$\approx \quad \underset{k}{\vee} \qquad (9.83)$$

In particular, this means that the ●-sum is (up to a number) a function map. So by Proposition 8.19 this implies that:

$$\approx \qquad \qquad \approx$$

Finally, any subgroup contains the unit of the group, so the classical subgroup must contain the unit of the ●-phase group. Thus, the unit of the grey phase group is a ○-basis state, and hence:

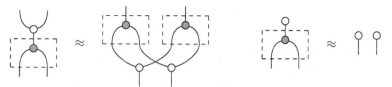

$$\approx$$

$\square$

Convention 9.78 We will use Greek letters κ, κ', etc. to denote phase states that are in the classical subgroup, whereas we stick to $\alpha, \beta, \gamma, \ldots$ for general phases. Hence the phase groups and their classical subgroups for the two colours are depicted respectively as:

$$\left\{ \vcenter{\hbox{○}} \right\}_{\vec{\kappa}} \subset \left\{ \vcenter{\hbox{○}} \right\}_{\vec{\alpha}}$$

and:

$$\left\{ \vcenter{\hbox{●}} \right\}_{\vec{\kappa}} \subset \left\{ \vcenter{\hbox{●}} \right\}_{\vec{\alpha}}$$

In particular, for a classical phase state the colour always indicates the spider for which it is a <u>phase</u>, as opposed to the colour of the spider that copies it.

From Theorem 9.77, we get another alternative characterisation of strong complementarity: the copiable states of one colour form a subgroup of the phase group of the other colour. Indeed, since basis states and copiable states are one and the same, the following corollary is an immediate result.

Corollary 9.79 For ○ and ● strongly complementary, and for:

$$\left\{ \vcenter{\hbox{●}} \right\}_{\vec{\kappa}} \qquad \text{and} \qquad \left\{ \vcenter{\hbox{○}} \right\}_{\vec{\kappa}}$$

the basis states for ○ and ●, respectively, we have:

$$\vcenter{\hbox{○}} \approx \bullet\bullet \qquad\qquad \vcenter{\hbox{●}} \approx \circ\circ \qquad\qquad (9.84)$$

We call this the *κ-copy rule*.

Example 9.80 In the particular case of qubits we have:

$$\vcenter{\hbox{○}} \approx \bullet\bullet \qquad\qquad (9.85)$$

This now allows us to give a diagrammatic derivation of the fact that:

$$\vcenter{\hbox{○—●}} \qquad\qquad (9.86)$$

is indeed the quantum CNOT-gate on ○ basis states. We have:

$$\vcenter{\hbox{○—●}} \approx \vcenter{\hbox{●—●}} = \bullet\bullet$$

hence:

So what is:

$$\text{(9.87)}$$

here? Since we have:

it follows that (9.87) is the *quantum NOT-gate* on ○ basis states, and so (9.86) is indeed the quantum CNOT-gate. The same argument with thin wires of course also does the trick for the classical CNOT-gate.

Equation (9.84) involves copy spiders and phase states. We can lift it to one involving copy spiders and phase maps.

Proposition 9.81 For ○ and ● strongly complementary, we have:

$$\text{(9.88)}$$

We call this the *κ-map-copy rule*.

Proof We have:

Note that this equation holds on the nose, because the doubled version of (9.67) introduces a factor of D, whereas the on-the-nose counterpart to (9.84) introduces a factor of $1/D$. □

Example 9.82 By (9.88) a NOT-gate can be pushed past a CNOT-gate:

Hence, in a circuit consisting only of NOT-gates and CNOT-gates, NOT-gates can all be pushed to one end of the circuit:

Equations (9.84) involve undecorated spiders of one colour and phase gates of the other colour. We can transform these into an equation just involving phase gates of two colours.

Proposition 9.83 For ○ and ● strongly complementary, we have:

$$
\tag{9.89}
$$

or equivalently, via doubling:

$$
\tag{9.90}
$$

We call this the κ-κ'-*commute rule*.

Proof First, note that:

$$
\tag{9.91}
$$

We rely on the undoubled version of (9.88), which can easily be shown to hold on the nose (i.e. with no global phase):

For the doubled version, we need to show that:

is a global phase, which is indeed the case:

$$\text{double}\left(\frac{1}{\sqrt{D}} \;\; \begin{matrix}\tilde{\kappa}\\ \tilde{\kappa}'\end{matrix} \right) = \frac{1}{D} \;\; \begin{matrix}\tilde{\kappa}\\ \tilde{\kappa}'\end{matrix} \;\; \overset{(9.40,\,9.42)}{=} \;\; D \;\; \begin{matrix}\triangle\\ \triangledown\end{matrix} \;\; \overset{(9.38)}{=} \;\; \boxed{}$$

$\square$

Exercise 9.84 Show that (9.64), (9.84), (9.88), (9.90), and (9.92) are all equivalent. (Hint and warning: for some of the proofs one will need to rely on the fact that when two maps agree on a basis, then they are equal. But, as explained in Remark 5.11, in this case numbers do matter!)

Corollary 9.85 Each of the equations (9.84), (9.88), (9.90), and (9.92) – when paired with the equations in (9.65) – provides an alternative definition for strong complementarity.

Remark* 9.86 Note that we chose to include equation (9.89) in undoubled form, which includes a non-trivial complex phase. The fact that the phase maps in (9.89) only commute up to a phase is a version of the *canonical commutation relations* (the *Weyl relations*, to be precise). The canonical commutation relations were the first tool that physicists used to probe complementarity in quantum theory.

In the proof of Proposition 9.83 we discovered another consequence of strong complementarity, namely that phase states in the classical subgroup are 'immune' to phase states of the other colour.

Proposition 9.87 For ○ and ● strongly complementary, we have:

$$\begin{matrix}\tilde{\alpha}\\ \tilde{\kappa}\end{matrix} = \begin{matrix}\\ \tilde{\kappa}\end{matrix} \qquad\qquad \begin{matrix}\tilde{\alpha}\\ \tilde{\kappa}\end{matrix} = \begin{matrix}\\ \tilde{\kappa}\end{matrix} \qquad (9.92)$$

We call this the *κ-eliminate rule*.

Proof Simply double and reflect equation (9.91). $\square$

Proposition 9.83 has a big brother, too, which states that phase maps in the classical subgroup can pass by phases of the other colour.

Exercise 9.88 Show that for $\bigcirc$ and $\bullet$ strongly complementary, we have:

$$
\begin{array}{ccc}
\overset{\vec{\alpha}}{\underset{\vec{\kappa}}{}} & = & \overset{\vec{\kappa}}{\underset{\vec{\kappa}(\vec{\alpha})}{}} \qquad\qquad
\overset{\vec{\alpha}}{\underset{\vec{\kappa}}{}} & = & \overset{\vec{\kappa}}{\underset{\vec{\kappa}(\vec{\alpha})}{}}
\end{array}
\tag{9.93}
$$

where:

$$
\widehat{\vec{\kappa}(\vec{\alpha})} := \overset{\vec{\kappa}}{\underset{\vec{\alpha}}{}} \qquad\qquad
\widehat{\vec{\kappa}(\vec{\alpha})} := \overset{\vec{\kappa}}{\underset{\vec{\alpha}}{}}
$$

are phase states for $\bullet$ and $\bigcirc$, respectively (as the notation suggests).

Exercise 9.89 For $\bigcirc$ and $\bullet$ strongly complementary, show that for any $\vec{\kappa}$, the function $\vec{\kappa}(-)$ defined in Exercise 9.88 is a *group homomorphism*, i.e.:

$$
\widehat{\vec{\kappa}(\vec{\alpha}+\vec{\beta})} = \widehat{\vec{\kappa}(\vec{\alpha})+\vec{\kappa}(\vec{\beta})} \qquad\qquad
\widehat{\vec{\kappa}(\vec{0})} = \widehat{\vec{0}}
$$

Furthermore, show that the map:

$$
\vec{\kappa} \mapsto \vec{\kappa}(-)
$$

is a *group action*, i.e.:

$$
\widehat{(\vec{\kappa} + \vec{\kappa}')(\vec{\alpha})} = \widehat{\vec{\kappa}(\vec{\kappa}'(\vec{\alpha}))} \qquad\qquad
\widehat{\vec{0}(\vec{\alpha})} = \widehat{\vec{\alpha}}
$$

Example 9.90 In the case of two dimensions, the classical subgroup has two elements, 0 and π. From Exercise 9.89, we know 0 acts trivially. We saw in Section 9.3.4 that π acts as a NOT-gate for $\bigcirc$-basis elements:

$$
\boxed{\pi} \;::\; \overset{}{\underset{0}{\triangledown}} \mapsto \overset{}{\underset{1}{\triangledown}} \;,\; \overset{}{\underset{1}{\triangledown}} \mapsto \overset{}{\underset{0}{\triangledown}}
$$

Hence, for $\bigcirc$-phase states:

$$
\boxed{\pi} \;::\; \begin{pmatrix} 1 \\ e^{i\alpha} \end{pmatrix} \mapsto \begin{pmatrix} e^{i\alpha} \\ 1 \end{pmatrix} = e^{i\alpha}\begin{pmatrix} 1 \\ e^{-i\alpha} \end{pmatrix}
$$

Thus, after doubling, π flips the phase, i.e.:

$$
\overset{\pi}{\underset{\alpha}{}} = \widehat{-\alpha} \qquad\qquad
\overset{\pi}{\underset{\alpha}{}} = \widehat{-\alpha}
\tag{9.94}
$$

where the second equation comes from interchanging the roles of $\bigcirc$ and $\bullet$.

9.3.5 Parity Maps from Spiders

Spiders were initially introduced in the previous chapter as classical data operations. We will see in this section that strong complementarity gives us many more classical data operations.

In Proposition 8.19 we encountered a characterisation of function maps as linear maps f satisfying:

In Section 9.3.1 we used these equations to motivate strong complementarity equations when taking f to be XOR:

which arises (up to a number) as an X-spider (cf. Proposition 5.88). It is easily seen that this connection to function maps applies for any strongly complementary pair of spiders.

Proposition 9.91 For $\bigcirc$ and $\bullet$ strongly complementary:

is a function map (up to a number) for $\bigcirc$.

In fact, recalling that function maps are particular examples of the more general classical maps (cf Section 8.2.1), the previous proposition generalises to all $\bullet$-spiders.

Proposition 9.92 For $\bigcirc$ and $\bullet$ strongly complementary:

is a classical process (up to a number) for $\bigcirc$.

Proof We have:

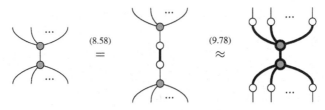

The rightmost diagram consists just of ◯-measure, ◯-encode, and pure quantum maps, so by Definition 8.3, it is a cq-map, and in particular it is a classical map (i.e. a cq-map with no quantum inputs/outputs). Hence the ●-spider is equal up to a number to a cq-map. Since this number is positive (cf. Remark 9.36) the ●-spider is itself a cq-map. Causality follows from the 'generalised copying' equation for strongly complementary spiders:

We can get a better idea of what these new classical processes do by looking at what they do on basis states and effects in two dimensions. For this it is helpful to treat basis states/effects as elements in the classical subgroup:

for $\kappa \in \{0, \pi\}$, so:

In order to figure out which of these are non-zero, we need to know which 'phase numbers' (i.e. phase spiders with no legs) taken from the classical subgroup are non-zero.

Lemma 9.93 For ◯ and ● strongly complementary:

$$\boxed{\kappa} \neq 0 \iff \kappa = 0$$

where the first 0 stands for the zero number while the second 0 stands for the unit of the phase group.

Proof Let the (classical) ●-phase state with phase 0 be equal, up to a number, to the first ◯-ONB state (which is also labelled '0' below). Then we have:

By orthonormality this number is only non-zero if $i = 0$, i.e. $\kappa = 0$. □

So, a matrix entry is non-zero if and only if we have:

$$\sum_i \kappa_i + \sum_i \kappa'_i = 0$$

In that case, it is equal to a fixed positive number p.

Exercise 9.94 What is the value of p, i.e. the number:

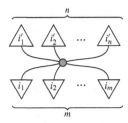

in terms of m and n?

For two dimensions, the classical subgroup is $\mathbb{Z}_2$ (cf. Example 9.76). In that case, the group-sum is equal to 0 if and only if there are an even number of 1s, so:

$$\approx \sum_{i_1 \ldots i_m i'_1 \ldots i'_n} \oplus (i_1 \ldots i_m i'_1 \ldots i'_n)$$

where $\oplus$ is the *even-parity function*, i.e.:

$$\oplus (i_1 \ldots i_m i'_1 \ldots i'_n) := \begin{cases} 1 & \text{if number of 1s is \underline{even}} \\ 0 & \text{if number of 1s is \underline{odd}} \end{cases}$$

So only those terms with an even number of 1-states occur in the sum. One example is the *even-parity state*:

$$\qquad\qquad (9.96)$$

Another example of this is the *parity map*, which returns the 0-state if the number of 1-states is even, and the 1-state if it is odd:

$$\qquad\qquad (9.97)$$

If we decorate $\bullet$-spiders with a π, the parity is reversed, so we obtain:

$$\approx \sum_{i_1 \ldots i_m i'_1 \ldots i'_n} \overline{\oplus} (i_1 \ldots i_m i'_1 \ldots i'_n)$$

where $\overline{\oplus}$ is the *odd-party function*, i.e.:

$$\overline{\oplus} (i_1 \ldots i_m i'_1 \ldots i'_n) := \begin{cases} 1 & \text{if number of 1s is \underline{odd}} \\ 0 & \text{if number of 1s is \underline{even}} \end{cases}$$

The (classical) NOT-gate is a special case:

since each term has precisely one 1. Another example is the *odd-parity state*:

$$\tag{9.98}$$

Example 9.95 In the case of three systems we have:

These two special cases, along with the three-system parity function:

will play an important role in derivation of quantum non-locality.

In addition to the new classical maps we can build out of ●-spiders, we can now build even more by combining ○-spiders and ●-spiders, for example, the classical CNOT-gate:

9.3.6 Classifying Strong Complementarity

In Example 9.44 we stressed how little is actually known about classifying complementary measurements. In fact establishing how many pairwise complementary measurements there are just in dimension 6 remains an open question (or, more accurately, a black hole that sucks in quantum information scientists).

So what about strong complementarity? What do we actually know? The answer is as satisfying as it ever gets: everything! We already know from Section 9.3.2 that sets of pairwise strongly complementary spiders are always of size 2. Hence, it only remains to classify these pairs. Since strong complementarity is really about the relationship between two families of spiders, we can furthermore assume that one of the families is fixed. Hence, classifying strongly complementary pairs amounts to answering the following question:

> *For a fixed family of spiders* ○*, can we classify all*
> *spiders* ● *that are strongly complementary to* ○ *?*

To attack this question, we should think about how much data we need to uniquely fix ●. But in fact, we have already answered this question! In (9.95), we saw that the ●-spiders, considered as parity maps, are all entirely fixed by the group-sum, which in the case of $\mathbb{Z}_2$ is XOR (hence 'parity'). Furthermore, for any commutative group G of size D, we can fix a D-dimensional system and label the ○-basis states by elements $g \in G$:

$$\left\{ \; \overset{|}{\underset{g}{\bigvee}} \; \right\}_{g \in G}$$

Then, the *generalised parity maps*:

$$
\begin{array}{c}
\overset{\triangle g_1' \; \triangle g_2' \; \cdots \; \triangle g_n'}{} \\
\underset{\triangledown g_1 \; \triangledown g_2 \; \cdots \; \triangledown g_m}{}
\end{array}
:= \begin{cases} \left(\dfrac{1}{\sqrt{D}} \right)^{m+n-2} & \text{if } \sum g_j = \sum g_j' \\ 0 & \text{otherwise} \end{cases} \qquad (9.99)
$$

(where '$\sum$' is the group-sum from G) yield a family of spiders. Furthermore, ○ and ● will always be strongly complementary! Hence:

> *Strongly complementary pairs of spiders are classified by commutative groups.*

So, why did we say we 'know everything' about classifying strongly complementary spiders? Well, because we know everything about classifying finite commutative groups, of course!

The simplest commutative groups are *cyclic groups* $\mathbb{Z}_k$, whose elements are $\{0, \ldots, k-1\}$, with group-sum given by addition modulo k. Then every other finite commutative group can be expressed uniquely (up to isomorphism) as a product of cyclic groups:

$$\mathbb{Z}_{k_1} \times \mathbb{Z}_{k_2} \times \cdots \times \mathbb{Z}_{k_n}$$

where each $k_i = p_i^{n_i}$ for some prime number p_i and some integer n_i; i.e. each of the k_i are *prime powers*. Using this characterisation, we know exactly how to build all of the strongly complementary pairs in every dimension.

Example 9.96 In dimension 2, there is only the Z/X pair, which corresponds to the 'parity' cyclic group $\mathbb{Z}_2$, whereas in dimension 36, there are four different strongly complementary pairs, corresponding to each of the ways to factor 36 into prime powers:

$$\mathbb{Z}_2 \times \mathbb{Z}_2 \times \mathbb{Z}_3 \times \mathbb{Z}_3 \qquad \mathbb{Z}_4 \times \mathbb{Z}_3 \times \mathbb{Z}_3 \qquad \mathbb{Z}_2 \times \mathbb{Z}_2 \times \mathbb{Z}_9 \qquad \mathbb{Z}_4 \times \mathbb{Z}_9$$

Okay, great, strongly complementary pairs are totally classified. Is this useful? Yes! Equation (9.99) implies in particular that:

where $g + h$ and 0 are the group-sum and unit in G, respectively. Hence G arises as the classical subgroup associated with the strongly complementary pair $\bigcirc/\bullet$. That is, we have a set of $\bullet$-phases:

$$\left\{ \begin{array}{c} \triangledown_g \end{array} \right\}_{g \in G} \cong \left\{ \begin{array}{c} \kappa_g \end{array} \right\}_{g \in G}$$

which is classical for $\bigcirc$ (up to a number):

and encodes G via $\bullet$:

$$ \overset{\kappa_g \quad \kappa_h}{\bullet} = \begin{array}{c} \kappa_{g+h} \end{array} \qquad\qquad \overset{}{\underset{\bullet}{\mid}} = \kappa_0 $$

And since strong complementarity is symmetric with respect to $\bigcirc/\bullet$, we also get an encoding of G in the classical subgroup of $\bigcirc$-phases:

$$ \overset{\kappa_g}{\bigcirc} \approx \kappa_g \quad \kappa_g \qquad\qquad \overset{\kappa_g \quad \kappa_h}{\bigcirc} = \kappa_{g+h} $$

So we can totally encode this group (in two ways) using a strongly complementary pair of spiders. Now, if we want to study this group (or, better, build some quantum processes that study the group for us!), we can use this pair of spiders. This is in fact precisely what we'll do in Section 12.2.4, when we provide a quantum algorithm for solving the *hidden subgroup problem*.

Remark 9.97 A careful reader might have noticed from the definition of $\bullet$ in (9.99) that the usual spider equation:

$$ \smile_{\bullet} = \smile $$

implies that $g = -g$ for all $g \in G$! This of course does not hold for all commutative groups, but holds only for those of the form:

$$\underbrace{\mathbb{Z}_2 \times \mathbb{Z}_2 \times \cdots \times \mathbb{Z}_2}_{N}$$

When $g \neq -g$, we still get spiders, but not necessarily ones corresponding to self-conjugate ONBs (cf. Section* 8.6.3). In this case, the linear map:

will be equal not to the plan wire, but rather to the function map that sends each group element to its inverse. This is discussed in detail in Section* 9.6.1.

9.4 ZX-Calculus

In this section, we will specialise the diagrammatic creatures and their interactions developed earlier in this chapter to the particular case of qubits. The two relevant questions in this context are the following:

1. Which cq-maps can we express using just phase Z- and X-spiders? In particular, can we express all of them?
2. Which equations between cq-maps can we prove using a *graphical calculus*, i.e. a fixed set of diagrammatic equations, picked from those we've already seen and maybe some new ones?

The answer to the first question is a resounding Yes! More specifically, we can build <u>any</u> linear map from m copies of $\mathbb{C}^2$ to n copies of $\mathbb{C}^2$ just using phase Z- and X-spiders. Hence, in particular, we can build any quantum map on qubits just by doubling, and, since we have spiders around, we can also build any cq-map on bits and qubits.

The answer to the second question, somewhat embarrassingly, is that we aren't really sure yet. But then again, it turns out to be an extremely hard question. However, if we restrict our phases to multiples of $\frac{\pi}{2}$, we obtain an important subtheory of **pure quantum maps** called **Clifford maps**. As we will show in Chapter (11), this subtheory of **Clifford maps** already provides enough quantum maps to prove that quantum theory is non-local.

For this process theory just four equations (or technically, four families of equations) are sufficient to prove everything! The first two equations tell us how spiders of the same colour <u>combine</u>. These are of course just the (undoubled) spider-fusion rules we are now very familiar with:

The third equation tells us how spiders of different colours can <u>commute</u> past each other:

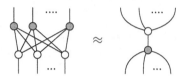

which via Theorem 9.71 is equivalent to strong complementarity. The fourth equation is new. It tells us how to <u>convert</u> spiders of one colour into spiders of the other colour:

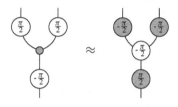

While the other equations are generic to strongly complementary pairs in any dimension, this equation is really saying something special about qubits. Indeed, we will see in the next section that this rule is intimately connected to the geometry of the Bloch sphere.

Interestingly, the keys to answering both of the questions above, which are all about quantum picturalism, come from the literature on quantum computation! Indeed, the first question turns out to be related to the quantum gate sets required to build a universal computing device, while for the second question the proof of completeness draws from results in measurement-based quantum computation.

Remark 9.98 We are working almost exclusively with single wires in this section. This gives us our most general-purpose rules, as special cases involving quantum and bastard spiders can all be obtained by folding/unfolding quantum wires. Therefore, we will regularly use undoubled versions (thanks to Corollary 6.18) of some of the doubled equations that we established earlier in this chapter. Note that, when there can be no confusion, we still refer to the single wires resulting from undoubling as qubits.

9.4.1 ZX-Diagrams Are Universal

In specialising to qubits, we define a new type of diagram.

Definition 9.99 A *ZX-diagram* is a string diagram consisting of just phase Z- and X-spiders:

$$
\vcenter{\hbox{\includegraphics{}}} := \vcenter{\hbox{\includegraphics{}}} + e^{i\alpha} \vcenter{\hbox{\includegraphics{}}} \tag{9.100}
$$

(9.101)

So for ZX-diagrams, we don't allow arbitrary processes, but rather just those built out of these two kinds of phase spiders. However, rather than thinking of this as removing all of the other boxes from our language, we can instead think of this as 'filling in the boxes':

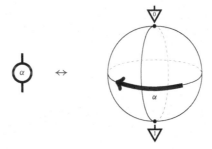

ZX-diagrams are significantly more expressive than plain string diagrams or dot diagrams, given that we have phases and two colours of spiders. In this section, we will see that this actually suffices to build any pure quantum map from qubits to qubits. If we additionally add discarding or (more generally) bastard spiders, we can build, respectively, any quantum map or any cq-map on (qu)bits.

First we show how ZX-diagrams can be used to construct arbitrary single-qubit unitaries. Recall that qubit unitaries correspond to rotations of the Bloch sphere. We already have two quite useful families of rotations: the Z-phase gates, which provide Z-axis rotations:

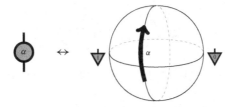

and the X-phase gates, which provide X-axis rotations:

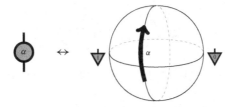

In fact, it is a standard property about spheres that any rotation can be decomposed as three rotations about a pair of orthogonal axes. Applying this to unitary quantum maps yields the following result.

Proposition 9.100 For any unitary quantum map $\widehat{U}$ on a single qubit there exist phases α, β, and γ such that $\widehat{U}$ can be written as:

This is called the *Euler decomposition* of $\widehat{U}$, and the phases α, β, and γ are called the *Euler angles*.

Since we can perform any unitary, it is possible to obtain any single-qubit state by just starting with some fixed state and transforming it into the state $\widehat{\psi}$ we want, now expressed as a ZX-diagram, e.g.:

By undoubling, we can see that any state in **linear maps** of type $\mathbb{C}^2$ can be expressed, up to a number (namely, a global phase), as a ZX-diagram. In fact, this generalises to many-qubit states.

Proposition 9.101 Any state in **linear maps** on n copies of $\mathbb{C}^2$ can be expressed, up to a number, as a ZX-diagram.

We will hold off on proving this theorem until Section 12.1.3 in the chapter on quantum computing, where we will borrow some results from the quantum circuits literature. More specifically, in Section 12.1.3 we will show that ZX-diagrams can be used to construct any unitary $\widehat{U}$ on n qubits, and just like above, we can use this fact to obtain any n-qubit quantum state:

Undoubling this gives us any state, up to a number (cf. Corollary 6.18):

Then, just by applying process–state duality to Proposition 9.101, we can also construct any linear map whose input/output wires all are of type $\mathbb{C}^2$:

Clearly if ψ is a ZX-diagram, then so too is f. Moreover, we can get all of the complex numbers back as ZX-diagrams.

Proposition 9.102 Any complex number can be expressed as a ZX-diagram of the form:

for some α, β, and k.

Proof First, note that we can obtain $\sqrt{2}$ times any complex phase via:

$$\text{(9.46)} \quad = \quad \sqrt{2} \quad \text{(9.11)} \quad = \quad \sqrt{2}\, e^{i\alpha}$$

So, it suffices to show that we can express any positive real number. First:

$$\text{(-β)} \; \text{(β)} \quad \overset{\text{(9.100)}}{=} \quad (1 + e^{i\beta})(1 + e^{-i\beta}) \; = \; 1 + e^{i\beta} + e^{-i\beta} + 1$$

Using the fact that $e^{i\beta} = \cos\beta + i\sin\beta$ (cf. Section 5.3.1), this reduces to:

$$\text{(-β)} \; \text{(β)} \; = \; 2\,(1 + \cos\beta)$$

which can be any real number between 0 and 2. To get larger numbers, we simply need to add more dots:

$$\text{(-β)} \; \text{(β)} \underbrace{\text{o} \cdots \text{o}}_{k} \; = \; 2^{k+1}(1 + \cos\beta)$$

Thus, for any positive real number r, we can fix some k such that $2^{k+1} \geq r$, then choose β accordingly. $\qquad\square$

Hence we can conclude the following.

Theorem 9.103 Any linear map whose input/output wires all are of type $\mathbb{C}^2$ can be expressed as a ZX-diagram:

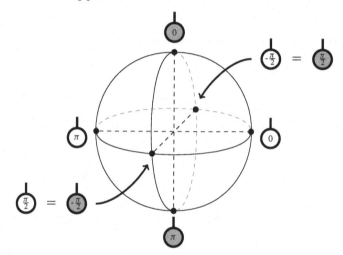

9.4.2 ZX-Calculus for Clifford Diagrams

Instead of considering the entire qubit, we can also construct lots of interesting states and processes (indeed most of those we've encountered so far!) by restricting to just six representative states on the Bloch sphere, namely, the Z-, X-, and Y-basis states, which correspond to the following phase states:

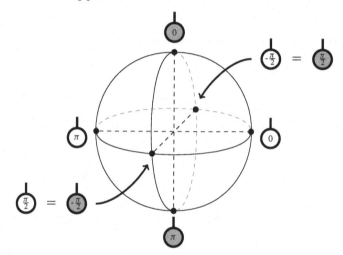

The phase groups are therefore also reduced to the four-element subgroup $\mathbb{Z}_4$ of $U(1)$:

Note that each of these states is representable as a Z- or X-phase state whose phase is a multiple of $\frac{\pi}{2}$. We can of course consider all such ZX-diagrams.

Definition 9.104 A *Clifford diagram* is a ZX-diagram where the phases are restricted to integer multiples of $\frac{\pi}{2}$.

In turn, these diagrams define a new process theory.

Definition 9.105 Let **Clifford maps**, be the subtheory of **pure quantum maps** obtained by doubling those linear maps that are expressible as Clifford diagrams.

As we claimed before, Clifford diagrams admit a graphical calculus for which **Clifford maps** are complete; that is, there exists a set of equations such that, whenever two Clifford maps are equal (up to a number), we can apply the equations of the graphical calculus to rewrite the diagram of one into the other.

In fact, the phase spider-fusion rules and strong complementarity already get us most of the way there. However, they cannot get us all the way there, since these rules are generic for all quantum systems, not just qubits. Hence, there has to be at least one more rule that clearly tells us that we are dealing with qubits. This final missing ingredient comes from the fact that, even though we only have Z-spiders and X-spiders in a ZX-diagram, somehow the Y-basis, and hence Y-spiders, are also hiding in there as well, by means of phases. Is there any way we can bring them out?

It turns out that just thinking about how to build a Y-copy spider will be enough to unlock the full power of the ZX-calculus for Clifford diagrams. In single form, the Y-basis can be written as follows:

$$
\begin{cases}
\sqrt{2}\;\blacktriangledown_{\!0} \;=\; \bigcirc_{\frac{\pi}{2}} \;=\; e^{i\frac{\pi}{4}}\;\bigcirc_{-\frac{\pi}{2}} \\[2mm]
\sqrt{2}\;\blacktriangledown_{\!1} \;=\; \bigcirc_{-\frac{\pi}{2}} \;=\; e^{-i\frac{\pi}{4}}\;\bigcirc_{\frac{\pi}{2}}
\end{cases}
\tag{9.102}
$$

(We include the complex phases explicitly, as they'll play a role shortly.)

Exercise 9.106 Using the concrete definitions of the Y-basis and the phase Z- and X-spiders, prove the equations in (9.102) and, in particular, the complex phases are as shown.

The two ways of expressing the Y-basis tell us two ways to copy it. First, a $-\frac{\pi}{2}$ rotation around the Z-axis will send the Y-basis to the X-basis:

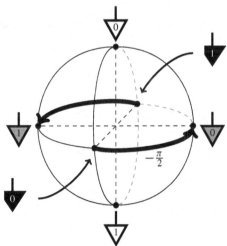

And a $\frac{\pi}{2}$ rotation will of course send the X-basis back to the Y-basis. Hence, one way to copy the Y-basis is:

send X-basis to Y-basis

copy X-basis

send Y-basis to X-basis

We can see this copying in action by plugging in the two Y-basis states, written as ○-phase states according to (9.102):

$$= \quad \overset{(9.68)}{=} \quad \frac{1}{\sqrt{2}} \quad = \quad \frac{1}{\sqrt{2}}$$

$$= \quad \overset{(9.85)}{=} \quad \frac{1}{\sqrt{2}} \quad = \quad \frac{1}{\sqrt{2}}$$

We can almost do the analogous thing with the colours reversed for ○-phases, but we need to correct the difference in the phases between the two ONB states. From (9.102), we can see these phases are $e^{i\frac{\pi}{4}}$ and $e^{-i\frac{\pi}{4}}$, respectively, so the overall difference is $\frac{\pi}{2}$. We can account for this by incorporating a $-\frac{\pi}{2}$ phase into the ○-spider:

send Z-basis to Y-basis

copy Z-basis (and fix the phase)

send Y-basis to Z-basis

Exercise 9.107 Prove by evaluating on Y-basis states that:

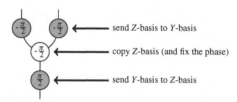

for some fixed global phase $e^{i\alpha}$.

Hence:

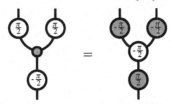

$$\approx \tag{9.103}$$

or, more accurately (since the LHS and RHS differ only by a global phase):

Adding this *Y-rule* to what we already know about strongly complementary pairs of spiders yields the following definition.

Definition 9.108 The *ZX-calculus for Clifford diagrams* consists of the following four rules:

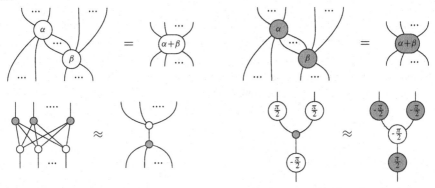

A first thing that may surprise some readers is the fact that the fourth rule, in contrast to all of the other rules we have encountered so far, is not symmetric in ○ and ●. This asymmetry is only an artefact of going for a minimal set of rules, and in the following section we will show that the ZX-calculus is indeed colour-symmetric. That is, any rule we prove can also be proven with the colours reversed. Clearly, the colour-reverse of the first three rules holds, so this is equivalent to showing that the colour-reverse of the *Y*-rule:

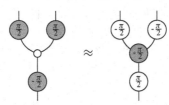

holds in the ZX-calculus.

Of course, from the previous sections we already know many other things about ZX-calculus. For example, since the third rule is equivalent to strong complementarity via Theorem 9.71, we know that the initial strong complementarity equations follow:

$$\approx \tag{9.104}$$

as well as the complementarity equation:

$$\approx \quad \leftarrow \text{disconnected} \tag{9.106}$$

These special cases also show explicitly that the ZX-calculus contains some equations that make diagrams disconnect. These equations are the most important ones, because ZX-diagrams, like all of the simpler kinds of diagrams we encountered before, are still most fundamentally about 'what is connected to what'.

As with strong complementarity, there isn't just one way to define the ZX-calculus, but many equivalent ways. In Section 9.4.4 below we will build a version of ZX-calculus with a substantially different fourth rule. In fact, the version we presented above has never been given before. However, it is to our knowledge the smallest possible set of rules. Moreover, it is very easy to remember, since each of the rules tells us one particular thing about spiders, namely:

- how spiders of the same colour <u>combine</u>
- how spiders of different colours can <u>commute</u> past each other
- how to <u>convert</u> spiders of one colour into another.

In Section 9.3.4, we derived a number of rules involving phases in the classical subgroup, which in the case of qubits are 0 and π. When we first derived those rules we took as given, simply by staring at the Bloch sphere, that π-phase states for one colour are basis states for the other colour (up to a number). However, to stay true to our conviction to use <u>only</u> the graphical calculus to prove everything, we should really derive these just using the four defining rules of ZX-calculus. This is indeed possible, as we will demonstrate in the next section.

In addition to the 'big four' rules of the ZX-calculus, we also need a few little rules for eliminating non-zero numbers:

$$\bigcirc \; = \; \bullet \; = \; \underset{\alpha}{\bullet} \; = \; \underset{\alpha}{\bullet} \; \approx \; \boxed{} \tag{9.107}$$

Using a technique similar to Exercise 9.62, we can indeed show that these numbers cannot be zero, unless every spider is already zero.

9.4.3 ZX for Dodos: Just Diagrams, Nothing Else

Now that we have fixed the ZX-calculus, our goal is to prove everything in the rest of this chapter just using those rules and nothing else. This will give a clear idea of what the ZX-calculus looks like 'in action'. First, we'll start with a little warmup.

Proposition 9.109 The ZX-calculus obeys:

$$\underset{\alpha}{} \approx \quad \qquad \underset{\alpha}{} \approx \quad \tag{9.108}$$

Proof The proof goes mostly just like (9.91), with a little 'non-zero' rule at the end:

$$\underset{\alpha}{} = \underset{\alpha}{} \overset{(9.105)}{\approx} \overset{(9.107)}{\approx} $$

The colour-reversed version is similar. As the use of phase spider fusion is usually self-evident, we won't explicitly indicate its use. □

Next we show that equations (9.102), which we used as a starting point to build the fourth rule, now pop out.

Proposition 9.110 The ZX-calculus obeys:

$$\underset{-\frac{\pi}{2}}{} \approx \underset{\frac{\pi}{2}}{} \qquad \qquad \underset{\frac{\pi}{2}}{} \approx \underset{-\frac{\pi}{2}}{} \tag{9.109}$$

Proof After taking the partial trace of both sides of the *Y*-rule:

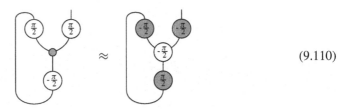

$$\tag{9.110}$$

the LHS reduces to:

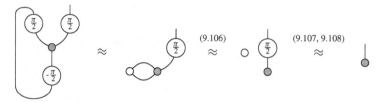

and the RHS to:

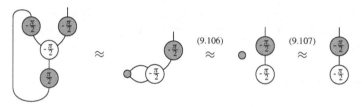

Then, applying a ○ $\frac{\pi}{2}$-phase to both sides yields the first equation in (9.109):

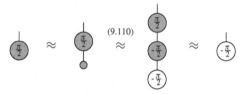

The second equation can then be obtained by conjugating both sides, which flips the signs.   □

We now show that the rules involving π-phases from Section 9.3.4 can all be derived in the ZX-calculus (as long as we restrict phases to multiples of $\frac{\pi}{2}$). Most importantly, we need to show that π-phase states are indeed basis states. We start with the π-commute rule.

Proposition 9.111 The ZX-calculus obeys the π-commute rule:

(9.111)

for $\alpha \in \left\{0, \frac{\pi}{2}, \pi, -\frac{\pi}{2}\right\}$.

Proof By evaluating the LHS of the Y-rule on the ○-state we get:

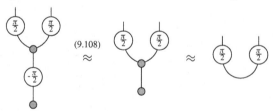

and doing the same with the RHS we get:

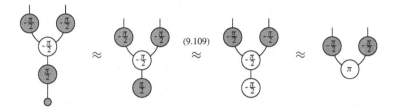

so we conclude:

$$\approx \qquad (9.112)$$

Unbending the wire yields:

$$\approx \qquad \overset{(9.112)}{\approx}$$

Then, applying a ⬤ $\frac{\pi}{2}$-phase gate to the output gives:

$$\approx \qquad (9.113)$$

i.e. (9.111) for $\alpha := \frac{\pi}{2}$. For the other angles, we can simply decompose as a series of $\frac{\pi}{2}$ gates; e.g. for $\alpha := \pi$ we have:

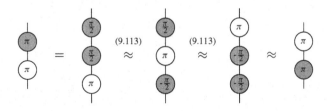

□

Exercise 9.112 Show that, if instead of on a ⬤-state, we evaluate on a ◯-state at the beginning of the proof above, the resulting equation is:

$$(9.114)$$

These rules give us enough to show that the ZX-calculus is colour-symmetric.

Theorem 9.113 The ZX-calculus obeys:

$$(9.115)$$

Hence any equation provable in the ZX-calculus also holds with the colours reversed.

Proof   Applying a series of phase gates to the LHS of the *Y*-rule gives:

Applying the same phases to the RHS gives:

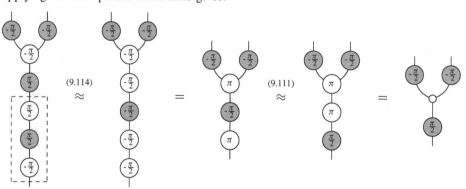

where for the first step we conjugated both sides of (9.114) to flip the signs. Hence we obtain:

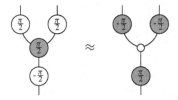

and conjugating both sides completes the proof. □

Exercise 9.114 Use the π-commute rule to prove the π-eliminate rule:

$$\approx \qquad (9.116)$$

for $\alpha \in \left\{0, \frac{\pi}{2}, \pi, -\frac{\pi}{2}\right\}$. Then prove its colour-reverse:

$$\approx \qquad (9.117)$$

The following rule is one that we have not seen yet, but it's an important one that also disconnects diagrams.

Proposition 9.115 The ZX-calculus obeys $\frac{\pi}{2}$-*supplementarity*:

$$\approx \qquad (9.118)$$

Proof First, we have:

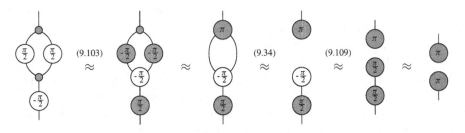

Then, applying a ○ $\frac{\pi}{2}$-phase gate gives:

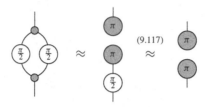

$$(9.117)$$

$\square$

Remark 9.116 The use of the term 'supplementarity' above refers to the fact that $\left(\frac{\pi}{2}, \frac{\pi}{2}\right)$ is a pair of *supplementary* angles, i.e. angles adding up to π (a.k.a. 180°). We'll see a generalisation of this rule in Section 9.4.6.

Finally, we reach our goal of showing that the ● π-phase state is a basis state for ○ (and vice versa via colour-reversal) using just the ZX-calculus.

Proposition 9.117 The ZX-calculus obeys the π-*copy* rule:

$$\text{(diagram)} \approx \text{(diagram)} \qquad\qquad (9.119)$$

Proof We have:

$$\text{(diagram)} \underset{(9.109)}{\approx} \text{(diagram)} \underset{(9.104)}{\approx} \text{(diagram)} \approx \text{(diagram)} \underset{(9.118)}{\approx} \text{(diagram)}$$

where you might need to stare at the fourth step for a bit to realise it's just an application of ○-phase spider fusion. $\square$

9.4.4 ZX for Pros: Build Your Own Calculus

We'll soon see that the four rules from Definition 9.108 suffice to prove everything for Clifford diagrams. However, that doesn't necessarily mean they are the most convenient set of rules for someone to work with. For example, a person who is used to more traditional algebraic structures will find the many-input many-output rules very awkward and may find the Frobenius algebra rules that we discussed in Section* 8.6.1 a lot more appealing than spider fusion. (You: Are you joking? Us: No, we're not.) That same person would also find the three defining equations of strong complementarity more appealing than the single rule in which we packaged them. Besides being corrupted by traditional algebra, one might find it more appealing simply to have a primitive rule for removing 4-cycles:

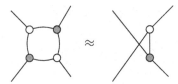

as we used to motivate strong complementarity in the first place. In a similar vein, one might wish to treat complementarity, and not strong complementarity, as a primitive:

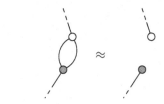

and come up with some new combination of rules that happens to imply strong complementarity. Why not? Everyone is different.

In this section, we'll derive an equivalent version of the ZX-calculus where the *Y*-rule:

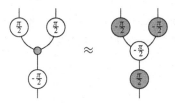

is replaced by something that is symmetric in the two colours. In the previous section, we did quite a bit of work to show that if any rule holds in the ZX-calculus, then so too does its colour-reversed version. Our new rule will show this colour-symmetry very explicitly by means of an explicit 'colour changer' that we construct out of phase gates.

First, with a bit of phase spider fusion, we have:

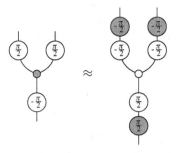

Then, by applying some $\frac{\pi}{2}$ phase gates to both sides, we can get rid of all the minus signs:

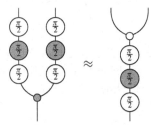

which at first seems like we've made everything worse. But then, if we let:

$$(9.120)$$

we get:

$$(9.121)$$

So we obtain a simple rule that tells us that the little white box passes through copy spiders and changes their colour. Aha! We seem to have found ourselves a candidate colour-changer!

But what is this mysterious box? One way to figure this out is by using some geometry. Let's have a look at the corresponding quantum map:

As rotations of the Bloch sphere, this gives:

Those who are particularly spacially gifted will see that this amounts to the following 180° rotation:

Those who are not particularly spacially gifted are encouraged to find something ball-shaped and give it a try, or, if you happen to lack opposable thumbs, like a dodo, you can use ZX-calculus to show that the little white box sends X-basis states to Z-basis states.

Proposition 9.118 The ZX calculus obeys:

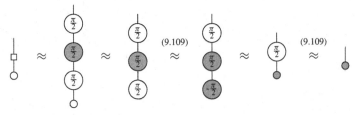 (9.122)

Proof We have:

$$
\begin{array}{ccccccccc}
\end{array}
$$

and similarly for the other basis state. □

Since the little white box yields a 180° rotation, if we do it twice we get back to where we started, so it is self-inverse. For the dodos following along, we can again check this with the ZX-calculus.

Proposition 9.119 The ZX calculus obeys:

$$
\quad = \quad (9.123)
$$

Proof One application of π-commute and some spider fusion yields:

□

So the little white box interchanges the Z- and X-basis states, and it is self-inverse. Sounds familiar? Indeed:

Our little white box and the Hadamard linear map that we first encountered back in Section 5.3.5 differ only by a global phase. Hence, we'll typically refer to either of them as the Hadamard gate, or *H-gate*. So, the punchline is as follows.

Theorem 9.120 The ZX-calculus can equivalently be presented as:

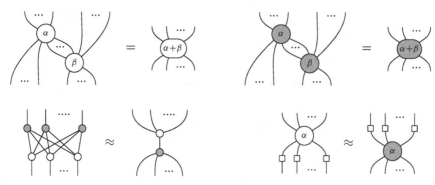

where:

Proof Since any ◯-phase spider in a Clifford diagram can be built from just these pieces:

   (9.124)

in order to change a whole ◯-phase spider to a ⬤-phase spider, all we need are rules to push *H*-gates through each of these pieces. Rules (9.121) and (9.122) together with their colour-reverses give us everything except:

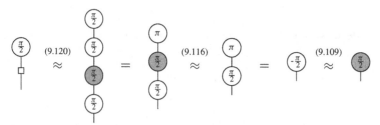

To prove this, first, we have:

then, using the above in the third step:

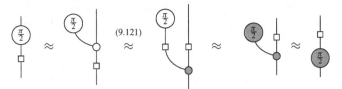

Conversely, equation (9.121) arises as a special case of the colour change rule. Just by unfolding the definition of the colour-changer and doing the phase-juggling from the beginning of this section backwards, (9.121) clearly implies the *Y*-rule.   □

The new rule:

$$\text{(9.125)}$$

will be referred to as *colour change*.

9.4.5 ZX for the God(esse)s: Completeness

The ultimate goal of life, the universe, and everything is of course to replace horrible symbolic manipulations by diagrams. So how far are ZX-diagrams getting us towards that Final Frontier? As we already mentioned, we don't know yet. What we do know is that in the restricted case of the ZX-calculus for Clifford diagrams, we actually achieve that goal. That is, as far as deriving equations between Clifford diagrams goes, we can forget entirely that these diagrams ever had anything to do with linear maps, and simply use graphical calculus. To warp us there, our starship *Enterprise* will be the following concept.

Definition 9.121 A *graph state* is a state whose ZX-diagram consists only of (undecorated) ○-spiders and *H*-gates where:

1. every ○-spider has exactly one output, and
2. all non-output wires connect two ○-spiders with a single *H*-gate.

Since any ○-spider has exactly one output, we can identify quantum systems with the ○-spiders. The edges then represent the way in which the systems are entangled with each other in the associated quantum state. We will see in Chapter 12 that *quantum graph states* obtained by doubling graph states form the basis of a *measurement-based* model of quantum computation.

Here are some examples of graph states:

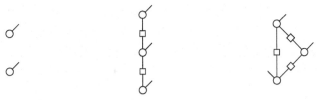

Note that, for clarity, we typically don't bother to extend output wires all the way to the top, nor do we always write them vertically. Also, since an *H*-gate is self-transposed, there is no need to distinguish its input from its output. Thus, we can write *H*-gates on wires going sideways without ambiguity:

This comes in handy when drawing more elaborate graph states:

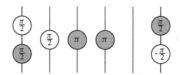

The reason we call these 'graph states' is that they are totally determined by the underlying (undirected) graph, i.e. the set of vertices connected by edges, which tells us where to add ○-spiders and wires with *H*-gates:

For our purposes, graph states are useful because they form the basis for a well-behaved canonical form for Clifford diagrams. This canonical form also involves *local Clifford unitaries*, i.e. untaries expressible as parallel compositions of single-system Clifford diagrams. For example:

Remark 9.122 Since they only act on systems separately, local Clifford unitaries will not effect the entanglement between systems. Hence, when we consider graph states as an *entanglement resource*, local Clifford unitaries do not alter it in any essential way. We'll discuss this concept in detail in Section 13.3.

To obtain the canonical form, we first apply process–state duality to turn any Clifford diagram into a state. To these Clifford states we then apply the following result.

Proposition 9.123 Every Clifford state can be transformed using the ZX-calculus into a graph state, followed by local Clifford unitaries, e.g.:

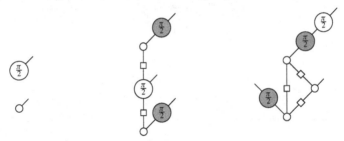

This is called the *graph-form* for a Clifford diagram.

We won't give the proof of this theorem here, but it essentially goes as follows. First, the spiders in a Clifford diagram are decomposed using spider fusion into a finite set of 'little' spiders, namely those depicted in (9.124) and their ⬤ counterparts. Then, the proof proceeds by induction on 'little' spiders. Whenever each type of 'little' spider is added to a Clifford diagram in graph form, the result can again be transformed into graph form.

Since we can convert every Clifford diagram into graph form, we only need to show the ZX-calculus suffices to prove equality between diagrams in graph form. Let's feed a bit of antimatter into the warp core.

Definition 9.124 For an undirected graph G and a vertex v of G, the *local complementation* of G at v, written $G \star v$, is the graph obtained by *complementing* all the pairs w, w' of vertices adjacent to v, where by complementing we mean:

- If there is an edge connecting w and w', remove it, and
- if there is not an edge connecting w and w', add one.

Okay, that's a pretty tricky definition, so let's look at an example. Take this graph, with the vertex v shown in white:

The vertices *adjacent* to v are those connected to v by an edge, i.e. the ones in white here:

To do the local complementation, we delete edges between white vertices where we see them, and add them where we don't:

And that's all there is to it:

Note in particular that none of the edges connecting the vertex v to its neighbours has changed, only the edges <u>between</u> its neighbours. This extends in the obvious way to an operation on graph states:

A remarkable property of graph states is that local complementation does nothing but introduce local Clifford unitaries.

Proposition 9.125 Let G be a graph state whose j-th output wire has associated graph node v_j. Then we have:

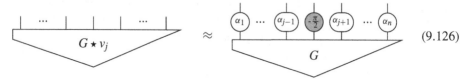
$$(9.126)$$

where:

$$\alpha_i = \begin{cases} \frac{\pi}{2} & \text{if } v_i \text{ is adjacent to } v_j \\ 0 & \text{otherwise} \end{cases}$$

In particular, applying the local complementation rule (9.126) will turn a Clifford diagram in graph form into another Clifford diagram in graph form representing the same state. Remarkably, we can produce <u>every</u> graph form representing that state in this way.

Proposition 9.126 Two Clifford diagrams in graph form represent the same state if and only if one can be transformed into the other using the local complementation rule (9.126), and applications of the ZX-calculus rules to local Clifford unitaries.

Again we'll omit the proof of this rather meaty theorem. However, the take-home message is: using the local complementation rule we can decide when two graph forms are equal by diagram rewriting. Combining this with Proposition 9.123, this means that if we can derive the local complementation rule in the ZX-calculus, then we can prove any equation between Clifford maps.

So, for our grand graphical finale, we'll derive the local complementation rule using the ZX-calculus. We begin with a little lemma.

Lemma 9.127 The ZX calculus obeys:

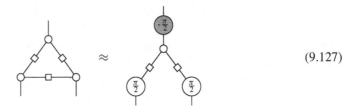

$$(9.127)$$

Proof First the *H*-gate goes up:

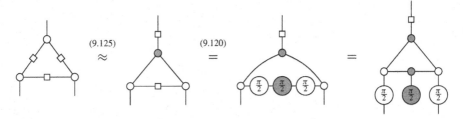

then we use some strong complementarity:

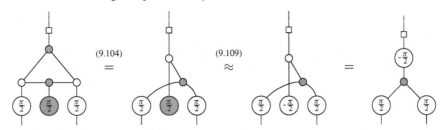

then the *H*-gate goes back down:

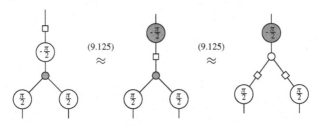

□

The next lemma generalises the one above. First, let K_n be the ZX-diagram defined recursively as follows:

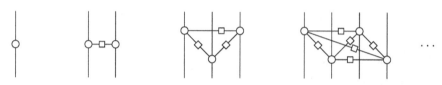

$$(9.128)$$

Thanks to spider fusion, this gives a bunch of vertical wires attached to a totally connected graph of H-edges:

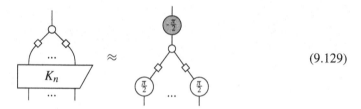

Lemma 9.128 The ZX calculus obeys:

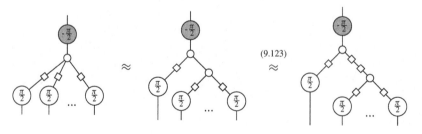

$$(9.129)$$

Proof   We prove this by induction on n. For $n = 0$, this is:

$$\begin{array}{ccc} & (9.108) & \\ \mid & \approx & \end{array}$$

For our induction hypothesis, assume (9.129) holds for some fixed n, which we indicate as (ih) below. Then – brace yourself – here comes the proof for $n + 1$:

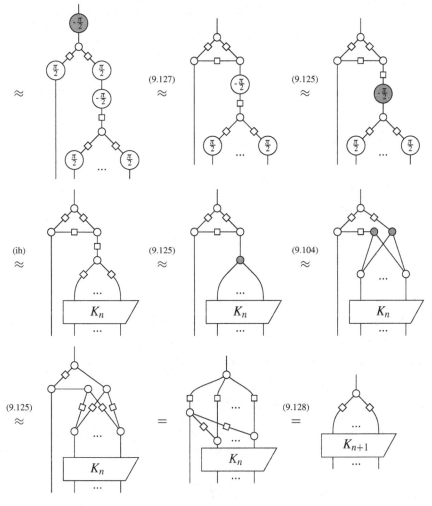

So, what does this lemma say? It says that by applying a ⬤-phase gate of $-\frac{\pi}{2}$ to a node v_j in the graph state, and ○-phase gates of $\frac{\pi}{2}$ to all of its neighbours, just like in the RHS of the local complementation rule (9.126), we introduce a new H-edge between every pair of neighbours of v_j:

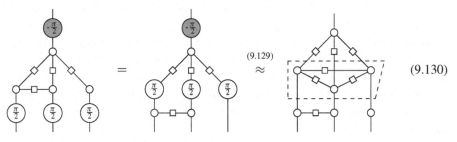

$$(9.130)$$

But then, just like with complementary spiders, pairs of *H*-edges in graph states cancel out:

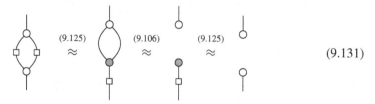

$$(9.131)$$

so the overall result is a local complementation around v_j:

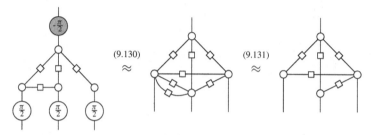

Bingo! We derived the local complementation rule using nothing but the four humble rules from Definition 9.108. All together we can now conclude the following.

Theorem 9.129 The ZX-calculus is complete for **Clifford maps**. That is, for any two Clifford diagrams D, E, the following are equivalent:

- $D = E$ can be derived in ZX-calculus, and
- the associated Clifford maps $[\![D]\!]$ and $[\![E]\!]$ are equal, up to a number.

We can thus make a statement analogous to the one about string diagrams and dot diagrams, provided we interpret two Clifford diagrams as being 'the same' when we can rewrite one diagram into the other by means of ZX-calculus:

> *An equation between Clifford maps holds if and only if the Clifford diagrams are the same.*

9.4.6 *Where We Stand with Full ZX-Calculus*

Even though Clifford maps already exhibit many quantum features, there is at least one reason one wants to consider more general ZX-diagrams. It is well known that Clifford diagrams (or, rather, their unitary cousins Clifford quantum circuits) are efficiently classically simulable. That is, if we input some fixed state into a Clifford map and measure the outputs, we can write a program on a classical computer that efficiently computes the Born-rule probabilities. Hence, if we want to build a quantum computer, Clifford maps don't really give us anything new. We'll discuss this more in Chapter 12.

On the other hand, if we add just one more phase to Clifford diagrams, we actually get a lot more.

Definition 9.130 A *Clifford+T diagram* is a ZX-diagram where the phases are restricted to integer multiples of $\frac{\pi}{4}$, and **Clifford+T maps** is the corresponding subtheory of **pure quantum maps**.

The funny name comes from the fact that, in the quantum computing literature, the ○ $\frac{\pi}{4}$-phase gate is often called the *T gate*. Adjoining the $\frac{\pi}{4}$ phase, for practical purposes pretty much gives us everything, in the sense that we can always get arbitrarily close to what we aim for.

Theorem 9.131 Clifford+T diagrams are *approximately universal*. That is, any linear map can be approximated up to arbitrary precision by a Clifford+T diagram.

We already know that ZX-diagrams let us build any quantum process, and now we know even Clifford+T diagrams let us pretty much build any process. So, how much can we say about these richer diagrams?

If we move from Clifford diagrams to arbitrary ZX-diagrams, the first thing we notice is several of the rules we proved for $\alpha \in \{0, \frac{\pi}{2}, \pi, -\frac{\pi}{2}\}$ extend to arbitrary phases. For example, clearly applying H-gates to every leg of a phase spider will change its colour, regardless of the value of α. Consequently:

$$ \tag{9.132} $$

holds concretely, for all α. Similarly:

$$ \tag{9.133} $$

also holds for all α.

In fact, if we add the two rules above to the ZX-calculus we get a bit closer completeness for all ZX-diagrams.

Theorem 9.132 The ZX-calculus, with the addition of the rules (9.132) and (9.133), is complete for **single qubit Clifford+T maps**, i.e. maps of the form:

Well, that might not look like much, but it's a start. Could it be the case that with this new extended version of the ZX-calculus we have completeness with respect to all Clifford+T diagrams?

Unfortunately, no. Like the other two rules above, the $\frac{\pi}{2}$ supplementarity rule has a big brother, which holds for (almost) all phases. For all α not equal to 0 or π, we have:

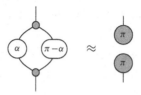

That is, for any (non-trivial) pair of supplementary angles $(\alpha, \pi - \alpha)$ the diagram above separates. It turns out, even when restricting to Clifford+T diagrams, that the equation:

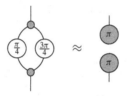

isn't provable from the existing rules. And that's all we know at the moment! Maybe there is a single magical rule that does the job, maybe not. So, alas, we must finish with an 'exercise'.

Exercise* 9.133 Find a complete set of rules for (Clifford+T) ZX-diagrams.

As with Exercise 7.39, if you find a solution, we'd love to hear about it! All the relevant publications will be discussed at the end of this chapter, but by the time you get to read this, probably there will be more.

9.5 Summary: What to Remember

1. The *unbiased* states for spiders are those states that satisfy:

That is, unbiased states for spiders ○ give the uniform probability distribution for ○-measurements. In terms of the Born rule, this means for all i:

Dropping normalisation gives us *phase states*:

The *phases* that decorate these states have a clear interpretation as:

the data destroyed by the classical-quantum passage

In other words, phases constitute the stuff that is genuinely quantum. In the case of a qubit they take the following form:

2. *Phase spiders* arise as follows:

and *phase spider fusion* is:

where:

(9.134)

This operation provides phases with the structure of a commutative group, the *phase group*, for which the unit and the inverse, respectively, are:

An important example of phase spiders are *phase gates*:

3. Spiders ○ and ● are *complementary* if:

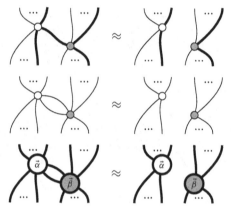

Complementarity admits the following interpretation:

(encode in ○) THEN (measure in ●) = (no data flow)

Complementary is equivalent to all of the ONB-states of ○ being unbiased for ● and vice versa, which can be expressed in two ways:

4. Complementarity induces *spider detachement rules*:

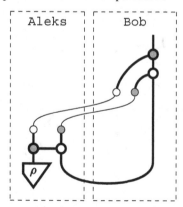

5. A complementary pair yields generalised CNOT-gates:

$$\sqrt{D}\ \ \circ\!\!-\!\!\bullet \qquad\qquad D\ \ \circ\!\!-\!\!\bullet$$

It also provides all of the pieces needed for teleportation:

and for proving its correctness:

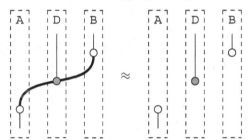

We can also do a protocol called *quantum key distribution*:

which makes it possible to detect eavesdropping on a quantum channel:

6. Strong complementarity:

strictly implies complementarity:

It also gives us much more rewriting power and implies, for instance:

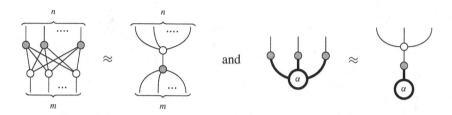

as well as:

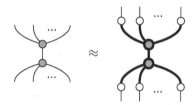

which shows that ● spiders are classical maps for ○. Some examples of classical maps for ○ in terms of ● spiders are the *parity map*, the *even-parity state*, and the *odd-parity state*:

7. Strong complementarity is equivalent to the basis states of ○ forming a *subgroup* of the phase group of ●, and vice versa. That is, phase states satisfying:

form the following subgroups:

$$\left\{\bigodot_{\vec{\kappa}}\right\}_{\vec{\kappa}} \subset \left\{\bigodot_{\vec{\alpha}}\right\}_{\vec{\alpha}} \qquad \left\{\bigodot_{\vec{\kappa}}\right\}_{\vec{\kappa}} \subset \left\{\bigodot_{\vec{\alpha}}\right\}_{\vec{\alpha}}$$

8. Strongly complementary pairs of spiders are *classified* by commutative groups. That is, for a family of spiders ○ and any finite commutative group G, there exists a unique family of spiders ● such that ○/● is strongly complementary and:

Conversely, all strongly complementary pairs arise in this manner.

9. *ZX-diagrams*, i.e. diagrams made up of ○ and ● phase spiders:

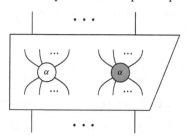

are *universal for qubits*, that is, we can express any classical-quantum map on qubits as a ZX-diagram.

10. *Clifford diagrams* are ZX-diagrams where the phases are restricted to integer multiples of $\frac{\pi}{2}$. The *ZX-calculus*, a graphical calculus for Clifford diagrams, consists of the following rules:

1. Two rules to <u>combine</u> spiders of the same colour:

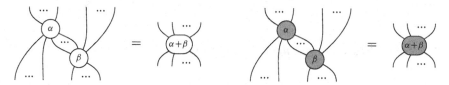

2. One rule to <u>commute</u> spiders of different colours past each other:

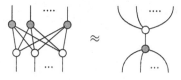

3. One rule to <u>convert</u> spiders of one colour into another:

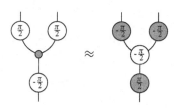

Equivalently, we can replace the last rule with the *colour-change* rule:

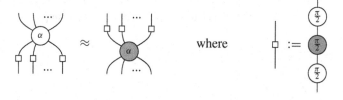

The ZX-calculus is complete for **Clifford maps**, i.e. pure quantum maps expressible as Clifford diagrams. Hence, any equation between Clifford maps can be proven just using the rules of the ZX-calculus.

9.6 Advanced Material*

9.6.1 Strongly Complementary Spiders Are Hopf Algebras*

In Section* 8.6.1 we saw that spiders have been more commonly known as (†-special commutative) Frobenius algebras. Strongly complementary pairs of spiders have also been around for a long time.

Definition 9.134 A *bialgebra* on a vector space V consists of an associative algebra ($\bigtriangleup$, $\bullet$) and a coassociative coalgebra ($\bigtriangledown$, $\circ$) satisfying:

A bialgebra is called a *Hopf algebra* if it additionally has a linear map:

$$\boxed{\iota} \ : V \to V$$

called the *antipode* such that:

$$(9.135)$$

That indeed looks pretty familiar. If we add the $\frac{1}{\sqrt{D}}$-factors and let the antipode be trivial:

$$\boxed{\iota} \ := \ |$$

then we get exactly the (strong) complementarity equations.

So, what's the deal with this antipode thing anyway? Let's have a look again at the $\circ$-spiders from Section 9.3.6 defined in terms of a commutative group G:

$$(9.136)$$

Plugging a ◯ ONB-state corresponding to a group element $g \in G$ into equation (9.135) yields:

$$\begin{array}{ccccccccc} & \overset{(8.21)}{=} & & \overset{(9.135)}{=} & & \overset{(8.13)}{=} & & \overset{(9.136)}{\approx} & \end{array}$$

So, if we take the group-sum of g and ι applied to g, we get 0. This means that ι encodes the group-inverse:

$$ \quad = \quad $$

The antipode law of a Hopf algebra captures the fact that $g - g = 0$:

$$ \left.\begin{array}{c} \text{multiply by its inverse} \end{array}\right\{ \qquad = \qquad \right\} \text{produce unit} $$
$$ \left.\begin{array}{c} \text{copy group element} \end{array}\right\{ \qquad \right\} \text{delete group element} $$

Hopf algebras that come from a group in this way are called *group algebras*. A group algebra has trivial antipode precisely when G consists only of self-inverse elements $g = -g$, like for example the parity group $\mathbb{Z}_2$.

Remark 9.135 Even though we wrote the group operation as '+' (which is usually done only for commutative groups), this construction works just as well for non-commutative groups. In that case the algebra ⬥ becomes non-commutative, but ⋎ (which is still just copying) remains co-commutative. A large literature exists studying certain kinds of Hopf algebras that are neither commutative nor co-commutative, called *quantum groups*.

At this point, you might be thinking, 'Hang on, doesn't strong complementarity (the bialgebra equations) imply complementarity (the extra Hopf algebra equation)?' The answer is of course Yes, but ◯ and ● need to be spiders (a.k.a. †-special commutative Frobenius algebras), not just plain old (co)algebras.

Moreover, having a careful look at the proof that strong complementarity implies complementarity:

$$ \quad = \quad = \quad = \quad \approx \quad \approx \quad \approx $$

we see that the second step relies crucially on the fact that:

$$\underset{\circ}{\smile} = \underset{\bullet}{\smile} = \smile \qquad \text{and} \qquad \overset{\circ}{\frown} = \overset{\bullet}{\frown} = \frown$$

which is only true for families of spiders that come from self-conjugate ONBs, an assumption that we make throughout this book.

In fact, we could drop this assumption, provided that we modify complementarity to:

$$\approx \tag{9.137}$$

where we take the following antipode:

$$:= \tag{9.138}$$

Then we indeed have:

$$\overset{(9.138)}{=} \quad = \quad \approx \quad \overset{(*)}{\approx}$$

Exercise 9.136 Prove the last step of this derivation marked (∗). This, in particular, will require proving:

$$\approx \qquad\qquad \approx$$

Much of what we did in this book can be extended to this more general setting, although diagrams become a bit more complicated.

Exercise 9.137 Adapt the description of quantum teleportation using complementary spiders of Section 9.2.7 to this more general setting. Bonus points: use the 'hairy spiders' from Section* 8.6.3 for expressing non-self-conjugate spiders.

9.6.2 Strong Complementarity and Normal Forms*

We have already seen the following consequence of strong complementarity:

$$\approx \tag{9.139}$$

whereby a complete bipartite graph of spiders can be replaced by a single ●-spider followed by a single ○-spider. But in fact, there are many more equations of this kind one can derive; for example, this equation:

$$\begin{array}{c}\end{array} \approx \begin{array}{c}\end{array} \qquad (9.140)$$

lets us turn large ($2N$-long) cycles of alternating ○- and ●-spiders into a bunch of connected six-cycles.

Exercise* 9.138 Prove equation (9.140).

Equations (9.139) and (9.140) are both instances of the following, much more general result.

Theorem 9.139 An equation:

where Γ and Λ are both circuits consisting only of spiders of the form:

is provable using spider fusion and strong complementarity if and only if the number of (forward-directed) paths, modulo 2, connecting each input to each output is the same in Γ and Λ.

In other words, we can prove any equation that follows from strong complementarity just by path-counting. Let's see this in action for the simplest example, which is just the first strong complementarity equation itself:

If we count the number of paths from the first input to the first output, we see there is just one:

In fact, there is exactly one path for every combination of input/output:

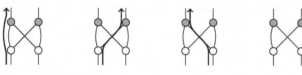

and the same is true on the RHS:

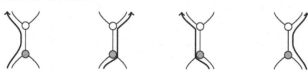

Hence we can conclude that the LHS and the RHS are equal. Checking the other two strong complementarity rules, we see they also respect the number of paths from inputs to outputs (which in those cases is always 0).

Exercise 9.140 Prove equation (9.140) using Theorem 9.139.

It is convenient to collect all of this path-counting information into the *path matrix* of a diagram. That is, a matrix **m** where each entry $\mathbf{m}_i^j$ gives the number of paths (modulo 2) from input i to output j. For instance, the path matrices of the diagrams above are both:

$$\rightsquigarrow \begin{pmatrix} 1 & 1 \\ 1 & 1 \end{pmatrix} \leftsquigarrow$$

The reason we count modulo 2 is that pairs of paths can be eliminated using the complementarity rule:

$$\rightsquigarrow \begin{pmatrix} 1 & 0 \\ 0 & 2 \end{pmatrix} = \begin{pmatrix} 1 & 0 \\ 0 & 0 \end{pmatrix} \rightsquigarrow$$

This clearly gives an equivalent statement to Theorem 9.139.

Corollary 9.141 Diagrams Γ and Λ (as in Theorem 9.139) are equal whenever they have the same path matrix.

So, how do we prove it? First, note that all of the strong complementarity rules and spider fusion respect the path matrix. So, it suffices to show that we can use these rules to rewrite any diagram into a *normal form*, which is uniquely fixed by a given path matrix. These normal forms are described as follows:

(i) No spiders of the same colour are touching;
(ii) any pair of spiders is connected by at most 1 edge; and
(iii) all ○-spiders occur before ●-spiders.

Pictorially, these normal forms look like this:

From such a normal form, we can immediately read off the path matrix just by putting a 1 whenever we see a wire connecting the appropriate spiders. For example:

$$
\rightsquigarrow \quad
\begin{pmatrix}
1 & 0 & 0 \\
1 & 1 & 0 \\
0 & 0 & 1 \\
0 & 1 & 1
\end{pmatrix}
$$

Conversely, for any path matrix, there is a unique normal form that has the correct paths from its inputs to its outputs.

So, it only remains to show that any diagram can be put into normal form. If our diagram doesn't satisfy (i) or (ii) above, we can always apply spider fusion or complementarity until it does. So, (iii) is the only tricky condition. This is where restricting to these spiders:

plays an important role. The only way a diagram will not satisfy (iii) is if it contains the RHS of (9.139). But then, if we just apply this equation backwards:

$$ \approx $$

we can push the $\bigcirc$-spider past the $\bullet$-spider. Since we restrict to circuit diagrams, if we do this repeatedly, all the $\bullet$-spiders will float to the top, while all the $\bigcirc$-spiders will sink to the bottom, giving us a normal form.

Interestingly, path matrices themselves form a process theory. First note that o-composing two matrices yields the same result as counting paths on the composed diagram:

$$
\rightsquigarrow \quad
\begin{pmatrix}
1 & 0 \\
0 & 1 \\
0 & 1
\end{pmatrix}
\begin{pmatrix}
1 & 1 \\
1 & 1
\end{pmatrix}
=
\begin{pmatrix}
1 & 1 \\
1 & 1 \\
1 & 1
\end{pmatrix}
\quad \leftsquigarrow
$$

The ⊗-composition of diagrams does not yield the Kronecker product of path matrices, but rather the *direct sum*:

$$\begin{pmatrix} 1 & 1 \\ 1 & 1 \end{pmatrix} \oplus \begin{pmatrix} 1 \\ 1 \end{pmatrix} = \begin{pmatrix} 1 & 1 & 0 \\ 1 & 1 & 0 \\ 0 & 0 & 1 \\ 0 & 0 & 1 \end{pmatrix}$$

This is a perfectly reasonable way to compose matrices in parallel, so let **matrices**$_\oplus(\mathbb{Z}_2)$ be just like the process theory of matrices defined in Section 5.2.5, except that parallel composition is $\oplus$, not Kronecker product.

The special 'one-to-many-legged' spiders from Theorem 9.139 both live in this process theory:

$$\rightsquigarrow \begin{pmatrix} 1 \\ 1 \\ \vdots \\ 1 \end{pmatrix} \qquad \rightsquigarrow \begin{pmatrix} 1 & 1 & \cdots & 1 \end{pmatrix}$$

and the only equations that hold between diagrams of these spiders are those that come from spider fusion and strong complementary (or, equivalently, the Hopf algebra equations). This process theory is 'the walking Hopf algebra', in the sense that it contains a Hopf algebra and nothing else. Hence, if we really wish to study the essence of Hopf algebras, we should study this process theory. To category theorists, this is known as the *PROP for Hopf algebras*.

9.7 Historical Notes and References

The bulk of this chapter, most notably the diagrammatic notions of phase states, phase spiders, the phase group, complementarity (i.e. mutual unbiasedness), and strong complementarity, as well as all of the equivalent characterisations of Section 9.3.4, are taken from Coecke and Duncan (2008, 2011). However, in Coecke and Duncan (2008) there is a void statement, namely that, under the 'mild' assumption of classical subgroup closure, complementarity and strong complementarity are equivalent. Only later the authors realised that classical subgroup closure is already equivalent to strong complementarity, so it's not so mild after all!

The usual notion of mutual unbiasedness was first introduced by Schwinger (1960). An extensive survey of what is known about them is in Durt et al. (2010), including problems concerning their classification. On the other hand, the classification of strongly complementary pairs is due to Kissinger (2012a).

The ZX-calculus was also introduced in Coecke and Duncan (2008, 2011). However, the version presented there wasn't enough for the completeness theorem of Section 9.4.5. While the *Y*-rule is new (besides a more complex version of it having appeared in talks by Ross Duncan), the equivalent version in terms of Euler-angle decomposition of the

H-gate is due to Duncan and Perdrix (2009). Most versions of the ZX-calculus that have been around contained many redundancies, but a minimal version (from which our presentation is derived) was recently given by Backens et al. (2016).

The completeness theorem with respect to Clifford maps was proved by Backens (2014a), and the completeness theorem with respect to single qubit Clifford+T maps was also proved by Backens (2014b). A related theorem is the complete characterisation for *n*-qubit Clifford circuits in terms of generators and relations given by Selinger (2015).

Schröder de Witt and Zamdzhiev (2014) showed that completeness cannot be achieved with the current rules for arbitrary quantum maps on qubits, and Perdrix and Wang (2015) showed that this is also the case for Clifford+T maps. The supplementarity equation that was used for that purpose first appeared in Coecke and Edwards (2010).

Graph states, which played an important role in the completeness proof, were introduced by Hein et al. (2004). Proposition 9.125, which is due to Backens, builds further on a powerful theorem by van den Nest that states that graph states are equivalent up to local Clifford unitaries if and only they can be turned into each other via local complementation (Van den Nest et al., 2004).

The first person to realise that strong complementarity could be used to model how classical systems and quantum data interact as in example 9.74 was Quanlong (Harny) Wang, as part of the team that produced the paper by Coecke et al. (2012).

Quantum circuits were introduced by Deutsch (1989). Quantum key distribution was first introduced by Bennett and Brassard (1984), and the version obtained by 'bending the wire' (i.e. using an entangled state instead of sending a quantum system) is due to Ekert (1991). The protocol in Example 9.61 is due to Perdrix (2005).

The use of ZX-calculus to model quantum circuits is from Coecke and Duncan (2008), and its use for quantum key distribution as in Section 9.2.6 is from Coecke and Perdrix (2010) and Coecke et al. (2011a). Building controlled operations in the ZX-calculus as in Section 9.2.7 comes from Coecke and Duncan (2011). A similar construction, just in terms of CNOT and phase gates, appeared in Barenco et al. (1995).

A good resource for Hopf algebras and quantum groups is Street (2007). Other standard references are Kassel (1995) and Majid (2000). The 'path counting' characterisation, as well as the normal form for bialgebras from Section* 9.6.2, was given by Lack (2004), as an example of systematically 'composing' two diagrammatic theories (a.k.a. *PROPs*), namely, algebras and coalgebras. This same technique was used to compose a pair of bialgebras to obtain the entire phase-free fragment of the ZX-calculus in Bonchi et al. (2014b). Interestingly, the theory has much the same characterisation as the one we gave for bialgebras in Section* 9.6.2, but with matrices generalised to 'linear relations'. An amusing and pedagogical account of this result involving football, LEGO, and dividing by zero is available as a blog (Sobocinski, 2015).

The phrase 'unreasonable effectiveness of diagrammatic reasoning with spiders' is stolen from Wigner (1995b), who argues the 'unreasonable effectiveness of mathematics in the natural sciences'.

10

Quantum Theory: The Full Picture

Philosophy [i.e. physics] is written in this grand book – I mean the universe – which stands continually open to our gaze, but it cannot be understood unless one first learns to comprehend the language and interpret the characters in which it is written. It is written in the language of mathematics, and its characters are triangles, circles, and other geometrical figures, without which it is humanly impossible to understand a single word of it; without these, one is wandering around in a dark labyrinth.

–*Galileo Galilei, Il Saggiatore, 1623*

In the previous chapters we constructed diagrammatic representations of the key ingredients of quantum theory and related them to the usual quantum formalism in terms of Hilbert spaces and linear maps. However, now it is time to forget about the latter and do pure quantum picturalism! Since it has quite some time to get to this point, in this chapter we give the whole quantum story, as a tale of diagrams and diagrams only.

10.1 The Diagrams

Diagrams consist of boxes and wires, which represent *processes* and *systems*, respectively:

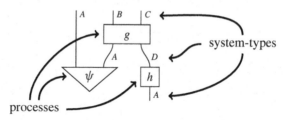

The golden rule of diagrams is:

Only connectivity matters!

That is, two diagrams are equal whenever they contain the same boxes, connected in the same way, regardless of how we write them on the page:

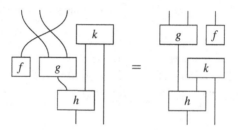

Underlying the story of this book is the story of an evolving diagrammatic language. In Section 2.2.1, we organised this evolution into layers of increasing expressiveness. Now we know all about these layers.

10.1.1 Circuit Diagrams

Circuit diagrams are diagrams that contain no directed cycles:

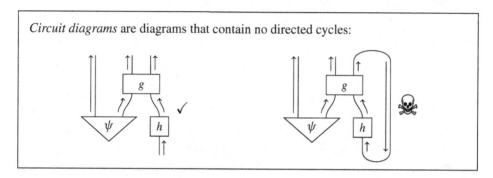

These diagrams are characterised by the fact that they give a clear notion of future and past. They can always be organised into time steps:

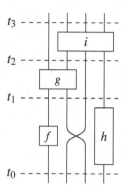

though not necessarily in a unique way.

Processes with no inputs are called *states*, and processes with no outputs are called *effects*. When a state hits an effect, a number pops out:

$$\left. \begin{array}{l} \text{effect} \left\{ \begin{array}{c} \text{\includegraphics{pi}} \end{array} \right. \\ \\ \text{state} \left\{ \begin{array}{c} \text{\includegraphics{psi}} \end{array} \right. \end{array} \right\} \text{number} \qquad (10.1)$$

which we can interpret as the probability (or sometimes just the possibility) that π will happen, given state ψ. This is called the *generalised Born rule*.

In general, a number is just a process with no inputs or outputs:

which we usually write simply as λ. We always have one special number around, the empty diagram, a.k.a. 1:

Sometimes, we also have 0, the number that 'eats everything'. Just as you would expect from these two numbers, 1 'times' something is that thing again, and 0 'times' something is 0:

 $\qquad (10.2)$

See also *circuits* in Section 3.2; *states, effects,* and *numbers* in Section 3.4.1; and *zero* in Section 3.4.2.

10.1.2 String Diagrams

String diagrams do away with the ban on directed cycles and even allow inputs to be connected to inputs and outputs to outputs:

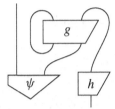

Equivalently, these come from circuit diagrams by appending special processes called:

$$cups := \quad \bigcup \qquad \text{and} \qquad caps := \quad \bigcap$$

which satisfy the *yanking equations*:

$$\bigcap \bigcup = \mid \qquad \bigotimes = \bigcup \qquad \bigotimes = \bigcap \qquad (10.3)$$

The existence of cups and caps explicitly witness *non-separability* in the following sense:

> *For any (non-trivial) system, the cup/cap is <u>never</u> ⊗-separable.*

However, rather than dwelling on what we <u>can't</u> do with cups/caps (namely, separate them), it's much more interesting to see what we <u>can</u> do! They allow us to encode processes and bipartite states and to go back without losing anything:

This *process–state duality* yields an isomorphism:

$$\left\{ \begin{array}{c} \big|^{B} \\ \boxed{f} \\ \big|_{A} \end{array} \right\}_{f} \cong \left\{ \begin{array}{c} \big|^{A} \quad \big|^{B} \\ \overline{\psi} \end{array} \right\}_{\psi} \qquad (10.4)$$

Cups and caps also induce a 180° rotation of boxes, called the *transpose*:

$$(10.5)$$

On the other hand, *adjoints* reflect boxes vertically:

Combining these two operations gives us four incarnations for any box:

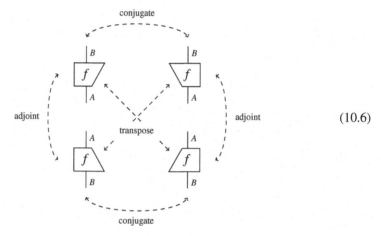

$$(10.6)$$

In particular, the adjoint of a state is the effect that tests for that state:

We expect testing a state for itself to give 1:

$$(10.7)$$

so by default, we expect states to be *normalised*. However, unnormalised states do naturally occur when expressing non-determinism; for example:

can be interpreted as 'ψ happens with probability p'.

Adjoints allow for defining special processes, for example, *isometries*:

$$\tag{10.8}$$

unitaries, which are two-sided isometries, and *positive processes* f, i.e. those for which there exists some other process g such that:

$$\tag{10.9}$$

See also *(non)separability* in Section 4.1.1; *process–state duality* in Section 4.1.2; *transpose* in Section 4.2.1; *adjoints* in Section 4.3.1; *conjugates* in Section 4.3.2; *isometries and unitaries* in Section 4.3.4; and *positive processes* in Section 4.3.5.

10.1.3 Doubled Diagrams

Doubled diagrams are diagrams of the form:

$$\tag{10.10}$$

where we interpret the effect:

$$\tag{10.11}$$

as *discarding* some outputs of a process.

They arise from a two-step construction. First, we double systems:

$$\mathrm{double}\left(\Big| \right) = \Big| \; := \; \big\| $$

and processes:

This operation preserves diagrams, and hence equations between diagrams:

Therefore, anything we can prove using diagrams of single processes also holds doubled. The converse, that any equation between doubled processes also holds non-doubled, is almost true. The only thing that doesn't survive the transition are certain numbers, called *global phases*:

but since these won't have any effect on probabilities, good riddance!

So if working with doubled processes is the same as just working with single processes (up to a global phase), why bother doubling at all? The crucial feature of doubling is that it makes room for something new. The second step in the doubling construction is we adjoin discarding, which represents the act of literally throwing a system away (or simply ignoring it). It works by connecting the two copies of a normalised state together, letting them annihilate:

(10.12)

Since discarding doesn't arise from doubling something, it is called *impure*. More generally, by discarding outputs of *pure* (i.e. doubled) processes, we obtain lots of other impure processes of the form (10.10).

By doubling all wires, we have also created a vacancy for a different type of system, which we can represent as plain old single wires:

So how do these two types of systems interact? Via *spiders*.

See also *doubling* in Section 6.1; *global phases* in Section 6.1.2; *discarding* in Section 6.2.1; *doubled process theory* in Section 6.2.4; and *classical wires* in Section 8.1.

10.1.4 Spider Diagrams

Spider diagrams consist of boxes and 'generalised wires', which are allowed to connect any number of inputs to any number of outputs:

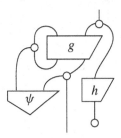

Again these can be equivalently presented as circuit diagrams by appending special processes called:

$$spiders \; := \qquad\qquad (10.13)$$

The only rule that governs them is that adjacent spiders fuse together:

$$= \qquad\qquad (10.14)$$

In particular, it follows that any two <u>connected</u> diagrams made up of spiders with the same number of inputs and outputs are equal:

Given that two-legged spiders are just 'plain old' wires:

$$
\phi = \Big| \qquad \smile = \smile \qquad \frown = \frown \tag{10.15}
$$

spider diagrams subsume string diagrams, and spider fusion generalises the yanking equations, e.g.:

$$
\boxed{\quad} = \phi = \Big| \tag{10.16}
$$

Doubling gives us two extra species of spiders for free. We obtain new pure processes by doubling the whole spider:

$$
\bowtie \; := \; \text{double}\left(\bowtie \right) = \bowtie \tag{10.17}
$$

We also obtain *bastard spiders* by interpreting some pairs of legs together as doubled systems (a.k.a. *folding*), while leaving others single:

doubled systems   single systems

$$
\tag{10.18}
$$

These satisfy their own fusion laws:

$$
\tag{10.19}
$$

and additionally, bastard spiders can fuse with the other two kinds of spiders:

$$
\tag{10.20}
$$

Discarding is itself a bastard spider:

$$\overline{\overline{\top}} \;=\; \varphi \tag{10.21}$$

whence:

$$\overset{\circ}{\underset{\displaystyle\varphi}{}} \;=\; \overline{\overline{\top}} \qquad\qquad \overline{\overline{\underset{\varphi}{\top}}} \;=\; \varphi \tag{10.22}$$

To every family of spiders, we can associate *copiable states*:

$$\left\{ \left.\bigtriangledown_{\!i}\;\right|\; \overset{\displaystyle\curlyvee}{\bigtriangledown_{\!i}} \;=\; \bigtriangledown_{\!i}\;\bigtriangledown_{\!i} \right\} \tag{10.23}$$

which are moreover orthonormal:

$$\frac{\bigtriangleup^{\,j}}{\bigtriangledown_{\!i}} \;=\; \delta_i^j \tag{10.24}$$

From this and spider fusion, we get a *generalised copy rule*:

$$\;=\; \delta_{i_1\ldots i_m}^{j_1\ldots j_n} \;\; \bigtriangledown_{\!i_1}\!\cdots\bigtriangledown_{\!i_1}\;\bigtriangleup_{i_1}\!\cdots\bigtriangleup_{i_1} \tag{10.25}$$

By doubling or folding parts of this equation, quantum and bastard versions also emerge, e.g.:

$$\;=\; \delta_{i_1\ldots i_m}^{j_1\ldots j_n} \tag{10.26}$$

See also *spiders* in Section 8.2.3; *quantum and bastard spiders* in Section 8.3.3; and *copiable (a.k.a. ONB) states* in Section 8.2.2.

10.1.5 ZX-Diagrams

ZX-*diagrams* consist entirely of two kinds of *phase spiders*:

where $\alpha \in [0, 2\pi)$. When phase spiders of the same colour meet, they fuse and their *phases* add (modulo 2π):

ZX-diagrams are the richest diagrams encountered in this book. They have two important ingredients: the diagrammatic structure of spiders and the group structure of the phases. In fact, the second arises from the first!

Like their undecorated cousins, phase spiders too can be doubled:

$$\text{double}\left(\text{⊠}_\alpha \right) = \text{⊠}_\alpha \tag{10.27}$$

When a phase spider meets a bastard spider, i.e. when it comes into contact with the single-world, the phase is destroyed:

$$\text{⊠} = \text{⊠} \tag{10.28}$$

This property of not surviving the passage from double to single totally characterises phases. To see this, consider all states where:

$$\bigtriangledown_\psi = \; |$$

Then, as if by magic, a commutative *phase group* emerges. Setting:

$$\bigcirc_\alpha := \bigtriangledown_\psi$$

we have:

$$\text{\alpha+\beta} := \quad \alpha \quad \beta \qquad 0 := \qquad -\alpha := \psi$$

Then, by letting:

$$\alpha := \alpha$$

we recover all phase spiders and their associated fusion law:

$$\alpha \quad \beta = \alpha+\beta \qquad (10.29)$$

Actually, it wasn't so magic. The property of 'not surviving the passage from double to single' is precisely what does the non-trivial work, namely giving this group its inverses:

$$\alpha = \quad \Longrightarrow \quad -\alpha \quad \alpha = \quad \Longrightarrow \quad -\alpha \quad \alpha =$$

In the case of ZX-diagrams, this emergent group is the circle group $U(1)$:

and in the case of *Clifford ZX-diagrams*, we restrict just to a four-element subgroup $\mathbb{Z}_4$ of $U(1)$:

$$0 \qquad \frac{\pi}{2} \qquad \pi \qquad -\frac{\pi}{2}$$

In either case, we might think of these as rotations of a sphere of some kind. But let's not get ahead of ourselves.

See also *phase spiders* in Section 9.1.2; *phase group* in Section 9.1.4; *ZX-diagrams* in Section 9.4.1; and *Clifford diagrams* in Section 9.4.2.

10.2 The Processes

A *process theory* is a collection of processes that make sense to plug together. In other words, it is *closed under forming diagrams*. Here's a process theory we really like, called *quantum theory*:

There are two kinds of systems, quantum and classical systems:

The processes that <u>may</u> be realised (not necessarily with certainty) are:

$$(10.30)$$

where the f-labeled box is made up of phase spiders:

Processes that can be realised with certainty furthermore obey *causality*:

$$(10.31)$$

10.2.1 Causality

Causality is an extremely important postulate for quantum theory, which nevertheless has an extremely simple interpretation:

> *If the output of a process is discarded, it may as well have never happened.*

For a quantum process (10.30) causality is equivalent to f being an *isometry*:

It guarantees that quantum theory is compatible with special relativity; i.e. it is *non-signalling*. Non-signalling says that the flow of information must respect the *causal structure*. So, if Aleks and Bob are very far apart, but possibly share some correlation from the past:

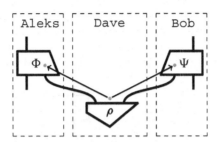

then it is impossible for Aleks to communicate directly to Bob. This is witnessed by the fact that if Bob doesn't know the output of Aleks' process:

then he can't possibly know the input, either.

See also *causality* in Section 6.2.5; *causal structure* in Section 6.3.1; and *non-signalling* in Section 6.3.2.

10.2.2 Process Decomposition and No-Broadcasting

A process theory has *spectral decompositions* if any positive process can be written in this form:

$$ \tag{10.32} $$

A process theory furthermore has *singular value decompositions* if <u>any</u> process can be decomposed as:

$$\boxed{f} \quad = \quad \begin{array}{c} \boxed{V} \longleftarrow \text{isometry} \\ \triangledown_{p} \longleftarrow \text{classical state} \\ \boxed{U} \longleftarrow \text{adjoint of isometry} \end{array} \qquad (10.33)$$

As a consequence of 10.32, quantum states ρ encode classical states:

$$\triangledown_{\rho} \quad = \quad \begin{array}{c} \boxed{\widehat{U}} \;\}\; \text{unitary} \\ \;\}\; \text{encoding} \\ \triangledown_{p} \;\}\; \text{classical state} \end{array}$$

From the form (10.32), using properties of spiders, it also follows that:

$$\left(\exists \psi, \phi : \quad \begin{array}{c} \boxed{f} \\ \boxed{f} \end{array} = \begin{array}{c} \triangledown_{\phi} \\ \triangle_{\psi} \end{array} \right) \quad \Longleftrightarrow \quad \left(\exists \psi', \phi' : \quad \boxed{f} = \begin{array}{c} \triangledown_{\phi'} \\ \triangle_{\psi'} \end{array} \right)$$

From this we can show that if the *reduced map* of a cq-map is pure:

$$\begin{array}{c} \overline{\underline{\underline{\equiv}}} \\ \boxed{\Phi} \end{array} = \boxed{\widehat{f}} \qquad \begin{array}{c} \text{\large ?} \\ \boxed{\Phi} \end{array} = \boxed{\widehat{f}} \qquad (10.34)$$

then the process Φ separates as follows:

$$\boxed{\Phi} = \triangledown_{\rho} \, \boxed{\widehat{f}} \qquad \boxed{\Phi} = \triangledown_{p} \, \boxed{\widehat{f}} \qquad (10.35)$$

for (causal) states ρ and p.

From these separation results, we immediately arrive at *no-broadcasting*; that is, there exists no quantum process such that:

$$\begin{array}{c} \overline{\underline{\underline{\equiv}}} \\ \boxed{\Delta} \end{array} \quad \overset{(l)}{=} \quad \Big| \quad \overset{(r)}{=} \quad \begin{array}{c} \overline{\underline{\underline{\equiv}}} \\ \boxed{\Delta} \end{array} \qquad (10.36)$$

The identity is pure, so if such a process existed, it would separate as:

$$\boxed{\Delta} \quad = \quad \boxed{\rho} \quad \Big| \tag{10.37}$$

which yields a contradiction:

$$\Big| \quad \overset{(10.36\text{r})}{=} \quad \boxed{\Delta} \quad \overset{(10.37)}{=} \quad \boxed{\rho} \quad \overset{\text{\raisebox{2pt}{\rule{10pt}{1pt}}}}{\rule{10pt}{0pt}}$$

See also *spectral theorem* in Section 5.3.3.1; *spectral and singular value decompositions* in Section 8.2.5; and *no-broadcasting* in Section 6.2.8.

10.2.3 Examples

The processes in quantum theory that may be released non-deterministically are called **cq-maps**, whereas causal cq-maps are called simply **quantum processes**. In this section, we'll give some important examples.

10.2.3.1 Classical Maps

Classical maps are cq-maps with no quantum inputs or outputs:

$$\boxed{f} \quad = \quad \boxed{\Phi} \quad = \quad \Big(\boxed{f} \quad \boxed{f} \Big) \tag{10.38}$$

and *classical processes* are causal classical maps:

$$\boxed{f} \quad = \quad \overset{\circ}{\Big|} \tag{10.39}$$

Consequently, classical maps are automatically self-conjugate:

$$\boxed{f} \quad = \quad \boxed{f} \tag{10.40}$$

The simplest examples of classical maps are ○-copiable states and effects, which represent:

$$classical\ values\ :=\ \text{∇}_i \qquad\qquad classical\ tests\ :=\ \text{△}^i$$

A system that admits just two such values/tests, 0 and 1, is called a *bit*.

Examples of classical processes include:

- classical values, because:

$$\text{∇}_i\ =\ \boxed{} \tag{10.41}$$

- *function maps*, i.e. processes that encode a function:

$$f : \{1, \ldots, m\} \to \{1, \ldots, n\}$$

via:

$$\boxed{f}\ \text{∇}_i\ :=\ \text{∇}_{f(i)} \tag{10.42}$$

which are characterised by this equation:

$$\boxed{f}\ =\ \boxed{f}\ \boxed{f} \tag{10.43}$$

and consequently satisfy:

$$\boxed{\widehat{f}}\ =\ \boxed{f} \tag{10.44}$$

- ○-spiders with precisely one input:

$$delete\ :=\ \text{⦿} \qquad copy\ :=\ \text{Y}_○ \qquad n\text{-}copy\ :=\ \text{Y}_○^{\cdots}$$

- ○-spiders with no inputs (after renormalising):

$$perfect\ correlation\ :=\ \tfrac{1}{D}\ \text{⌣}_○^{\cdots}$$

and as a special case:

$$uniform\ distribution\ :=\ \tfrac{1}{D}\ \text{|}_○$$

- ⊙-spiders with more than one output (after renormalising):

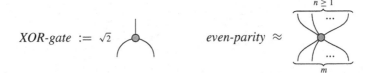

$$XOR\text{-}gate := \sqrt{2} \qquad\qquad even\text{-}parity \approx$$

- the same, but with π phases:

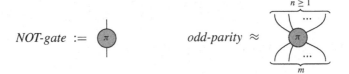

$$NOT\text{-}gate := \qquad\qquad odd\text{-}parity \approx$$

- and combinations of the above, e.g.:

$$CNOT\text{-}gate := \sqrt{2}$$

The last three examples are given in the special case of bits, but also generalise to other classical systems.

10.2.3.2 Quantum Maps

Quantum maps are cq-maps with no classical inputs or outputs:

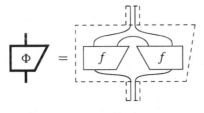

Some examples of *quantum states* are:

$$maximally\ mixed\ state := \frac{1}{D}$$

$$Bell\ state := \frac{1}{D}$$

$$GHZ\ state := \frac{1}{D}$$

Restricting to *qubits*, i.e. two-dimensional quantum systems, we have:

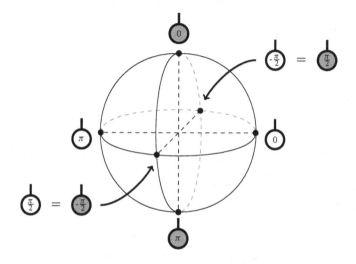

which can be depicted on the *Bloch sphere*:

On a pair of qubits, the Bell state extends to the *Bell basis*:

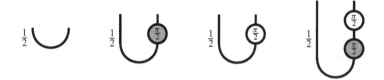

which is expressed in terms of the Bell state and the four *Bell maps*:

These are all unitary quantum processes, or *quantum gates*. Other important single-qubit quantum gates are:

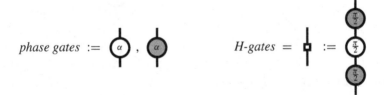

and two-qubit quantum gates are:

10.2.3.3 Classical-Quantum Interactions

Many classical-quantum interactions arise as special cases of bastard spiders:

- *encode* and *measure* for all colours:

the corresponding *non-demolition measurements*:

and *decoherence*:

More general *demolition* and *non-demolition ONB measurements* arise from combining bastard spiders with a unitary $\widehat{U}$:

Important examples are the *Y-measurements*:

and the non-separable *Bell measurement*:

While measurements extract classical data from a quantum system, *controlled unitaries*:

use classical data to change a quantum system. They are characterised by the following equations:

$$\qquad\qquad = \qquad\qquad\qquad = \qquad\qquad\qquad (10.45)$$

An example is the *correction* used in quantum teleportation:

$$\qquad\qquad := \qquad\qquad\qquad (10.46)$$

which further decomposes as:

$$Z\text{-correction} := \qquad\qquad X\text{-correction} := \qquad\qquad (10.47)$$

So, that gives us a pretty good collection of parts. Now let's plug them together and turn this thing on!

See also *classical maps and function maps* in Section 8.2.1; *classical logic gates* in Section 5.3.4; *parity maps* in Section 9.3.5; *maximally mixed state* in Section 6.2.2; *Bloch sphere* in Section 6.1.2; *Bell maps/basis* in Section 5.3.6; *phase gates* in Section 9.1.5; *H-gate* in Section 5.3.5; *measure and encode* in Section 8.1.3; *decoherence* in Section 8.3.2; *ONB-measurements* in Section 8.4.1; *controlled unitaries* in Section 8.4.2; and *Bell measurement/correction* in Section 9.2.7.

10.3 The Laws

We now turn to the most important laws governing quantum processes. We can already get a bit of mileage out of the fact that spiders of the same colour fuse together. However, things start to get really interesting when spiders of different colours start to fight.

10.3.1 Complementarity

Spiders are *complementary* if:

$$\begin{array}{c} \text{(diagram)} \end{array} = \frac{1}{D} \begin{array}{c} \text{(diagram)} \end{array} \qquad (10.48)$$

or, equivalently, if:

$$\begin{array}{c} \text{(diagram)} \end{array} = \frac{1}{D} \begin{array}{c} \text{(diagram)} \end{array} \qquad (10.49)$$

or in words:

$$(\text{encode in } \bigcirc) \text{ THEN } (\text{measure in } \bullet) = (\text{no data flow})$$

From the simple form (10.48), we can derive lots of other equations. Notably, a version for bastard spiders:

$$\begin{array}{c} \text{(diagram)} \end{array} = \frac{1}{D} \begin{array}{c} \text{(diagram)} \end{array} \qquad (10.50)$$

which unfolded yields:

$$\text{(figure)} = \frac{1}{D} \text{(figure)} \tag{10.51}$$

which doubled gives:

$$\text{(figure)} = \frac{1}{D^2} \text{(figure)} \tag{10.52}$$

and which combined with the bastard/quantum spider fusion (10.20) yields yet another variation:

$$\text{(figure)} = \frac{1}{D^2} \text{(figure)} \tag{10.53}$$

Complementarity is kind of a big deal. For example, it explains the behaviour of the Stern–Gerlach device:

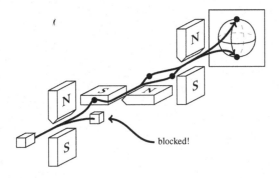

blocked!

as:

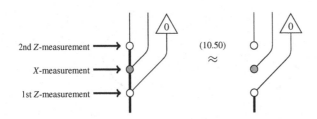

2nd Z-measurement ⟶

X-measurement ⟶

1st Z-measurement ⟶

$$\begin{array}{cc} (10.50) \\ \approx \end{array}$$

it provides the diagrammatic magic for quantum teleportation:

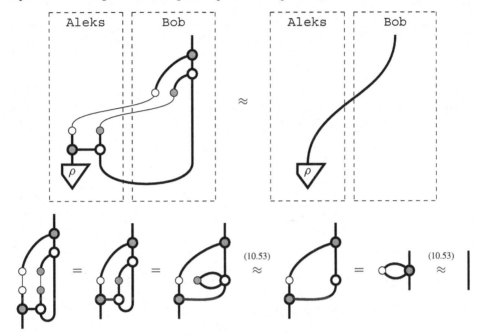

and it induces basic properties of classical and quantum gates:

$$(10.54)$$

For complementary spiders, the copiable states of one colour are, up to a number, phase states for the other colour:

$$(10.55)$$

Hence complementarity gives us some new equations, the κ-*copy rules*:

$$(10.56)$$

Up to a global phase, these copiable phase states pass right through phase gates of the other colour:

$$(10.57)$$

See also *complementarity* in Section 9.2.1; *Stern–Gerlach* in Section 9.2.5; and *teleportation via complementarity* in Section 9.2.7.

10.3.2 Strong Complementarity

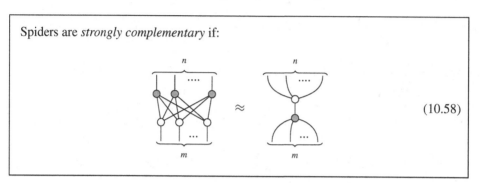

$$\tag{10.58}$$

Equivalently, strong complementarity can be expressed as three rules:

$$= \frac{1}{\sqrt{D}} \tag{10.59}$$

$$= \frac{1}{\sqrt{D}} \qquad \qquad = \frac{1}{\sqrt{D}} \tag{10.60}$$

While these rules don't have a (single) natural interpretation as in the case of complementarity, they do <u>imply</u> complementarity:

$$= \quad = \quad = \quad \overset{(10.59)}{\approx} \quad \overset{(10.60)}{\approx} \quad \approx$$

One way of interpreting equation (10.58) is to treat some parts of the equation as doubled, e.g.:

$$\approx \tag{10.61}$$

This consequence of strong complementarity is used to show that correlations for the GHZ state take a particular form:

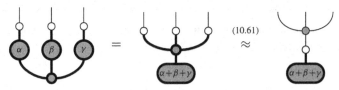

which will play a crucial role in establishing *quantum non-locality*.

It also furnishes a plethora of new equations concerning copiable phases, stemming from the fact that, for strongly complementary spiders, the group-sum of two copiable phases is again a copiable phase. So, copiable phases actually form a subgroup of the phase group, called the *classical subgroup*:

$$\left\{ \raisebox{-3pt}{\includegraphics{}} \right\}_\kappa \subset \left\{ \raisebox{-3pt}{\includegraphics{}} \right\}_\alpha \qquad\qquad \left\{ \raisebox{-3pt}{\includegraphics{}} \right\}_\kappa \subset \left\{ \raisebox{-3pt}{\includegraphics{}} \right\}_\alpha$$

So, in addition to the κ-copy rules (10.56), strong complementarity furthermore implies:

- that the unit is a copiable state:

$$\raisebox{-3pt}{\includegraphics{}} \approx \raisebox{-3pt}{\includegraphics{}} \qquad\qquad \raisebox{-3pt}{\includegraphics{}} \approx \raisebox{-3pt}{\includegraphics{}} \qquad\qquad (10.62)$$

- the copiability of the corresponding phase gates:

$$\raisebox{-3pt}{\includegraphics{}} = \raisebox{-3pt}{\includegraphics{}} \qquad\qquad \raisebox{-3pt}{\includegraphics{}} = \raisebox{-3pt}{\includegraphics{}} \qquad (10.63)$$

- and the commutation of classical phase gates, up to a global phase:

$$\raisebox{-3pt}{\includegraphics{}} = \raisebox{-3pt}{\includegraphics{}} \qquad\qquad (10.64)$$

For ZX-diagrams, the classical subgroup is:

$$\left\{ \raisebox{-3pt}{\includegraphics{}} \, , \, \raisebox{-3pt}{\includegraphics{}} \right\} \subseteq \left\{ \raisebox{-3pt}{\includegraphics{}} \right\}_\alpha$$

because:

$$\raisebox{-3pt}{\includegraphics{}} \approx \raisebox{-3pt}{\includegraphics{}} \qquad\qquad \raisebox{-3pt}{\includegraphics{}} \approx \raisebox{-3pt}{\includegraphics{}} \qquad (10.65)$$

and similarly, with the colours reversed. But you don't have to take our word for it, you can prove it using the *ZX-calculus*!

See also *strong complementarity* in Section 9.3; *generalised form and doubling* in Section 9.3.3; and *classical subgroup* in Section 9.3.4.

10.3.3 ZX-Calculus

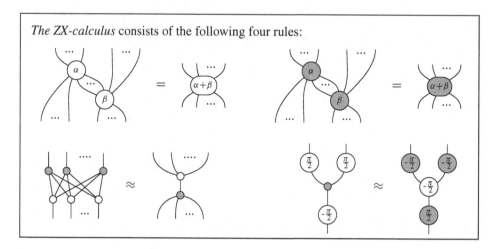

The ZX-calculus is a complete graphical calculus for Clifford ZX-diagrams. That is, if an equation holds between two ZX-diagrams, it is provable in the ZX-calculus. It consists of three kinds of rules, which tell us:

1. how spiders of the same colour <u>combine</u>,
2. how spiders of different colours can <u>commute</u> past each other, and
3. how to <u>convert</u> spiders of one colour into another.

The first kind of rule is phase spider fusion, whereas the second comes from strong complementarity. Hence, the first two kinds of rules hold for arbitrary strongly complementary spiders. On the other hand, the third kind of rule, the *Y-rule*, tells us something specifically about qubits and the Bloch sphere, namely that the *Y*-basis states can be copied in two equivalent ways.

The rules above are the most succinct way to give the calculus we know. However, a convenient, alternative presentation replaces the *Y*-rule with the *colour-change rule*:

$$\tag{10.66}$$

From the ZX-calculus, we can also derive:

- the 0-*copy* rule:

(10.67)

and the π-*copy* rule:

(10.68)

which confirm that $\{0, \pi\}$ indeed forms the classical subgroup:

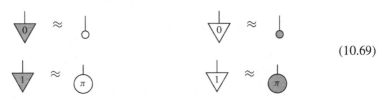

(10.69)

- the π-*commute* rule:

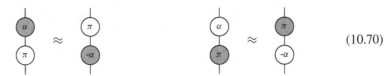

(10.70)

- the *phase eliminate* rule:

(10.71)

as well as its π-*eliminate* counterpart, specialising (10.57):

(10.72)

- Together (10.70) and (10.71) yield:

(10.73)

- the $\frac{\pi}{2}$-*supplementarity* rule:

$$(10.74)$$

- two equivalent expressions of the Y-basis:

$$(10.75)$$

- equations between bastard spiders, obtained from (un)folding:

$$(10.76)$$

$$(10.77)$$

$$(10.78)$$

$$(10.79)$$

- self-inverseness of the H-gate:

$$(10.80)$$

Many of these $\approx$-equations hold up to a global phase, so doubling yields:

$$(10.81)$$

$$(10.82)$$

$$(10.83)$$

$$(10.84)$$

$$(10.85)$$

$$(10.86)$$

$$(10.87)$$

So, from the humble four equations of the ZX-calculus, we can actually derive much more. Of course, we didn't do this just because it was fun (actually, it was kind of fun). In the next four chapters, we will see how these equations, along with the graphical presentation of quantum theory in general, can be exploited in quantum foundations, quantum computation, quantum resource theories and automating proofs in all of the above.

See also ZX-*calculus* in Section 9.4 and the *colour-change rule* in Section 9.4.4.

10.4 Historical Notes and References

While we have already given all the relevant references for the development of quantum picturalism in the previous chapters, here we give a (somewhat idiosyncratic) timeline of the events leading up to here. The use of diagrams to discuss quantum teleportation first appeared in Coecke (2003, 2014a). This paper was not published until 2014. From then onwards, the further refinement of the diagrammatic language was tightly intertwined with the development of the corresponding category-theoretic axiomatics. Part of the reason for this is that diagrams look a bit silly, and if a paper contains no hard mathematics there is no chance to get the work published in a prestigious venue. The categorical axiomatisation of the aforementioned paper was given in Abramsky and Coecke (2004). Basically, all that was used here were string diagrams. Of course, Penrose had already been drawing string diagrams since the 1970s (Penrose, 1971). However, they weren't used to describe quantum features such as quantum teleportation, for the simple reason that they weren't known to anyone yet. Penrose himself even said that 'the notation seems to be of value mainly for private calculations because it cannot be printed in the normal way', in a book coincidentally also published by Cambridge University Press (Penrose, 1984).

Doubling popped up quite soon after the 2004 paper, despite the fact that the relevant papers only went to press a bit later. In Coecke (2007) doubling was proposed in order to get the correct Born rule, and independently Selinger (2007) moreover adjoined impure processes. Also, the asymmetric boxes that allow one to clearly distinguish adjoints, transpose, and conjugate were introduced in Selinger (2007). The idea of a discarding process was put forward in Coecke and Perdrix (2010).

Spiders required an evolutionary process starting in 2006 and spanning 6 years (Coecke and Pavlovic, 2007; Coecke and Paquette, 2008; Coecke et al., 2010a, 2012) in order to adequately account for classical systems and processes. Completeness of spider diagrams is in Kissinger (2014b). In contrast, phases and (strong) complementarity in terms of spiders popped out all at once (Coecke and Duncan, 2008, 2011). Completeness is still an unfinished story, but the strongest current results for ZX-diagrams appear in Backens (2014a,b).

Causality, although it plays a very central role in this book, was the last one to enter the picture. Its importance became clear from the information-theoretic axiomatization of Chiribella et al. (2010, 2011).

11

Quantum Foundations

Mermin once summarized a popular attitude towards quantum theory as 'Shut up and calculate.' We suggest a different slogan: 'Shut up and contemplate!'

– Lucien Hardy and Rob Spekkens, 2010

This chapter is dedicated to the foundations of quantum theory, or as it's more fashionably called these days, quantum foundations. Here, we will use all of the things we've learned so far to probe some very deep questions:

1. What features of nature are imposed on us by quantum theory?
2. Conversely, what features of a physical theory are imposed on us by (our current understanding of) nature?
3. Which of these features are 'properly quantum', in the sense that they have no counterpart in any classical physical theory?

We'll address these questions by looking at one of those most celebrated (and historically controversial) properties of quantum theory: *quantum non-locality*. First, we will give a precise definition of non-locality and prove that it exists within the theory of **quantum processes** and in fact already within the comparatively tiny subtheory of causal **Clifford maps**. Then, we will present a new process theory called **spek**, which has locality built right in. A remarkable thing is that the two theories of **Clifford maps** and **spek** are identical in every respect except one: the phase group of a single system. And (another spoiler alert!) it is indeed this one difference that kills the proof of non-locality that works for quantum theory.

11.1 Quantum Non-locality

Quantum non-locality is probably still the least understood of all the new quantum features, in both philosophical and structural terms. Our upbringing in a seemingly 'classical' world, and especially our undeniably corrupting 'classical' scientific education, tends to make us expect two things from a physical theory:

1. *Realism*: physical systems have real pre-existing properties, and hence the outcome of 'measuring' such a property is fixed in some way prior to the measurement.
2. *Locality*: it is impossible for one system to affect another distant system instantaneously.

Very early on, something made Einstein, the father of relativity theory, extremely uncomfortable with quantum theory. He realised that something really weird was going on:

Quantum theory is <u>not</u> a local-realistic theory.

It was the first reaction of many (including Einstein) to think this simply meant quantum theory was 'incomplete'. Since of course any theory should be local and realistic, the failure of quantum theory to be so was simply a bug that needed to be fixed.

But, as we'll see shortly, we will have to learn to live in absence of localrealism. The failure of <u>any</u> local-realistic theory to reproduce the predictions of quantum theory is what we call *quantum non-locality*.

11.1.1 Refinements of Quantum Theory

Typically we consider physical systems to have certain properties even if they are not observed, and when we observe the system, it is these properties that we witness. For example, the colour of a pencil won't change when we don't look at it. Realism stands for the assumption that something like this is true in quantum theory, namely, that what we learn in quantum measurements is not just created out of the blue during the measurement process but has some cause in the past.

Of course, when we started this book with Dave's travels to the North and South Poles, we made it clear that measurements change the state of the system non-deterministically and that the measurement outcome does not faithfully reflect what the state of the system was. We know now that this is just what the standard quantum formalism tells us.

However, there is no reason a priori that we couldn't refine quantum theory in such a way that each measurement outcome can be traced back to something pre-existing. It could be the case that somebody had already put Dave in a rocket aimed at the South Pole, and they were just waiting for us to ask where he was. That is, there could be some *hidden variables* at work here, and it is our ignorance of them that leads to the apparent non-determinism. We can thus refine our theory, putting those extra variables in, and poof! – no non-determinism.

Such a refinement is sometimes called a *hidden variable model*, or more recently, an *ontological model*. To avoid any philosophical baggage that comes with each of these terms, we will stick to *refinement*.

The crucial feature of a refinement is that even though it may be adding additional variables to account for measurement outcomes, it should retain the predictions of quantum theory. A famous example of such a refinement is *de Broglie–Bohm theory*, which postulates the existence of particles flying around that always have precise positions and momenta. The catch is, they are being pushed around by something else spread out in space, called the *pilot wave*, which explains the characteristically quantum behaviour we saw in Section 7.1.4 whenever we try to measure those particles. Hence the theory keeps realism, at the cost of dropping locality.

On the other hand, as we saw in Section 6.3, relativity theory is derived from the principle that nothing travels faster than the speed of light. Therefore, many would consider dropping locality too high of a price to pay for realism.

In that case, we should require that any refinement of quantum theory that provides it with predetermined measurement outcomes should be compatible with relativity theory. Quantum theory is of course already compatible with relativity theory, thanks to the causality postulate (cf. Section 6.3.2). To keep this compatibility intact, any newly introduced variables should not travel faster than the speed of light.

So in particular, any correlation that may occur for spatially separated systems must have some common cause in the past. In other words, they should respect *Reichenbach's common cause* principle. This principle states that every correlation is either a consequence of a direct causal link or due to a common cause. An example of the first is that being shot by a gun causes pain (or death). An example of the second is the strong correlation in the previous century between the spread of televisions in households and the death of hedgehogs. The common cause is the spread of wealth, which caused people not only to buy televisions, but also to buy cars, and these cars killed hedgehogs. Bummer.

In order to establish a contradiction between quantum theory and local realism, we will proceed as follows. We will consider carefully chosen *measurement scenarios*; that is, we fix a particular quantum state and measure each of its systems in a number of different ways. We compute the probabilities for each scenario, and hence the *correlations* between the outcomes in each measurement, and study the properties these correlations obey. For example, if we were to consider two systems in a Bell state and measure each system with the following measurements:

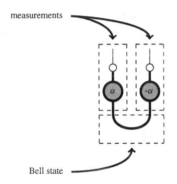

simply using phase spider fusion we learn that the outcomes for the measurements on each of the systems will always be the same; i.e. they are perfectly correlated:

Despite the fact that this scenario involves quantum processes, we could of course also produce these same correlations by some totally classical (and hence local) process. In order to establish a contradiction with local realism, we will need to consider a scenario consisting of several different choices of measurement on the same quantum state.

11.1.2 GHZ-Mermin Scenarios

Consider again the following measurements on a GHZ-state:

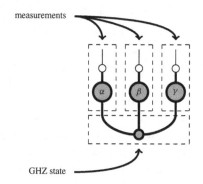

Each of the measurements seems to depend (locally) on a phase, but by using phase spider fusion and strong complementarity we obtain:

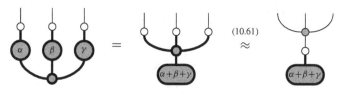

In the particular cases that $\alpha + \beta + \gamma$ is either 0 or π, then the phase state:

is in the classical subgroup for $\bigcirc$, and hence, by (10.69) and (10.26):

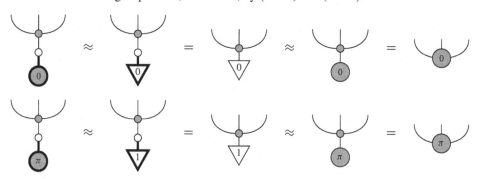

That is, we obtain the even-parity state and odd-parity state, respectively (cf. Section 9.3.5).

Now, let's consider some fixed choices for the phases α, β, and γ. If we let the phase be 0, we obtain the usual Z-measurement, whereas if it is $\frac{\pi}{2}$, the result is a Y-measurement:

In order to produce a contradiction with local realism, we will need to consider one choice of measurements that yields the even-parity state, namely:

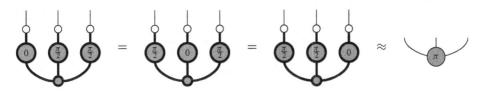

as well as three choices that yield the odd-parity state, namely:

In other words, we consider the following measurement choices:

	system A	system B	system C
scenario 1	Z	Z	Z
scenario 2a	Z	Y	Y
scenario 2b	Y	Z	Y
scenario 2c	Y	Y	Z

We'll use a particular property of all of these scenarios together that allows us to draw a contradiction with local realism. That property is the 'overall parity':

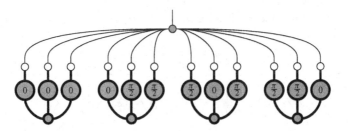

Substituting the parities for the individual scenarios:

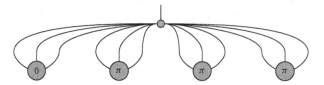

by phase spider fusion we obtain:

So the overall parity is <u>odd</u>.

11.1.3 Drawing a Contradiction

Local realism assumes that all measurement outcomes have some common cause in the past. So, we construct a refined model, whose classical values already 'know in advance' what outcome they will provide for either measurement:

$$\text{(11.1)}$$

So, for instance, if we measure Z on the first system, we will get the outcome $z^A \in \{0, 1\}$; if we measure Y on the third system, we will get y^C; and so on. A generic state in this model is then a probability distribution over these classical values:

$$\sum_i p^i \qquad \text{(11.2)}$$

Now, we know that quantum theory predicts that the overall parity for the four measurement choices is always odd. Hence, to be consistent with quantum theory, it should be the case that every possible value in the above probability distribution yields an odd overall parity. But, as we shall now see, not one of them does!

By definition, each value in (11.2) gives the following outcomes for each of the four measurement choices:

Combining duplicate states via copy spiders, we obtain:

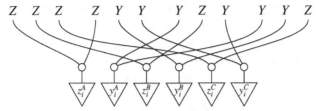

Then, the overall parity is given by:

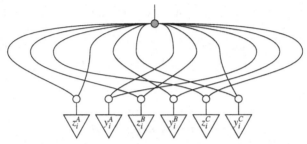

Looking closely at this 'locality Spaghetti Monster', we see that exactly two legs connect each ○-spider to the ◉-spider. So, applying the complementarity equation (10.49), we obtain:

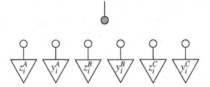

which is equal to:

$$\overset{\mid}{\underset{0}{\bigcirc}} \approx \overset{\mid}{\underset{0}{\triangledown}}$$

That is, for every state, the overall parity is <u>even</u>. Since even is not equal to odd:

$$\overset{\mid}{\underset{0}{\triangledown}} \neq \overset{\mid}{\underset{1}{\triangledown}}$$

quantum theory is non-local.

11.2 Quantum-like Process Theories

One way to understand a thing is by understanding how it relates to similar things. For example, suppose we wished to list the most remarkable traits of a dodo. If we compared dodos with people, we would find lots of totally uninteresting distinctions, like 'dodos don't have fingers'. However, if we compare dodos with other birds, we find their unique characteristics start to stand out more. For instance, we will immediately note their inability to fly and their legendary tastiness (which gives us some clues toward their extinction!).

Something similar has been done for quantum theory, by considering it as part of a broader class of theories. This can be done in many different ways, depending on what features of the theory one wishes to study and contrast with others. For example, a lot of effort has been put into understanding quantum theory as an instance of a *generalised probabilistic theory*. In this general setting, systems always have convex sets of states (e.g., the Bloch ball in the case of qubits), but many of the other characteristics of quantum theory start to break down.

In another direction, one can look at process theories that admit similar diagrammatic (i.e compositional) behaviour to quantum theory, which is what we (of course!) will focus on here. We saw in Chapter 9 that many quantum features can be expressed in terms of spiders, which are purely diagrammatic creatures. Hence, just by reinterpreting these spider diagrams in other process theories, we can see how things such as measurement, complementarity, and even non-locality arguments look in 'quantum-like' process theories.

11.2.1 Complementarity in relations

In Chapter 5, we learned that **relations** were totally boring when it comes to ONBs. Namely, each system has exactly one ONB (cf. Example 5.6), given just by the set of singletons. However, by Chapter 8, ONBs were superseded by spiders for our purposes, and indeed in the theory of **linear maps**, fixing an ONB is exactly the same as fixing a family of spiders. Of course, if that were true for **relations**, then spiders would be just as boring as ONBs, since there would only be spiders of one colour around.

However, it turns out that the theory of **relations** is a lot wilder than one might imagine at first, and there are lots of things that behave like spiders that do not arise from an ONB. For example, even for the system $\mathbb{B}$ there are already spiders of two colours. Among many other things, this means that the diagram:

makes perfect sense within the theory of **relations** (or, more precisely, the theory of 'cq-relations' built in the analogous way to **cq-maps**), and we can calculate with these spiders just as we have been doing all along.

Spiders of the first colour are indeed the ones that arise from the unique ONB for $\mathbb{B}$:

$$\ :: \ \begin{cases} (0,\ldots,0) \mapsto (0,\ldots,0) \\ (1,\ldots,1) \mapsto (1,\ldots,1) \end{cases}$$

and it's straightforward to see that these creatures indeed fuse in the appropriate manner. But what about the spiders of the second colour?

Recall from Section 9.3.5 that we gave a characterisation of ○-spiders in terms of the parity of ○-basis states. While the X-basis doesn't carry over into **relations** (thanks to that pesky minus sign in the second basis state), ○-spiders do! That is, we let:

$$:: (b_1, \ldots, b_m) \mapsto (b'_1, \ldots, b'_n)$$

if and only if the number of 1s in $b_1, \ldots, b_m, b'_1, \ldots, b'_n$ is even. Particular cases of these parity spiders are the relational versions of the examples we saw in Section 9.3.5, e.g. relational XOR:

$$:: (i,j) \mapsto i \oplus j$$

and the relational three-system parity state:

$$:: * \mapsto \{(0,0,0),(0,1,1),(1,0,1),(1,1,0)\}$$

Exercise 11.1 Show that the ○ spiders defined above indeed behave like spiders, that is, that we have:

$$=$$

as well as invariance under leg-swapping and conjugation and that the pair ○/○ is strongly complementary:

$$= \qquad = \qquad = $$

11.2.2 Spekkens' Toy Quantum Theory

While it has non-separable states, and even strongly complementary spiders, the theory of **relations** doesn't really look much like quantum theory. With this in mind, it may come as a surprise that there exists a subtheory of **relations** that does exhibit many quantum features. For instance, the states of the smallest non-trivial system do organise themselves into a sphere, which looks a whole lot like the six-state restriction of the Bloch sphere we encountered in Section 9.4.2 with the **Clifford maps**. We'll first define this theory concretely (i.e. 'the hard way'), and later show that it can be obtained equivalently as a small modification to ZX-calculus.

The basic system in this new theory, called **spek,** is the four-element set:

$$IV := \{1, 2, 3, 4\}$$

Since this will be a subtheory of **relations**, the states of IV are given by subsets. Rather than taking all $2^4 = 16$ subsets of IV, we'll take our (non-zero) states to be just the six two-element subsets:

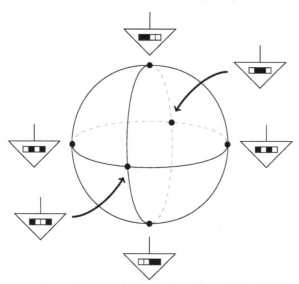

Notably, these organise themselves into three pairs of orthogonal states:

so we can, somewhat suggestively, place them on the 'spek sphere':

By analogue to the Bloch sphere, we will refer to the states on the Z-axis as the spek-Z states, those on the X-axis as spek-X states, and those on the Y-axis as spek-Y states. These are the states of our process theory, but what about processes? On a single system, we allow arbitrary permutations of IV, e.g.:

$$\sigma_{(23)} \quad :: \quad 1 \mapsto 1, \ 2 \mapsto 3, \ 3 \mapsto 2, \ 4 \mapsto 4$$

Since permutations cannot change the number of filled-in boxes, they will always send two-element subsets to other two-element subsets. For example:

Furthermore, they will always send orthogonal states to orthogonal states; i.e. they preserve antipodes on the 'spek sphere'. Hence, these permutations, which are easily seen to be unitary, are the analogues to the one-qubit unitaries in **Clifford maps**.

That's it for processes on a single system. To get all the processes in **spek**, we just add a single family of spiders, namely the spiders that copy the **spek** spek-Z states:

$$(11.3)$$

We call this family of spiders the $\{1, 2\}$-*parity spiders*. Explicitly, a $\{1, 2\}$-parity spider is given by the relation:

$$:: \quad \begin{cases} (a_1, \ldots, a_m) \mapsto (b_1, \ldots, b_n) \\ (a_1 + 2, \ldots, a_m + 2) \mapsto (b_1 + 2, \ldots, b_n + 2) \end{cases}$$

for all $a_i, b_i \in \{1, 2\}$ such that the number of 2s in $a_1, \ldots, a_m, b_1, \ldots, b_n$ is even. In fact, this is the unique family of a spiders whose copy relation:

$$:: \quad \begin{cases} 1 \mapsto \{(1, 1), (2, 2)\} \\ 2 \mapsto \{(1, 2), (2, 1)\} \\ 3 \mapsto \{(3, 3), (4, 4)\} \\ 4 \mapsto \{(3, 4), (4, 3)\} \end{cases}$$

satisfies equations (11.3). Furthermore:

$$\begin{matrix} \vdots \\ \end{matrix} \quad :: \quad * \mapsto \{1, 3\}$$

is one of the six states on the 'spek-sphere', so we can reach any other state by means of permutations, e.g.:

$$\sigma_{(3\,4)} \quad = \quad \sigma_{(3\,4)} \quad = \quad$$

Since we can recover the 'spek-sphere' using just spiders and permutations, we can define the full process theory **spek** as follows.

Definition 11.2 The theory **spek** is the following subtheory of **relations**:

- The systems consist of n copies of IV.
- The processes are string diagrams made up of:

  - $\{1, 2\}$-parity spiders:

  - all permutations on a single system:

Now compare this with the following characterisation of **Clifford maps**.

Proposition 11.3 The theory **Clifford maps** can equivalently be defined as the following subtheory of **pure quantum maps**:

- The systems consist of n copies of $\widehat{\mathbb{C}^2}$.
- The processes are string diagrams made up of:

  - the Z quantum spiders:

- all Clifford unitaries on a single system:

Proof Clearly any string diagram consisting of ◯-spiders and Clifford unitaries is a Clifford diagram (cf. Definition 9.104), and hence gives a Clifford map. Conversely, the following are Clifford unitaries on a single system:

From these and ◯-spiders, we can obtain:

which gives us all the pieces we need to construct an arbitrary Clifford diagram. □

Not too bad, eh?

Remark 11.4 One apparent difference between **Clifford maps** and **spek** is that the numbers are different; namely, in **Clifford maps** these are the positive real numbers, while in **spek** these are the booleans. We say 'apparent', since this can be easily adjusted. A simple way to do this is to take 'equality up to a number' as equality in **Clifford maps**, so that only two non-equal numbers (namely, 0 and 1) remain. It's a bit more tedious, but also possible, to augment the numbers in **spek** with the positive real numbers and adjust the definition of 'wiring processes together' a bit, so that, for example:

And the analogy between **Clifford maps** and **spek** goes even further. Just like we constructed the ◯ spiders (a.k.a. spek-Z spiders), we can construct two other families of spiders (a.k.a. spek-X spiders and spek-Y spiders) that copy the other two orthogonal pairs of states.

Exercise 11.5 Using ◯-spiders and permutations, define new spiders:

such that:

copy the spek-X and spek-Y states, respectively.

And of course these interact the way one would hope.

Exercise 11.6 Show that the spek $\bigcirc$-spiders and the spek $\bullet$-spiders are strongly complementary.

Let's see how far this analogy goes.

11.2.3 Phases in spek

Since we laid out our six states on a sphere it is tempting to try to think of the points on the equator as phases:

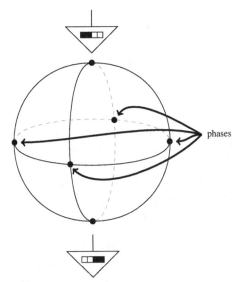

and try to fit them together into a (commutative) phase group. In fact, as we'll soon see, this is possible. And furthermore, once we start decorating a pair of strongly complementary spiders with these phases, we get universality for **spek**, a Y-rule that relates two different ways to copy the spek-Y basis, and even a corresponding completeness theorem! How on Earth could there be any difference between **spek** and **Clifford maps**?

Those who know a bit of group theory (or who have read Section* 9.3.6) may have guessed the answer. There are precisely two commutative groups with four elements, and as it turns out, **Clifford maps** has one and **spek** has the other! In the case of **Clifford maps**, the phase group is $\mathbb{Z}_4$, the four-element cyclic group. That is, it has four elements $\{0, 1, 2, 3\}$, where the group-sum is addition modulo 4.

'Hang on!' you might say, 'I thought the phase group consisted of rotations around the Bloch sphere equator':

– and you'd be right! But just by giving these group elements different names:

$$0 \leftrightarrow 0 \qquad 1 \leftrightarrow \frac{\pi}{2} \qquad 2 \leftrightarrow \pi \qquad 3 \leftrightarrow -\frac{\pi}{2}$$

we see that this is in fact the same group, e.g.:

$$1 + 2 = 3 \qquad \leftrightarrow$$

In order to be a phase for the spek-Z states, it must be the case that:

So, for a phase state, we must pick one element from $\{1, 2\}$ and one element from $\{3, 4\}$, which indeed gives the four remaining spek states. As with qubits, we can make this into a group by picturing little wheels. However, this time, we should picture two wheels rather than one, which can each be set to 0 or π:

To see which state each of these corresponds to, just colour in the boxes where the black dots land:

hence the four phase states are:

The group-sum is just adding the angles element-wise, e.g.:

$$\pi \quad 0 \quad + \quad \pi \quad \pi \quad = \quad 0 \quad \pi$$

and indeed we can check that:

$$
\begin{array}{c}
\vcenter{\hbox{a,b}} \quad \vcenter{\hbox{c,d}}
\end{array}
\quad = \quad
\boxed{a+c,\,b+d}
$$

As in the quantum case, we can use these new phase states to decorate a new breed of spiders:

$$
\cdots \quad (a,b) \quad := \quad (a,b) \quad \cdots
$$

which satisfy 'spek spider fusion':

$$
\begin{array}{c}
\boxed{a_1,b_1} \\
\boxed{a_2,b_2} \\
\boxed{a_3,b_3} \\
\boxed{a_4,b_4} \\
\boxed{a_5,b_5}
\end{array}
\quad = \quad
\boxed{\textstyle\sum a_i,\,\sum b_i}
$$

Whereas the phase group for **Clifford maps** is $\mathbb{Z}_4$, this new phase group is called $\mathbb{Z}_2 \times \mathbb{Z}_2$. Instead of a four-element group $\{0,1,2,3\}$ where we do addition modulo 4, this is a pair of identical two-element groups $\{0,1\}$ where we do addition modulo 2.

To form phase states for the spek-X states, we should choose one element from $\{1,3\}$ and one element from $\{2,4\}$. To get this from an element of the phase group, we just realign the little wheels accordingly:

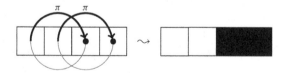

which gives us enough phase states to fill up the spek sphere:

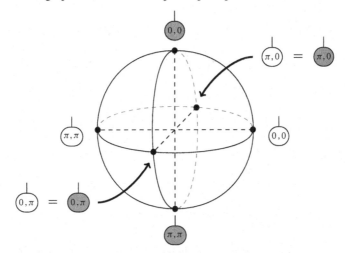

Let's just have a quick peek at the Bloch sphere again, shall we?

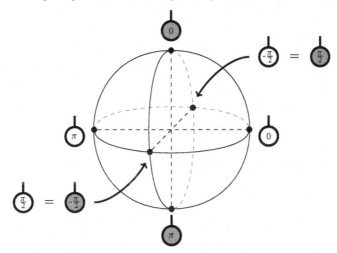

Wow! Like brothers from another mother! It's really starting to look like the only difference between these two theories is the phase group. In fact, this is indeed the case. But what precisely does it mean to 'change the phase group' of a whole process theory? We can make this precise with the help of our old friend, the ZX-calculus.

11.2.4 ZX-Calculus for spek

We now know that **spek** admits a phase group and a version of phase spider fusion. We furthermore know that ○ and ◉ are strongly complementary. Hence, we have nearly established a fully fledged ZX-calculus for **spek**. The only thing missing is the Y-rule, which we will add now.

Definition 11.7 The *spek ZX-calculus* consists of the following rules:

where $a, b, c, d \in \{0, \pi\}$ and $\bar{a}$ is shorthand for $\pi + a$.

This definition is almost identical to the definition of the ZX-calculus for **Clifford maps**. Even the 'extended' Y-rule, which in the case of **spek** relates four variations of the Y-copy, rather than two:

is in fact perfectly analogous to the Y-rule for **Clifford maps**. That is, they both arise, for chosen elements y_1, y_2 of the phase group, as:

for all $i, j \in \{1, 2\}$. If we assume $y_1 \neq y_2$ and that spider decorations are unique, i.e.:

then y_1 and y_2 are already uniquely fixed by the calculus (up to possibly renaming some elements of the phase group). For **Clifford maps**, they must be $y_1 := \frac{\pi}{2}$ and $y_2 := -\frac{\pi}{2}$ and for **spek**, $y_1 := (0, \pi)$ and $y_2 := (\pi, 0)$.

Exercise* 11.8 In **Clifford maps**, the extra variations on the Y-rule are redundant. Are they indeed necessary in the case of **spek**?

Example 11.9 Since doubling the processes in any process theory just eliminates global phases (cf. Remark 6.19) and the only global phase in **spek** is 1, doubling the processes in **spek** just gives **spek** again:

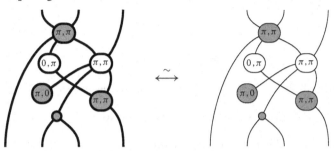

However, doubling enables us to encode measurements:

whose outcomes we can compute using the phase group. For example, if we ○-measure each of two spek-Z states, we get:

which stand for 'definitely in $\{0, 1\}$' and 'definitely in $\{2, 3\}$', respectively. The other states on the spek sphere are ○-phase states, so if we ○-measure them we get:

which stands for 'I have no idea'.

Just as with the ZX-calculus for **Clifford maps**, the first thing we want to do is derive some convenient rules.

Exercise 11.10 Assuming the analogous 'little rules' to equation (9.107):

show that the spek ZX-calculus obeys its own versions of the 'dodo rules' derived in Section 9.4.3:

and the colour-change rule:

Like its Clifford cousin, the following is also true for the spek ZX-calculus.

Theorem 11.11 The spek ZX-calculus is complete for **spek**.

In fact, the proof (which we won't go into here) works in much the same way as the proof for **Clifford maps**. Namely, the processes in **spek** admit a natural notion of *graph form*, into which any process can be translated. Then, using the spek version of the local complementation rule (9.129), it is possible to translate two graph-form diagrams into one another if and only if they are equal as processes in **spek**.

That means, just as in the case of **Clifford maps**, instead of defining **spek** concretely, we could just as well define it as a theory whose processes are diagrams of phase spiders, modulo the equations in Definition 11.7. Hence we have the following theorem.

Theorem 11.12 The only difference between **Clifford maps** and **spek** is the choice of phase group: $\mathbb{Z}_4$ versus $\mathbb{Z}_2 \times \mathbb{Z}_2$.

So, is this one little difference such a big deal?

11.2.5 Non-locality in spek?

Is this difference in phase groups a big deal? Yes, it is! Let's try to reproduce the GHZ-Mermin scenario from Section 11.1.2 in **spek** by doubling and fixing some measurements (cf. Example 11.9):

$$Z\text{-measurement} := \quad Y\text{-measurement} :=$$

Previously, a local-realistic theory predicted that the overall parity:

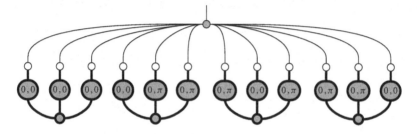

should be even, i.e. a ◉ phase of 0, in **spek** a.k.a. $(0,0)$. Since in **Clifford maps** we obtained a ◉ phase of π, quantum theory is non-local. However, since for **spek** the phase group is $\mathbb{Z}_2 \times \mathbb{Z}_2$, all pairs of $(0,\pi)$-phases all cancel out, e.g.:

So we obtain:

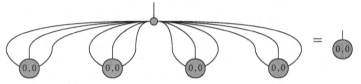

and the contradiction vanishes! In fact, a non-locality argument is doomed in **spek** no matter what choice of measurements we make. This is because **spek** is, <u>by construction</u>, a locally realistic theory!

To see this, all we need to do is think about what the states in **spek** actually mean. As we saw in Section 3.4.1, states in **relations** represent non-determinism. That is, the set IV can be thought of as a (classical) system, which has one of four possible states. For instance, it could be a collection of four boxes, where Dave is hiding in exactly one box:

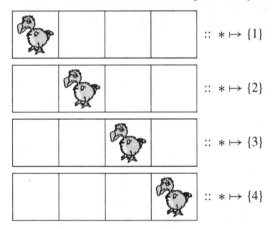

Since this is representing some actual state of affairs, these are called *ontic* states. On the other hand, it could be that we don't know exactly where Dave is, but we do know he's in one of the first two boxes:

Since this doesn't represent the real state of the system, but merely our <u>knowledge</u> about that system, this is called an *epistemic* state.

The crucial thing about **spek** is we put certain limitations on which epistemic states are allowed, so we never have perfect knowledge about which box Dave is in. Out of this comes many seemingly quantum features, such as (strong) complementarity, unseparability of systems, and something that looks a whole lot like the Bloch sphere. But, ultimately, this theory was designed from the start to always admit a refinement into a local realistic theory, by simply admitting the underlying ontic states as hidden variables.

Exercise* 11.13 Transform the ZX-calculus into a graphical calculus for a quantum system other than qubits. What are the theories that one can obtain by changing the phase group?

11.3 Summary: What to Remember

1. Quantum theory is not a *local-realistic theory*; that is, any refinement of quantum theory in which all measurement outcomes have some common cause in the past must violate locality.
2. This fact can be established by drawing a contradiction diagrammatically:

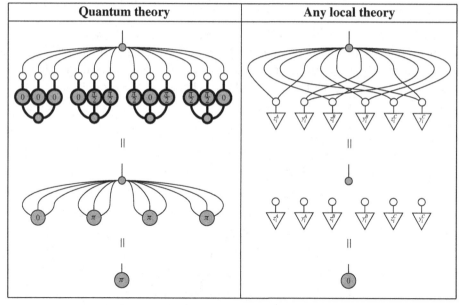

3. In the theory of **relations** there are complementary spiders even on the two-element set $\mathbb{B}$.
4. Spekkens' toy theory, a theory that closely resembles qubit quantum theory, can be formulated as a process theory **spek**, which is a subtheory of **relations**. It can also be formulated entirely diagrammatically, as a modification of ZX-calculus simply by replacing the group $\mathbb{Z}_4$ by the group $\mathbb{Z}_2 \times \mathbb{Z}_2$.
5. Quantum non-locality, and the GHZ-Mermin scenario in particular, is tightly intertwined with the fact that the phase group for qubits includes $\mathbb{Z}_4$. In contrast, the phase group in Spekkens' toy theory does <u>not</u> contain $\mathbb{Z}_4$.

6. More generally, studying **quantum theory** within a wider space of process theories teaches us which ingredients of quantum theory cause its remarkable features.

11.4 Historical Notes and References

The GHZ-Mermin argument was initially proposed in a somewhat more complicated form by Greenberger et al. (1990) and was put in its present form by Mermin (1990). A first overly complicated diagrammatic proof was produced by Coecke et al. (2012). The same authors did sober up not too long after and produced the proof presented here (Coecke et al., 2016). Generalisations of this argument can be found in Gogioso and Zeng (2015).

The first non-locality proof was due to John Bell (1964), and Einstein's related concern, which instigated the discussions leading to Bell's theorem, appeared in (Einstein et al., 1935; Einstein, 1936). The first experimental verification was due to Aspect et al. (1981), but only recently an experiment took place that is widely accepted to be 'loophole-free' (Hensen et al., 2015). David Bohm's hidden variable model was first published in (Bohm, 1952a,b).

The fact that one can represent complementarity in **relations** was observed by Coecke and Edwards (2011), where also **spek** was first presented as a subtheory of **relations**. Pavlovic (2009) classified all spiders in **relations**, and all pairs of strongly complementary spiders were classified by Evans et al. (2009) and, independently, also by Edwards. For a bestiary of sets and relations, see Gogioso (2015a).

Spekkens presented his toy theory in Spekkens (2007). An earlier very similar but less developed toy theory was presented by Hardy (1999). Rather than a full presentation of a theory, Spekkens gave a recipe to produce all states and processes, without evidence that a consistent theory would emerge in this manner. Key to that recipe was a so-called knowledge-balance principle, which restricted the amount of knowledge one could ever have about a system. That Spekkens' recipe produced a consistent theory was established in Coecke and Edwards (2012), by relating Spekkens' recipe to **spek**. That paper also contained a picture of process-theoretic graffiti under an Oxford bridge:

The two camps joined forces to figure out that it is the phase group that captures the true difference between quantum theory and the toy theory (Coecke et al., 2011b). This body of work then became the content of the DPhil thesis of Edwards (2009), which also contains some further elaborations. Edwards' current whereabouts are unknown to us.

Backens and Nabi Duman (2015) realised that one could adjust the ZX-calculus in order to obtain **spek** and also provided a corresponding completeness theorem, hence establishing an entirely diagrammatic presentation of Spekkens' toy theory.

The quote at the beginning of this chapter is taken from Hardy and Spekkens (2010). See also the discussion of Mermin's quote in Section 1.3.

12

Quantum Computation

In the Name of the Pasta, and of the Sauce, and of the Holy Meatballs ...
– *Bobby Henderson,* The Gospel of the Flying Spaghetti Monster, 2006

After the conceptual comes the practical. While the study of quantum foundations is as old as quantum theory itself, the field of quantum computing is relatively new. So new in fact that large-scale, practical quantum computing is still not a reality. A typical 'quantum computer' takes many months to set up before performing such astounding tasks as factoring 6 into 3 × 2. Nonetheless, if those machines would exist, we know that we would gain amazing speed-ups in solving some hard (classical) computational problems, such as those involved in breaking a huge portion of cryptographic systems in use today.

Before we get into 'quantum computing', we should say a couple of things about 'computing'. What is computing? Our answer by now probably won't come as such a shock: it's a process theory! Computation is indeed all about wiring the inputs and outputs of small processes together to make bigger processes. More specifically, a *computation* consists of a (finite) set of basic processes, which are wired together according to some (also finite) instructions, which we refer to as an *algorithm* or simply a *program*.

The only essential difference between classical and quantum computation is the contents of the basic processes. For classical computation, these operations consist of things like logical operations (e.g. XOR) or reading/writing locations in memory. For quantum computation, we can extend this with quantum processes and classical-quantum interactions such as measurements. So quantum computing is all about figuring out how to write new kinds of programs that exploit these new building blocks to build faster algorithms or accomplish new kinds of tasks that aren't possible classically.

The first quantum algorithms were 'proofs of concept', in the sense that they solved some problem much faster than a classical computer, but the kinds of problems they solved were not particularly interesting in their own right. However, this changed drastically with the advent of Grover's *quantum search* and Shor's *factoring* algorithms. The latter, which demonstrated the application of quantum computers to efficiently factor large numbers, is enormously interesting for one big reason: a huge percentage of the current cryptography in use today relies on a cryptographic system called RSA, which in turn depends on the fact that it is computationally infeasible to factor large numbers. What's more, pretty much every alternative to RSA (except for a handful of so-called post-quantum systems) relies on the closely related problem of computing 'discrete logarithms'. But both factoring and this problem can be solved efficiently as special cases of the *hidden subgroup problem*, which we lay out in Section 12.2.4. So, once there are quantum computers around, the padlock on your web browser meaning your bank details are being encrypted becomes effectively meaningless (though a quantum hacker is probably more interested in other things besides the size of your overdraft).

In this chapter, we'll combine the classical-quantum building blocks we have already seen into computations. There is in fact not just one way to do this, but many possible *models of quantum computation*. We will focus on two such models, the first being the *quantum circuit model*, which extends the classical concept of computing with circuits to quantum processes in the more or less obvious way. Rather than starting with a bunch of bits and performing classical logic gates, we start with a bunch of qubits, perform a bunch of quantum gates, then measure what comes out.

The second and markedly funkier model is called *measurement-based quantum computation* (MBQC). Rather than relying on anything like logic gates, in MBQC, measurements do all of the work. MBQC relies on the uniquely quantum feature that measurements enable the kind of drastic changes to the state required to perform any computation, and hence this model is nothing like anything we've seen in classical computation.

Both of these models can be expressed using the ZX-calculus, which enables us to reason and prove things about them diagrammatically, as well as translate computations from one model to the other. In fact, this is a two-way street. Two of the most important theorems from Chapter 9 (universality and completeness of the ZX-calculus) come from the encoding of ZX-diagrams into the circuit model and MBQC, respectively.

12.1 The Circuit Model

The *quantum circuit model* is a straightforward extension of computing with circuits of classical logic gates (e.g. AND, OR, NOT) to quantum processes. All computations in the circuit model are performed in three steps:

1. Prepare some qubits in a certain fixed state.
2. Perform a circuit consisting of basic quantum gates.
3. Measure (some of the) resulting qubits.

As in the case of classical circuits, we assume that the set of basic quantum gates is fixed in advance. Typical examples are phase gates:

and the CNOT gate:

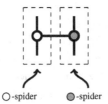

So, in general, spiders should be thought of not as gates, but rather as 'gate pieces'. For instance, neither of the quantum spiders in the CNOT gate is a unitary on its own, but together they make up a unitary quantum gate. The ZX-calculus rules governing these spiders then yield equations between the resulting gates. In fact, one can even do the following (although we don't really recommend it).

Exercise* 12.1 Give an equivalent presentation of the ZX-calculus using only equations between phase gates and the CNOT-gate.

12.1.1 Quantum Computing as ZX-Diagrams

The three steps making up a quantum computation in the circuit model together yield a ZX-diagram like this one:

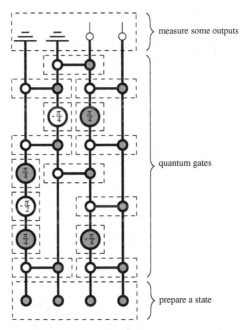

We can now use ZX-calculus for establishing equations between computations. If such a diagram only involves phases that are multiples of $\frac{\pi}{2}$, or only involves single qubit gates with phases that are multiples of $\frac{\pi}{4}$, then we know, by Theorem 9.129 and Theorem 9.132, respectively, that any equation that holds for circuits consisting of those gates can be derived in ZX-calculus. However, the ZX-calculus is still very useful even for more general circuits.

For example, let's see what the ZX-calculus tells us about the fairly complicated circuit depicted above. If we try to keep the quantum gates intact, we're pretty much stuck, but what if we temporarily forget that the chunk in the middle is made up of quantum gates, and just treat it as any ZX-diagram? That's when some ZX-magic happens!

First, consider the following chunks of the diagram:

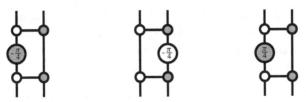

Forgetting that we are dealing with gates, we can identify a 4-cycle:

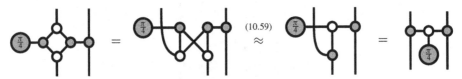

and make use of the strong complementarity rules:

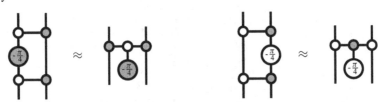

Similarly we have:

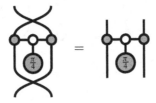

One immediate consequence, which wasn't visible before, is that these chunks are symmetric in their two inputs:

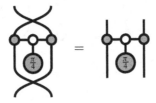

More strikingly, after substituting in the big circuit we obtain:

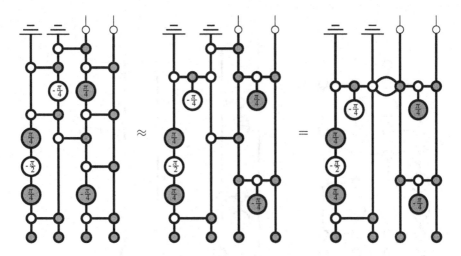

By complementarity, the left half of the circuit separates from the right half, so the (discarded) left half has no effect:

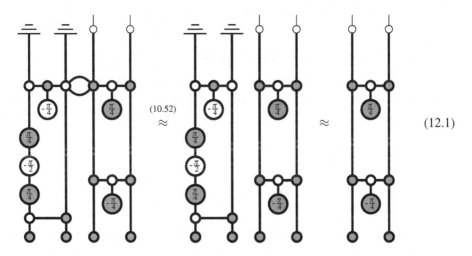

(12.1)

Since we moreover have:

(12.2)

(which is easy to show if we put these back in gate form), then our complicated circuit wasn't so complicated after all:

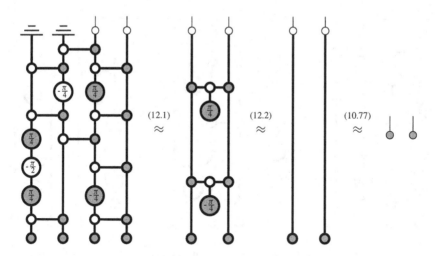

Exercise 12.2 Rather than reverting back to the gate form, prove equation (12.2) directly using strong complementarity.

12.1.2 Building Quantum Gates as ZX-Diagrams

Quantum gates are simple unitary quantum processes that can be used to construct more complicated ones. We have already seen some good candidates for quantum gates, such as phase gates and CNOT. We will now construct some more sophisticated ones using phase spiders as 'gate pieces'.

Recall from Example 9.80 that we can use:

to derive the fact that the CNOT gate uses the left qubit, called the *control qubit*, to decide whether to do nothing to the right qubit or to apply a ○-phase of π:

Naturally, we can ask if it is possible to construct controlled versions of other stuff such as, for example, a controlled ○-phase of π. The equations above suggest an easy way to do this, using the colour-change rule (10.81):

Simplifying this a bit gives:

$$\text{(diagram)}$$

Definition 12.3 The *CZ-gate* is:

$$\text{(diagram)}$$

What if, instead of selectively applying a ○-phase of π, we want to selectively apply any phase α? How could we build such a thing? Looking more closely at how the CZ-gate works gives us a clue:

$$\text{(diagram)} \overset{(10.67)}{\approx} \text{(diagram)} \overset{(*)}{=} \text{(diagram)} = \text{(diagram)}$$

$$\text{(diagram)} \overset{(10.68)}{\approx} \text{(diagram)} \overset{(*)}{=} \text{(diagram)} = \text{(diagram)}$$

The crucial step in both derivations, marked $(*)$, is where the H-gate turns a ●-phase state into a ○-phase state:

$$\text{(diagram)}$$

If we could generalise this by replacing π with any phase α:

$$\text{(diagram)} \tag{12.3}$$

then we would be there:

With a bit of ZX-trickery, building (12.3) isn't too hard.

Proposition 12.4 For:

$$
\boxed{\alpha} := \quad \text{(12.4)}
$$

we have:

$$
\text{(12.5)}
$$

Proof On the ◯-phase of 0 we have:

i.e. the phase $\frac{\alpha}{2}$ and $-\frac{\alpha}{2}$ cancel out, while on the ◯-phase of π:

the phase $\frac{\alpha}{2}$ and $\frac{\alpha}{2}$ add up.   □

Remark 12.5 When just looking at how the α-box acts on Z-basis states, the $\bigcirc$-phase at the bottom of the α-box doesn't play a role, i.e. this process:

does the same job. The important feature of the α-box is that, depending on the input, the two phases will either add up or cancel out. However, including this extra phase is convenient because it makes the α-box self-transposed and makes several other things work out more nicely (cf. Proposition 12.7 below).

Like the H-gate, this α-box is self-transposed, so we can still ignore the direction of the wire without ambiguity:

$$
\vcenter{\hbox{\includegraphics{placeholder}}}
$$

However, unlike the H-gate, this α-box isn't unitary for all α. For example, it can even separate:

$$
\vcenter{\hbox{\includegraphics{placeholder}}}
$$

So, like the spiders in the CNOT gate, it should be treated as a 'gate piece' from which we can build interesting quantum gates, most notably the one we just saw.

Proposition 12.6 The following $CZ(\alpha)$-gate is unitary:

$$
\vcenter{\hbox{\includegraphics{placeholder}}} \tag{12.6}
$$

Proof We will use the same trick as we used in Section 12.1.1 in order to simplify the big diagram. Since we can identify a 4-cycle:

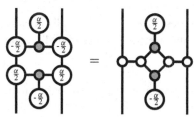

it is clear which rule we should use:

$$= \quad \text{[diagram]} \quad \approx \quad \text{[diagram]} \quad = \quad \text{[diagram]} \quad \approx \quad \text{[diagram]} \quad = \quad || \quad$$

and similarly for the other composition.   □

One would expect to recover the CZ-gate in the case where $\alpha := \pi$. Indeed this does look promising, as specialising the equations from (12.5) yields:

$$\text{[diagram]} \qquad\qquad \text{[diagram]} \qquad\qquad (12.7)$$

So, this map sends states of the (doubled) Z-basis to states of the (doubled) X-basis. In other words, it seems to behave like a H-gate. However, we've known since Section 6.1.5 that a doubled ONB is not an ONB, so we still need to check that this actually <u>is</u> an H-gate.

Proposition 12.7 The π-box is equal to the H-gate:

$$\text{[diagram]} \qquad\qquad (12.8)$$

Proof   We have:

$$\text{[diagram]}$$

□

Proposition 12.6 yields a simple form for the CZ(α)-gate, which we built up as a ZX-diagram using 'gate pieces'. Now, is this really a new quantum gate, or can it be built from the basic gates that we already had around? Again, ZX-calculus provides the answer.

Exercise 12.8 Show using ZX-calculus that CZ(α)-gate can be built from CNOTs and ○ phase gates as follows:

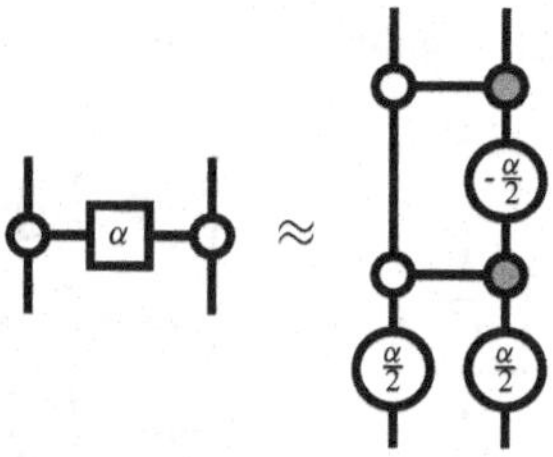

$$(12.9)$$

(Hint: the RHS contains a 4-cycle.)

We can pass back to an X version of this gate, called the $CX(\alpha)$-*gate*, by again pre- and post-composing H-gates on the second qubit:

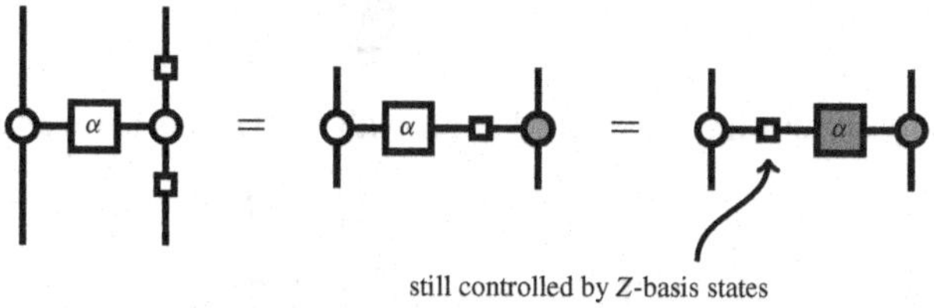

still controlled by Z-basis states

where:

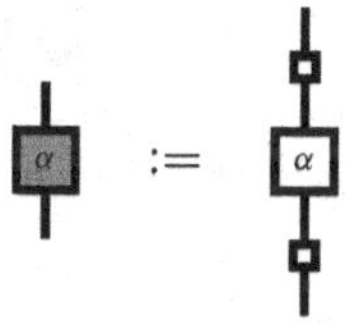

Now CNOT, (a.k.a. CX) arises as a special case where $\alpha = \pi$:

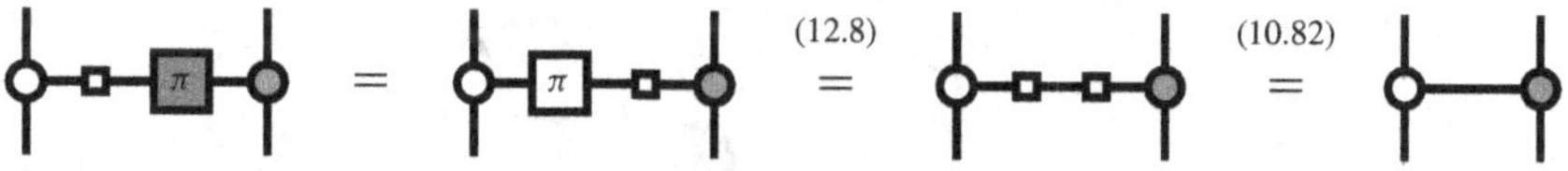

Since we can construct controlled phases of both colours, it is also possible to construct controlled unitarities, by exploiting the Euler angle decomposition from Proposition 9.100. If $\widehat{U}$ decomposes as:

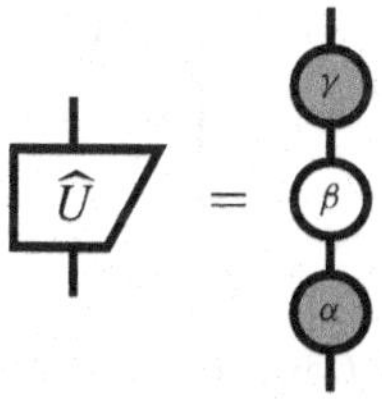

then we can construct a controlled-$\widehat{U}$ gate as follows:

$$(12.10)$$

The unitary $\widehat{U}$ only 'fires' if the control qubit is '1':

When pre- and post-composing the control qubit with NOTs, this is reversed:

We can now combine these two to selectively perform $\widehat{U}_0$ or $\widehat{U}_1$, depending on the value of the control qubit:

$$(12.11)$$

Exercise 12.9 Verify that the matrix of a *multiplexed unitary*, i.e. a gate of the form (12.11), is a block-diagonal matrix of the form:

$$(12.12)$$

where $\mathbf{U}_0$ and $\mathbf{U}_1$ are the matrices of U_0 and U_1, respectively.

12.1.3 Circuit Universality

We now have enough tools to show that we can express any pure quantum map from qubits to qubits as a ZX-diagram. As we first noted in Section 9.4.1, it suffices to show that we can realise any unitary as a ZX-diagram. Once we have any unitary, we can obtain any state as:

which can be transformed into an arbitrary map by process–state duality.

The reason for passing via unitaries is twofold. First, we will prove along the way that any unitary quantum map can be constructed using just a fixed set of unitary quantum gates, i.e that we have a *universal set of gates* for quantum computation. Second, since interest in universal sets of gates pre-dates the birth of ZX-diagrams, most of the hard work has already been done by someone else.

We'll show that we can realise any unitary as a ZX-diagram in three steps:

1. Construct the *Toffoli gate*:

   using just the CNOT-gate and phase gates, where:

2. Use the Toffoli gate to show that, whenever we can construct an n-qubit gate $\widehat{U}$ using just the CNOT-gate and phase gates, we can also construct a controlled-$\widehat{U}$ gate:

3. Use this fact to show that whenever we can build arbitrary $n - 1$ qubit unitaries, we can also build all n qubit unitaries.

12.1.3.1 Constructing Toffoli

We depicted a Toffoli gate in terms of an AND gate above, to indicate that it applies a NOT gate to the third qubit precisely when both of the first two qubits are '1' (i.e. both ⊙ phases of π):

It turns out an AND gate is pretty tricky to build (cf. Exercise 12.10 below), so we'll go straight for the Toffoli gate. First, realise that another way to think of Toffoli is as a 'controlled-CNOT':

So, how can we 'switch a CNOT off and on'? First, we'll write CNOT in terms of the $CZ(\pi)$-gate, which itself can be written in terms of CNOT gates and Z-phase gates:

Then simply turn all of the phase gates into controlled-phase gates:

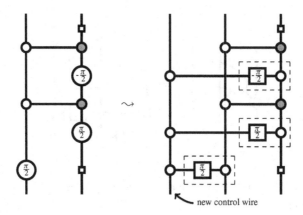

new control wire

Now we have a brand new control wire. If we input '0' into the control qubit, nothing happens to the other two qubits:

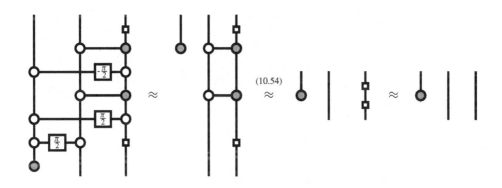

whereas if we input '1', a CNOT-gate is applied:

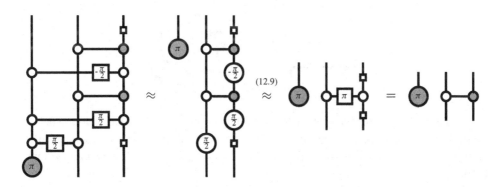

Nailed it! So let:

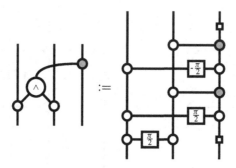

Since each of the $CZ(\alpha)$-gates above can be written in terms of CNOT and phase gates, so too can the Toffoli gate.

Exercise 12.10 Give an alternative construction of the Toffoli gate by first proving that the following linear map gives the AND gate:

(Hint: first evaluate '/' on Z-basis states, then show that '\' is its inverse.)

12.1.3.2 Constructing Controlled Unitaries

Now that step 1 is complete, we can accomplish step 2 using pretty much the same trick. Suppose $\widehat{U}$ is an n-qubit unitary built out of CNOTs and phase gates, for example:

Then, we can construct controlled-$\widehat{U}$ by adding a new control line that 'switches' every gate in the circuit off or on. That is, every phase gate becomes a controlled-phase, and every CNOT becomes a 'controlled-CNOT', i.e. a Toffoli gate. Our example becomes:

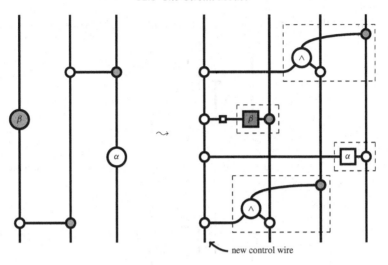

new control wire

The resulting circuit will therefore, by construction, be a controlled-$\widehat{U}$ gate:

Once we have controlled operations, we can build a multiplexor just as we did in (12.11):

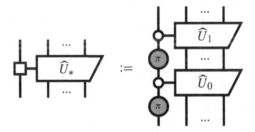

Furthermore, since the controlled-$\widehat{U}$ circuit is itself composed of CNOT and phase gates, we can repeat the whole process of adding a control line to obtain a 'controlled-controlled-$\widehat{U}$' gate:

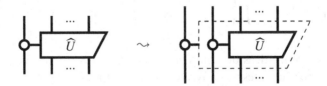

We can also repeat this n times to obtain an n-controlled-$\widehat{U}$ gate, which we could write as:

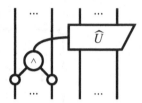

Exercise 12.11 Following (12.11), use n-controlled unitaries to construct an n-qubit multiplexed unitary, that is, a gate:

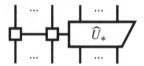

such that for any bit string $\vec{i} := i_1 i_2 \ldots i_n$, the gate applies a distinct unitary on the rightmost qubit:

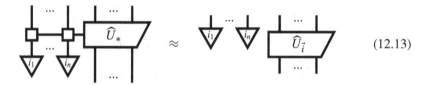 $\hspace{2cm}$ (12.13)

12.1.3.3 Putting the Pieces Together

This section could also be titled 'The Ugly Matrix Part'.

The main theorem depends on a generalisation of the Euler angle decomposition for larger unitary matrices, called the *cosine-sine decomposition*. As with the Euler decomposition, we'll omit the proof, which is essentially just a lot of matrix manipulation.

Proposition 12.12 The matrix of any unitary can be decomposed as:

$$
\begin{pmatrix} \mathbf{U}_0 & 0 \\ 0 & \mathbf{U}_1 \end{pmatrix}
\begin{pmatrix} \mathbf{C} & -\mathbf{S} \\ \mathbf{S} & \mathbf{C} \end{pmatrix}
\begin{pmatrix} \mathbf{V}_0 & 0 \\ 0 & \mathbf{V}_1 \end{pmatrix}
\hspace{1cm} (12.14)
$$

where $\mathbf{U}_i$ and $\mathbf{V}_i$ are matrices of unitary maps, and $\mathbf{C}$ and $\mathbf{S}$ are matrices whose (i, i)-th entries are $\cos \theta_i$ and $\sin \theta_i$, respectively, and all other entries are 0.

Now, we'll see how to construct these matrices from the gates we've built. Undoubing (12.13) gives:

Assuming the unitaries $U_{\vec{i}}$ are arbitrary, we can absorb the number (which is in fact just a global phase) into them, obtaining equality on the nose:

We already saw in Exercise 12.9 that the two block-diagonal matrices can be realised using multiplexors with single control wire:

The middle matrix is a bit trickier.

Exercise* 12.13 Show that for the following unitary matrix:

$$
\boxed{Y_{\vec{i}}} \quad \leftrightarrow \quad \begin{pmatrix} \cos\theta_{\vec{i}} & -\sin\theta_{\vec{i}} \\ \sin\theta_{\vec{i}} & \cos\theta_{\vec{i}} \end{pmatrix}
$$

whose angles $\theta_{\vec{i}}$ depend on a bit string $\vec{i}$, we have:

for an arbitrary matrix of sines and cosines as in Proposition 12.12.

We can therefore build arbitrary n-qubit unitaries inductively. We already know how to build arbitrary one-qubit unitaries. Assuming we can build arbitrary unitaries on $< n$ qubits, thanks to Proposition 12.12, we can construct n-qubit unitaries by means of:

This at last gives us the following theorem.

Theorem 12.14 Any n-qubit unitary can be constructed out of the CNOT-gate and phase gates:

12.2 Quantum Algorithms

Now that we know how to build any unitary out of quantum gates, let's start to put those unitaries to work in some quantum algorithms.

Before we dig in, some disclaimers are in order. While quantum algorithms are really the 'killer app' that has generated the most excitement about new quantum features over the past two decades, all quantum algorithms that are currently known span only a fairly limited range of problems. So by no means is it the case that every problem can be solved more efficiently if we had quantum computers at hand.

Second, the analysis of quantum algorithms with diagrams is still pretty young. As you'll see, there has been some progress in this direction, but rather than a complete story, this section should be read more as a preview of (or better, an invitation to take part in!) things to come. In fact, the diagrammatic presentation of the hidden subgroup problem given in Section 12.2.4 (of which quantum factoring is an instance) and its connection to strong complementarity was only discovered in the last stages of writing this book.

12.2.1 A Quantum Oracle's (False?) Magic

We already encountered the general form that a quantum computation takes in the circuit model:

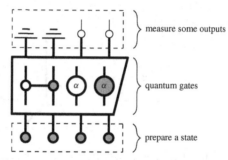

If we want to use quantum computation to beat a classical computer, we need some way or another to encode classical problems into quantum circuits. Clearly, the place to do so is the big unitary in the middle. Now, virtually any computational problem can be reduced to learning something about a function like this:

$$f : \{0, 1\}^n \to \{0, 1\}$$

For instance, the *satisfiability* problem asks whether there exists some bit string i such that $f(i) = 1$.

So, how do we turn f into a unitary? Well, first we can encode it as a linear map, as we did for classical logic gates back in Section 5.3.4:

but this linear map (and hence the quantum map $\widehat{f}$) will be unitary only if f is a bijection from its inputs to its outputs. That's no good! Especially when $N > 1$, that's just not going to happen.

But maybe not all is lost! Remember the trick we used to turn our non-unitary α-box into a two-qubit unitary? We did it like this:

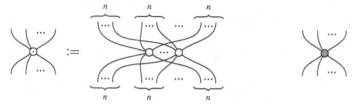

We can try a similar trick for $\widehat{f}$. First, fix spiders for both the input and the output type of f:

Instead of choosing $\bigcirc$-spiders for the output system, we take a complementary spider. Why? Since this is vital for establishing unitarity.

Proposition 12.15 The quantum map:

$$\widehat{U}_f \;\; := \;\; D \;\; \widehat{f} \tag{12.15}$$

is unitary for any function f if and only if ○ and ● are complementary.

Proof   First assume that ○ and ● are complementary. Then, also using the fact that f is a function map we have:

$$\widehat{U}_f \;\approx\; \cdots \;=\; \cdots \;\overset{(10.40)}{=}\; \cdots$$

$$\overset{(10.43)}{=}\; \cdots \;\overset{(10.52)}{\approx}\; \cdots \;\overset{(10.43)}{=}\; \cdots \;=\; \cdots$$

The proof for the other composition is similar. Conversely, the fact that unitarity of (12.15) implies complementarity comes from taking $\widehat{f}$ to be a plain wire. Then:

$$D \;\; \left[\;\cdots\;\right] \;\; = \;\; D \;\; \multimap$$

is unitary, which by Proposition 9.50 implies complementarity of ○ and ●.   □

So what does $\widehat{U}_f$ do? It turns the quantum map $\widehat{f}$ into a unitary by adjoining an extra input matching $\widehat{f}$'s output, and extra outputs matching $\widehat{f}$'s input:

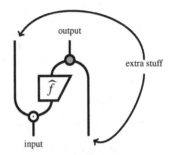

or more specifically:

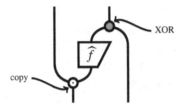

So a ○ basis state at the left input of $\widehat{U}_f$ gets copied. The first of these copies is fed to the first output of $\widehat{U}_f$, while the second one is fed into the input of $\widehat{f}$, and the corresponding output is then XOR'ed with the second input of $\widehat{U}_f$. If we take the second input to $\widehat{U}_f$ to be:

$$ \underset{0}{\triangledown} \approx \; \mathbf{|} $$

then we can evaluate the function f for any input:

$$ \text{[diagram]} = \text{[diagram]} = \underset{i}{\triangledown}\;\underset{f(i)}{\triangledown} $$

Pretty cool, right? Now, check out what happens when we put something that isn't a classical bit string into the first input:

$$ \text{[diagram]} = \text{[diagram]} $$

Rather than getting as output the value of f at one particular input, we get the entire function f, encoded as a state. In other words, we have a <u>superposition</u> of the values of f at <u>every</u> possible input:

$$
\left\lfloor \widehat{f} \right. = \text{double} \left(\sum_{i \in \{0,1\}^n} \bigvee_i \bigvee_i \boxed{f} \right) = \text{double} \left(\sum_i \bigvee_i \bigvee_{f(i)} \right)
$$

Whoa! Now you probably get what all the fuss about quantum computing is, namely:

A single quantum process can evaluate a function at all inputs simultaneously!

This single process $\widehat{U}_f$ with seemingly god-like knowledge of f is called a *quantum oracle*. But then, just how god-like is it?

Sure, we now have heaps of information about f, but how are we going to get to it? It's encoded in a quantum state, so the only thing we can do is measure it. If we're a bit thick and decide to measure the first system with respect to $\odot$, all that wonderful information becomes just <u>one</u> measurement outcome:

$$
\bigvee_i \bigvee_{f(i)}
$$

and we don't even get to choose *i*! In other words, after spending 10 million dollars to build a fancy quantum computer, all it does is evaluate f once at some random *i*. What a shockingly monumental waste of money!

12.2.2 The Deutsch–Jozsa Algorithm

So is quantum computing just one big scam? Of course not! But one has to be really clever to circumvent all the harm caused by quantum measurements.

While we won't ever get access to all of those input-output pairs of f, there are lot of other, highly non-trivial things we can ask about f that still have single answers. With a bit of cleverness, we could hope to extract with a single measurement the answer to a question that classically requires knowing many (if not all) input-output pairs of f. This would genuinely allow us to drastically out-perform a classical computer for certain tasks.

The new bad news (yes, there is more of it) is that the type of questions with single answers that can be obtained from a single measurement is very limited. For example, the 'satisfiability' problem we mentioned in the last section is provably <u>not</u> a question that can be answered by a single measurement.

Another question we could ask is whether a function f always returns the same answer for any input, i.e. whether a function is *constant*. Of course, if we knew the answer to this question, the 'satisfiability' problem becomes trivial (why?). Hence, we won't be able to find a solution to this problem either. However, maybe we can get some traction by assuming something about f.

The types of questions that lend themselves well to solutions by quantum algorithms are certain *promise problems*, i.e. problems where we know some piece of information about a function in advance (a 'promise'), and we are trying to learn something additional. In the case of the Deutsch–Jozsa algorithm, the question is whether a function either is constant or satisfies some other property 'X'. What makes it a promise problem is that 'X' is more specific than just 'not constant'. To figure out what 'X' should be, let's go ahead and start to build the algorithm and see what we need.

We'll start very generically by assuming that we apply the quantum oracle for a given function f on a superposition of all inputs:

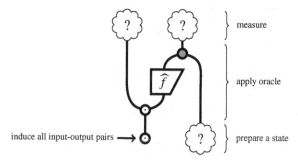

If f is constant, or equivalently, if $\widehat{f}$ is constant, then it has to be of the following form:

$$\widehat{f} = \bigtriangledown^i$$

That is, it simply deletes its input and provides the constant value i as output. Substituting this in our algorithm (to be) gives:

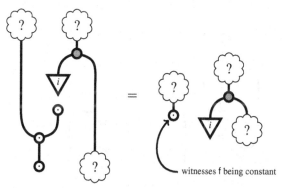

witnesses f being constant

So the diagram disconnects and the chunk on the left no longer has any dependency on the values of f. Instead, it contains a state that precisely witnesses the fact that f is constant. This state moreover happens to be an eigenstate of the ○-measurement:

(10.77)
≈

Hence:

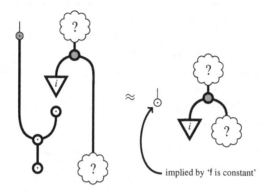

implied by 'f is constant'

Great! So we now know that a constant f will guarantee that:

$$ \overset{\cdot}{\odot} \; = \; \nabla_0 \cdots \nabla_0 \tag{12.16} $$

will be the outcome for a ⊙-measurement at the oracle's first output.

Remark 12.16 Note that here, we are not measuring the output of the function, but rather the 'extra stuff' that we added to maintain unitarity of the quantum oracle. So this 'extra stuff' has become very important in its own right!

So we now know that if f is constant, we always get that outcome. However, for a generic function f, we might also get that outcome if f is not constant. This is where our 'X' comes in. The property 'X' of a function should be chosen such that outcome (12.16) can never occur, which will guarantee that, whenever we do see this outcome, f must be constant. 'Never occurring' means probability zero, so f should satisfy 'X' if and only if:

testing for the 'f is constant' outcome

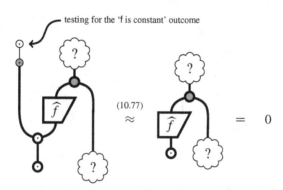

(10.77)
≈   =   0

We still have some freedom in choosing {?} appropriately. Thanks to causality, the cloud on top isn't of any use, so we can treat it as discarding. This just leaves the state that we feed into the second input of the oracle to play with. First, we try the ●-state we used in the previous section. But by causality we have:

$$
\hat{f} \quad = \quad \hat{f} \quad \approx \quad \Box
$$

So that doesn't give us anything. Next, let's try the other colour. Unfortunately, since $\hat{f}$ comes from a function on the ○-basis, this goes just as poorly:

$$
\hat{f} \quad \approx \quad \hat{f} \quad \overset{(10.43)}{\approx} \quad \Box
$$

We're running out of options here! We give it one more go:

$$
\hat{f} \quad \approx \quad \hat{f} \quad \approx \quad \hat{f}
$$

Aha! This seems like it at least might go to zero for some functions f. So what does this:

$$
\overset{\pi}{f} = 0 \tag{12.17}
$$

mean for the function f? The ○-effect with a π phase is almost like deleting, except we pick up a -1 phase when we 'delete' the second basis state:

$$
\overset{\pi}{\underset{0}{\nabla}} = \Box \qquad\qquad \overset{\pi}{\underset{1}{\nabla}} = -1\,\Box
$$

So, if:

$$\raisebox{0.5em}{\includegraphics{}} \;=\; \raisebox{0.5em}{\includegraphics{}} \;=\; \sum_i \raisebox{0.5em}{} \;=\; \sum_{\mathsf{f}(i)=0} 1 \;+\; \sum_{\mathsf{f}(i)=1} (-1) \;=\; 0$$

then the number of values i such that $\mathsf{f}(i) = 0$ must be the same as the number where $\mathsf{f}(i) = 1$. This means the function f is *balanced*.

So we found ourselves a genuine quantum algorithm! Namely, the following one.

Given. A function:

$$\mathsf{f} : \{0, 1\}^n \to \{0, 1\}$$

with the promise that it is either *constant* :=

- f either always returns 0 or always returns 1,

or *balanced* :=

- f returns 0 for the same number of inputs as it returns 1.

Problem. Is f constant or balanced?

Quantum algorithm. Perform:

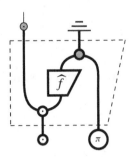

and if the outcome is:

then f is constant. Otherwise it is balanced.

Classically, to be totally sure, we will need to examine at least half of the outputs, plus 1, to determine if f is constant or balanced. In other words, we will need to *query* f at least $2^{n-1} + 1$ times. Shockingly, the quantum version requires only one query!

Remark 12.17 Note that, for this to be efficient, we must also assume that the unitary oracle can be implemented efficiently, e.g. by using quantum gates. This of course depends on the function f. However, if there exists an efficient way to implement f on a classical computer, it can be efficiently implemented using quantum gates. We refer the interested reader to further reading in the historical notes and references at the end of this chapter.

Generalising (12.17), one could say that the number:

can be used to gauge 'how balanced' f is. That is, it becomes positive if f produces more 0s than 1s, and negative if there are fewer 0s than 1, and the further the number is from 0, the greater the difference. We will make use of this fact in the next algorithm.

12.2.3 Quantum Search

We now know that quantum processes allow us to speed up at least one task. Unfortunately, there just aren't that many times when knowing if a function is constant or balanced is a matter of life and death. But maybe we can learn from the tricks we used and exploit these to do something useful.

The crux of the Deutsch–Jozsa algorithm is the fact that a $\bigcirc$-effect with a π phase deletes the result of f if it is 0, and introduces a -1 phase when it is 1:

$$\quad (12.18)$$

Plugging a $\bigcirc$-state with a π into the second input of our oracle gives:

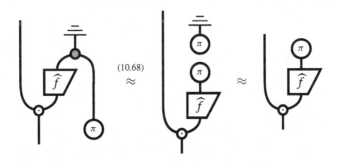

The (undoubled version of) this process acts on classical inputs as follows:

Hence by (12.18) we have:

$$
\begin{cases}
\text{if } f(i) = 0 \\
\text{if } f(i) = 1
\end{cases}
$$

So any i where $f(i) = 1$ gets 'marked' by flipping its sign. Hence, by applying this process to 'all inputs together' we obtain:

$$
= \sum_{f(i)=0} i - \sum_{f(i)=1} i \qquad (12.19)
$$

In the Deutsch–Jozsa algorithm, we exploited this state to detect constant versus balanced by means of a single ⊙-measurement. But what if, on the other hand, we could devise a measurement that gives us just one of the 'marked' bit strings i as an outcome? Well, this solves quite a useful problem indeed: the *search problem*. Many difficult computational problems boil down to search.

Suppose we have a set of things (e.g. apples), some of which are good (e.g. fresh) and some are bad (e.g. rotten). We want a fresh apple. This is easy if there are many apples and only one is rotten, but if it's the other way around, this is very hard, since we have to check the apples one by one. Putting this in terms of a function, suppose we have:

$$ f : \{0, 1\}^n \to \{0, 1\} $$

where 0 stands for rotten and 1 stands for fresh. Can we find a fresh apple, i.e. some i such that $f(i) = 1$?

Simplifying the LHS of (12.19) gives:

$$
= \sum_{f(i)=0} i - \sum_{f(i)=1} i
$$

All of the good stuff now carries a minus sign. Comparing this with the following state:

$$\downarrow = \sum_i \quad (12.20)$$

we see that subtracting these two states eliminates the rotten stuff:

$$\downarrow - \boxed{f}_\pi = 2 \sum_{f(i)=1}$$

Great! So we just measure the result and we'll get a fresh apple. So the search problem is now reduced to finding a unitary that does this:

$$\boxed{f}_\pi \;\mapsto\; \downarrow - \boxed{f}_\pi$$

Clearly it should consist of two terms, one that produces (12.20) as a constant, erasing the input, and one that does nothing except introducing a -1 phase. So let:

$$\boxed{d} := \lambda \;\; - \;\; |$$

Then:

$$\boxed{d} \;::\; \boxed{f}_\pi \;\mapsto\; \left(\lambda \boxed{f} \right) - \boxed{f}_\pi$$

Now we can choose λ to cancel out the extra number:

$$\boxed{f}_\pi$$

in the first term. In the previous section, we saw that this extra number measures the 'balance' of f:

$$\boxed{f}_\pi = \boxed{f} = \sum_{f(i)=0} 1 + \sum_{f(i)=1} (-1) = N_0 - N_1$$

where N_0 is the number of is where $f(i) = 0$, and N_1 is the number is where $f(i) = 1$. So if we take λ to be $\frac{1}{N_0 - N_1}$ then:

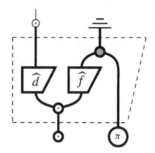

$$(12.21)$$

and hence:

only contains the good stuff. So:

will always give us a fresh apple!

But, as you are probably getting used to by now, there's a catch.

Exercise 12.18 Show that, for d to be unitary, it must be the case that:

$$\lambda = \frac{2}{N}$$

where λ is assumed to be a real number and $N := N_0 + N_1 = 2^n$.

Putting the two equations for λ together, we have:

$$\frac{1}{N_0 - N_1} = \frac{2}{N_0 + N_1}$$

which, after a bit of simple algebra, gives:

$$\frac{N_1}{N_0 + N_1} = \frac{1}{4}$$

Hence exactly one in four apples must be fresh. So we (re-)discovered the following quantum search algorithm.

Given. A function:

$$f : \{0, 1\}^n \to \{0, 1\}$$

with the promise that exactly one out of four bit strings are mapped to 1.

Problem. Find an bit string that is mapped to 1.

Quantum algorithm. Perform:

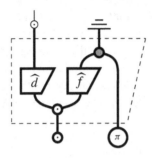

where:

$$\boxed{d} := \frac{2}{N} \, \overset{\odot}{\underset{\odot}{}} \, - \, \Big|$$

Then we always find such a bit string.

Classically, one has a probability $\frac{1}{4}$ to find a good bitstring each time one tries, and if one is really unlucky, one has to try $\frac{3N}{4} + 1$ times to find one. Again, in the quantum version it's bingo after the first try!

Of course, if fewer (or more!) than $D/4$ elements are marked, the probability of getting an unmarked element is non-zero. However, we can improve our chances if we simply iterate the unitary part of the protocol:

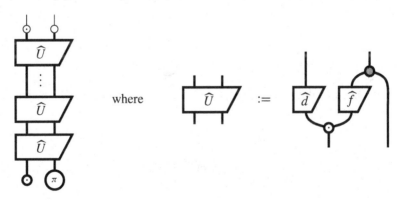

With each iteration, the probability of getting a marked element goes up, and one can show that for a <u>single</u> marked element, the probability of getting an unmarked outcome goes to zero after about $\sqrt{D}$ iterations. This multistep version is called *Grover's algorithm*.

12.2.4 The Hidden Subgroup Problem

To cap off our discussion of quantum algorithms, we'll now look at the *hidden subgroup problem*, whose solution is probably the most important quantum algorithm to date. Why is that? Since it's not clear why one should care about 'subgroups' or how they can 'hidden', it is of course not clear why one should care about the hidden subgroup problem. Well, at least one really good reason is that if we can efficiently solve the hidden subgroup problem, we can break lots of cryptography! That is, the factoring algorithm and discrete logarithm algorithms, which as we mentioned in the introduction to this chapter can be used to break many cryptographic systems, occur as special cases of the solution to the hidden subgroup problem. Who knew obscure problems about groups had anything to do with cryptography? (Answer: any cyptographer.)

We'll focus on commutative groups and make use of the classification of strongly complementary spiders from Section 9.3.6. For any commutative group G, there exists a system (which we'll also call G) and a pair of strongly complementary spiders $\bigcirc/\bullet$ that encode G. That is, for:

$$\left\{ \;\; \begin{matrix} | \; G \\ \triangledown \\ g \end{matrix} \;\; \right\}_{g \in G} \;\; \cong \;\; \left\{ \;\; \begin{matrix} | \; G \\ \kappa_g \end{matrix} \;\; \right\}_{g \in G}$$

we have:

$$(12.22)$$

Note that we carefully label wires above. This is because we now use a <u>different</u> pair of strongly complementary spiders to encode a subgroup $H \subseteq G$:

$$(12.23)$$

To witness the fact that this is a subgroup of G, we give an *inclusion map*, which embeds the group elements of H into G:

The unit and group-sum from H are preserved by inclusion, so i is a *group homomorphism*:

Note we again use wire labels to distinguish the spiders for G in (12.22) from the spiders for H in (12.23).

The way H is 'hidden' is via the *quotient group* G/H. This is a new group whose elements are sets of G elements, called *equivalence classes*. If we let:

$$[g] := \{ g' \in G \mid \exists h \in H . g' = g + h \}$$

then the set:

$$G/H := \{ [g] \mid g \in G \}$$

becomes a group, with unit $[0]$ and group-sum:

$$[g] + [g'] := [g + g']$$

As with the other two groups, we fix a strongly complementary pair of spiders to encode G/H:

This time we get a map coming out of G, called the *quotient map*:

which sends every $g \in G$ to $[g] \in G/H$. Just like with i, q is a group homomorphism:

(12.24)

Notably, if $h \in H$, the quotient map sends h to $[h] \in G/H$. But then, $h = 0 + h$, so $[h] = [0]$. Hence, every element in h gets sent to the unit. Diagrammatically, this means that if we compose i and q, this results in deleting h and sending out the unit:

$$(12.25)$$

Okay, we are ready to 'hide' H. Suppose we are given a function:

$$\mathsf{f} : G \to \{0, 1\}^N$$

with the promise that it decomposes as follows:

$$(12.26)$$

where 'injective function' just means f' is a function map and an isometry. Can we figure out what H is?

As usual, we use spiders to build the oracle for f:

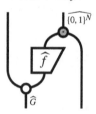

where $\odot/\bullet$ is any complementary pair of spiders for the bit-string system (last one, we promise!). Now, the first thing we thought was really cool about quantum oracles was that they let us prepare states like this:

$$(12.27)$$

Let's see what happens when we measure the left system using $\bullet$:

Individual measurement outcomes correspond to ●-ONB effects, which are the same as group elements, encoded as ○-phases:

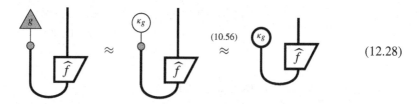

We can figure out which group elements we obtain this way with the help of a lemma about the (adjoint of the) quotient map.

Lemma 12.19 For i the subgroup map and q the quotient map we have:

$$(12.29)$$

Proof We have:

$$(10.40)$$

$$(12.24) \qquad (12.25)$$

$$(10.40)$$

$\square$

Looking at our promise for **f** in the context of (12.27), we have:

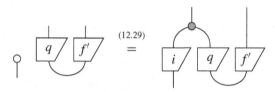

If we ⊗-compose with a ◯-effect on the left, we obtain:

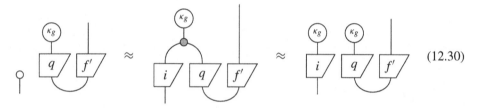

Now, plugging in a ◯-phase of κ_g gives:

So, it is either the case that:

$$\text{(diagram)} = 0$$

in which case the probability of getting the outcome corresponding to κ_g is:

$$\text{(diagram)} = \text{(diagram)} = 0$$

or we can cancel out this state from both sides of (12.30), giving:

$$\text{(diagram)} \approx \text{(diagram)}$$

In fact, it is not hard from here to show that the $\approx$-equation above actually holds on the nose. So, we can conclude that, by measuring the left system of the oracle as in (12.28), we will always obtain outcomes corresponding to group elements $g \in G$ where:

$$
\begin{array}{c}
\kappa_g \\
\big| G \\
i \diagup \\
\big| H
\end{array}
\quad = \quad
\begin{array}{c}
\circ \\
\big| H
\end{array}
\tag{12.31}
$$

This is all we need to solve the hidden subgroup problem. But to see this, we need to have a closer look at what the equation above means. Even though κ_g is <u>not</u> deleting:

$$
\kappa_g \quad \neq \quad \circ
$$

because of (12.31), it <u>acts</u> just like deleting when restricted to the subgroup H; that is, if $h \in H$ then:

$$
\frac{1}{\sqrt{D}} \;
\begin{array}{c}
\kappa_g \\
\kappa_h
\end{array}
\quad = \quad
\begin{array}{|c|}
\hline
 \\
\hline
\end{array}
$$

In group theory, the set of phases that 'locally' delete a subgroup H is called the *annihilator* of H:

$$
H' := \left\{ g \in G \;\middle|\; \forall h \in H : \quad \frac{1}{\sqrt{D}} \begin{array}{c} \kappa_g \\ \kappa_h \end{array} = \begin{array}{|c|} \hline \\ \hline \end{array} \right\}
$$

Now, if we have a subgroup, there exists an efficient <u>classical</u> algorithm for computing its annihilator, which basically amounts to solving some system of equations. But what if we only have the annihilator of a subgroup? In that case, we can exploit the following fact.

Exercise* 12.20 Assuming we have labelled classical phases such that:

$$
\begin{array}{c}
\kappa_h \\
\kappa_g
\end{array}
\quad = \quad
\begin{array}{c}
\kappa_g \\
\kappa_h
\end{array}
$$

show that $(H')' = H$. (Hint: prove that $H \subseteq (H')'$ and that H and $(H')'$ are the same size. For the latter, first show that H' is the same size as G/H.)

Voila! After not many uses of the oracle, our quantum measurement gives us the *generators* of H', i.e. enough elements to obtain any element in H' via group-sums. Then, we can use some classical post-processing to compute generators for $(H')'$, which by Exercise* 12.20, is H. In summary:

Given. A commutative group G and a function:

$$f : G \to \{0, 1\}^N$$

with a promise that there exists a subgroup $H \subseteq G$ such that:

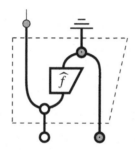

Problem. Find H.

Quantum algorithm. Perform:

and obtain outcome:

for $g \in H'$. Repeat until we obtain a generating set for H'. Use this set to (classically) compute $(H')' = H$.

So how do we go from here to factoring? Suppose we start with a function:

$$f : \mathbb{Z} \to \{0, 1\}^N$$

and we run the hidden subgroup algorithm to discover the subgroup H is:

$$\{kr \bmod D\}_k \subseteq \mathbb{Z}$$

that is, the group consisting of all the multiples of r, modulo D. Then, we have discovered that f has *period* r; i.e. for all x:

$$f(x) = f(x + r)$$

But why is that interesting? Well, if we can find the period of:

$$f(x) := a^x \bmod D$$

for randomly chosen values of a, then we can efficiently factor D! To see this, suppose the function above has period r:

$$a^x = a^{x+r} \pmod{D}$$

Then after a bit of algebra we have:

$$a^r - 1 = 0 \pmod{D}$$

Now, if we're only a little bit lucky, r is an even number, so letting $b := a^{r/2}$, we have:

$$b^2 - 1 = 0 \pmod{D}$$

Factoring the LHS gives:

$$(b + 1)(b - 1) = 0 \pmod{D}$$

So D divides the product of $b + 1$ and $b - 1$. This means one of two things is true: either D divides $b + 1$ or $b - 1$ or these both contain non-trivial factors of D. If the latter is true (which is at least as likely as the former), we can recover a factor efficiently as the greatest common divisor of $b + 1$ and D.

Remark 12.21 The careful reader will note that there was one small problem with this derivation: the commutative group $\mathbb{Z}$ is not finite! However, if we choose some q very large, then we do this algorithm with a function $f : \mathbb{Z}_q \to \{0, 1\}^N$ from the cyclic group $\mathbb{Z}_q$ and get the same result with high probability.

Exercise 12.22 Show that the one-bit Deutsch–Jozsa problem for a function:

$$f : \{0, 1\} \to \{0, 1\}$$

is an instance of the hidden subgroup problem for f. Show that for $N > 1$ and:

$$f : \{0, 1\}^N \to \{0, 1\}$$

there is no group-sum one can fix for the set $\{0, 1\}^N$ that makes the Deutsch–Joszagroup problem into an instance of the hidden subgroup problem.

12.3 Measurement-Based Quantum Computation

Measurement-based quantum computing (MBQC) is an alternative way to provide universal quantum computation. Rather than stuffing all the computational structure in unitaries, in MBQC all quantum processes are measurements. We first encountered something like this in Section 7.2.2 in the form of gate teleportation. Here we present an MBQC model based on *graph states* (cf. Section 9.4.5) and single-qubit measurements, also known as the *one-way model* of quantum computation. In this model, a computation consists of three steps:

1. Prepare a graph state.
2. Perform single-qubit measurements, where later measurements can be controlled by previous measurement outcomes, using *feed-forward*.
3. Possibly do some classical post-processing on measurement outcomes.

The key is to exploit the backaction (cf. Section 7.2.1) of single-qubit measurements to (non-deterministically) realise arbitrary quantum effects, which, when applied a graph state, are enough to realise any quantum computation. For example, if we choose measurements that give us the effects:

then we can turn this piece of a graph state:

into any single-qubit unitary:

$$\tag{12.32}$$

Pretty cool, right? In this section, we'll see not only how to recover arbitrary single- and multiqubit gates using measurements, but also how to do it deterministically, using a technique similar to the one we used for quantum gate teleportation.

The advantage of the MBQC paradigm over the circuit model is that, once we have a graph state, all subsequent computation is done using only single-qubit processes. With current technology, multiqubit operations like CNOT can be very tricky, and often introduce too much decoherence to be practical for quantum computing. This, combined with the fact that certain graph states are relatively straightforward to prepare in the lab has made MBQC a promising choice for actually implementing quantum computations.

12.3.1 Graph States and Cluster States

We introduced graph states:

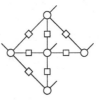

in Section 9.4.5 to prove completeness of the ZX-calculus. MBQC is based on *quantum graph states*, obtained by doubling:

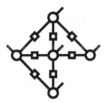

These many-qubit states can be realised using a simple circuit as follows:

1. Prepare n qubits in the $\bigcirc$-state:

2. To introduce an edge between the i-th and j-th qubit, apply a CZ-gate:

Since the spiders all fuse, it doesn't matter which order we apply CZ-gates. We will always get a graph state:

The most commonly studied type of graph state is a *cluster state*.

Definition 12.23 A *two-dimensional cluster state* is a graph state whose nodes form an $m \times n$ grid:

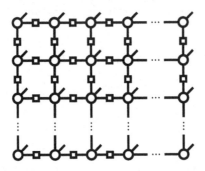

12.3.2 Measuring Graph States

In MBQC, all the magic comes from single-qubit measurements. In fact, it will suffice to consider just two kinds of measurements:

$$\begin{array}{cc} & := \text{ } Z\text{-measurement} \qquad\qquad := \text{ } X_\alpha\text{-measurement} \end{array}$$

Since an X_α-measurement consists of a $\bigcirc$-phase of α followed by an X-measurement, its associated quantum effects are:

$$\approx \quad \overset{(10.76)}{\approx} \quad = $$

$$\approx \quad \overset{(10.78)}{\approx} \quad = $$

Hence, an X_α-measurement amounts to a measurement for some ONB on the equator of the Bloch sphere:

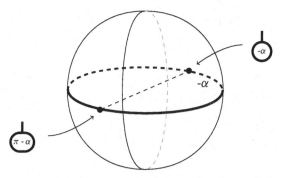

As a special case, X_0 is of course just a normal X-measurement.

The utility of an X_α-measurement is to introduce phases into a graph state. For example, if we measure a single qubit and get outcome 0, we will introduce a ○-phase of α:

$$\cdots \; \mathbin{\vdots} \; \alpha \; \cdots \;\; = \;\; \cdots \; \mathbin{\vdots} \; \alpha \; \cdots$$

If we get outcome 1, this produces a phase of $\alpha + \pi$ instead of α, which we treat as an error that will need to be corrected later. We will see how to do this in the next section. By carefully selecting where we perform X_α-measurements, we can produce ●-phases as well as ○-phases, as we did in (12.32) to realise an arbitrary single-qubit unitary.

Whereas X_α-measurements introduce phases, Z-measurements can be used to cut out unwanted qubits from a graph state. For example, if we perform a Z-measurement on a two-dimensional cluster state and get outcome 0, this leaves a hole:

$$\overset{(10.76)}{\approx} \qquad \overset{(10.58)}{\approx}$$

$$\overset{(10.66)}{=} \qquad =$$

Repeating this process many times, we can carve out arbitrary shapes:

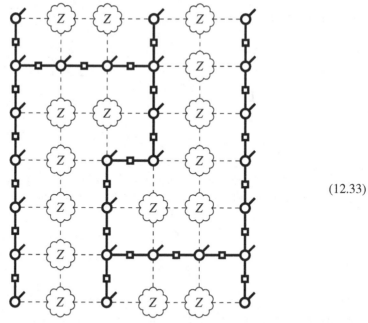

(12.33)

As in the case of X_α-measurements, we should treat outcome 1 as an error, which in this case will introduce extra π phases on all of the neighbours of the deleted qubit. We will now see how to correct both kinds of error using a technique called *feed-forward*.

12.3.3 Feed-Forward

In quantum teleportation (and quantum gate teleportation) we achieve an overall deterministic process by matching unitary corrections up with measurement outcomes. In some sense, MBQC is a vast, many-qubit generalisation of quantum gate teleportation. Thus, corrections are the key to eliminating errors and obtaining a deterministic computation. The ONB-measurements from the previous section will introduce the following effects:

Z-measurement: $\left\{ \begin{matrix} \kappa \\ \end{matrix} \right\}_\kappa$ X_α-measurement: $\left\{ \begin{matrix} \alpha+\kappa \\ \end{matrix} \right\}_\kappa$

where $\kappa \in \{0, \pi\}$. If $\kappa = \pi$, we treat this as an error, which we can correct by applying phase gates to qubits near the one we measured. For a Z-measurement, we obtain:

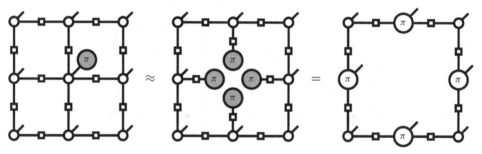

Hence we can correct the error by simply applying a ○-phase of π to all the neighbours of the qubit we measured.

In the case of an X_α-measurement, we can make the correction by 'pushing' the error along a graph state until it only appears on output wires, where it can be corrected. Let's see how this works for just a single measurement on a chain of three qubits. Measuring the first qubit produces an error of π. Using the spider-fusion rules, we can shift this error onto an edge of the graph state:

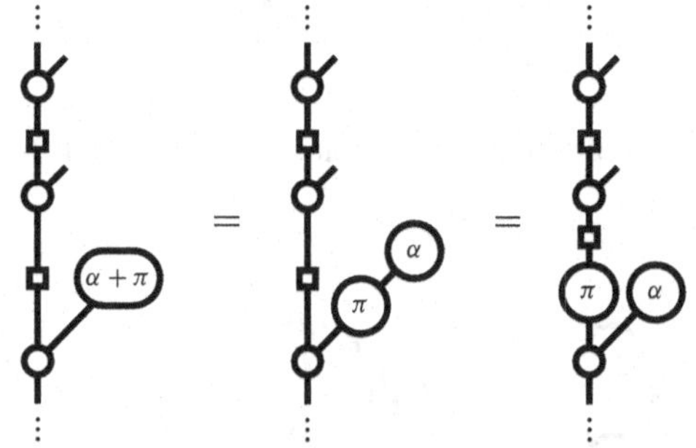

Now, using colour-change, π-copy, and spider rules, the π can be pushed upwards. After passing through the second qubit, an error still remains on the graph state:

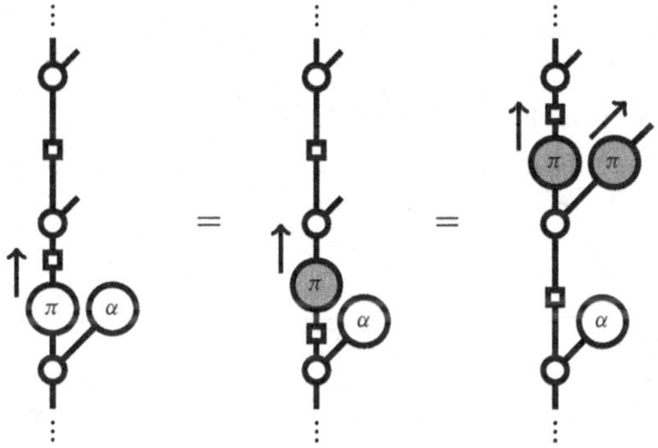

but after passing through the third qubit, the error only occurs on output wires:

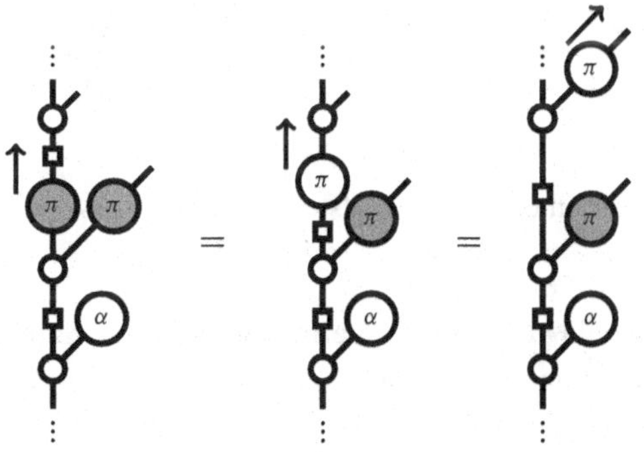

So, chaining everything together we obtain:

$$(12.34)$$

We can therefore apply corrections on the second and third qubits to correct the error from the first measurement:

$$(12.35)$$

which yields an α phase gate deterministically.

If we are also measuring the second and third qubits, we get these corrections for free, simply by adjusting our later measurement choices:

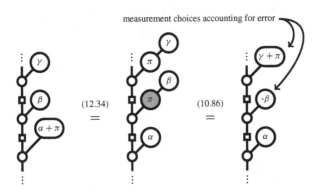

measurement choices accounting for error

(12.34)
$=$

(10.86)
$=$

Of course, these measurements could also produce errors, which also need to be corrected.

Exercise 12.24 How would you correct an error for an X_α measurement in a cluster state? What's the general rule for any graph state?

We can obviously adjust measurement choices only for measurements that haven't happened yet. So, the order in which we choose to perform measurements has an effect on when (and if) we make these corrections.

Exercise* 12.25 Can we measure the qubits in a cluster state in some order such that it is possible to feed-forward <u>all</u> corrections? How about for any graph state?

12.3.4 Feed-Forward with Classical Wires

Feed-forward seems to work pretty well, but somewhere along the way, we started working with effects rather than measurements, so we lost the classical wires and hence the explicit flow of classical data. Can we get them back? Of course! Here's one way to do this.

First, we make all classical data the same colour by making a minor modification to the X_α-measurement from before:

For corrections, we define cq-maps that, depending on a classical input bit, will either apply a π phase or do nothing. The two kinds of corrections we need are:

Note that these are nearly the same as the corrections (10.47), which were used for quantum teleportation before. The only differences are making the classical data the same colour to match the associated measurements and keeping a copy of the classical data, rather than deleting it. We'll see soon why we do this.

Now, the correction procedure given by (12.35) translates as follows:

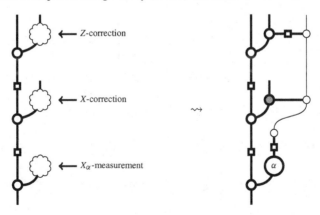

The measurement result is being fed-forward along a classical wire, and used twice to make the two corrections. Now, we will see that we can still reason in the same way as before, even with all the extra wires around. The only difference is, rather than pushing the π through the diagram as we did in (12.35), the thing we should now try to push through the diagram is ...

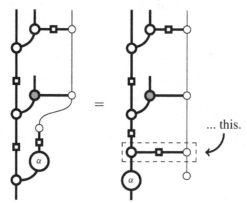

... this.

As before, there are a few crucial moves we use to push the error out. The first kind of move commutes a correction past an H-gate, which changes its (quantum) colour:

$$(12.36)$$

The second kind slides a Z-correction through a $\bigcirc$-spider:

$$(12.37)$$

which is just quantum spider fusion. The third kind copies an X-correction through a $\bullet$-spider:

$$(12.38)$$

This rule follows from strong complementarity:

The last type of rule cancels a correction with itself:

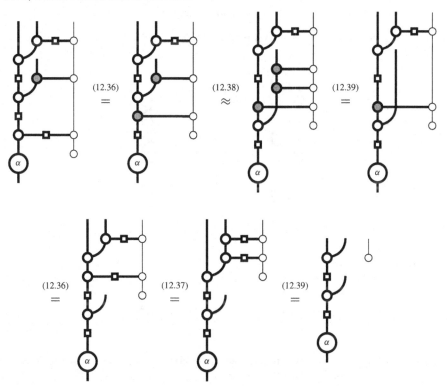

$$(12.39)$$

Both of these come from complementarity:

Now, let's see these moves in action:

Thus we get an α phase gate deterministically, just as expected. The only remnant of the measurement is the outcome itself, which as we can see, is a uniform probability distribution.

Exercise 12.26 Use the feed-forward rules to prove that the following diagram yields a phase of α on the bottom qubit, deterministically:

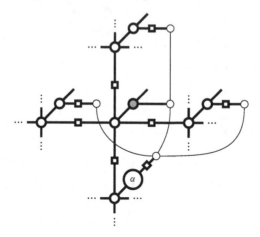

12.3.5 Universality

As we've seen, we can build arbitrary single-qubit unitaries like this:

$$\begin{array}{ccccc} & = & & = & \end{array}$$

and it's even easier to produce CZ-gates:

$$= $$

So we have, in principle, everything we need for universal quantum computation (cf. the construction throughout Section 12.1). But how do we put it all together into a circuit? One solution is to start with a big cluster state and 'carve out' the shape of the circuit we want by means of Z-measurements, as in (12.33). Then, we can use X_α-measurements to fill in all the phases. This will work, and in fact, it was how universality was originally shown.

However, if we are more clever about what kind of graph state we start with, we might be able to find something that's already universal using only X_α measurements. In other words: no carving necessary. This has the appealing feature that the entire computation now consists <u>only</u> of a list of angles. Such a state does exist, called the *brickwork state*:

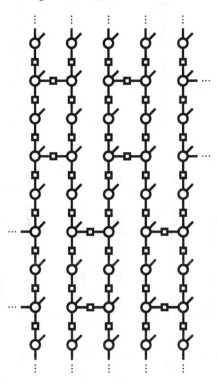

It consists of a repeating pattern of 'bricks':

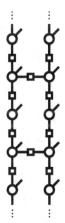

which can be made to do our bidding solely by choosing angles. If we choose 0 everywhere, a brick does nothing at all:

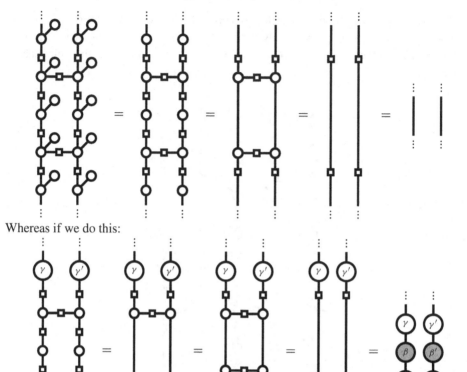

Whereas if we do this:

the result is two arbitrary single-qubit unitaries. And now to seal the deal. If we could turn a brick into a CZ-gate, we would of course be done already, since CZ-gates plus single-qubit gates are universal. But as we saw above, putting 0-phases everywhere will create <u>two</u> CZ-gates, which cancel out. However, if we do something a bit more clever:

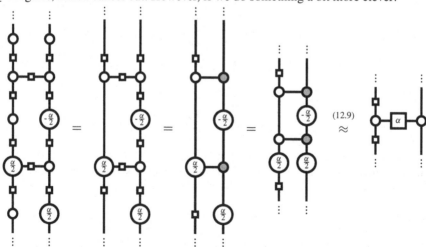

$$(12.9)$$

a controlled-phase gate appears. That should do the trick! Indeed, choosing $\alpha := \pi$, this particular brick gives us a CNOT:

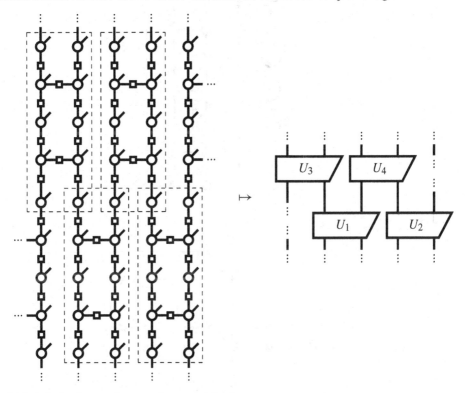

Hence we have a universal set of quantum gates.

In order to assemble these gates into arbitrary circuits, we can choose measurement angles for each of the bricks (and summing the angles when the bricks overlap). Then, the whole brickwork state can be turned into a circuit of interleaved quantum gates:

which is indeed enough to produce any circuit.

Exercise 12.27 Show that it is possible to turn an interleaved circuit consisting of single qubit unitaries and CNOT into an arbitrary circuit.

The final piece of the puzzle is that we should not only be able to recover any circuit, we should be able to do it deterministically. For this to be the case, there should exist some order for measuring the qubits such that we can always feed errors forward onto qubits we haven't measured yet.

Exercise 12.28 Show that any error on row k can be fixed by applying corrections only on rows $k + 1$ and $k + 2$.

Hence, it is possible to obtain any circuit deterministically by measuring qubits row by row, and therefore we have the following.

Theorem 12.29 MBQC with brickwork states and X_α-measurements is universal for quantum computation.

12.4 Summary: What to Remember

1. Two important models of quantum computation are the *quantum circuit model*:

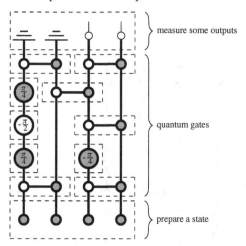

and *measurement-based quantum computation*:

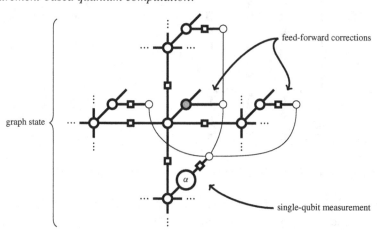

Both are *universal*, in the sense that they can implement any unitary from qubits to qubits.

2. Any function map can be turned into a unitary quantum map called a *quantum oracle*, using a complementary pair of spiders. These form the basis of most quantum algorithms:

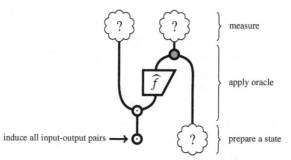

3. Most quantum algorithms solve *promise problems*. We covered these ones:

Algorithm	Promise	Problem	Diagram
Deutsch–Jozsa	f is constant or balanced.	Which is it?	
Quantum search	f maps one in four bit strings to 1.	Find i such that f(i) = 1.	
Hidden subgroup	f factors through G/H for some subgroup $H \subseteq G$.	Find H.	

A special case of the hidden subgroup problem is *period finding*, which allows efficient factorisation of large numbers.

12.5 Historical Notes and References

The first hints towards quantum computing were given by Paul Benioff (1980), where quantum mechanical models of computers were described, and by renowned mathematician

Yuri Manin (1980), who proposed the idea of quantum computing. This was, however, still at the time of the Cold War, and it was written in Russian. Therefore, many attribute quantum computing to Richard Feynman due to a talk he gave at MIT in 1981. (Note that this is the second time in this book that we find Feynman in a scooping mode.) The first universal quantum computer was described by David Deutsch (1985), using the notion of a *quantum Turing machine*. This has since largely been superceded by the (simpler) circuit model, also due to Deutsch (1989). The first proof of universality of the circuit model was given in Barenco et al. (1995). The proof we give is based (partially) on a substantially simpler proof from Shende et al. (2006).

The Deutsch–Jozsa algorithm appeared in Deutsch and Jozsa (1992); Shor's factoring algorithm appeared in Shor (1994, 1997); and the quantum search algorithm first appeared in Grover (1996). The hidden subgroup algorithm, along with its encoding of Shor's factoring algorithm and Simon's problem, was given by Jozsa (1997). A common misconception in the quantum computing community is that the Deutsch–Jozsa algorthm arises as a special case, but as highlighted in Exercise 12.22, this is only the case when $N = 1$. Recent surveys of quantum algorithms include Ambainis (2010) and Montanaro (2015).

Applications of quantum picturalism to quantum circuits include Boixo and Heunen (2012) and Ranchin and Coecke (2014), which includes the solution to Exercise* 12.1. The diagrammatic treatment of quantum algorithms was introduced by Vicary (2013) and further developed in Zeng and Vicary (2014) and Zeng (2015). The diagrammatic derivation of the hidden subgroup algorithm is given by Gogioso and Kissinger (2016).

The one-way MBQC model was first proposed by Raussendorf and Briegel (2001) and further elaborated on in Raussendorf et al. (2003). Graph states were first proposed in Hein et al. (2004). The diagrammatic treatment was first proposed in Coecke and Duncan (2008, 2011) and elaborated in Duncan and Perdrix (2010). Existing techniques in MBQC were improved upon using the ZX-calculus in Duncan and Perdrix (2010); Horsman (2011) gives a ZX-calculus-based presentation of *topological MBQC* (Raussendorf et al., 2007). A survey of MBQC in the ZX-calculus is Duncan (2012).

While practical quantum computers do not exist yet, there are notable achievements in the lab. The first implementation of a quantum algorithm on a three-qubit nuclear magnetic resonance quantum computer was realised in Oxford by Jones et al. (1998). In 2000, a five-qubit one was realised in Munich, as well as a seven-qubit one at Los Alamos. The fidelity of quantum computations has improved as well. For example, single- and double-qubit gates have been implemented up to a 99.9% fidelity using trapped ions by Ballance et al. (2016). The first realisation of MBQC was done in Vienna by Walther et al. (2005).

13

Quantum Resources

In my life, I have prayed but one prayer: oh Lord, make my enemies
ridiculous. And God granted it.
— Voltaire, letter to Étienne Noël Damilaville, 1767

Although bankers on Wall Street can have pretty much whatever they want in whatever
quantity they want, this chapter is for the rest of us cheapskates. Throughout this book, we
have assumed that we, like quantum bankers, have had access to whatever processes we
need in whatever quantity. For instance, in teleportation, we assumed we could obtain Bell
states at will and do joint measurements on pairs of systems. For non-locality, we assumed
we had lots of GHZ states around, so we could make enough measurements to see our
assumptions about locality cave in. And for MBQC, we assumed that we had huge graph
states around to implement universal quantum computation.

As you might expect, these things ain't cheap! Quantum processes involving multiple
systems typically take some very special equipment and many hours to implement. So,
when it comes to such *resources*, it behooves one to think about doing as much as possible
with as little as possible. This is of course true not just for quantum states, but for any kind
of resource: coal, oil, nuclear, wind, and solar energy; certain chemicals; or just a bit of
affection.

What is important about all resources is what we can do with them. If we think of
the benefits of resources (e.g. warm, comfy houses or secure communication) as being
resources themselves, we see that pretty much all questions about resources boil down to
the following two:

1. Can a given resource be <u>converted</u> into some other resource?
2. <u>How much</u> of resource X to do we need to obtain resource Y?

These are exactly the questions a *resource theory* aims to answer.

One place this idea of 'conversion of resources' appears very explicitly is in the study
of chemical reactions, where one finds expressions like these:

$$6CO_2 + 6H_2O + \text{light} \longrightarrow C_6H_{12}O_6 + 6O_2$$
$$C_6H_{12}O_6 \longrightarrow 2C_2H_5OH + 2CO_2$$

Such an expression tells us that we can convert the stuff on the LHS to the stuff on the RHS. Even though the expression doesn't provide any details about how this conversion is actually done, it does provide us with two very useful pieces of information. First, it specifies how much we need of one resource to get a certain amount of another, and second, it tells us all the resources a given resource can be converted into. Clearly, the more things it can be converted into, the more desirable it is.

This is true for any kind of resource, including the quantum examples we gave above. We listed several entangled states that were good for various tasks, but if some entangled state can be easily converted into any other entangled state, then clearly it should be at the top of the food chain. This idea is captured by the *resource theory of entanglement.*

Of course, entanglement is not the only quantum resource one may care about. Given that real-world devices tend to implement quantum processes in fairly noisy and unreliable ways, it is very natural to study *purity* of quantum states as a resource theory. Other resources are even more subtle. For instance, when doing quantum key distribution, sharing a reference frame, i.e. agreeing on what the Z- and the X-spiders are, is a key resource.

But what exactly is a resource theory? It's a process theory, of course! More specifically, it's a process theory with a very special interpretation: the types of the process theory represent resources and the processes represent conversions. From a process theory, we can derive a *convertibility relation*, which in turn tells us about the *conversion rates* of resources, what good (cost) *measures* should be, whether a theory has *catalysts*, etc.

Along the way, we'll make two interesting observations. When we study the resource theories of purity and entanglement, we'll see that, even though the conversion relations of these theories seem to be 'quantitative', we can still prove strong characterisation theorems diagrammatically. Moreover, when we study entanglement of three qubits, we'll see that each of the maximal *entanglement classes* is captured by some kind of spider. One of these spiders is very familiar by now, since it's been the main character of the past five chapters. However, the other kind of spider is quite different. Rather than happily fusing with its neighbours, it explodes:

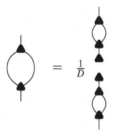

13.1 Resource Theories

A resource theory is really just a process theory, except for the wording.

Definition 13.1 A *resource theory* is a process theory where the types are called *resources*, and the processes are called *resource conversions*.

This actually makes it very different!

For example, we could define a resource theory called **food**, where the systems are ingredients and the processes are ways to prepare food, i.e. convert one food into an other food. A raw carrot can be converted into a boiled carrot, or a boiled carrot and a cup of broth can be converted into soup. Another example is **energy**, whose resources are forms of energy and whose resource conversions turn one form of energy into another: coal into heat, heat into electricity, electricity into work, etc.

13.1.1 Free Processes

For process theories such as **quantum processes**, all states of a given type have been treated on equal footing. However, clearly there is a big difference between the following two states in terms of what one can do with them:

For example, the first one allows one to do teleportation, while the second one doesn't. To capture this difference within a process theory we will treat the states themselves as different types. So, states in a process theory like **quantum processes** will become types in the resource theory we construct. Hence, we get some types (i.e. resources) that are useful for teleportation and others that are totally useless for teleportation.

They will be related by means of the processes (i.e. resource conversions) in our theory. But what should these be? One choice would be to take the conversions from a resource ρ to another resource ρ' to be all those processes in the original process theory that transform ρ into ρ':

$$\vcenter{\hbox{\includegraphics{fig1}}} = \vcenter{\hbox{\includegraphics{fig2}}} \qquad (13.1)$$

However, that wouldn't give us anything useful. For example, if we care about entanglement, we want to be able to learn from the structure of the resource theory **entanglement** that the Bell state is a much more valuable resource than any separable state. But if we include all processes as resource conversions, we could just use this process:

$$\vcenter{\hbox{\includegraphics{fig3}}} \quad :: \quad \vcenter{\hbox{\includegraphics{fig4}}} \mapsto \vcenter{\hbox{\includegraphics{fig5}}} \qquad (13.2)$$

to turn any state, including the separable ones, into a Bell state!

So, we should distinguish between those processes that allow us to 'create' more of our resource (in this case, entanglement) and those that do not. For example the quantum process:

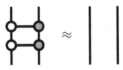

should be allowed, since it can only destroy entanglement, while the quantum process (13.2) should not be allowed, since it can create entanglement where there was none before.

A process that only decreases (or preserves) the resource of interest is called a *free process*, which is meant to suggest that the other processes are 'expensive', so we want to avoid using them. If two processes are free, then their compositions should also be free, so free processes always form a subtheory of a resource theory.

Remark 13.2 On the other hand, non-free processes typically do not form a process theory. For instance, the CNOT gate can introduce new entanglement, so it should not be free, but doing a CNOT-gate twice:

is just the same as doing nothing, which is always free!

Note that *free states* are processes in **F**:

that convert 'nothing' into that state itself. This perfectly matches the idea that 'free' means getting something for nothing.

We can capture all of this by specialising Definition 13.1 to those resource theories that arise from an underlying process theory.

Definition 13.3 Given a process theory **P** together with a subtheory **F** of *free processes*, we define the corresponding resource theory:

P-states/F

to be the process theory that has:

- the states of **P** as its types, and
- the processes in **F** that convert ρ into ρ' as its processes.

The beauty of such resource theories is that one never has to say exactly what it means for a state to have 'X amount of a resource'. For instance, as we'll see later, asking 'how much' entanglement a state has is not really a well-defined question, because states can be entangled in different, inequivalent ways. On the other hand, once we say what our free processes are, we can immediately start comparing resources.

13.1.2 Comparing Resources

Definition 13.4 For a resource theory **R**, resource A can be *converted* into resource B, which we write as follows:

$$A \succeq B$$

if and only if there exists a process in **R** from A to B:

Two resources A and B are *equivalent* if $A \succeq B$ and $B \succeq A$, in which case we write:

$$A \simeq B$$

This definition has three immediate consequences. First, that $\succeq$ is *reflexive*, i.e. $A \succeq A$, since the identity process always converts A to itself:

It is also *transitive*, i.e. if $A \succeq B$ and $B \succeq C$, then $A \succeq C$, by ∘-composition:

These two conditions make the relation $\succeq$ into a *preorder*. However, $\succeq$ does not need to be a *partial order*, which is a preorder that is additionally *anti-symmetric*, i.e. if $A \succeq B$ and $B \succeq A$ then $A = B$, because it's perfectly reasonable to have resources that are interconvertible but not equal. That is, there can exist:

In addition to being a preorder, $\succeq$ plays well with $\otimes$:

So in summary we have the following theorem.

Theorem 13.5 For R the set of resources in a resource theory:

$$(R, \succeq, \otimes, I)$$

forms a *preordered monoid*. That is, $(R, \otimes, I)$ forms a monoid:

$$(A \otimes B) \otimes C = A \otimes (B \otimes C) \qquad A \otimes I = A = I \otimes A$$

$(R, \succeq)$ forms a preorder:

$$A \succeq A \qquad\qquad A \succeq B \text{ and } B \succeq C \implies A \succeq C$$

and these two structures are compatible:

$$A_1 \succeq B_1 \text{ and } A_2 \succeq B_2 \implies A_1 \otimes A_2 \succeq B_1 \otimes B_2$$

Remark 13.6 Equivalently, a preordered monoid can be defined as a process theory where there exists at most one process of any given type, which we could write like this:

The presence of such a process then simply witnesses the fact that $A \succeq B$. So, the passage from a resource theory to a preordered monoid $(R, \succeq, \otimes)$ can be seen as passing from a big process theory, with lots of different processes, to a much smaller one, where we only remember whether there were any processes of a given type or not.

Very early on in this book we saw that process theories put $\circ$-composition and $\otimes$-composition of processes on the same footing. Similarly, convertibility of resources is tightly intertwined with $\otimes$-composition of systems. While one may not be able to convert a single pound coin into a home in Oxford, one million or so of those would get you something:

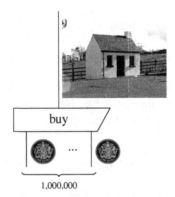

Hence, the addition of ⊗ gives us the ability to trade quantities of one resource for quantities of another. In other words, we can express the rate at which one resource can be converted into another.

Definition 13.7 For resources A and B in a resource theory, the *conversion rate* is given by:

$$r(A \succeq B) := \mathsf{supremum} \left\{ \frac{N}{M} \;\middle|\; \underbrace{A \otimes \cdots \otimes A}_{M} \succeq \underbrace{B \otimes \cdots \otimes B}_{N} \right\}$$

If we consider a resource theory just containing the process above, the conversion rate from pounds to Oxford houses is $\frac{1}{1,000,000}$. So, a pound can be converted into one one-millionth of an Oxford house. What a bargain!

Remark* 13.8 Since the conversion rate is computed as a supremum, this can in general be irrational. Take, for example, a theory where resources are little strings that we can cut and arrange into shapes. Then, we need 4 strings of length 1 to make a circle of diameter 1:

but only 7 strings to make 2 circles:

and 10 strings to make 3 circles, 13 to make 4, and so on. If we carry on to infinity, we approach the optimal rate, which is given by the supremum:

$$\text{supremum} \left\{ \frac{1}{4}, \frac{2}{7}, \frac{3}{10}, \frac{4}{13}, \cdots \right\} = \frac{1}{\pi}$$

So, we can optimally produce one circle for every π strings.

The convertibility relation totally forgets what the processes in a resource theory are and remembers only if such a process exists. Nevertheless, it already contains a great deal of information about the structure of a resource theory. For example, it tells us whether our resource theory has *catalysts*, that is, resources C for which:

$$A \otimes C \succeq B \otimes C \text{ while } A \not\succeq B$$

It also tells us if resources are *quantity-like*:

$$\left. \begin{array}{r} A_1 \otimes A_2 \simeq B_1 \otimes B_2 \\ \\ A_1 \succeq B_1 \end{array} \right\} \implies B_2 \succeq A_2$$

i.e. resources behave like 'quantities of stuff', where the whole is just the sum of the parts. Similarly, it tells us if resources are *non-interacting*:

$$A \succeq B_1 \otimes B_2 \implies \exists A_1, A_2 : \left\{ \begin{array}{l} A \simeq A_1 \otimes A_2 \\ A_1 \succeq B_1 \\ A_2 \succeq B_2 \end{array} \right.$$

which means that whenever a resource gives us multiple things, we can 'cut it up' and produce each of those things independently.

Exercise 13.9 Prove that if a resource theory is quantity-like and non-interacting, then it is also *catalysis-free*:

$$A \otimes C \succeq B \otimes C \implies A \succeq B$$

13.1.3 Measuring Resources

Physicist like real numbers, and so do many others. There is nothing wrong with that, as long as it doesn't become an obsession. When it comes to gauging the value of a resource, numbers might be part of the story, but they can never be the whole story, because they often miss vital information. This stems from the fact that real numbers form not only a partial order, but in fact a *total order*. That is, for any two real numbers a and b we have either:

$$a \geq b \qquad\qquad \text{or} \qquad\qquad b \geq a$$

Hence, if we want to compare two resources by assigning them real numbers, it will always be the case that:

$$A \succeq B \qquad \text{or} \qquad B \succeq A$$

But what if neither of these things is true?

A key example comes from entanglement. While using numbers to measure entanglement works okay for a pair of qubits, we'll see in Section 13.3.2 that as soon as we have a slightly more complicated system (e.g. three qubits), this breaks down precisely because incomparable states start to appear. Rather than giving us 'more' or 'less' entanglement, these states give us different kinds of entanglement. However, this didn't stop people from coming up with a whole zoo of ways to measure entanglement with real numbers, which work well for comparing some states, but not others.

On the other hand, when it comes to measuring impurity of quantum states, there is a very famous number called *entropy*, which does a pretty good job of telling us how pure a state is.

But before we get there, how do we go about assigning numbers to resources in a consistent way? We begin by making the positive real numbers $\mathbb{R}_{\geq 0}$ into a preordered monoid. We always let $\succeq$ be the usual ordering on positive real numbers, but then for $\otimes$, we have many choices. For instance, we can take $\otimes$ to be the sum of two positive real numbers. Then clearly:

$$a \geq a' \text{ and } b \geq b' \quad \Longrightarrow \quad a + b \geq a' + b'$$

But we can also use:

$$\max(a, b) := \begin{cases} a & \text{if } a \geq b \\ b & \text{if } b \geq a \end{cases}$$

Exercise 13.10 Show that max is associative:

$$\max(a, \max(b, c)) = \max(\max(a, b)c)$$

and compatible with $\geq$, that is:

$$a \geq a' \text{ and } b \geq b' \quad \Longrightarrow \quad \max(a, b) \geq \max(a', b')$$

These two choices give us two different preordered monoids:

$$\mathcal{R}^+ := (\mathbb{R}_{\geq 0}, \geq, +, 0) \qquad \mathcal{R}^{\max} := (\mathbb{R}_{\geq 0}, \geq, \max, 0)$$

and two corresponding kinds of measures.

Definition 13.11 Given a resource theory, an *additive measure* is a function:

$$M : (R, \succeq, \otimes) \to \mathcal{R}^+$$

where:

$$A \succeq B \implies M(A) \geq M(B) \qquad M(A \otimes B) = M(A) + M(B)$$

Similarly, a *supremal measure* is a function:

$$M : (R, \succeq, \otimes) \to \mathcal{R}^{\max}$$

where:

$$A \succeq B \implies M(A) \geq M(B) \qquad M(A \otimes B) = \max(M(A), M(B))$$

A measure gives us only partial information about a resource theory and its convert-ibility relation $\succeq$. Notably, measures can be used to prove that a given resource <u>cannot</u> be converted into another, for if $M(A) < M(B)$, then it must not be the case that $A \succeq B$. However, the converse isn't true. It could still be the case that $M(A) \geq M(B)$ even when no conversion $A \succeq B$ exists.

On the other hand, numbers are (usually) a lot easier to compare than the resources they came from, so in the rare cases that a measure does in fact totally capture $\succeq$, we are happy.

13.2 Purity Theory

Most of the time in quantum theory, we value states and processes that are as pure as possible. For example, the Bell state in quantum teleportation is valuable not just because it is entangled, but also because it is pure. If instead we use an impure state in teleportation, then also the resulting state will inherit that impurity:

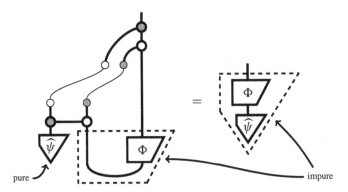

Introducing a bit of impurity like this is okay, and in fact unavoidable, since pure states represent an ideal that is impossible to achieve in practice. However, if we introduce too much impurity like this, pretty soon we'll just have garbage, e.g. a maximally mixed state. Hence purity is a vital resource. In this section, we will define a resource theory for purity and show how the purity of states can be compared and quantified.

13.2.1 Comparing Purity

If it is purity that we cherish, then the free processes should be precisely those that cannot create new purity. In particular, they should preserve maximal impurity:

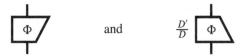

or equivalently:

$$\frac{D'}{D} \ \Phi \ = \ \perp \tag{13.3}$$

where D is the dimension of Φ's input and D' the dimension of its output.

Definition 13.12 Let **unital quantum maps** be the subtheory of **causal quantum maps** obtained by restricting to processes satisfying (13.3). Then:

purity := **causal quantum states/unital quantum maps**

The following is an alternative characterisation of unital quantum maps.

Proposition 13.13 A quantum map Φ is unital if and only if:

$$\Phi \qquad \text{and} \qquad \frac{D'}{D} \ \Phi$$

are both causal.

Proof Unital quantum maps are causal by definition, and if we take the adjoint of (13.3), the resulting equation is causality of $\frac{D'}{D} \ \Phi$. □

So in particular, if a quantum map Φ has input and output types with the same dimension, Φ is unital precisely when Φ and the adjoint of Φ are both causal. Hence, if a quantum map is pure and unital, the map and its adjoint must both be isometries, so the following characterisation is immediate.

Corollary 13.14 A pure quantum map is unital if and only if it is unitary.

As a result, the only pure unital maps must go between systems of the same dimension, so in particular there are no pure unital states or effects. Going beyond pure maps, there is precisely one unital state:

$$\frac{1}{D} \ \perp$$

This follows from the fact that discarding is the unique causal effect. Consequently, the maximally mixed state is the unique free state in **purity**. One therefore expects that it can be obtained via conversion from any other state. This is indeed possible, since the process that discards a state and replaces it with the maximally mixed state:

$$\frac{1}{D} \; \overline{\overline{\equiv}} \; \overline{\overline{\top}} \quad :: \quad \overset{\bigtriangledown}{\rho} \; \mapsto \; \frac{1}{D} \; \perp$$

is clearly unital, so it is a free process in **purity**. Hence:

$$\overset{\bigtriangledown}{\rho} \; \succeq \; \frac{1}{D} \; \underline{\perp}$$

Less drastically, the process that adds some noise to a given state:

$$(1-p) \Bigg| \; + \; \frac{p}{D} \; \overline{\overline{\underset{\top}{\equiv}}} \tag{13.4}$$

is also a free process for **purity**, which we'll call a *noise map*. Since:

$$(1-p) \Bigg| \; + \; \frac{p}{D} \; \overline{\overline{\underset{\top}{\equiv}}} \quad :: \quad \overset{\bigtriangledown}{\rho} \; \mapsto \; (1-p) \overset{\bigtriangledown}{\rho} \; + \; \frac{p}{D} \; \underline{\perp}$$

it follows that:

$$\overset{\bigtriangledown}{\rho} \; \succeq \; (1-p) \overset{\bigtriangledown}{\rho} \; + \; \frac{p}{D} \; \underline{\perp}$$

Remark 13.15 In quantum information literature, and especially in quantum optics, a noise map is often referred to as a *depolarizing channel*.

So, which states can be converted to others via unital quantum maps? If we initially restrict to qubits, we can start to get a picture of this by looking at the geometry of the Bloch ball. We already know that unitaries and noise maps are free processes, so we can see how these act on the Bloch ball. Since a noise map simply mixes its input state with the maximally mixed state, it shrinks any point in the Bloch ball towards the centre:

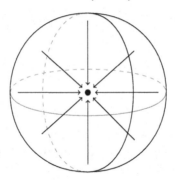

We already know that unitaries give rotations of the Bloch ball, so by composing a noise map with a unitary, any point on the Bloch ball can be taken to any other point closer (or equally close) to the centre of the Bloch ball:

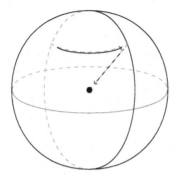

Hence, if ρ' is not further away from the centre of the Bloch sphere than ρ, then $\rho \succeq \rho'$ in **purity**. In fact, the converse is also true. Rather than proving this directly, we can actually give a characterisation for this convertibility relation that works in all dimensions. We will make use of the following preorder.

Definition 13.16 For probability distributions:

$$
\overset{|}{\underset{p}{\bigtriangledown}} \;\leftrightarrow\; \begin{pmatrix} p^1 \\ \vdots \\ p^n \end{pmatrix}
\qquad\qquad
\overset{|}{\underset{q}{\bigtriangledown}} \;\leftrightarrow\; \begin{pmatrix} q^1 \\ \vdots \\ q^n \end{pmatrix}
$$

we say that p *majorizes* q, written:

$$
\overset{|}{\underset{p}{\bigtriangledown}} \;\succeq\; \overset{|}{\underset{q}{\bigtriangledown}}
$$

if, after rearranging the numbers in each probability distribution in decreasing order:

$$
p^1 \geq p^2 \geq \cdots \geq p^n
\qquad\qquad
q^1 \geq q^2 \geq \cdots \geq q^n
$$

we have:

$$
\begin{cases}
p^1 & \geq & q^1 \\
p^1 + p^2 & \geq & q^1 + q^2 \\
& \vdots & \\
p^1 + \cdots + p^n & \geq & q^1 + \cdots + q^n
\end{cases}
\tag{13.5}
$$

It is straightforward to check that this gives a preorder, which we call the *majorization order*. It is not a partial order since the same elements in different order, e.g. point distributions, are equivalent with respect to majorization:

$$
\begin{pmatrix} 1 \\ 0 \\ \vdots \\ 0 \end{pmatrix} \simeq \begin{pmatrix} 0 \\ 1 \\ \vdots \\ 0 \end{pmatrix} \simeq \cdots \simeq \begin{pmatrix} 0 \\ 0 \\ \vdots \\ 1 \end{pmatrix}
$$

while obviously being non-equal. Furthermore, even if we consider probability distributions up to reordering of elements, this does not give a total order, since, e.g.:

$$
\begin{pmatrix} \frac{1}{2} \\ \frac{1}{2} \\ 0 \end{pmatrix} \not\succeq \begin{pmatrix} \frac{3}{4} \\ \frac{1}{8} \\ \frac{1}{8} \end{pmatrix}
\qquad\qquad
\begin{pmatrix} \frac{3}{4} \\ \frac{1}{8} \\ \frac{1}{8} \end{pmatrix} \not\succeq \begin{pmatrix} \frac{1}{2} \\ \frac{1}{2} \\ 0 \end{pmatrix}
$$

Exercise 13.17 Probability distributions:

$$
\overset{|}{\underset{p}{\triangledown}} \quad \leftrightarrow \quad \begin{pmatrix} p^1 \\ p^2 \\ p^3 \end{pmatrix}
$$

can be represented on a triangle by taking the p^is to be coordinates:

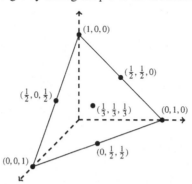

For an arbitrary probability distribution p', depict the region on the triangle of all probability distributions p for which we have:

$$
\overset{|}{\underset{p}{\triangledown}} \succeq \overset{|}{\underset{p'}{\triangledown}}
$$

and the region for which we have:

$$\text{\raisebox{-0.5em}{$\bigtriangledown_{p'}$}} \succeq \text{\raisebox{-0.5em}{$\bigtriangledown_{p}$}}$$

Majorization gives us a preorder, which we can turn into a preordered monoid by taking $\otimes$ to be the usual parallel composition of probability distributions:

$$\text{\raisebox{-0.5em}{$\bigtriangledown_{p}$}} \; \text{\raisebox{-0.5em}{$\bigtriangledown_{p'}$}}$$

To show that this is compatible with the majorization ordering, it is helpful to give an alternative characterisation of majorization. This alternative characterisation is closer to the soul of resource theories in that it presents majorization as a conversion relation, namely, convertibility by means of doubly stochastic maps.

Definition 13.18 *Doubly stochastic maps* are classical processes f (cf. Definition 8.11) such that:

$$\boxed{f} \quad = \quad \frac{1}{D'} \, \circ \tag{13.6}$$

Or equivalently, they are classical maps f where:

$$\boxed{f} \qquad \text{and} \qquad \frac{D'}{D}\boxed{f}$$

are both causal.

Proposition 13.19 The following are equivalent:

• p majorizes q:

$$\text{\raisebox{-0.5em}{$\bigtriangledown_{p}$}} \succeq \text{\raisebox{-0.5em}{$\bigtriangledown_{q}$}}$$

• There exists a doubly stochastic map such that:

$$\text{\raisebox{0em}{$\boxed{f}$}}\,\text{\raisebox{-0.5em}{$\bigtriangledown_{p}$}} \quad = \quad \text{\raisebox{-0.5em}{$\bigtriangledown_{q}$}}$$

Proof (sketch) The majorization preorder admits a 'tower of Hanoi' interpretation, where $p^1, \ldots, p^n$ represent different stacks:

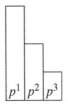

To realise conversion, one first 'moves' $p^1 - q^1$ from p^1 to p^2, so that p^1 becomes q^1, then one moves $(p^1 + p^2 - q^1) - q^2$ from $p^1 + p^2 - q^1$ to p^3, so that p^2 has become q^2, and so on, e.g.:

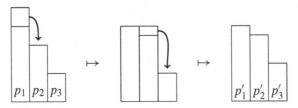

The fact that one can do so requires:

$$p^1 - q^1 \geq 0 \quad p^1 + p^2 - q^1 - q^2 \geq 0 \quad \ldots \quad p^1 + \cdots + p^{n-1} - q^1 - \cdots - q^{n-1} \geq 0$$

which, after moving all the numbers q^i to the RHS of the inequalities, exactly recovers the majorization condition. Since the composition of doubly stochastic maps is again doubly stochastic, it suffices to show the 'Hanoi moves' can be realised by doubly stochastic maps, and vice versa. We leave this as an exercise.   □

Exercise 13.20 Show that whenever p majorizes q, there exists a doubly stochastic map sending p to q by giving the matrices that realise the 'Hanoi moves' in the above proof. Hint: use the fact that a matrix for a generic 2×2 doubly stochastic map is of the form:

$$\begin{pmatrix} 1 - x & x \\ x & 1 - x \end{pmatrix}$$

where $0 \leq x \leq 1$. Conversely, show that for f a doubly stochastic map and p a probability distribution, $p \succeq f \circ p$ in the majorization preordering as defined in (13.5).

Thus we have the following proposition.

Proposition 13.21 Majorization, together with $\otimes$-composition, makes the set of probability distributions for a classical system into a preordered monoid.

Proof If for doubly stochastic maps f and g we have:

i.e. $p \succeq p'$ and $q \succeq q'$, then:

$$\frac{f}{p} \; \frac{g}{q} \;\; = \;\; \frac{}{p'} \; \frac{}{q'}$$

so since $f \otimes g$ is evidently also doubly stochastic, $p \otimes q \succeq p' \otimes q'$. □

Unital quantum maps are characterised by the fact that they as well as their adjoints are causal, and doubly stochastic maps are characterised by the fact that they as well as their adjoints are causal. This looks like a theorem waiting to happen, and indeed these two concepts are mutually related by encoding/measurement.

Theorem 13.22 If a classical map f is doubly stochastic, then:

$$\raisebox{0pt}{f} \tag{13.7}$$

is a unital quantum map, and if a quantum map Φ is unital, then:

$$\raisebox{0pt}{Φ}$$

is a doubly stochastic (classical) map.

Proof The proof directly follows from the fact that measure and encode both are causal, and each other's adjoints. For example, unitality of (13.7) can be established as follows:

□

Remark 13.23 We have tried to conform with standard terminology, but it would of course have made perfect sense to call doubly stochastic maps, instead, 'unital classical maps' or to call unital quantum maps, instead, 'doubly causal quantum maps'.

This correspondence between unital quantum maps and doubly stochastic maps translates into a correspondence between the convertibility of quantum states and the convertibility of probability distributions. The key bridge here is the spectral theorem, which, thanks to Proposition 8.56, lets us express a quantum state in terms of a probability distribution.

Lemma 13.24 For quantum states ρ and ρ', which decompose as:

$$\tag{13.8}$$

for unitaries $\widehat{W}$ and $\widehat{W}'$ and probability distributions p and p', the following are equivalent:

- there exists a unital quantum map Φ such that:

$$\tag{13.9}$$

- there exists a doubly stochastic map f such that:

$$\tag{13.10}$$

Proof Assume first that ρ and ρ' are related by a unital quantum map Φ as in (13.9). Then, by (13.8):

We can then use unitarity to move $\widehat{W}'$ to the LHS:

$$\tag{13.11}$$

Then, measuring both sides yields:

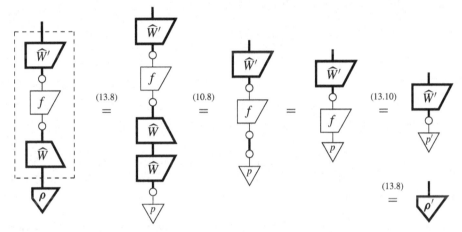

By Proposition 13.22, the map applied in the LHS to p is doubly stochastic, since all quantum maps involved are unital. Conversely, assuming p and p' are related by a doubly stochastic map as in (13.10), it follows that:

where, again by Proposition 13.22, the quantum map applied in the LHS to ρ is unital. □

Using more traditional notation, we refer to the probability distribution p related to the state ρ in equation (13.8) as the *spectrum* of ρ, written $\mathsf{spec}(\rho)$ (cf. Definition 5.73). Together, Proposition 13.19 and Lemma 13.24 yield the following characterisation of convertibility in **purity** in terms of majorization.

Theorem 13.25 The conversion relation of **purity** for the states of a fixed quantum system of arbitrary dimension is given by:

$$\overset{\rho}{\triangledown} \succeq \overset{\rho'}{\triangledown} \quad \Longleftrightarrow \quad \mathsf{spec}(\rho) \succeq \mathsf{spec}(\rho')$$

where the second $\succeq$ is the majorization preordering.

13.2.2 Measuring (Im)purity

A typical measure of purity is the *von Neumann entropy* of a quantum state. This is computed as follows. First, use Corollary 6.68 to decompose ρ over an ONB of pure states:

$$\bigtriangledown_{\!\rho} = \sum_i p^i \; \bigtriangledown_{\!i} \tag{13.12}$$

Then, the von Neumann entropy is computed as:

$$S\left(\bigtriangledown_{\!\rho}\right) := -\sum_i p^i \log_D(p^i)$$

where $\log_D(p^i)$ is the logarithm of p^i for some fixed base D, which is typically taken to be the dimension of a single system, e.g. 2 for qubits. In that case, the entropy varies from 0 for pure states to 1 for the maximally mixed state.

Remark 13.26 For a classical probability distribution:

$$\bigtriangledown_{\!p} = \sum_i p^i \; \bigtriangledown_{\!i}$$

the quantity:

$$S\left(\bigtriangledown_{\!p}\right) := -\sum_i p^i \log_D(p^i)$$

is the *Shannon entropy*. Since every quantum state diagonalises, it encodes a probability distribution with respect to <u>some</u> ONB. The von Neumann entry is then the Shannon entropy of that encoded probability distribution.

Now, this almost gives us an additive measure, in the sense of Definition 13.11. Indeed, it satisfies:

$$S\left(\bigtriangledown_{\!\rho_1}\,\bigtriangledown_{\!\rho_2}\right) = S\left(\bigtriangledown_{\!\rho_1}\right) + S\left(\bigtriangledown_{\!\rho_2}\right) \tag{13.13}$$

which follows straightforwardly from the fact that:

$$\log_D(p^i q^j) = \log_D(p^i) + \log_D(q^j)$$

Moreover, for general states of two systems, we have:

$$S\left(\bigtriangledown_{\!\rho}\right) \le S\left(\bigtriangledown_{\!\rho}\right) + S\left(\bigtriangledown_{\!\rho}\right)$$

with equality if and only if ρ is separated. However, von Neumann entropy is not a measure of *purity*, but rather a measure of *impurity*. That is, for $\succeq$ the purity preorder, we have:

Remark 13.27 Given that mixing stands for introducing a lack of knowledge, this means that the maximally mixed state is the 'least informative' state, in contrast to pure quantum states, which are 'maximally informative'. Thus there is a close connection between mixedness/entropy of quantum state and its information content. Studying this and related issues is an important part of *quantum information theory*.

13.3 Entanglement Theory

While entanglement is a very different concept from purity, involving at least two systems, when characterising the conversion relation of the resource theory of entanglement, we'll encounter many of the same ingredients.

We'll actually study two distinct resource theories of entanglement, one that is roughly on par with **purity** in terms of how fine-grained the conversion relation becomes, and one that is much coarser. The latter is so coarse in fact, that for two- and three-qubit states, it yields only a finite number of equivalent resources. Along the way, we will see that, that while 'entanglement' is a single word, it can stand for very different things.

Throughout this section we consider pure states of the form:

$$\overbrace{}^{\hat{A}\quad\hat{A}\quad\cdots\quad\hat{A}}_{\hat{\psi}}$$

i.e. states whose type consists of two or more copies of the same system. This is not an essential assumption, but it simplifies things a bit.

13.3.1 LOCC Entanglement

If it is entanglement that we cherish, then the free processes should be the ones that create no new entanglement. We already saw that disentangled states are of the form:

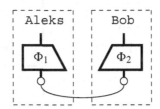

and given the discussion in Section 8.3.5 one may be tempted to think that the kind of processes that don't create any entanglement are those of the form:

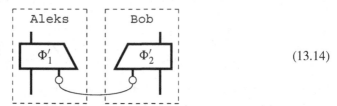

(13.14)

However, in that discussion we ignored causality, for the simple reason that states are always causal, up to a number. However, for processes, causality is a non-trivial requirement. Within the context of resource theories, it is important that one can actually realise 'free' processes, since otherwise there isn't much free about them! Non-causal quantum maps can only be realised non-deterministically, and if we're unlucky, rather than converting our resource we may as well have lost it all together. Hence, we'll restrict to only causal free processes for the time being.

Once causality enters the picture, we should distinguish whether (i) Aleks and Bob share a classical cup; (ii) Aleks sends some classical data to Bob; or (iii) Bob sends some classical data to Aleks. So, in addition to (13.14), we should also consider processes of the form:

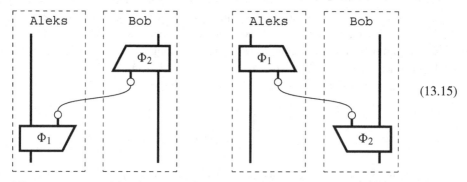

(13.15)

Luckily, once we consider these two forms, we no longer need to think about (13.14), because this arises as a special case of Aleks sending classical data to Bob (or vice versa):

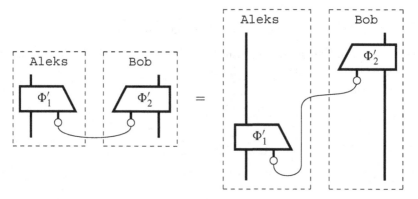

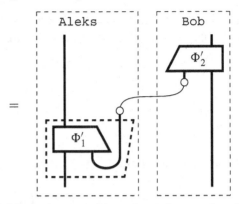

Quantum processes of the forms (13.15), and compositions thereof, are referred to as *LOCC-operations*, for the reason that they can be decomposed in two kinds of basic processes, namely:

- Local (causal) Operations :=

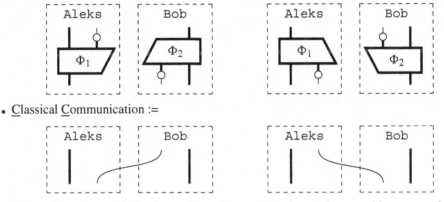

- Classical Communication :=

Now, throughout this book we have drawn dashed grey boxes with names of people (or extinct birds) attached to them. We never really said what this means formally. However, this is really simple. We can capture this by creating a new process theory, called **quantum processes**[2], which has exactly the same processes, but two copies of each (classical and quantum) type, one for Aleks and one for Bob:

$$\widehat{A} \rightsquigarrow \left(\widehat{A}_{\text{Aleks}}, \widehat{A}_{\text{Bob}}\right) \qquad X \rightsquigarrow \left(X_{\text{Aleks}}, X_{\text{Bob}}\right)$$

In this new process theory, local operations are then just the quantum processes that only connect Aleks' types to Aleks' types and Bob's to Bob's, e.g.:

Then, classical communication is just a classical wire, where we label one end by Aleks' type and the other by the corresponding type for Bob (or vice versa):

and **pure quantum states**[2] is the set of states of the form:

Of course, the same can be done for any number of agents.

Definition 13.28 Let **locc**[2] be the subtheory of **quantum processes**[2] obtained by restricting to processes corresponding to local operations, classical communication, and compositions thereof. Then:

$$\textbf{LOCC entanglement}^2 := \textbf{pure quantum states}^2/\textbf{locc}^2$$

So general quantum processes that don't create entanglement involve a game of ping-pong of classical communication between the two agents:

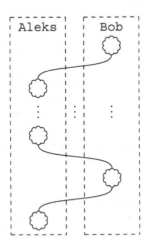

How long does this game have to go on in order to obtain the most general kind of conversions? Fortunately, one single use of classical communication already does the job, since the two forms (13.15) are in fact interchangeable. To show this, we first observe that we can interchange the roles of the two systems of a bipartite state via unitaries.

Lemma 13.29 For any pure bipartite state $\widehat{\psi}$ there exist unitaries $\widehat{U}$ and $\widehat{V}$ that swap the two systems:

$$ \text{(13.16)} $$

Proof Applying the singular value decomposition from Exercise 8.50 (which relies on the fact that the two systems are the same) to ψ gives:

for unitaries U' and V'. Then:

Since the RHS is invariant under swapping, we have:

Moving all of the unitaries to the LHS then completes the proof:

$\square$

Importantly, to realise this swapping we need only local operations. As a consequence of this 'local swapping', we get the following.

Proposition 13.30 There exist Φ_1 and Φ_2 such that:

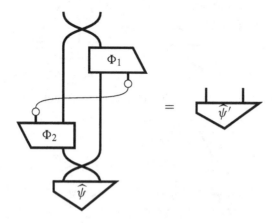

$$(13.17)$$

if and only if there exist Φ_1' and Φ_2' such that:

$$(13.18)$$

Proof Let Φ_1 and Φ_2 satisfy (13.17). Deforming the LHS gives:

Then, since $\widehat{\psi}$ and $\widehat{\psi}'$ are both pure bipartite states, we can apply Lemma 13.29 to remove the two swaps:

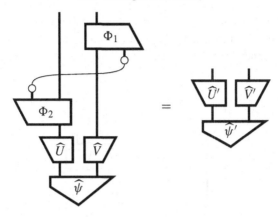

Then, moving the unitaries to the LHS:

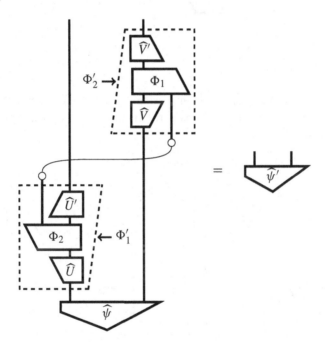

we obtain processes Φ_1' and Φ_2' satisfying (13.18). The converse follows symmetrically.

$\square$

For the sake of simplicity, assume that we have a LOCC protocol where each step deterministically yields a pure state:

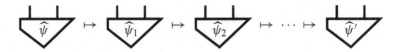

Then, this consists of a ∘-composition of processes satisfying (13.17) or (13.18). So, by applying Proposition (13.30), we can make all of the classical wires go from Aleks to Bob and bundle them together into a single classical wire. Hence we obtain a process of the form:

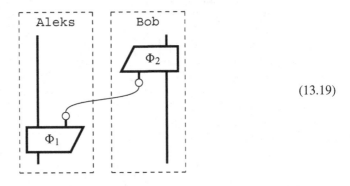

(13.19)

Exercise 13.31 Show that (by possibly increasing the size of the classical system) we can furthermore assume that Φ_1 in (13.19) is pure. That is, for all (causal) Φ_1 and Φ_2 there exist (causal) $\widehat{f}$ and Φ such that:

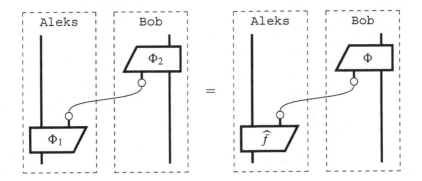

Exercise* 13.32 Show that any LOCC protocol can be rewritten in the form (13.19) without assuming each step deterministically yields a pure state. Hint: first purify Aleks' and Bob's processes such that each step yields a 'non-deterministic' pure state:

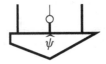

and use the following generalisation of Lemma 13.29: For any non-deterministic state $\widehat{\psi}$, there exist <u>controlled</u> unitaries $\widehat{U}$, $\widehat{V}$ such that:

$$(13.20)$$

In the proof of the characterisation of the conversion relation for this resource theory we will make use of the following fact that tells us how the reduced states when discarding a different system are related.

Lemma 13.33 For every non-deterministic bipartite state $\widehat{\psi}$ (i.e. a pure bipartite state with a classical output) there exists a controlled unitary $\widehat{U}$ such that its two reduced states are related as follows:

In the special case where the classical system is trivial, this becomes:

for some unitary $\widehat{U}$.

Proof We can rewrite equation (13.20) of Exercise* 13.32 as follows:

$$(13.21)$$

Then, using causality of the controlled unitary $\widehat{V}$ (in the second step):

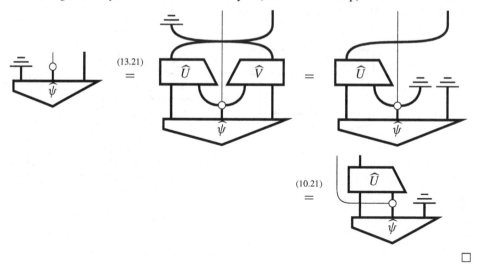

We also will rely on the following fact.

Lemma 13.34 Any mixture of unitaries:

$$ \tag{13.22} $$

is a unital quantum map.

Proof From equations (10.45) it follows that for any controlled unitary:

its 'controlled inverse':

is also a controlled unitary, and in particular, it is causal:

$$ \tag{13.23} $$

The adjoint of this equation then yields unitality of (13.22):

$$
\left[\widehat{U}\, p \right] \;=\; \left[p \; \widehat{U} \right] \;\overset{(13.23)}{=}\; \left[p \right] \;=\; \bot
$$

□

We now have enough ingredients to characterise the conversion relation for **LOCC entanglement**[2], which looks a lot like the one for **purity**.

Theorem 13.35 If a bipartite state $\widehat{\psi}$ can be converted into a bipartite state $\widehat{\psi}'$ in **LOCC entanglement**[2], then there exists a unital quantum map Ψ that converts the reduced state of $\widehat{\psi}'$ into the reduced state of $\widehat{\psi}$:

$$
\Psi \, \widehat{\psi}' \;=\; \widehat{\psi}
\tag{13.24}
$$

Hence, the conversion relation of **LOCC entanglement**[2] for bipartite states of a pair of the same fixed quantum system is given by:

$$
\widehat{\psi} \;\succeq\; \widehat{\psi}' \quad\Longrightarrow\quad \mathrm{spec}\!\left(\widehat{\psi}' \right) \;\succeq\; \mathrm{spec}\!\left(\widehat{\psi} \right)
$$

where the second $\succeq$ is the majorization preordering.

Proof Assume $\widehat{\psi} \succeq \widehat{\psi}'$. That is, thanks to Proposition 13.30 and Exercise 13.31, there exist quantum processes $\widehat{f}$ and Φ such that:

$$
\Phi \;\; \widehat{f} \;\; \widehat{\psi} \;=\; \widehat{\psi}'
\tag{13.25}
$$

We will start with the RHS of (13.24), and work our way to the LHS. For this purpose, let us first focus just on local operation $\widehat{f}$ on the LHS of (13.25). Applying $\widehat{f}$ to $\widehat{\psi}$ yields a non-deterministic pure state; hence, by Lemma 13.33, there exists a controlled unitary such that:

Then, deleting the classical output on both sides yields this equation:

$$(13.26)$$

where we used causality to eliminate $\widehat{f}$ on the LHS. Next, by the second part of Lemma 13.33 there exists a unitary $\widehat{V}$ such that:

$$(13.27)$$

Combining equations (13.26) and (13.27), then moving $\widehat{V}$ to the RHS we obtain:

$$(13.28)$$

Now, we will modify equation (13.25) a bit so that we can plug it into the RHS of (13.28). By Proposition 8.59 we know that if deleting a classical system yields a pure state, then the classical system separates. Applying this to (13.25), which we can rewrite as:

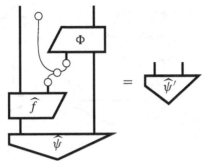

yields the following separation:

$$(13.29)$$

for some probability distribution p. By causality of Φ we also have:

$$(13.30)$$

Hence:

$$(13.31)$$

This equation can now be plugged into (13.28), which yields:

By Lemma 13.34 the quantum map:

is unital, so we indeed obtain equation (13.24). □

Exercise* 13.36 Prove the converse to Theorem 13.35. Namely, show:

Comparing Theorem 13.25 and Theorem 13.35 we see that less purity of the reduced state means more entanglement.

Example 13.37 Since the reduced state of the Bell state is the maximally mixed state:

it can be converted into any other bipartite pure state via LOCC. In other words, it is *LOCC-maximal*.

13.3.2 SLOCC Entanglement

We started the previous section with an argument that the very nature of 'freeness' requires causality of all processes involved in conversion of resources. However, the situation changes if instead of one pair of systems in state $\widehat{\psi}$, one has an unlimited supply of systems in that state. Then, if we try to convert to a state $\widehat{\psi}'$ by some non-deterministic LOCC process, we only need at least one branch to succeed:

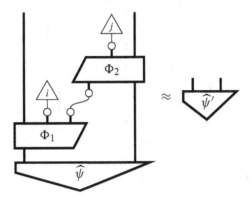

Using terminology of Definition 13.7, this amounts to having a non-zero conversion rate from $\widehat{\psi}$ to $\widehat{\psi}'$:

$$r(\widehat{\psi} \succeq \widehat{\psi}') > 0$$

Since by Theorem 6.94) any quantum map can be realised as a branch of a (causal) quantum process, passing to this more liberal kind of local operation is equivalent to allowing local operations to be any cq-map, not just the causal ones. These new 'free' processes are called *SLOCC-operations*, as they can be decomposed in two kinds of basic processes:

- Stochastic Local Operations (a.k.a. possibly non-causal cq-maps) and
- Classical Communication

Definition 13.38 Let **slocc²** be the subtheory of **quantum maps²** obtained by restricting to processes realising stochastic local operations, classical communication, and compositions thereof. Then:

$$\textbf{SLOCC entanglement}^2 := \textbf{pure quantum states}^2/\textbf{slocc}^2$$

Though it looks pretty similar to **LOCC entanglement²**, this resource theory is actually a lot easier to work with. For one thing, classical communication now becomes irrelevant.

Theorem 13.39 A bipartite state $\widehat{\psi}$ can be converted into a bipartite state $\widehat{\psi}'$ in **SLOCC entanglement²** if and only if there exist quantum maps Φ_1 and Φ_2 such that:

$$\text{(13.32)}$$

Proof   Assume $\widehat{\psi} \succeq \widehat{\psi}'$; then there exist cq-maps Φ_1' and Φ_2' such that:

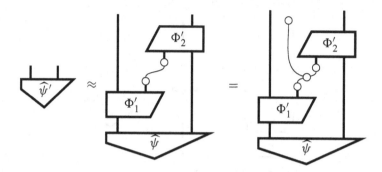

so by Proposition 8.59 we have:

$$ \tag{13.33} $$

for some p. Since p is a (causal) probability distribution, there must be some ONB effect i such that:

$$ \tag{13.34} $$

Hence:

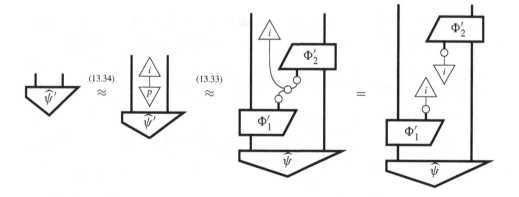

Letting:

(up to an appropriately chosen number) yields equation (13.32). □

Exercise 13.40 Show, using a similar technique to the proof above, that it suffices to consider only pure quantum maps as local operations:

(13.35)

In order to characterise convertibility in **SLOCC entanglement**[2] we make use of the following standard notion from linear algebra, adopted to pure quantum maps.

Definition 13.41 For a pure quantum map or bipartite state, represented in terms of its singular value decomposition:

its *rank*, respectively:

is the number of non-zero entries in the matrix of p:

Remark 13.42 Just as the 'sideways' singular value decomposition of a bipartite state is often called the Schmidt decomposition, the 'sideways' rank of a bipartite state is often called the *Schmidt rank* in the literature.

We have already encountered the extreme cases of rank.

Exercise 13.43 First, show that 'maximum rank' is the same as 'non-degenerate' as in Definition 4.75, that the cup has maximum rank, and that 'rank 1' is the same as separable. Next, show that the conversion relation $\succeq$ for **SLOCC entanglement**[2] is given by:

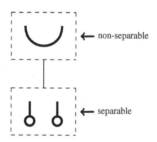

where you can make use of the fact that for pure quantum maps $\widehat{f}$ and $\widehat{g}$:

$$\mathrm{rank}\left(\boxed{\widehat{f}}\right) \geq \mathrm{rank}\left(\begin{array}{c}\boxed{\widehat{g}}\\\boxed{\widehat{f}}\end{array}\right) \qquad \mathrm{rank}\left(\boxed{\widehat{g}}\right) \geq \mathrm{rank}\left(\begin{array}{c}\boxed{\widehat{g}}\\\boxed{\widehat{f}}\end{array}\right)$$

Thus, two bipartite states are equivalent (cf. Definition 13.4) if and only if they have the same rank. Since every bipartite qubit state must have either rank 1 or rank 2, for example:

$$\mathrm{rank}\left(\bigcup\right) = 2 \qquad\qquad \mathrm{rank}\left(\mathbf{\phi}\ \mathbf{\phi}\right) = 1$$

there are only two *equivalence classes*: the class of states equivalent to the Bell state and the class of those equivalent to a separable state. Furthermore, any separable state can be obtained from a non-separable one via SLOCC-operations, hence:

$$\bigcup \;\succeq\; \mathbf{\phi}\ \mathbf{\phi}$$

Since there are finitely many equivalence classes, we can depict this conversion relation as a *convertibility diagram*:

where the boxes represent equivalence classes, and the <u>downward</u> edge(s) represent(s) when one class is convertible to other.

Whereas LOCC-convertibility captures quantitative differences in entanglement, SLOCC-convertibility is much better at capturing purely 'qualitative' differences, e.g. 'separable' versus 'non-separable'. In the bipartite case, as we just saw, this creates a very simple total ordering on SLOCC-equivalence classes, dictated by the rank. So for all bipartite states $\widehat{\psi}$ and $\widehat{\psi}'$, either:

$$\text{(diagram)} \succeq \text{(diagram)} \qquad \text{or} \qquad \text{(diagram)} \succeq \text{(diagram)}$$

However, once we go beyond two systems, the situation changes. We start to get states that are SLOCC-maximal, but in inequivalent ways.

Theorem 13.44 The conversion relation for **SLOCC entanglement**[3], when restricting to qubits, is given by:

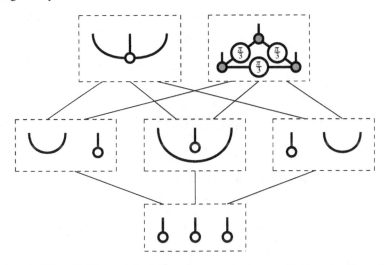

The key feature of this convertibility diagram is that at the top level we have two incomparable classes, respectively, witnessed by the GHZ state that we have encountered before, and something new, called the *W state*:

Exercise 13.45 Show that:

$$\text{(diagram)} \approx \text{(diagram)} + \text{(diagram)} + \text{(diagram)}$$

The qualitatively different entanglement properties between the GHZ state and the W state are a bit more subtle than, say, separable versus non-separable. For instance, if for a GHZ state we discard a system, the remaining two systems disentangle:

whereas for a W state they do not. In the next section, we see an even more striking difference between 'GHZ-spiders' and this new kind of 'arachnid'.

Remark 13.46 Once one goes beyond three qubits, it is no longer the case that we get finitely many SLOCC-equivalence classes. There are simply too many free parameters for a state for four or more systems to cancel out via SLOCC-operations. One can still study parametrized SLOCC-equivalence classes (a.k.a. 'SLOCC super-classes') for four or more systems, but these are not nearly as well understood.

13.3.3 Exploding Spiders

Rather than looking like a spider, W states look more like a spider orgy:

So what comes out? It cannot be just an ordinary spider, because of the following.

Proposition 13.47 For any spiders ● on a two-dimensional system, the state:

is SLOCC-equivalent to a GHZ state.

Proof From Theorem 8.41, any spider corresponds to an ONB. In particular:

Then, for U the unitary that sends the ●-ONB to the ○-ONB, we have:

$\square$

Since we can't obtain the W state using normal spiders, we will obtain it using a different kind of spider-like arachnid. That is, we will define a family:

such that:

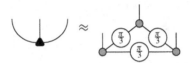

Following Exercise 13.45 we let:

We can generalise this straightforwardly to produce n-partite states. First, consider all bit strings of length n containing a single 0:

$$C_n := \{011...1 , 101...1 , ... , 11...10\}$$

Then, to form the n-partite state, we sum over all bit strings $\vec{i} \in C_n$:

Similarly, we can obtain input legs, with a little modification: the input-ONB states should be negated. That is, for $\bar{i} := 1 - i$, we set:

$$\tag{13.36}$$

These new arachnids behave much like spiders, except they like each other a bit too much. For instance, standard spider behaviour includes obeying the leg-swapping equations:

$$\tag{13.37}$$

Also, if they are connected by a <u>single</u> leg, they fuse as expected:

$$\tag{13.38}$$

from whence we can derive most of usual stuff that holds for spiders. For instance, like spiders, these new things give us cups and caps:

$$\text{(13.39)}$$

However, if these arachnids shake <u>two</u> (or more) legs, they become over-excited and explode:

$$\approx \qquad \text{(13.40)}$$

two wires

leaving a bunch of smoking stubs of legs behind:

Definition 13.48 A family ▲ of linear maps is called a family of *antispiders* if it satisfies equations (13.37), (13.38), and (13.40).

Exercise 13.49 Show (13.36) indeed defines a family of antispiders.

We can isolate the key difference between spiders and antispiders by looking at a simpler version of spider fusion, as compared with the 'antifusion' equation (13.40); namely, what happens to a single 'loop':

$$= \qquad \text{vs.} \qquad \approx \qquad \text{(13.41)}$$

While in the case of spiders we get a plain wire, antispiders do the exact opposite: they separate. Although it looks like they separate in a very specific way, in fact it is already enough to say just that they separate:

Proposition 13.50 A family of linear maps satisfying (13.37) and (13.38) are antispiders if and only if:

$$= \qquad \text{(13.42)}$$

Proof Clearly antispiders satisfy (13.42), setting:

Conversely, assume (13.42). First, we have:

Since the LHS is non-zero, the two numbers on the RHS must also be non-zero, that is:

Now we can learn what ψ and π are, up to a number:

Substituting into (13.42), we obtain:

$$(13.43)$$

From this, it is straightforward to show that (13.40) follows. We leave this as an exercise for the reader. □

Exercise* 13.51 Show that for any antispider on a non-trivial ($D > 1$) system there cannot be 'double loops':

$$\phi = 0 \qquad\qquad (13.44)$$

(Hint: first compute the number involved in equation (13.43).)

Just as with normal spiders and the GHZ state, the antispider equations totally characterise the SLOCC-equivalence class of W.

Theorem 13.52 Let ▲ be a family of antispiders for a two-dimensional system. Then:

is SLOCC-equivalent to the W state.

Proof First, note that:

$$\blacktriangle \not\approx \begin{array}{c}\blacktriangle\\\blacktriangle\end{array} \qquad\qquad (13.45)$$

since otherwise the plain wire separates:

$$\left|\quad \overset{(13.38)}{=} \quad \bigwedge \quad \approx \quad \bigwedge \quad \overset{(13.38)}{=} \quad \bigcirc \quad \overset{(13.40)}{\approx} \quad \begin{array}{c}\blacktriangle\\\blacktriangle\end{array}\right.$$

Hence the two states (13.45) form a basis for the two-dimensional system, so it is possible to define an invertible linear map f such that:

$$\boxed{f} \;::\; \underset{0}{\bigtriangledown} \;\mapsto\; \begin{array}{c}\blacktriangle\\\blacktriangle\end{array}\;,\quad \underset{1}{\bigtriangledown} \;\mapsto\; \blacktriangle$$

If we additionally let:

$$\boxed{g} \;:=\; \boxed{f}$$

then by plugging in ○-ONB effects, it is straightforward to check that:

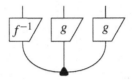

indeed gives a W state. □

The bottom line is: once again we find the distinction between 'separable' and 'non-separable' playing a crucial role, this time in highlighting the qualitative difference between the two SLOCC-maximal states on three qubits.

13.3.4 Back to Basics: Arithmetic

Theorem 13.44 tells us that there really are only two kinds of non-separable qubit states on three systems; that is, up to local operations, each such state is equivalent to either:

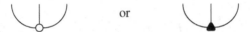

By bending some wires, we could just as easily say that any non-separable, linear map from two qubits to a qubit, up to local operations, must be equivalent to either:

or any one-to-two map must be locally equivalent to either:

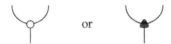

Because of (anti)spider fusion, an entire family of (anti)spiders is determined by these three-legged spiders:

so one would expect spiders to reduce to just two cases as well, up to some local linear maps. This is indeed the case.

Theorem 13.53 Let:

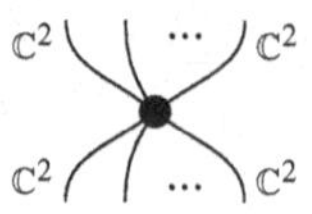

$$\mathbb{C}^2 \diagdown \quad \cdots \quad \diagup \mathbb{C}^2$$
$$\mathbb{C}^2 \diagup \quad \cdots \quad \diagdown \mathbb{C}^2$$

be a family of linear maps satisfying:

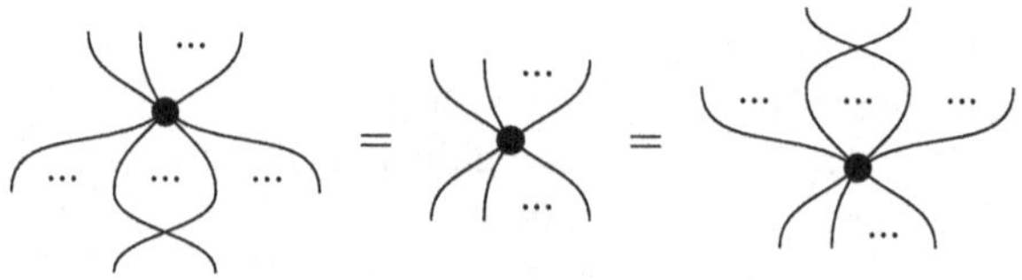

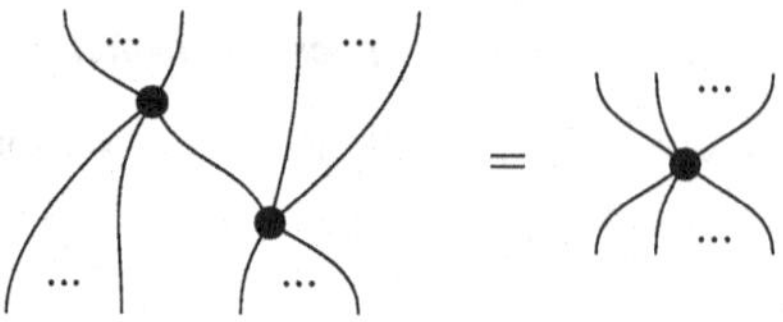

Then it must be the case that ● is *isomorphic* to either ○ or ▲. That is, there exists some isomorphism (i.e. invertible linear map) f such that either:

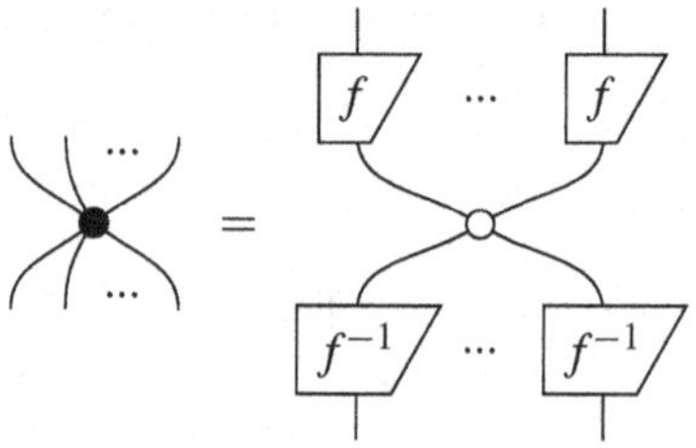

or

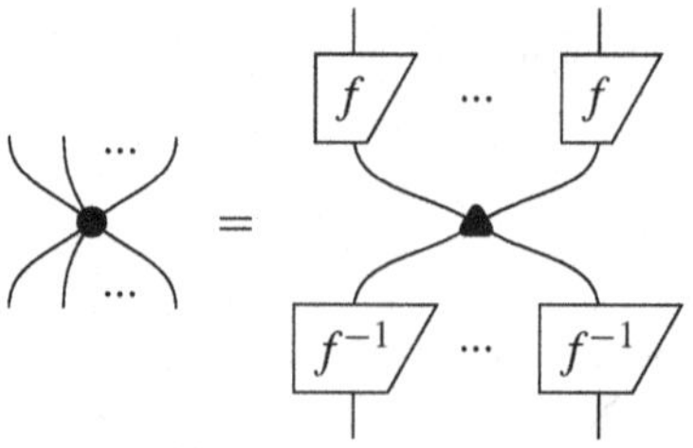

So spiders and antispiders are, in a strong sense, the <u>only</u> choice we have for a 'fusing' family of qubit operations. In fact, even without assuming we have a whole family of spiders, any pair of maps:

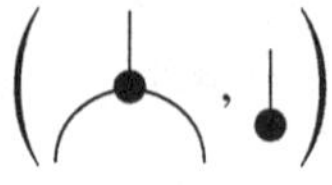

satisfying:

must be isomorphic to either:

$$\left(\begin{array}{c} \end{array}, \begin{array}{c} \end{array} \right) \qquad \text{or} \qquad \left(\begin{array}{c} \end{array}, \begin{array}{c} \end{array} \right)$$

But what are these two operations ○ and ▲ ? What do they do? Let's start with something we know about ○-spiders: how they interact with phase states:

$$\begin{array}{c} \alpha \quad \beta \end{array} = \boxed{\alpha+\beta}$$

A phase state 'encodes' a complex number $e^{i\alpha}$, and when a pair of phase states meets a ○-spider, these numbers are multiplied:

$$\begin{pmatrix} 1 \\ e^{i\alpha} \end{pmatrix} \star \begin{pmatrix} 1 \\ e^{i\beta} \end{pmatrix} = \begin{pmatrix} 1 \\ e^{i(\alpha+\beta)} \end{pmatrix} = \begin{pmatrix} 1 \\ e^{i\alpha} \cdot e^{i\beta} \end{pmatrix}$$

In fact, as we saw back in Section 8.2.2.4 this will still be true if we replace complex numbers of the form $e^{i\alpha}$ with arbitrary complex numbers:

$$\begin{pmatrix} 1 \\ \lambda \end{pmatrix} \star \begin{pmatrix} 1 \\ \lambda' \end{pmatrix} = \begin{pmatrix} 1 \\ \lambda \cdot \lambda' \end{pmatrix}$$

That is, by letting:

$$\boxed{\lambda} \leftrightarrow \begin{pmatrix} 1 \\ \lambda \end{pmatrix}$$

We have:

$$\begin{array}{c} \lambda \quad \lambda' \end{array} = \boxed{\lambda \cdot \lambda'}$$

Since these 'generalised phases' form a basis, this totally characterises ○-matching for qubits. Hence:

> *Qubit spiders correspond to multiplication.*

If we plug those same generalised phase states into the ▲-matching, something surprising happens:

$$\begin{array}{c} \lambda \quad \lambda' \end{array} = \boxed{\lambda+\lambda'}$$

Hence:

Qubit antispiders correspond to addition.

Now, if you have two numbers and you want to make one number, the first two things you would try are addition and multiplication. Surprisingly, when it comes to qubits, these are our only choices!

From here, we can start to play a similar game to that of the ZX-calculus and start to find a series of graphical rules governing the interaction of the ○-spider with the ▲-antispider. For instance, by letting:

we obtain generalised phase gates. Then, we can use a copy law:

to capture the fact that 'times' distributes over 'plus':

which symbolically we know as:

$$\lambda \cdot (\mu + \mu') = \lambda \cdot \mu + \lambda \cdot \mu'$$

Just taking ○/▲, along with the usual π phase:

we obtain a language that is universal for linear maps whose matrices are restricted to integers, i.e. the process theory **matrices**($\mathbb{Z}$) $\subseteq$ **linear maps**. There even exists a graphical calculus that is complete for this theory. What is it? Well, we need to leave something for *Picturing Even More Quantum Processes*, don't we?!

13.4 Summary: What to Remember

1. A *resource theory* is a kind of process theory where the types are called *resources* and the processes are called *resource conversions*. It captures the idea that resources (e.g. states) of some types are more valuable than others; e.g. entangled states are more desirable than separable states:

Given any process theory **P** (e.g. **quantum processes**) and a sub-theory of *free processes* **F**, a resource theory **P/F** arises whose:

- types are the states of **P** and
- processes from a type ρ to ρ' are those Φ in **F** that convert ρ into ρ':

$$\vcenter{\hbox{Φ on ρ}} = \vcenter{\hbox{ρ'}}$$

2. For a resource theory **R**, resource A can be *converted* into resource B:

$$A \succeq B$$

if and only if there exists a process in **R** that realises this conversion:

$$\frac{B}{\boxed{f}}\Big|_{A}$$

Denoting the resources by R, for any resource theory:

$$(R, \succeq, \otimes)$$

forms a *preordered monoid*, i.e. a preorder $(R, \succeq)$ for which we have:

$$A_1 \succeq B_1 \quad \text{and} \quad A_2 \succeq B_2 \quad \Longrightarrow \quad A_1 \otimes A_2 \succeq B_1 \otimes B_2$$

3. An *additive measure* for a resource theory is a function:

$$M : (R, \succeq, \otimes) \rightarrow \left(\mathbb{R}_{\geq 0}, \leq, +\right)$$

that preserves the preordered monoid structure; and a *supremal measure* is a function:

$$M : (R, \succeq, \otimes) \rightarrow \left(\mathbb{R}_{\geq 0}, \leq, \mathsf{max}\right)$$

that again preserves the preordered monoid structure.

4. Unital quantum map are causal quantum maps Φ that also satisfy:

$$\vcenter{\hbox{Φ with $\tfrac{1}{D}$}} = \tfrac{1}{D}\ \vcenter{\hbox{}}$$

The resource theory **purity** arises from **quantum maps** by taking unital quantum maps as the free processes:

$$\textbf{purity} := \textbf{causal quantum states/unital quantum maps}$$

The conversion relation for **purity** is characterised by:

$$\vcenter{\hbox{ρ}} \;\succeq\; \vcenter{\hbox{ρ'}} \quad\Longleftrightarrow\quad \mathsf{spec}(\rho) \;\succeq\; \mathsf{spec}(\rho')$$

where the second $\succeq$ is the *majorization preordering*:

$$\begin{cases} \begin{aligned} p^1 &\geq q^1 \\ p^1 + p^2 &\geq q^1 + q^2 \\ &\vdots \\ p^1 + \cdots + p^n &\geq q^1 + \cdots + q^n \end{aligned} \end{cases}$$

and the p^i are assumed to be in decreasing order. The *von Neumann entropy*:

$$S\left(\vcenter{\hbox{ρ}}\right) := -\sum_i p^i \log_D(p^i)$$

gives an additive measure for (im)purity, namely:

$$\vcenter{\hbox{ρ}} \;\succeq\; \vcenter{\hbox{ρ'}} \;\Longrightarrow\; S\left(\vcenter{\hbox{ρ}}\right) \leq S\left(\vcenter{\hbox{ρ'}}\right)$$

5. The subtheory **locc²** of *LOCC-operations* (local operations and classical communication) consists of quantum processes of the form:

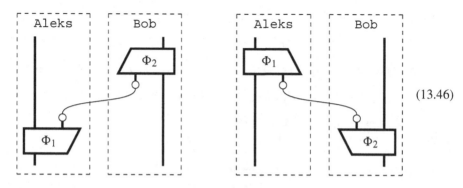

(13.46)

Then:

$$\textbf{LOCC entanglement}^2 := \textbf{pure quantum states}^2\textbf{/locc}^2$$

The conversion relation of **LOCC entanglement**[2] is characterised by:

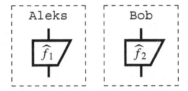

where the second $\succeq$ is again the majorization preordering.

6. The subtheory **slocc**[2] of *SLOCC-operations* (stochastic local operations and classical communication) consists of (possibly non-causal) cq-maps of the form of (13.46). Then:

$$\textbf{SLOCC entanglement}^2 := \textbf{pure quantum states}^2/\textbf{slocc}^2$$

SLOCC-convertibility can furthermore always be realised by separable pure quantum maps:

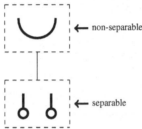

that is, classical communication is not necessary. For two qubits, there are only two *SLOCC-equivalence classes*:

but for three qubits the story gets more interesting:

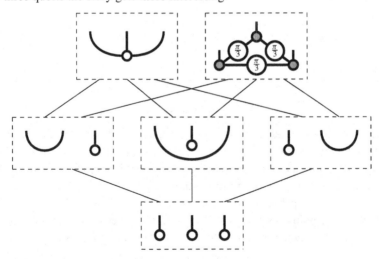

7. The states:

are the GHZ state and the *W state*. For any family of spiders over $\mathbb{C}^2$, the associated tripartite state is always SLOCC-equivalent to the GHZ state. On the other hand, for any family of *antispiders*, which still fuse when there is one wire:

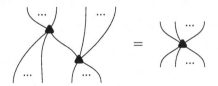

but explode when there are two:

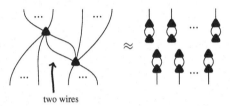

the associated tripartite state:

is SLOCC-equivalent to the W state.

8. Spiders and antispiders are the only two possibilities for spider-like families of linear maps on $\mathbb{C}^2$, and they correspond to 'plus' and 'times':

13.5 Historical Notes and References

The concept of a resource theory that we have adopted mainly emerged within the quantum information community. The process-theoretic formulation was put forward by Coecke et al. (2014). The notion of free processes first appeared in Horodecki et al. (2002). Many resource theories had already been proposed, e.g. entanglement (Horodecki et al., 2009), symmetry (Gour and Spekkens, 2008; Marvian and Spekkens, 2013), purity (Horodecki et al., 2003), non-equilibrium (Gour et al., 2013), and athermality (Brandão et al., 2013). Elaborations on the framework presented here include Fong and Nava-Kopp (2015) and

Fritz (2015). As resource theory is currently taking off as a subject, many more papers will have seen daylight by the time you read this.

A similar analysis to the one performed here, but within the framework of operational probabilistic theories (a hybrid of generalised probabilistic theories and process theories), can be found in Chiribella and Scandolo (2015). In that paper, the results presented here are referred to as an axiomatisation of thermodynamics. Also, our Lemma 13.29 is taken to be an axiom and is referred to as the *local exchangeability axiom*.

A discussion of the entropy associated to mixed quantum states was already in von Neumann (1927), which was in fact a long time before Shannon's seminal paper on entropy (Shannon, 1948), which started the field of information theory.

The majorization preordering dates back to Robert Franklin Muirhead (1903), who made several important contributions to mathematics but never held a faculty position. Theorem 13.25 is taken from Alberti and Uhlmann (1982). The fact that a mixture of unitaries yields a unital quantum map is a generalisation of one direction of Birkhoff's theorem, which states that mixtures of permutations and doubly stochastic maps are one and the same thing. Proposition 13.30 is taken from Lo and Popescu (2001) and Theorem 13.35 is taken from Nielsen (1999).

The SLOCC-classification of three qubits is taken from Dür et al. (2000). From four qubits onward there is an uncountably infinite set of SLOCC classes. One can nevertheless still identify finitely many parametrised 'super-classes' (see e.g. Verstraete et al., 2002; Lamata et al., 2007). The form of the W state in ZX-calculus is taken from Coecke and Edwards (2010).

The treatment of the W state as antispiders was introduced in Coecke and Kissinger (2010), and so was the interaction of spiders and antispiders. Further elaborations are in Herrmann (2010) and Kissinger (2012a). An extension of these ideas to qutrits can be found in Honda (2012). The encoding of spiders (a.k.a. the GHZ state) and antispiders (a.k.a. the W state) as 'times' and 'plus' first appeared in Coecke et al. (2010b). The completeness for the corresponding calculus is due to Hadzihasanovic (2015).

14

Quantomatic

We will encourage you to develop the three great virtues of a programmer: laziness, impatience, and hubris.

– Larry Wall, Programming Perl, 1st edition

So there you have it. Hundreds of pages and more than 3000 diagrams later, we've told you (pretty much) everything we know about quantum theory and diagrammatic reasoning. So, where to now? Is it time for everyone to start covering blackboards and filling papers with diagrams? Of course!

But even better, what if someone else did all the diagrammatic proving for you while you sit back, relax, and have a beer? That's even better! The fact that the diagrams we use are essentially made up of a finite number of ingredients (namely, spiders) means they are particularly well suited to *automated reasoning*. In this subfield of artificial intelligence, one develops software that allows a computer to do a whole range of things often associated with human mathematicians: from simply checking mathematical proofs for correctness to automatically searching for new proofs or even new and interesting conjectures to then try and prove automatically.

In the past, automated reasoning has typically concerned traditional, formula-based mathematics built on formal logics and set-based algebraic structures, and there it has been very successful. It has provided us with tools called *proof assistants*, which allow us to automatically construct proofs of mind-bending results, such as Gödel's incompleteness theorems, and rigorously check proofs that are way too big for a human mathematician to get totally correct, like Kepler's conjecture, the four-colour theorem, and the Feit–Thompson theorem (famously massive proofs in geometry, graph theory, and group theory, respectively).

In addition to serving essentially as 'robot teaching assistants', which tirelessly check the work of human mathematicians, techniques from automated reasoning can actually tell us something new, via *conjecture synthesis*. Much as a human mathematician would discover features and behaviours of an unfamiliar mathematical creature by 'poking at it' (i.e. making educated guesses about how it will behave and trying to prove them), there exist automated techniques that do this at high speed. When it succeeds in a proof, the result is a freshly minted theorem that no human has ever seen or even thought to ask about.

In fact, if you are a bit stuck for what to buy someone for Christmas, such theorems can be bought (and named after that special someone) online at theorymine.co.uk.

'Okay, okay,' you say, 'but what does this have to do with diagrams?' In this chapter, we will talk about how automated reasoning has made it into the diagrammatic realm and can be applied to all of the stuff we've done in this book, via a diagrammatic proof assistant called Quantomatic. This tool lets one create string diagrams, and string diagram equations, then apply those equations to prove theorems. This can be done either in a step-by-step fashion, as we've been doing in this book all along, or in an automated way, via *proof strategies*. In this way, we can easily handle string diagrams and diagram proofs that get way too big to work with by hand. We also see powerful theorems such as the completeness theorem for the ZX-calculus from Section 9.4.5 start to pay some serious dividends, as they can be translated into strategies that totally automate graphical proofs.

Diagrams are particularly well suited for conjecture synthesis, as we often have a collection of basic 'generators' we are interested in (e.g. spiders) and wish to discover their graphical calculus. At the end of this chapter, we will discuss how conjecture synthesis works in Quantomatic and examine the progress made towards automatic discovery of new theorems about physical theories.

14.1 Taking Quantomatic for a Spin

We'll now give some idea of what it's like to work with Quantomatic and the type of calculations it can do. We will avoid going into too many details, as these will inevitably become out of date as the tool grows and changes over time. Of course, the best thing you can do is download the latest version from quantomatic.github.io and follow along for yourself. On the website, you will also find sample projects and tutorials.

A good candidate for automation is MBQC computations, given how big the diagrams can become. Consider this computation:

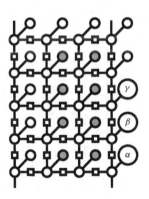

Here it is in Quantomatic:

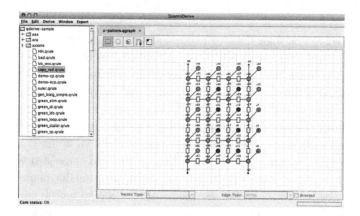

Now, let's see what this simplifies to by starting a *derivation*:

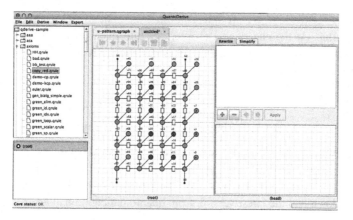

We add a few rules and start doing some rewriting:

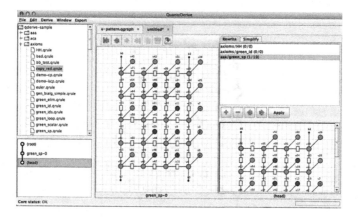

and keep rewriting:

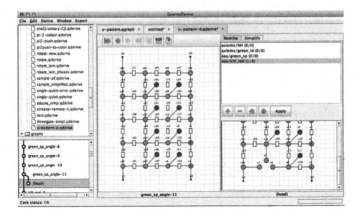

... still going:

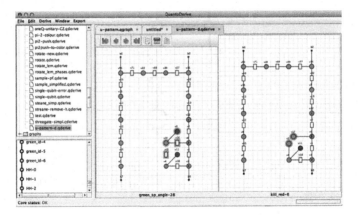

... almost there ...

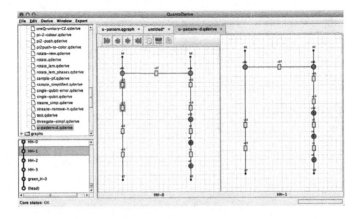

... and voila:

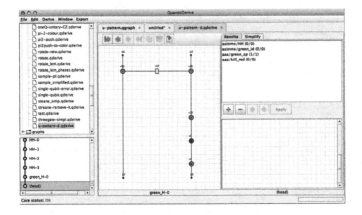

After a little over 50 steps, we end up with phase gates and a CZ-gate:

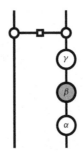

Wow, that seems like a lot of work! Actually, it only took about 15 minutes to do in Quantomatic. However, we would prefer to do this in 15 seconds, so let's switch to the *simplifier*:

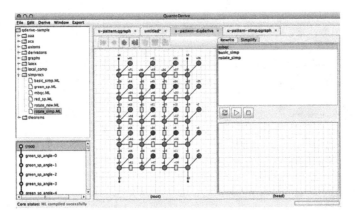

The Quantomatic's simplifier does automated rewriting, using one or more programmable proof strategies. These are little programs that tell Quantomatic how to choose the next rule to apply. We pick the mbqc strategy, and click ▷. After about 15 seconds and a trippy light show, we get:

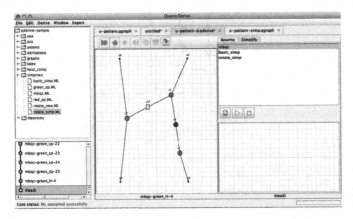

which is exactly what we expected, modulo a bit of extra spider fusion:

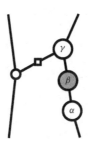

Despite having lots of steps, this calculation was really pretty straightforward. After a bit of spider fusion, the only thing left to do was eliminate the ●-spiders coming from Z-measurements, as we explained in Section 12.3.2, and eliminate the remaining H-gates with the colour-change rule.

So, let's try something a bit less obvious, coming from the circuit model. We already saw that three CNOT-gates equals swap:

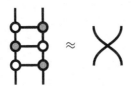

But how about this much more complicated configuration of CNOT gates?

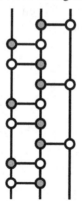

Well, we can start to do some spider fusion, or maybe use strong complementarity to get rid of the four-cycles, but even after we do that, it's far from obvious how to proceed. So, let's feed it to Quantomatic's simplifier:

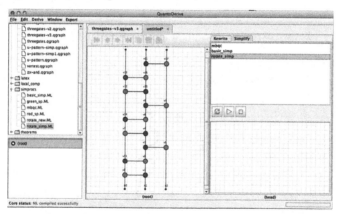

This time, we choose `rotate_simp`, which is a proof strategy that always reduces any ZX-diagram without phases into canonical form:

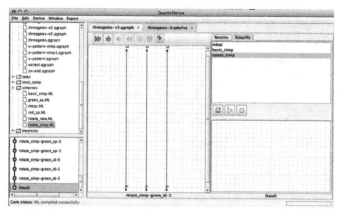

So, this is in fact just the identity. To see how Quantomatic got there, we can export the proof that the simplifier came up with:

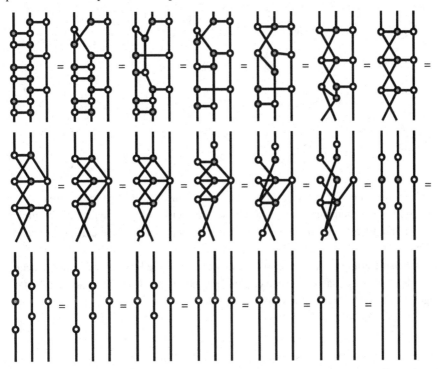

Even though we were a bit stuck on how to proceed, Quantomatic happily reduced this to the identity. And we get to take all the credit!

14.2 !-Boxes: Replacing the 'Dot, Dot, Dot'

One thing we've happily skirted over so far is what rules look like in Quantomatic. Simple rules are basically what you'd expect: a pair of graphs with identical inputs/outputs on each side:

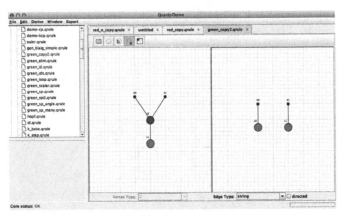

but looking at the ZX-calculus:

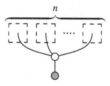

three out of the four rules are actually whole families of rules, as evident in the use of '…' in the LHS and RHS. When we give such a rule, we implicitly rely on the fact that a human with a brain is reading it, so either it is obvious what we mean or we can pretty easily explain it in words.

Clearly this isn't going to work for Quantomatic, so we need to formalise the concept of 'dot, dot, dot' in a way that a machine can understand. To see how this works, let's start with a slightly simpler '…' rule, the n-fold copy rule:

The LHS and the RHS both have some subdiagram that has been copied n times. On the LHS, this subdiagram just consists of a single output wire:

where every copy is connected to the same ○-spider, whereas the RHS contains n copies of an entire ●-spider with one output:

We can indicate this repetition by enclosing part of a diagram in a *!-box*:

(which, incidentally is pronounced 'bang-box'). A diagram with !-boxes is interpreted as a family of diagrams obtained by copying the diagram inside the !-box n times, while retaining any wires into or out of the !-box:

$$\left[\!\!\left[\begin{array}{c}\end{array}\right]\!\!\right] := \left\{\begin{array}{c}\end{array}, \begin{array}{c}\end{array}, \mathsf{Y}, \mathsf{Y}, \ldots \right\}$$

Hence an equation between diagrams with !-boxes gives an entire family of diagram equations, where each !-box on the LHS and its corresponding !-box on the RHS are copied n times:

$$\left[\!\!\left[\begin{array}{c}\end{array} = \begin{array}{c}\end{array}\right]\!\!\right] := \left\{ \begin{array}{c}\end{array} = \begin{array}{c}\end{array}, \begin{array}{c}\end{array} = \begin{array}{c}\end{array}, \mathsf{Y} = \begin{array}{c}\end{array}, \ldots \right\}$$

This !-box rules look like this in Quantomatic:

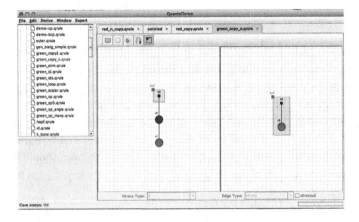

which can be applied to produce, e.g.:

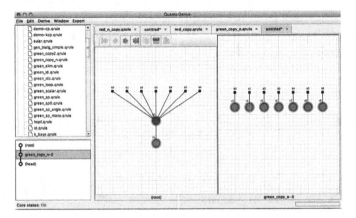

Quantomatic

Using !-box rules, we can express the whole ZX-calculus in Quantomatic as follows:

Remarkably, the fully general strong complementarity rule, which was a bit difficult to describe in words, is very easy to write down with !-boxes.

14.3 Synthesising Physical Theories

Can a machine come up with interesting new theorems about physics? More specifically, can it take the ingredients of a diagrammatic theory (e.g. spiders or antispiders) and start plugging them together to discover automatically how they interact? This is what the next feature of Quantomatic, called *conjecture synthesis*, is all about.

Rather than starting with a fixed set of diagram rewrite rules and trying to derive new rules, it starts with a concrete model and tries to discover which rules should hold. In other words, suppose we have a collection of generators, given concretely, e.g. as matrices:

but we have no idea what sorts of equations hold between diagrams of these generators. Are they spiders? Do they satisfy something like strong complementarity? Or are they something else altogether?

The way human scientists (e.g. us) would try to figure this out is simply to start plugging these things together and see which equations arise. This process is totally algorithmic, so in principal, a machine could do it. Furthermore, a machine can do it in a it would be much faster and more systematic way.

This is where conjecture synthesis comes in. This is a routine to effectively enumerate diagrams up to some given size while using the rules we discover along the way to eliminate redundancies and speed up the search for new, interesting rules. After setting up a list of generators (called 'gens' below), this routine is invoked in Quantomatic with an incantation like this:

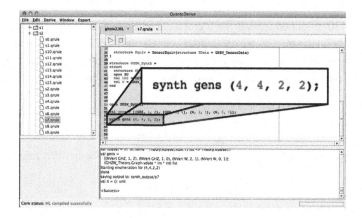

which performs the synthesis up to size $(4, 4, 2, 2)$; that is, it will generate rules involving at most four generators, four wires between them, two inputs, and two outputs. With those parameters, it found about 170 rules. Amongst them are lots of versions on spider fusion, our friends the exploding antispiders (shown here in black):

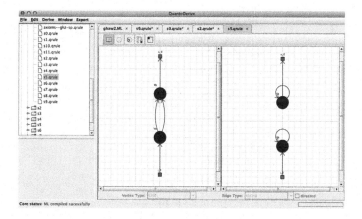

and also a variation with the GHZ-spider on top:

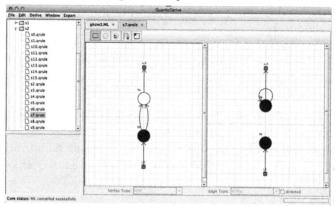

Of course, 170 unrelated rules is not a graphical calculus. This is the point where things start to get really interesting, as we try to discover relationships between these rules. Which rules are the most important? Which give the most 'proving power'? The search procedure is designed to eliminate rules that are 'obviously' provable from others; e.g. if we know this rule:

$$\vcenter{\hbox{\includegraphics{}}} = \vcenter{\hbox{\includegraphics{}}}$$

conjecture synthesis is smart enough not to bother checking whether this rule holds:

$$\vcenter{\hbox{\includegraphics{}}} = \vcenter{\hbox{\includegraphics{}}}$$

But what about the less obvious stuff? For this, one can use automated proof strategies such as the ones we described in Section 14.1 to try to build up a dependency relation between rules, i.e. a graph of which rules can be proven using others. For instance, the ZX-calculus would provide something like this:

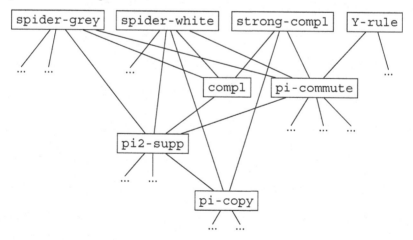

from which one could very quickly discover what are the most interesting basic rules and non-trivial theorems that exist in this theory.

This is very much ongoing work, and many aspects about what makes such a procedure effective are still unclear. The high-water mark is of course discovering theorems that are interesting in their own right, quite apart from how they were found. Of course, this has the unfortunate side effect of putting us mere human scientists out of work!

14.4 Historical Notes and References

The theory behind Quantomatic Kissinger and Zamdzhiev (2015) was developed by Dixon and Kissinger (2013), following from earlier results by Dixon and Duncan (2010) and Dixon et al. (2010). !-boxes were first proposed in Dixon and Duncan (2009) and formalised in Kissinger et al. (2014).

One of the first interactive theorem provers was Stanford LCF, developed by Robin Milner (1972) and named after the logic of computable functions by Dana Scott (1993). This was succeeded by Edinburgh LCF (Gordon et al., 1979), which originated the *LCF paradigm* of a core logical 'kernel' driven by various (semi-)automated 'tactics'. A short history can be found in Gordon (2000). Other notable provers to adopt this paradigm are Isabelle (Paulson et al., 1986) and Coq (Coqand et al., 1984). The latter has gained significant attention lately due to its use in homotopy type theory and the univalent foundation of mathematics (Shulman et al., 2013).

In recent years, several enormous proofs were fully formalised in theorem provers. The FlysPecK project, led by Hales et al. (2015), succeeded in giving a formal proof of the Kepler conjecture in 2015. A formal proof of the four-colour theorem was given by Gonthier (2008) and the Feit–Thompson theorem by Gonthier et al. (2013).

The method for synthesising diagrammatic theories was proposed by Kissinger (2012b), based on techniques introduced by Johansson et al. (2011). The latter has been used to (among other things!) automatically prove little theorems that can be bought as gifts (Bundy et al., n.d.).

The second example in Section 14.1 is equivalent to rule (C14) from (Selinger, 2015), one of the 15 rules used to construct a complete calculus for Clifford circuits.

Appendix
Some Notations

The following notations are used throughout this book.

General

- $X := Y$ means *X is defined to be (or, is interpreted as) Y*
- $X \leftrightarrow Y$ means *X corresponds to Y*
- $X \Leftrightarrow Y$ means *X if and only if Y*
- LHS means *left-hand-side*
- RHS means *right-hand-side*
- $\exists$ means *there exists*
- $\forall$ means *for all*
- $\checkmark$ means *ok*
- ☠ means *not ok*

Concerning Sets

- $\emptyset$ means *the empty set*
- $\mathbb{R}$ means *the set of all real numbers*
- $\mathbb{R}^+$ means *the set of all positive real numbers*
- $[0, 1]$ means *the set of real numbers between 0 and 1 (inclusive)*
- $x \in X$ means *x is an element of X*
- $X \subseteq Y$ means *X is a subset of Y*
- $X - Y$ means *the set of all elements of X that are not in Y*
- $X \cong Y$ means *X is isomorphic to Y*
- $\{X \mid Y\}$ means *the set of all X such that Y holds*

804

Concerning Functions and Relations

- $x \mapsto y$ means *x is mapped to y*
- $f : X \to Y$ means *f is a function from set X to set Y*
- $f :: x \mapsto y$ means *function or relation f maps x to y*

- $f ::$ $\begin{cases} x_1 \mapsto y_1 \\ x_2 \mapsto y_2 \\ \vdots \\ x_n \mapsto y_n \end{cases}$ means *function (or relation) f maps ..., x_i to y_i, ...*

- $R ::$ $\begin{cases} x_1 \mapsto Y_1 \\ x_2 \mapsto Y_2 \\ \vdots \\ x_n \mapsto Y_n \end{cases}$ means *relation R maps ..., x_i to all $y_i \in Y_i$, ...*

- $f(a)$ means *those elements function (or relation) f maps to*

Concerning Diagrams

- means *the empty diagram*

- 0 means *any zero diagram*

References

Abramsky, S. 2010. No-cloning in categorical quantum mechanics. Pages 1–28 of: Gay, S., and Mackie, I. (eds), *Semantic Techniques in Quantum Computation*. Cambridge University Press. Arxiv preprint arXiv:0910.2401.

Abramsky, S., and Coecke, B. 2004. A categorical semantics of quantum protocols. Pages 415–425 of: *Proceedings of the 19th Annual IEEE Symposium on Logic in Computer Science (LICS)*. arXiv:quant-ph/0402130.

Abramsky, S., and Coecke, B. 2005. Abstract physical traces. *Theory and Applications of Categories*, **14**(6), 111–124. arXiv:0910.3144.

Abramsky, S., and Heunen, C. 2012. Operational theories and categorical quantum mechanics. In: *Logic and Algebraic Structures in Quantum Computing*. Cambridge University Press. arXiv:1206.0921.

Abramsky, S., and Jagadeesan, R. 1994. New foundations for the geometry of interaction. *Information and Computation*, **111**, 53–119.

Abramsky, S., and Tzevelekos, N. 2011. Introduction to categories and categorical logic. Pages 3–94 of: Coecke, B. (ed), *New Structures for Physics*. Lecture Notes in Physics. Springer-Verlag.

Alberti, P. M., and Uhlmann, A. 1982. *Stochasticity and Partial Order*. Mathematics and Its Applications, vol. 9. Reidel.

Ambainis, A. 2010. *New developments in quantum algorithms*. arXiv:1006.4014.

Aspect, A., Grangier, P., and Roger, G. 1981. Experimental tests of realistic local theories via Bell's theorem. *Physical Review Letters*, **47**(7), 460.

Aspect, A., Dalibard, J., and Roger, G. 1982. Experimental test of Bell's inequalities using time-varying analyzers. *Physical Review Letters*, **49**(25), 1804.

Awodey, S. 2010. *Category Theory*. Oxford University Press.

Backens, M. 2014a. The ZX-calculus is complete for the single-qubit Clifford+T group. Pages 293–303 of: Coecke, B., Hasuo, I. and Panangaden, P. (eds), *Proceedings of the 11th Workshop on Quantum Physics and Logic*. Electronic Proceedings in Theoretical Computer Science, vol. 172. Open Publishing Association.

Backens, M. 2014b. *The ZX-calculus is complete for the single-qubit Clifford+T group*. arXiv:1412.8553.

Backens, M., and Nabi Duman, A. 2015. A complete graphical calculus for Spekkens' toy bit theory. *Foundations of Physics*. arXiv:1411.1618.

Backens, M., Perdrix, S., and Wang, Q. 2016. A simplified stabilizer ZX-calculus. In: *Proceedings of the 13th International Conference on Quantum Physics and Logic*. arXiv:1602.04744.

Baez, J. C. 1993–2010. *This week's finds in mathematical physics.* math.ucr.edu/home/baez/TWF.html.

Baez, J. C. 2006. Quantum quandaries: a category-theoretic perspective. Pages 240–266 of: Rickles, D., French, S., and Saatsi, J.T. (eds), *The Structural Foundations of Quantum Gravity.* Oxford University Press. arXiv:quant-ph/0404040.

Baez, J. C., and Dolan, J. 1995. Higher-dimensional algebra and topological quantum field theory. *Journal of Mathematical Physics,* **36,** 6073. arXiv:q-alg/9503002.

Baez, J. C., and Erbele, J. *Categories in control.* arXiv:1405.6881.

Baez, J. C., and Fong, B. 2015. A compositional framework for passive linear networks. *arXiv:1504.05625.*

Baez, J. C., and Lauda, A. 2011. A prehistory of n-categorical physics. Pages 13–128 of: Halvorson, H. (ed), *Deep Beauty: Understanding the Quantum World through Mathematical Innovation.* Cambridge University Press.

Baez, J. C., and Stay, M. 2011. Physics, topology, logic and computation: a Rosetta Stone. Pages 95–172 of: Coecke, B. (ed), *New Structures for Physics.* Lecture Notes in Physics. Springer.

Balkir, E., Sadrzadeh, M., and Coecke, B. 2016. *Distributional Sentence Entailment Using Density Matrices.* Cham: Springer International Publishing. Pages 1–22.

Ballance, C. J., Harty, T. P., Linke, N. M., Sepiol, M. A., and Lucas, D. M. 2016. High-Fidelity Quantum Logic Gates Using Trapped-Ion Hyperfine Qubits. *Physical Review Letters,* **117**(6), 060504.

Baltag, A., and Smets, S. 2005. Complete axiomatizations for quantum actions. *International Journal of Theoretical Physics,* **44,** 2267–2282.

Bankova, D., Coecke, B., Lewis, M., and Marsden, D. 2016. Graded entailment for compositional distributional semantics. In: *Proceedings of the 13th International Conference on Quantum Physics and Logic.* arXiv:1601.04908.

Barenco, A., Bennett, C. H., Cleve, R., DiVincenzo, D. P., Margolus, N., Shor, P. W., Sleator, T., Smolin, J. A., and Weinfurter, H. 1995. Elementary gates for quantum computation. *Physical Review A,* **52,** 3457–3467.

Barnum, H., Caves, C. M., Fuchs, C. A., Jozsa, R., and Schumacher, B. 1996. Noncommuting mixed states cannot be broadcast. *Physical Review Letters,* **76,** 2818.

Barnum, H., Barrett, J., Leifer, M., and Wilce, A. 2007. A generalized no-broadcasting theorem. *Physical Review Letters,* **99**(24), 240501.

Barr, M., and Wells, C. 1990. *Category Theory for Computing Science.* New York: Prentice Hall.

Barrett, J. 2007. Information processing in generalized probabilistic theories. *Physical Review A,* **75,** 032304.

Belinfante, F. J. 1973. *Survey of Hidden-Variables Theories.* Pergamon Press.

Bell, J. S. 1964. On the Einstein–Podolsky–Rosen paradox. *Physics,* **1**(3), 195–200.

Benabou, J. 1963. Categories avec multiplication. *Comptes Rendus des Séances de l'Académie des Sciences. Paris,* **256,** 1887–1890.

Benioff, P. 1980. The computer as a physical system: a microscopic quantum mechanical Hamiltonian model of computers as represented by Turing machines. *Journal of Statistical Physics,* **22,** 563–591.

Bennett, C. H., and Brassard, G. 1984. Quantum cryptography: public key distribution and coin tossing. Pages 175–179 of: *Proceedings of IEEE International Conference on Computers, Systems and Signal Processing.* IEEE.

Bennett, C. H., and Wiesner, S. 1992. Communication via one- and two-particle operators on Einstein–Podolsky–Rosen states. *Physical Review Letters,* **69,** 2881–2884.

Bennett, C. H., Brassard, G., Crepeau, C., Jozsa, R., Peres, A., and Wootters, W. K. 1993. Teleporting an unknown quantum state via dual classical and Einstein–Podolsky–Rosen channels. *Physical Review Letters*, **70**(13), 1895–1899.

Birkhoff, G., and von Neumann, J. 1936. The logic of quantum mechanics. *Annals of Mathematics*, **37**, 823–843.

Bloch, F. 1946. Nuclear induction. *Physical Review*, **70**, 460–474.

Blute, R. F., Ivanov, I. T., and Panangaden, P. 2003. Discrete quantum causal dynamics. *International Journal of Theoretical Physics*, **42**(9), 2025–2041.

Bohm, D. 1952a. A suggested interpretation of the quantum theory in terms of hidden" variables. I. *Physical Review*, **85**(2), 166.

Bohm, D. 1952b. A suggested interpretation of the quantum theory in terms of 'hidden' variables. II. *Physical Review*, **85**(2), 180.

Bohm, D. 1986. Time, the implicate order and pre-space. Pages 172–208 of: Griffin, D. R. (ed), *Physics and the Ultimate Significance of Time*. SUNY Press.

Bohm, D., and Peat, F. D. 1987. *Science, Order, and Creativity*. Routledge.

Bohr, N. 1931. *Atomtheorie und Naturbeschreibung*. Springer.

Bohr, N. 1935. Quantum mechanics and physical reality. *Nature*, **136**, 65.

Bohr, N. 1961. *Atomic Physics and Human Knowledge*. Science Editions.

Boixo, S., and Heunen, C. 2012. Entangled and sequential quantum protocols with dephasing. *Physical Review Letters*, **108**, 120402. arXiv:1108.3569.

Bonchi, F., Sobocinski, P., and Zanasi, F. 2014a. A categorical semantics of signal flow graphs. Pages 435–450 of: *CONCUR'14: Concurrency Theory*. Lecture Notes in Computer Science, vol. 8704. Springer.

Bonchi, F., Sobocinski, P., and Zanasi, F. 2014b. Interacting bialgebras are Frobenius. Pages 351–365 of: *17th International Conference on Foundations of Software Science and Computation Structures (FOSSACS)*.

Borceux, F. 1994a. *Handbook of Categorical Algebra: Basic Category Theory*. Cambridge University Press.

Borceux, F. 1994b. *Handbook of Categorical Algebra: Categories and Structures*. Cambridge University Press.

Born, M. 1926. Quantenmechanik der stoßvorgänge. *Zeitschrift für Physik*, **38**(11–12), 803–827.

Born, M., and Jordan, P. 1925. Zur Quantenmechanik. *Zeitschrift für Physik*, **34**, 858–888.

Bourbaki, N. 1959–2004. *Éléments de mathématique*. CCLS & Editions Masson.

Bourbaki, N. 1981. *Espaces vectoriels topologiques*. Springer.

Bourbaki, N. 1987. *Topological Vector Spaces*. Springer.

Bouwmeester, D., Pan, J.-W., Mattle, K., Eibl, M., Weinfurter, H., and Zeilinger, A. 1997. Experimental quantum teleportation. *Nature*, **390**(6660), 575–579.

Brandão, F. G. S. L., Horodecki, M., Oppenheim, J., Renes, J. M., and Spekkens, R. W. 2013. The resource theory of quantum states out of thermal equilibrium. *Physical Review Letters*, **111**, 250404.

Brauer, R., and Nesbitt, C. 1937. On the regular representations of algebras. *Proceedings of the National Academy of Sciences of the United States of America*, **23**(4), 236.

Bub, J. 1999. *Interpreting the Quantum World*. Cambridge University Press.

Buchsbaum, D. 1955. Exact categories and duality. *Transactions of the American Mathematical Society*, **80**, 1–34.

Bundy, A., Cavallo, F., Dixon, L., Johansson, M., and McCasland, R. N.d. 2015. The theory behind TheoryMine. *IEEE Intelligent Systems*, **30**(4), 64–69.

Carboni, A., and Walters, R. F. C. 1987. Cartesian bicategories I. *Journal of Pure and Applied Algebra*, **49**, 11–32.

Carroll, L. 1942. *Alice in Wonderland*. Pelangi Publishing Group Bhd.

Chiribella, G. 2014. *Distinguishability and copiability of programs in general process theories*. arXiv:1411.3035.

Chiribella, G., and Scandolo, C. M. 2015. Entanglement and thermodynamics in general probabilistic theories. *New Journal of Physics*, **17**, 103027.

Chiribella, G., D'Ariano, G. M., and Perinotti, P. 2010. Probabilistic theories with purification. *Physical Review A*, **81**(6), 062348.

Chiribella, G., D'Ariano, G. M., and Perinotti, P. 2011. Informational derivation of quantum theory. *Physical Review A*, **84**(1), 012311.

Choi, M.-D. 1975. Completely positive linear maps on complex matrices. *Linear Algebra and Its Applications*, **10**, 285–290.

Clark, S., Coecke, B., Grefenstette, E., Pulman, S., and Sadrzadeh, M. 2014. A quantum teleportation inspired algorithm produces sentence meaning from word meaning and grammatical structure. *Malaysian Journal of Mathematical Sciences*, **8**, 15–25. arXiv:1305.0556.

Clifton, R., Bub, J., and Halvorson, H. 2003. Characterizing quantum theory in terms of information-theoretic constraints. *Foundations of Physics*, **33**, 1561–1591.

Coecke, B. 2000. Structural characterization of compoundness. *International Journal of Theoretical Physics*, **39**, 585–594.

Coecke, B. 2003. *The logic of entanglement. An invitation*. Tech. rept. RR-03-12. Department of Computer Science, Oxford University.

Coecke, B. 2005. Kindergarten quantum mechanics. Pages 81–98 of: Khrennikov, A. (ed), *Quantum Theory: Reconsiderations of the Foundations III*. AIP Press. arXiv:quant-ph/0510032.

Coecke, B. 2007. De-linearizing linearity: projective quantum axiomatics from strong compact closure. *Electronic Notes in Theoretical Computer Science*, **170**, 49–72. arXiv:quant-ph/0506134.

Coecke, B. 2008. Axiomatic description of mixed states from Selinger's CPM-construction. *Electronic Notes in Theoretical Computer Science*, **210**, 3–13.

Coecke, B. 2009. Quantum picturalism. *Contemporary Physics*, **51**, 59–83. arXiv:0908.1787.

Coecke, B. 2011. A universe of processes and some of its guises. Pages 129–186 of: Halvorson, H. (ed), *Deep Beauty: Understanding the Quantum World through Mathematical Innovation*. Cambridge University Press. arXiv:1009.3786.

Coecke, B. 2013. An alternative Gospel of structure: order, composition, processes. Pages 1–22 of: Heunen, C., Sadrzadeh, M., and Grefenstette, E. (eds), *Quantum Physics and Linguistics: A Compositional, Diagrammatic Discourse*. Oxford University Press. arXiv:1307.4038.

Coecke, B. 2014a. *The Logic of Entanglement*. Cham: Springer International Publishing. Pages 250–267.

Coecke, B. 2014b. *Terminality implies non-signalling*. arXiv:1405.3681.

Coecke, B. 2016. The logic of quantum mechanics – take II. Pages 174–198 of: Chubb, J., Eskandarian, A., and Harizanov, V. (eds), *Logic and Algebraic Structures in Quantum Computing*. Cambridge University Press. arXiv:1204.3458.

Coecke, B., and Duncan, R. 2008. Interacting quantum observables. In: *Proceedings of the 37th International Colloquium on Automata, Languages and Programming (ICALP)*. Lecture Notes in Computer Science.

Coecke, B., and Duncan, R. 2011. Interacting quantum observables: categorical algebra and diagrammatics. *New Journal of Physics*, **13**, 043016. arXiv:quant-ph/09064725.

Coecke, B., and Edwards, B. 2010. Three qubit entanglement within graphical Z/X-calculus. *Electronic Proceedings in Theoretical Computer Science*, **52**, 22–33.

Coecke, B., and Edwards, B. 2011. Toy quantum categories. *Electronic Notes in Theoretical Computer Science*, **270**(1), 29–40. arXiv:0808.1037.

Coecke, B., and Edwards, B. 2012. Spekkens's toy theory as a category of processes. In: Abramsky, S., and Mislove, M. (eds), *Mathematical Foundations of Information Flow*. Proceedings of symposia in applied mathematics. American Mathematical Society. arXiv:1108.1978.

Coecke, B., and Heunen, C. 2011. Pictures of complete positivity in arbitrary dimension. *Quantum Phsyics and Logic, Electronic Proceedings in Theoretical Computer Science*, **95**, 27–35. arXiv:1210.0298.

Coecke, B., and Kissinger, A. 2010. The compositional structure of multipartite quantum entanglement. Pages 297–308 of: *Automata, Languages and Programming*. Lecture Notes in Computer Science. Springer. arXiv:1002.2540.

Coecke, B., and Lal, R. 2013. Causal categories: relativistically interacting processes. *Foundations of Physics*, **43**, 458–501. arXiv:1107.6019.

Coecke, B., and Paquette, É. O. 2008. POVMs and Naimark's theorem without sums. *Electronic Notes in Theoretical Computer Science*, **210**, 15–31. arXiv:quant-ph/0608072.

Coecke, B., and Paquette, É. O. 2011. Categories for the practicing physicist. Pages 167–271 of: Coecke, B. (ed), *New Structures for Physics*. Lecture Notes in Physics. Springer. arXiv:0905.3010.

Coecke, B., and Pavlovic, D. 2007. Quantum measurements without sums. Pages 567–604 of: Chen, G., Kauffman, L., and Lamonaco, S. (eds), *Mathematics of Quantum Computing and Technology*. Taylor and Francis. arXiv:quant-ph/0608035.

Coecke, B., and Perdrix, S. 2010. Environment and classical channels in categorical quantum mechanics. Pages 230–244 of: *Proceedings of the 19th EACSL Annual Conference on Computer Science Logic (CSL)*. Lecture Notes in Computer Science, vol. 6247. Extended version: arXiv:1004.1598.

Coecke, B., and Smets, S. 2004. The Sasaki hook is not a [static] implicative connective but induces a backward [in time] dynamic one that assigns causes. *International Journal of Theoretical Physics*, **43**, 1705–1736.

Coecke, B., and Spekkens, R. W. 2012. Picturing classical and quantum Bayesian inference. *Synthese*, **186**, 651–696. arXiv:1102.2368.

Coecke, B., Moore, D. J., and Wilce, A. 2000. Operational quantum logic: an overview. Pages 1–36 of: Coecke, B., Moore, D. J., and Wilce, A. (eds), *Current Research in Operational Quantum Logic: Algebras, Categories and Languages*. Fundamental Theories of Physics, vol. 111. Springer-Verlag. arXiv:quant-ph/0008019.

Coecke, B., Moore, D. J., and Stubbe, I. 2001. Quantaloids describing causation and propagation of physical properties. *Foundations of Physics Letters*, **14**, 133–146. arXiv:quant-ph/0009100.

Coecke, B., Paquette, É. O., and Perdrix, S. 2008a. Bases in diagrammatic quantum protocols. *Electronic Notes in Theoretical Computer Science*, **218**, 131–152. arXiv:0808.1029.

Coecke, B., Paquette, É. O., and Pavlović, D. 2008b. *Classical and quantum structures*. Tech. rept. RR-08-02. Department of Computer Science, Oxford University.

Coecke, B., Paquette, É. O., and Pavlović, D. 2010a. Classical and quantum structuralism. Pages 29–69 of: Gay, S., and Mackie, I. (eds), *Semantic Techniques in Quantum Computation*. Cambridge University Press. arXiv:0904.1997.

Coecke, B., Kissinger, A., Merry, A., and Roy, S. 2010b. The GHZ/W-calculus contains rational arithmetic. *Electronic Proceedings in Theoretical Computer Science*, **52**, 34–48.

Coecke, B., Sadrzadeh, M., and Clark, S. 2010c. Mathematical foundations for a compositional distributional model of meaning. Pages 345–384 of: van Benthem, J., Moortgat, M., and Buszkowski, W. (eds), *A Festschrift for Jim Lambek*. Linguistic Analysis, vol. 36. arxiv:1003.4394.

Coecke, B., Wang, Q., Wang, B., Wang, Y., and Zhang, Q. 2011a. Graphical calculus for quantum key distribution (extended abstract). *Electronic Notes in Theoretical Computer Science*, **270**(2), 231–249.

Coecke, B., Edwards, B., and Spekkens, R. W. 2011b. Phase groups and the origin of non-locality for qubits. *Electronic Notes in Theoretical Computer Science*, **270**(2), 15–36. arXiv:1003.5005.

Coecke, B., Duncan, R., Kissinger, A., and Wang, Q. 2012. Strong complementarity and non-locality in categorical quantum mechanics. In: *Proceedings of the 27th Annual IEEE Symposium on Logic in Computer Science (LICS)*. arXiv:1203.4988.

Coecke, B., Heunen, C., and Kissinger, A. 2013a. *Categories of quantum and classical channels*. arXiv:1305.3821.

Coecke, B., Heunen, C., and Kissinger, A. 2013b. Compositional quantum logic. Pages 21–36 of: *Computation, Logic, Games, and Quantum Foundations: The Many Facets of Samson Abramsky*. Springer.

Coecke, B., Pavlović, D., and Vicary, J. 2013c. A new description of orthogonal bases. *Mathematical Structures in Computer Science*, **23**, 555–567. arXiv:quant-ph/0810.1037.

Coecke, B., Fritz, T., and Spekkens, R. W. 2016. A mathematical theory of resources. *Information and Computation*.

Coecke, B., Duncan, R., Kissinger, A., and Wang, Q. 2016. Generalised compositional theories and diagrammatic reasoning. In: Chiribella, G., and Spekkens, R. W. (eds), *Quantum Theory: Informational Foundations and Foils*. Fundamental Theories of Physics. Springer. arXiv:1203.4988.

Coqand, T., Heut, G., et al. 1984. *Coq theorem prover*. https://coq.inria.fr/.

Cunningham, O., and Heunen, C. 2015. Axiomatizing complete positivity. Pages 148–157 of: Heunen, C., Selinger, P., and Vicary, J. (eds), *Proceedings of the 12th International Workshop on Quantum Physics and Logic*. Electronic Proceedings in Theoretical Computer Science, vol. 195. Open Publishing Association.

Davies, E. B. 1976. *Quantum Theory of Open Systems*. Academic Press.

Davies, E. B., and Lewis, J. T. 1970. An operational approach to quantum probability. *Communications in Mathematical Physics*, **17**, 239–260.

Deutsch, D. 1985. Quantum theory, the Church–Turing principle and the universal quantum computer. *Proceedings of the Royal Society of London. A. Mathematical and Physical Sciences*, **400**(1818), 97–117.

Deutsch, D. 1989. Quantum computational networks. *Proceedings of the Royal Society of London*, **425**.

Deutsch, D. 1991. Quantum mechanics near closed timelike lines. *Physical Review D*, **44**, 3197.

Deutsch, D., and Jozsa, R. 1992. Rapid solution of problems by quantum computation. *Proceedings of the Royal Society of London. Series A: Mathematical and Physical Sciences*, **439**(1907), 553–558.

Dieks, D. G. B. J. 1982. Communication by EPR devices. *Physics Letters A*, **92**(6), 271–272.

Dijkstra, E. W. 1968. A constructive approach to the problem of program correctness. *BIT Numerical Mathematics*, **8**, 174–186.

Dirac, P. A. M. 1926. On the theory of quantum mechanics. *Proceedings of the Royal Society A*, **112**, 661–677.

Dirac, P. A. M. 1939. A new notation for quantum mechanics. Pages 416–418 of: *Proceedings of the Cambridge Philosophical Society*, vol. 35. Cambridge University Press.

Dixon, L., and Duncan, R. 2009. Graphical reasoning in compact closed categories for quantum computation. *Annals of Mathematics and Artificial Intelligence*, **56**(1), 23–42.

Dixon, L., and Duncan, R. 2010. Extending graphical representations for compact closed categories with applications to symbolic quantum computation. *Intelligent Computer Mathematics*, 77–92.

Dixon, L., and Kissinger, A. 2013. Open-graphs and monoidal theories. *Mathematical Structures in Computer Science*, **23**(2), 308–359.

Dixon, L., Duncan, R., and Kissinger, A. 2010. Open graphs and computational reasoning. Pages 169–180 of: Cooper, S. B., Panangaden, P. and Kashefi, E. (eds), *Proceedings of the Sixth Workshop on Developments in Computational Models: Causality, Computation, and Physics*. Electronic Proceedings in Theoretical Computer Science, vol. 26. Open Publishing Association.

Dixon, L., Duncan, R., Merry, A., Kissinger, A., Soloviev, M., and Zamzhiev, V. 2011. `quantomatic`. http://quantomatic.github.io.

Duncan, R. 2006. *Types for quantum computation*. DPhil Thesis, Oxford University.

Duncan, R. 2012. *A graphical approach to measurement-based quantum computing*. arXiv:1203.6242.

Duncan, R., and Lucas, M. 2013. Verifying the Steane code with Quantomatic. In: *Proceedings of the 10th International Workshop on Quantum Physics and Logic*. arXiv:1306.4532.

Duncan, R., and Perdrix, S. 2009. Graph states and the necessity of Euler decomposition. *Mathematical Theory and Computational Practice*, 167–177.

Duncan, R., and Perdrix, S. 2010. Rewriting measurement-based quantum computations with generalised flow. Pages 285–296 of: *Proceedings of ICALP*. Lecture Notes in Computer Science. Springer.

Duncan, R., and Perdrix, S. 2013. Pivoting makes the ZX-calculus complete for real stabilizers. In: *Proceedings of the 10th International Workshop on Quantum Physics and Logic*. arXiv:1307.7048.

Dür, W., Vidal, G., and Cirac, J. I. 2000. Three qubits can be entangled in two inequivalent ways. *Physical Review A*, **62**(062314).

Durt, T., Englert, B.-G., Bengtsson, I., and Życzkowski, K. 2010. On mutually unbiased bases. *International Journal of Quantum Information*, **8**, 535–640.

Eckmann, B., and Hilton, P. J. 1962. Group-like structures in general categories. I. Multiplications and comultiplications. *Mathematische Annalen*, **145**(3).

Edwards, B. 2009. *Non-locality in categorical quantum mechanics*. PhD thesis, University of Oxford.

Eilenberg, S., and Mac Lane, S. 1945. General theory of natural equivalences. *Transactions of the American Mathematical Society*, **58**(2), 231.

Einstein, A. 1936. Physics and reality. *Journal of the Franklin Institute*, **221**(3), 349–382.

Einstein, A., Podolsky, B., and Rosen, N. 1935. Can quantum-mechanical description of physical reality be considered complete? *Physical Review*, **47**(10), 777.

Ekert, A. K. 1991. Quantum cryptography based on Bell's theorem. *Physical Review Letters*, **67**(6), 661–663.

Evans, J., Duncan, R., Lang, A., and Panangaden, P. 2009. *Classifying all mutually unbiased bases in Rel*. arXiv:0909.4453.

Everett, H. III. 1957. "Relative state" formulation of quantum mechanics. *Reviews of Modern Physics*, **29**(3), 454.

Faure, C.-A., Moore, D. J., and Piron, C. 1995. Deterministic evolutions and Schrödinger flows. *Helvetica Physica Acta*, **68**(2), 150–157.

Fauser, B. 2013. Some graphical aspects of Frobenius structures. Pages 23–48 of: Heunen, C., Sadrzadeh, M., and Grefenstette, E. (eds), *Quantum Physics and Linguistics: A Compositional, Diagrammatic Discourse*. Oxford University Press. arXiv:1202.6380.

Feynman, R. P. 1982. Simulating physics with computers. *International Journal of Theoretical Physics*, **21**, 467–488.

Fong, B., and Nava-Kopp, H. 2015. Additive monotones for resource theories of parallel-combinable processes with discarding. *Electronic Proceedings in Theoretical Computer Science*, **195**, 170–178. arXiv:1505.02651.

Fort, C. 1931. *Lo!* Cosimo Books.

Foulis, D. J., and Randall, C. H. 1972. Operational statistics. I. Basic concepts. *Journal of Mathematical Physics*, **13**(11), 1667–1675.

Freyd, P. 1964. *Abelian Categories*. New York: Harper and Row.

Freyd, P., and Yetter, D. 1989. Braided compact closed categories with applications to low-dimensional topology. *Advances in Mathematics*, **77**, 156–182.

Fritz, T. 2014. *Beyond Bell's theorem II: scenarios with arbitrary causal structure*. arXiv:1404.4812.

Fritz, T. 2015. Resource convertibility and ordered commutative monoids. *Mathematical Structures in Computer Science*, **10**, 1–89.

Fuchs, C. A. 2002. *Quantum mechanics as quantum information (and only a little more)*. arXiv: quant-ph/0205039.

Fuchs, C. A., Mermin, N. D., and Schack, R. 2014. An introduction to QBism with an application to the locality of quantum mechanics. *American Journal of Physics*, **82**, 749–754. arXiv:1311.5253.

Ghirardi, G.-C., Rimini, A., and Weber, T. 1980. A general argument against superluminal transmission through the quantum mechanical measurement process. *Lettere Al Nuovo Cimento*, **27**(10), 293–298.

Gilbreth, F. B., and Gilbreth, L. M. 1922. Process charts and their place in management. *Mechanical engineering*, **70**, 38–41.

Girard, J.-Y. 1989. Towards a geometry of interaction. *Contemporary Mathematics*, **92**, 69–108.

Gleason, A. M. 1957. Measures on the closed subspaces of a Hilbert space. *Journal of Mathematics and Mechanics*, **6**, 885–893.

Gogioso, S. 2015a. *A bestiary of sets and relations*. arXiv:1506.05025.

Gogioso, S. 2015b. *Categorical semantics for Schrödinger's equation*. arXiv:1501.06489.

Gogioso, S. 2015c. *Monadic dynamics*. arXiv:1501.04921.

Gogioso, S., and Genovese, F. 2016. Infinite-dimensional categorical quantum mechanics. In: *Proceedings of QPL*. arXiv:1605.04305.

Gogioso, S., and Kissinger, A. 2016. *Fully graphical treatment of the Hidden Subgroup Problem*. Unpublished.

Gogioso, S., and Zeng, W. 2015. *Mermin non-locality in abstract process theories*. arXiv:1506.02675.

Gonthier, G. 2008. *The Four Colour Theorem: Engineering of a Formal Proof*. Berlin and Heildelberg: Springer. Page 333.

Gonthier, G., Asperti, A., Avigad, J., Bertot, Y., Cohen, C., Garillot, F., Le Roux, S., Mahboubi, A., O'Connor, R., Biha, S. O., et al. 2013. A machine-checked proof of the odd order theorem. Pages 163–179 of: *Interactive Theorem Proving*. Springer.

Gordon, M. 2000. From LCF to HOL: a short history. Pages 169–186 of: *Proof, Language, and Interaction*.

Gordon, M. J., Milner, A. J., and Wadsworth, C. P. 1979. *Lecture Notes in Computer Science*. Vol. 78. Berlin: Springer-Verlag.

Gottesman, D., and Chuang, I. L. 1999. Demonstrating the viability of universal quantum computation using teleportation and single-qubit operations. *Nature*, **402**(6760), 390–393.

Gour, G., and Spekkens, R. W. 2008. The resource theory of quantum reference frames: manipulations and monotones. *New Journal of Physics*, **10**, 033023.

Gour, G., Müller, M. P., Narasimhachar, V., Spekkens, R. W., and Yunger Halpern, N. 2013. *The resource theory of informational nonequilibrium in thermodynamics*. arXiv:1309.6586.

Greenberger, D. M., Horne, M. A., Shimony, A., and Zeilinger, A. 1990. Bell's theorem without inequalities. *American Journal of Physics*, **58**, 1131–1143.

Grefenstette, E., and Sadrzadeh, M. 2011. Experimental support for a categorical compositional distributional model of meaning. Pages 1394–1404 of: *The 2014 Conference on Empirical Methods on Natural Language Processing*. arXiv:1106.4058.

Gröblacher, S., Paterek, T., Kaltenbaek, R. R., Brukner, Č., Żukowski, M., Aspelmeyer, M., and Zeilinger, A. 2007. An experimental test of non-local realism. *Nature*, **446**, 871–875.

Grothendieck, A. 1957. Sur quelques points d'algèbre homologique. *Tohoku Math J.*, 119–221.

Grover, L. K. 1996. A fast quantum mechanical algorithm for database search. Pages 212–219 of: *Proceedings of the Twenty-eighth Annual ACM Symposium on Theory of Computing*. STOC '96. New York: ACM.

Hadzihasanovic, A. 2015. A diagrammatic axiomatisation for qubit entanglement. In: *Proceedings of the 30th Annual IEEE Symposium on Logic in Computer Science (LICS)*. arXiv:1501.07082.

Hales, T., Adams, M., Bauer, G., Dang, D. T., Harrison, J., Hoang, T. L., Kaliszyk, C., Magron, V., McLaughlin, S., Nguyen, T. T., et al. 2015. A formal proof of the Kepler conjecture. *arXiv preprint arXiv:1501.02155*.

Harding, J. 2009. A link between quantum logic and categorical quantum mechanics. *International Journal of Theoretical Physics*, **48**(3), 769–802.

Hardy, L. N.d. *Disentangling nonlocality and teleportation*. arXiv:quant-ph/9906123.

Hardy, L. 2001. Quantum theory from five reasonable axioms. *arXiv:quant-ph/0101012*.

Hardy, L. 2011. Foliable operational structures for general probabilistic theories. Pages 409–442 of: Halvorson, H. (ed), *Deep Beauty: Understanding the Quantum World through Mathematical Innovation*. Cambridge University Press. arXiv:0912.4740.

Hardy, L. 2012. *The operator tensor formulation of quantum theory.* arXiv:1201.4390.

Hardy, L. 2013a. A formalism-local framework for general probabilistic theories, including quantum theory. *Mathematical Structures in Computer Science*, **23**(2), 339–440.

Hardy, L. 2013b. On the theory of composition in physics. Pages 83–106 of: *Computation, Logic, Games, and Quantum Foundations: The Many Facets of Samson Abramsky.* Springer. arXiv:1303.1537.

Hardy, L., and Spekkens, R. W. 2010. Why physics needs quantum foundations. *Physics in Canada*, **66**, 73–76.

Harrigan, N., and Spekkens, R. W. 2010. Einstein, incompleteness, and the epistemic view of quantum states. *Foundations of Physics*, **40**, 125–157.

Hasegawa, M., Hofmann, M., and Plotkin, G. D. 2008. Finite dimensional vector spaces are complete for traced symmetric monoidal categories. Pages 367–385 of: Avron, A., Dershowitz, N., and Rabinovich, A. (eds), *Pillars of Computer Science*. Lecture Notes in Computer Science, vol. 4800. Springer.

Hedges, J., Shprits, E., Winschel, V., and Zahn, P. 2016. *Compositionality and string diagrams for game theory.* arXiv:1604.06061.

Hein, M., Eisert, J., and Briegel, H. J. 2004. Multiparty entanglement in graph states. *Physical Review A*, **69**, 062311.

Heisenberg, W. 1925. Über quantentheoretische Umdeutung kinematischer und mechanischer Beziehungen. *Heisenberg (1925)*, **33**, 879–893.

Heisenberg, W. 1930. *Die physikalischen Prinzipien der Quantentheorie.* Leipzig: S. Hirzel.

Hensen, B., Bernien, H., Dreau, A. E., Reiserer, A., Kalb, N., Blok, M. S., Ruitenberg, J., Vermeulen, R. F. L., Schouten, R. N., Abellan, C., Amaya, W., Pruneri, V., Mitchell, M. W., Markham, M., Twitchen, D. J., Elkouss, D., Wehner, S., Taminiau, T. H., and Hanson, R. 2015. Loophole-free Bell inequality violation using electron spins separated by 1.3 kilometres. *Nature*, **526**(10), 682–686.

Henson, J., Lal, R., and Pusey, M. F. 2014. *Theory-independent limits on correlations from generalised Bayesian networks.* arXiv:1405.2572.

Herrmann, M. 2010. *Models of multipartite entanglement.* MSc Thesis, Oxford University.

Heunen, C., and Jacobs, B. 2010. Quantum logic in dagger kernel categories. *Order*, **27**(2), 177–212.

Heunen, C., and Kissinger, A. 2016. *Can quantum theory be characterized in information-theoretic terms?* arXiv:1604.05948.

Heunen, C., Contreras, I., and Cattaneo, A.o S. 2012b. Relative Frobenius algebras are groupoids. *Journal of Pure and Applied Algebra*, **217**, 114–124.

Heunen, C., Sadrzadeh, M., and Grefenstette, E. (eds). 2012a. *Quantum Physics and Linguistics: A Compositional, Diagrammatic Discourse.* Oxford University Press.

Hinze, R., and Marsden, D. 2016. Equational reasoning with lollipops, forks, cups, caps, snakes, and speedometers. *Journal of Logical and Algebraic Methods in Programming*.

Hoare, C. A. R., and He, J. 1987. The weakest prespecification. *Information Processing Letters*, **24**, 127–132.

Honda, K. 2012. Graphical classification of entangled qutrits. *Electronic Proceedings in Theoretical Computer Science*, **95**, 123–141.

Horodecki, M., Oppenheim, J., and Horodecki, R. 2002. Are the laws of entanglement theory thermodynamical? *Physical Review Letters*, **89**, 240403.

Horodecki, M., Horodecki, P., and Oppenheim, J. 2003. Reversible transformations from pure to mixed states and the unique measure of information. *Physical Review A*, **67**, 062104.

Horodecki, R., Horodecki, P., Horodecki, M., and Horodecki, K. 2009. Quantum entanglement. *Reviews of Modern Physics*, **81**, 865–942. arXiv:quant-ph/0702225.

Horsman, C. 2011. Quantum picturalism for topological cluster-state computing. *New Journal of Physics*, **13**, 095011. arXiv:1101.4722.

Jacobs, B. 2010. Orthomodular lattices, Foulis semigroups and Dagger kernel categories. *Logical Methods in Computer Science*, **6**(2), 1.

Jamiołkowski, A. 1972. Linear transformations which preserve trace and positive semidefiniteness of operators. *Reports on Mathematical Physics*, **3**, 275–278.

Jammer, M. 1974. *The Philosophy of Quantum Mechanics*. John Wiley & Sons.

Jauch, J. M. 1968. *Mathematical Foundations of Quantum Mechanics*. Addison-Wesley.

Jauch, J. M., and Piron, C. 1963. Can hidden variables be excluded in quantum mechanics? *Helvetica Physics Acta*, **36**, 827–837.

Johansson, M., Dixon, L., and Bundy, A. 2011. Conjecture synthesis for inductive theories. *Journal of Automated Reasoning*, **47**, 251–289.

Jones, J. A., Mosca, M., and Hansen, R. H. 1998. Implementation of a quantum search algorithm on a quantum computer. *Nature*, **393**(6683), 344–346.

Jones, V. F. R. 1985. A polynomial invariant for knots via von Neumann algebras. *Bulletin of the American Mathematical Society*, **12**, 103–111.

Joyal, A., and Street, R. 1991. The geometry of tensor calculus I. *Advances in Mathematics*, **88**, 55–112.

Joyal, A., Street, R., and Verity, D. 1996. Traced monoidal categories. Pages 447–468 of: *Mathematical Proceedings of the Cambridge Philosophical Society*, vol. 119. Cambridge University Press.

Jozsa, R. 1997. Quantum algorithms and the Fourier transform. In: *Proceedings of the Santa Barbarba Conference on Coherence and Decoherence*. Proceedings of the Royal Society of London.

Kartsaklis, D., and Sadrzadeh, M. 2013. Prior disambiguation of word tensors for constructing sentence vectors. Pages 1590–1601 of: *The 2013 Conference on Empirical Methods on Natural Language Processing*. ACL.

Kassel, C. 1995. *Quantum Groups*. Vol. 155. Springer.

Kauffman, L. H. 1987. State models and the Jones polynomial. *Topology*, **26**, 395–407.

Kauffman, L. H. 1991. *Knots and Physics*. World Scientific.

Kauffman, L. H. 2005. Teleportation topology. *Optics and Spectroscopy*, **99**, 227–232.

Kelly, G. M. 1972. Many-variable functorial calculus I. Pages 66–105 of: Kelly, G. M., Laplaza, M., Lewis, G., and Mac Lane, S. (eds), *Coherence in Categories*. Lecture Notes in Mathematics, vol. 281. Springer-Verlag.

Kelly, G. M., and Laplaza, M. L. 1980. Coherence for compact closed categories. *Journal of Pure and Applied Algebra*, **19**, 193–213.

Kissinger, A. 2012a. *Pictures of processes: automated graph rewriting for monoidal categories and applications to quantum computing*. PhD thesis, University of Oxford.

Kissinger, A. 2012b. Synthesising graphical theories. arXiv:1202.6079.

Kissinger, A. 2014a. Abstract tensor systems as monoidal categories. In: Casadio, C., Coecke, B., Moortgat, M., and Scott, P. (eds), *Categories and Types in Logic, Language, and Physics: Festschrift on the Occasion of Jim Lambek's 90th Birthday*. Lecture Notes in Computer Science, vol. 8222. Springer. arXiv:1308.3586.

Kissinger, A. 2014b. *Finite matrices are complete for (dagger-)hypergraph categories*. arXiv:1406.5942 [math.CT].

Kissinger, A., and Zamdzhiev, V. 2015. *Quantomatic: a proof assistant for diagrammatic reasoning.* arXiv:1503.01034.

Kissinger, A., Merry, A., and Soloviev, M. 2014. Pattern graph rewrite systems. Pages 54–66 of: Löwe, B., and Winskel, G. (eds), *Proceedings 8th International Workshop on Developments in Computational Models.* Electronic Proceedings in Theoretical Computer Science, vol. 143. Open Publishing Association.

Kochen, S., and Specker, E. P. 1967. The problem of hidden variables in quantum mechanics. *Journal of Mathematics and Mechanics*, **17**(1), 59–87.

Kock, J. 2004. *Frobenius Algebras and 2D Topological Quantum Field Theories.* Vol. 59. Cambridge University Press.

Kraus, K. 1983. *States, Effects and Operations.* Springer.

Lack, S. 2004. Composing PROPs. *Theory and Applications of Categories*, **13**, 147–163.

Laforest, M., Baugh, J., and Laflamme, R. 2006. Time-reversal formalism applied to maximal bipartite entanglement: theoretical and experimental exploration. *Physical Review A*, **73**(3), 032323.

Lamata, L., León, J., Salgado, D., and Solano, E. 2007. Inductive entanglement classification of four qubits under stochastic local operations and classical communication. *Physical Review A*, **75**, 022318.

Lambek, J., and Scott, P. J. 1988. *Introduction to Higher-Order Categorical Logic.* Cambridge University Press.

Leinster, T. 2004. *Higher Operads, Higher Categories.* Cambridge University Press.

Lemmens, P. W. H., and Seidel, J. J. 1973. Equiangular lines. *Journal of Algebra*, **24**(3), 494–512.

Lloyd, S., Maccone, L., Garcia-Patron, R., Giovannetti, V., Shikano, Y., Pirandola, S., Rozema, L. A., Darabi, A., Soudagar, Y., Shalm, L. K., and Steinberg, A. M. 2011. Closed timelike curves via postselection: theory and experimental test of consistency. *Physical Review Letters*, **106**(4), 040403.

Lo, H.-K., and Popescu, S. 2001. Concentrating entanglement by local actions: beyond mean values. *Physical Review A*, **63**, 022301.

Ludwig, G. 1985. *An Axiomatic Basis of Quantum Mechanics*, volume 1: *Derivation of Hilbert Space.* Springer-Verlag.

Mac Lane, S. 1950. Duality for groups. *Bull. Am. Math. Soc.*, **56**, 485–516.

Mac Lane, S. 1963. Natural associativity and commutativity. *The Rice University Studies*, **49**(4), 28–46.

Mac Lane, S. 1998. *Categories for the Working Mathematician.* Springer-Verlag.

Mackey, G. W. 1963. *The Mathematical Foundations of Quantum Mechanics.* New York: W. A. Benjamin.

Macrakis, K. 1993. *Surviving the Swastika: Scientific Research in Nazi Germany.* Oxford University Press.

Majid, S. 2000. *Foundations of Quantum Group Theory.* Cambridge University Press.

Manin, Y. I. 1980. *Vychislimoe i Nevychislimoe.* Sovetskoye Radio.

Markopoulou, F. 2000. Quantum causal histories. *Classical and Quantum Gravity*, **17**(10), 2059.

Marvian, I., and Spekkens, R. W. 2013. The theory of manipulations of pure state asymmetry: I. Basic tools, equivalence classes and single copy transformations. *New Journal of Physics*, **15**(3), 033001.

Mehra, J. 1994. *The Beat of a Different Drum: The Life and Science of Richard Feynman.* Clarendon Press.

Mellies, P.-A. 2012. Game semantics in string diagrams. Pages 481–490 of: *Proceedings of the 27th Annual IEEE Symposium on Logic in Computer Science (LICS)*. IEEE Computer Society.

Mermin, N. D. 1990. Quantum mysteries revisited. *American Journal of Physics*, **58**(Aug.), 731–734.

Mermin, N. D. April 1989. What's wrong with this pillow? *Physics Today*.

Mermin, N. D. May 2004. Could Feynman have said this? *Physics Today*.

Milner, R. 1972. *Logic for computable functions; description of a machine implementation.* Tech. rept. STAN-CS-72-288. Stanford University.

Montanaro, A. 2015. *Quantum algorithms: an overview.* arXiv:1511.04206.

Moore, D. J. 1995. Categories of representations of physical systems. *Helvetica Physica Acta*, **68**, 658–678.

Moore, D. J. 1999. On state spaces and property lattices. *Studies in History and Philosophy of Modern Physics*, **30**(1), 61–83.

Muirhead, R. F. 1903. Some methods applicable to identities and inequalities of symmetric algebraic functions of *n* letters. *Proceedings of the Edinburgh Mathematical Society*, **21**, 144–157.

Neumark, M. A. 1943. On spectral functions of a symmetric operator. *Izvestiya Rossiiskoi Akademii Nauk. Seriya Matematicheskaya*, **7**(6), 285–296.

Nielsen, M. A. 1999. Conditions for a class of entanglement transformations. *Physical Review Letters*, **83**(2), 436–439.

Nielsen, M. A., and Chuang, I. L. 2010. *Quantum Computation and Quantum Information.* Cambridge University Press.

Ozawa, M. 1984. Quantum measuring processes of continuous observables. *Journal of Mathematical Physics*, **25**(1), 79–87.

Pan, J.-W., Bouwmeester, D., Daniell, M., Weinfurter, H., and Zeilinger, A. 2000. Experimental test of quantum nonlocality in three-photon Greenberger–Horne–Zeilinger entanglement. *Nature*, **403**, 515–519.

Panangaden, P., and Paquette, É. O. 2011. A categorical presentation of quantum computation with anyons. Pages 983–1025 of: Coecke, B. (ed), *New Structures for Physics*. Lecture Notes in Physics. Springer.

Paquette, É. O. 2008. *Categorical quantum computation.* PhD thesis, University of Montreal.

Paulsen, V. 2002. *Completely Bounded Maps and Operator Algebras.* Cambridge University Press.

Paulson, L., et al. 1986. *Isabelle theorem prover.* https://isabelle.in.tum.de/.

Pavlovic, D. 2009. Quantum and classical structures in nondeterminstic computation. Pages 143–157 of: *Proceedings of the 3rd International Symposium on Quantum Interaction*. QI '09. Berlin and Heidelberg: Springer-Verlag.

Pavlovic, D. 2013. Monoidal computer I: basic computability by string diagrams. *Information and Computation*, **226**, 94–116.

Pearl, J. 2000. *Causality: Models, Reasoning and Inference.* Cambridge University Press.

Penrose, R. 1971. Applications of negative dimensional tensors. Pages 221–244 of: *Combinatorial Mathematics and Its Applications*. Academic Press.

Penrose, R. 1984. *Spinors and Spacetime*, vol. 1. Cambridge University Press.

Penrose, R. 2004. *The Road to Reality: A Complete Guide to the Physical Universe.* Jonathan Cape.

Perdrix, S. 2005. State transfer instead of teleportation in measurement-based quantum computation. *International Journal of Quantum Information*, **3**(1), 219–223.

Perdrix, S., and Wang, Q. 2015. *The ZX calculus is incomplete for Clifford+T quantum mechanics.* arXiv:1506.03055.

Piedeleu, R., Kartsaklis, D., Coecke, B., and Sadrzadeh, M. 2015. Open system categorical quantum semantics in natural language processing. In: *CALCO 2015.* arXiv:1502.00831.

Pierce, B. C. 1991. *Basic Category Theory for Computer Scientists.* MIT Press.

Piron, C. 1976. *Foundations of Quantum Physics.* W. A. Benjamin.

Piron, Constantin. 1964. Axiomatique quantique. *Helvetia Physica Acta,* **37**, 439–468.

Planck, M. 1900. Zur Theorie des Gesetzes der Energieverteilung im Normalspektrum. *Verhandlungen der Deutschen Physikalischen Gesellschaft,* **2**, 237–245.

Poincaré, H. 1902. *La science et l'hypothèse.* Flammarion.

Pusey, M. F., Barrett, J., and Rudolph, T. 2012. On the reality of the quantum state. *Nature Physics,* **8**(6), 475–478.

Ranchin, A., and Coecke, B. 2014. Complete set of circuit equations for stabilizer quantum mechanics. *Physical Review A,* **90**, 012109.

Rauch, H., Zeilinger, A., Badurek, G., Wilfing, A., Bauspiess, W., and Bonse, U. 1975. Verification of coherent spinor rotation of fermions. *Physics Letters A,* **54**, 425–427.

Raussendorf, R., and Briegel, H. J. 2001. A one-way quantum computer. *Physical Review Letters,* **86**, 5188.

Raussendorf, R., Browne, D. E., and Briegel, H. J. 2003. Measurement-based quantum computation on cluster states. *Physical Review A,* **68**(2), 22312.

Raussendorf, R., Harrington, J., and Goyal, K. 2007. Topological fault-tolerance in cluster state quantum computation. *New Journal of Physics,* **9**, 199.

Redei, M. 1996. Why John von Neumann did not like the Hilbert space formalism of quantum mechanics (and what he liked instead). *Studies in History and Philosophy of Modern Physics,* **27**(4), 493–510.

Redhead, Michael. 1987. *Incompleteness, Nonlocality, and Realism: A Prolegomenon to the Philosophy of Quantum Mechanics.* Clarendon Press.

Rickles, D. 2007. *Symmetry, Structure, and Spacetime.* Elsevier.

Roddenberry, G. 1966. *Star Trek* (television series). NBC.

Rowe, M. A., Kielpinski, D., Meyer, V., Sackett, C. A., Itano, W. M., Monroe, C., and Wineland, D. J. 2001. Experimental violation of a Bell's inequality with efficient detection. *Nature,* **409**, 791–794.

Sadrzadeh, M., Clark, S., and Coecke, B. 2013. The Frobenius anatomy of word meanings I: subject and object relative pronouns. *Journal of Logic and Computation,* **23**, 1293–1317. arXiv:1404.5278.

Sadrzadeh, M., Clark, S., and Coecke, B. 2014. The Frobenius anatomy of word meanings II: possessive relative pronouns. *Journal of Logic and Computation,* exu027.

Schröder de Witt, C., and Zamdzhiev, V. 2014. *The ZX calculus is incomplete for quantum mechanics.* arXiv:1404.3633.

Schrödinger, E. 1926. An undulatory theory of the mechanics of atoms and molecules. *Physical Review Letters,* **28**(6), 1049–1070.

Schrödinger, E. 1935. Die gegenwärtige Situation in der Quantenmechanik. *Naturwissenschaften,* **23**, 823–828.

Schrödinger, E. 1935. Discussion of probability relations between separated systems. *Mathematical Proceedings of the Cambridge Philosophical Society,* **31**, 555–563.

Schumacher, B. 1995. Quantum coding. *Physical Review A,* **51**, 2738.

Schwinger, J. 1960. Unitary operator bases. *Proceedings of the National Academy of Sciences of the U.S.A.,* **46**, 570–579.

Scott, D. S. 1993. A type-theoretical alternative to ISWIM, CUCH, OWHY. *Theoretical Computer Science*, **121**(1), 411–440.

Selinger, P. 2007. Dagger compact closed categories and completely positive maps. *Electronic Notes in Theoretical Computer Science*, **170**, 139–163.

Selinger, P. 2011a. Finite dimensional Hilbert spaces are complete for dagger compact closed categories (extended abstract). *Electronic Notes in Theoretical Computer Science*, **270**(1), 113–119.

Selinger, P. 2011b. A survey of graphical languages for monoidal categories. Pages 275–337 of: Coecke, B. (ed), *New Structures for Physics*. Lecture Notes in Physics. Springer-Verlag. arXiv:0908.3347.

Selinger, P. 2015. Generators and relations for n-qubit Clifford operators. *Logical Methods in Computer Science*, **11**.

Shannon, C. E. 1948. A mathematical theory of communication. *The Bell System Technical Journal*, **27**, 379–423.

Shende, V. V., Bullock, S. S., and Markov, I. L. 2006. Synthesis of quantum-logic circuits. *IEEE Transactions on Computer-Aided Design of Integrated Circuits and Systems*, **25**(6), 1000–1010.

Shor, P. W. 1994. Algorithms for quantum computation: discrete logarithms and factoring. Pages 124–134 of: *Proceedings of the 35th Annual Symposium on Foundations of Computer Science*. IEEE.

Shor, P. W. 1997. Polynomial-time algorithms for prime factorization and discrete logarithms on a quantum computer. *SIAM Journal on Computing*, **26**(5), 1484–1509.

Shulman, M., et al. 2013. *Homotopy type theory: univalent foundations of mathematics*. https://homotopytypetheory.org/book/.

Simon, D. R. 1997. On the power of quantum computation. *SIAM Journal on Computing*, **26**(5), 1474–1483.

Sobocinski, P. 2015. *Graphical linear algebra*. http://graphicallinearalgebra.net.

Spekkens, R. W. 2007. Evidence for the epistemic view of quantum states: a toy theory. *Physical Review A*, **75**(3), 032110.

Stay, M., and Vicary, J. 2013. Bicategorical semantics for nondeterministic computation. *Electronic Notes in Theoretical Computer Science*, **298**, 367–382. arXiv:1301.3393.

Stinespring, W. F. 1955. Positive functions on C*-algebras. *Proceedings of the American Mathematical Society*, **6**(2), 211–216.

Street, R. 2007. *Quantum Groups: A Path to Current Algebra*. Cambridge University Press.

Stubbe, I., and van Steirteghem, B. 2007. Propositional systems, Hilbert lattices and generalized Hilbert spaces. Pages 477–524 of: Gabbay, D., Lehmann, D., and Engesser, K. (eds), *Handbook Quantum Logic*. Elsevier Publ.

Sudarshan, E. C. G., Mathews, P. M., and Rau, J. 1961. Stochastic dynamics of quantum-mechanical systems. *Physical Review*, **121**(3), 920.

Svetlichny, G. 2009. *Effective quantum time travel*. arXiv:0902.4898.

Tull, S. 2016. *Operational theories of physics as categories*. arXiv:1602.06284.

Turing, A. M. 1937. On computable numbers, with an application to the Entscheidungsproblem. *Proceedings of the London Mathematical Society*, **42**, 230–265.

Van den Nest, M., Dehaene, J., and De Moor, B. 2004. Graphical description of the action of local Clifford transformations on graph states. *Physical Review A*, **69**(2), 9422.

Verstraete, F., Dehaene, J., De Moor, B., and Verschelde, H. 2002. Four qubits can be entangled in nine different ways. *Physical Review A*, **65**(052112). arXiv:quant-ph/0109033.

Vicary, J. 2011. Categorical formulation of finite-dimensional quantum algebras. *Communications in Mathematical Physics*, **304**(3), 765–796.

Vicary, J. 2013. The topology of quantum algorithms. Pages 93–102 of: *Proceedings of the 28th Annual IEEE Symposium on Logic in Computer Science (LICS)*. IEEE Computer Society.

Von Neumann, J. 1927a. Thermodynamik quantenmechanischer Gesamtheiten. *Nachrichten von der Gesellschaft der Wissenschaften zu Göttingen, Mathematisch-Physikalische Klasse*, **1**, 273–291.

Von Neumann, J. 1927b. Wahrscheinlichkeitstheoretischer aufbau der quantenmechanik. *Nachrichten von der Gesellschaft der Wissenschaften zu Göttingen, Mathematisch-Physikalische Klasse*, **1**, 245–272.

Von Neumann, J. 1932. *Mathematische Grundlagen der quantenmechanik*. Springer-Verlag. Translation, *Mathematical Foundations of Quantum Mechanics*, Princeton University Press, 1955.

Walther, P., Resch, K. J., Rudolph, T., Schenck, E., Weinfurter, H., Vedral, V., Aspelmeyer, M., and Zeilinger, A. 2005. Experimental one-way quantum computing. *Nature*, **434**, 169–176.

Wedderburn, J. H. M. 1906. On a theorem in hypercomplex numbers. *Proceedings of the Royal Society of Edinburgh*, **26**, 48–50.

Weihs, G., Jennewein, T., Simon, C., Weinfurter, H., and Zeilinger, A.n. 1998. Violation of Bell's inequality under strict Einstein locality conditions. *Physical Review Letters*, **81**, 5039.

Werner, R. F. 2001. All teleportation and dense coding schemes. *Journal of Physics A: Mathematical and General*, **34**(35), 7081.

Whitehead, A. N. 1957. *Process and Reality*. Harper & Row.

Wigner, E. P. 1931. *Gruppentheorie und ihre Anwendung auf die Quanten mechanik der Atomspektren*. Friedrich Vieweg und Sohn.

Wigner, E. P. 1995a. *Remarks on the Mind-Body Question*. Springer. Pages 247–260.

Wigner, E. P. 1995b. The unreasonable effectiveness of mathematics in the natural sciences. Pages 534–549 of: *Philosophical Reflections and Syntheses*. Springer.

Wilce, A. 2000. Test spaces and orthoalgebras. Pages 81–114 of: Coecke, B., Moore, D. J., and Wilce, A. (eds), *Current Research in Operational Quantum Logic: Algebras, Categories and Languages*. Fundamental Theories of Physics, vol. 111. Springer.

Wittgenstein, L. 1953. *Philosophical Investigations*. Basil & Blackwell.

Wood, C. J., and Spekkens, R. W. 2012. *The lesson of causal discovery algorithms for quantum correlations: causal explanations of Bell-inequality violations require fine-tuning*. arXiv:1208.4119.

Wootters, W., and Zurek, W. 1982. A single quantum cannot be cloned. *Nature*, **299**, 802–803.

Zeilinger, A. 1999. Experiment and the foundations of quantum physics. *Reviews of Modern Physics*, **71**, S288.

Zeng, W. 2015. *The abstract structure of quantum algorithms*. PhD thesis, University of Oxford. arXiv:1512.08062.

Zeng, W., and Vicary, J. 2014. *Abstract structure of unitary oracles for quantum algorithms*. arXiv:1406.1278.

Zukowski, M., Zeilinger, A., Horne, M. A., and Ekert, A. K. 1993. "Event-ready-detectors" Bell experiment via entanglement swapping. *Physical Review Letters*, **71**, 4287–4290.

Index

Printed in the United States
By Bookmasters